Optical Techniques for Industrial Inspection

Optical Techniques for Industrial Inspection

P. Cielo

National Research Council of Canada
Industrial Materials Research Institute
Boucherville, Quebec, Canada

ACADEMIC PRESS, INC.
Harcourt Brace Jovanovich, Publishers

Boston San Diego New York
Berkeley London Sydney
Tokyo Toronto

ACADEMIC PRESS, INC.
1250 Sixth Avenue, San Diego, 92101

United Kingdom Edition published by
ACADEMIC PRESS, INC. (LONDON) LTD.
24-28 Oval Road, London NW1 7DX

Library of Congress Cataloging-in-Publication Data

Cielo, Paolo G.
Optical techniques for industrial inspection.

Bibliography: p.
Includes index.
1. Engineering inspection. 2. Quality control—Optical methods. I. Title.
TS156.2.C55 1987 670.42′5 87-12957
ISBN 0-12-174655-0

88 89 90 91 9 8 7 6 5 4 3 2 1
Printed in the United States of America

Contents

Preface

On-line and unsupervised industrial inspection for quality control and process monitoring is increasingly required in the modern automated factory. Optical techniques are particularly well suited to industrial inspection needs because of their noncontact nature and imaging capabilities. Optical sensors can be used for remote inspection of high temperature products or otherwise inaccessible parts, provided they are in a line-of-sight relation with the sensor. Moreover, the imaging capabilities of optical systems enable the simultaneous collection of large amounts of information and hence the possibility of compensating for variable part position and orientation with respect to the sensor. This is an important requirement in a flexible automation industrial environment.

Electro-optical technology is now mature for operation in hostile industrial environments. One of the main reasons for the growing interest for optical inspection systems in the industrial world is the recent appearance of rugged, reliable and affordable optoelectronic components. The development of such components is mainly a spin-off of the rapid adoption of optical technology in fields such as fiber optic communications and optical disk devices. One example is the laser diode, a coherent solid state source which is extremely compact and surprisingly powerful as compared to the earlier gas laser devices. Such diodes have a lifetime ten times longer than an ordinary tungsten lamp at prices which are rapidly approaching those of inexpensive incoherent light sources. Other important advances in optical technology which are directly relevant to optical inspection requirements are in the fields of optical detectors, compact staring-array video devices and signal processing capabilities.

This book is organized in such a way that optical techniques are introduced as an answer to industrial inspection needs. Rather than describing each technique and its applications separately, the industrial requirements are first outlined in Chapter 1, while Chapter 2 gives an overview of the competitive possible approaches using techniques other than optical. Optical methods are then described in the subsequent chapters, each dealing with a specific industrial inspection field. Comparisons are thus favored between different optical as well as nonoptical methods, and development efforts are justified in terms of the basic industrial needs. In an effort to make each chapter as self-contained as possible, some specific optical techniques relating to different inspection problems are discussed in several chapters.

The book is directed to an audience encompassing optical scientists who are aiming at the industrial inspection sector, practicing engineers in search of suitable sensors for industrial control requirements, as well as students specializing in industrial inspection technology or in electro-optic instrumentation. Chapters 1 and 2 are particularly directed to optical scientists who are not familiar with industrial needs and nonoptical inspection systems, while Chapter 3 reviews modern optical technology for the benefit of engineers having a limited optoelectronics background. As a general rule, only a knowledge of introductory college level optical physics and of general instrumentation is assumed.

I am grateful to many colleagues for their help and encouragement in this endeavor. I wish to thank in particular G. Bata and J.F. Bussière for their support, as well as J. Domey, M. Dufour, J.-P. Monchalin and M. Rioux for many useful discussions. L.A. Utracki provided early encouragement and advice on the book organization. K.C. Cole competently reviewed the chapter on spectroscopic techniques for the analysis of industrial materials. I am particularly grateful to G. Birnbaum of the National Bureau of Standards who kindly reviewed the whole manuscript and made several appropriate and greatly appreciated suggestions. Finally, my deepest debt is to my wife, Claudia, without whose patience and moral support this book would not have become a reality.

P. Cielo

Chapter 1

Need and Functions of Industrial Inspection

1.1 Definition and Scope of Inspection Activities in Industry

In the broadest industrial sense, inspection may be defined as the function of determining the conformance of the product to the requirements. Similar expressions which are also used in this context are *quality control, statistical process control, industrial sensing and gaging* or *nondestructive testing.*

In the modern materials processing and manufacturing industry, the main inspection requirements are in the fields of product specification and industrial process monitoring. The term *quality control* refers to the first of these two requirements. Such controls are usually performed on the finished product and are mainly devoted to verifying product quality as it may be assessed by the customer. Visual inspection of the product appearance, color, surface finish and geometrical characteristics are typical examples of controls which are being increasingly performed by means of a quantitative machine inspection approach.

The present trend toward automated production has intensified the need for on-line inspection at each production stage to remove faulty products from the production line before incurring additional product transformation expenses in subsequent manufacturing steps. The detection through a *statistical process control*, of repetitive faults in the product after a given step usually indicates a malfunction of the manufacturing process. On-line inspection tools are desirable to control the automated production process, a function which was carried out by the operator in the nonautomated process. A further step consists of using the information continuously supplied by the inspection sensors to feed the decision making unit of a Computer-Integrated-Manufacturing system. In this case, the sensor acts as the essential portion of a feedback loop which finely modifies the process parameters as soon as minute departures from product specifications are detected. In a Flexible-Manufacturing-Technology environment, the choice of the process to execute relies on sensor information about the workpiece. The term *sensing and gaging* is a general expression which encompasses both quality control and process monitoring activities.

The concept of *testing* refers to the procedure of submitting the product to the function for which it has been designed in order to verify its suitability to this function. For example, an insulated high voltage cable may be submitted to a voltage level higher than the maximum level expected during service or a pressure vessel may be raised to a pressure 50% higher than in the typical in-service operation. Such tests are often too expensive to be performed on 100% of the production. Moreover, a static initial test is usually not sufficient to assure that the product will not fail during the predicted operation life. A pressure vessel may survive a sudden pressure burst to 150% of the operation pressure but may fail after 1,000 pressure cycles not exceeding the normal operation pressure because of cumulative fatigue effects during service. Destructive fatigue tests are thus routinely conducted on products such as nuclear pressure vessels or aircraft structures to evaluate the expected lifetime of such products under normal loading conditions. Safety sensitive structures are automatically withdrawn from service after a fraction of their lifetime in order to avoid catastrophic failures. Destructive test procedures required to obtain statistically significant lifetime predictions are extremely expensive and by definition cannot be applied to 100% of the production. *Nondestructive-testing*, or *-evaluation*, or *-inspection* procedures, are thus widely used both in the production process and during service to collect indirect information on product integrity and suitability. Typical physical properties which may provide such information are thickness and uniformity of structure components and coatings or the presence of surface irregularities and cracks. Optical inspection techniques such as fluorescent-penetrant inspection, holographic pressure testing or thermographic depth-imaging are an expanding subfield of nondestructive testing techniques.

1.2 Human Versus Automated Inspection

Inspection is an integral aspect of manufacturing activity, and human inspection dates back to the stone age craftsman examining his workpiece during carving. Operator-performed manual, and particularly visual, inspection is still by far the most widespread product control activity in the modern production plant. It has been indicated (Geverter, 1982; Miller, 1985) that as much as 90% of all industrial visual inspection activities might be performed by computer vision within the next decade.

Human inspection, as compared to automated inspection, is much simpler and faster to set up, while human versatility and judgement make a strict and detailed specification of the product requirements and tolerances unnecessary. Conversely, inspection machines are not affected by fatigue or task monotony, and they provide reliable decisions based on objective criteria which are not expected to change, as in the case of human inspection, when the inspector changes. Verifying the repetitivity of a given color in a garment manufacturing line may prove a difficult task for the quality control operator, while color quantification is very easily performed by a chromaticity measuring optical system.

The effect of human fatigue and psychological effects in human inspection may be appreciated from a typical rate-of-inspection curve of the kind shown in Figure 1-1 (Kennedy and Andrews, 1977). Such a curve is the result of a series of studies whereby an inspector undertook a 100 percent visual inspection task of the same item at increasing production rates. The fraction of defects escaping detection was plotted versus the production rate. As the chart in Figure 1-1 shows, practically no defects escaped the inspector when the inspection rate was very slow (at a rate smaller than A. As the pace was stepped up, more defects were missed (range B). However, as the rate was further increased, an increase in inspection efficiency could be observed (range C). This is attributed to less inspection monotony at rate C compared to rate B. Only at higher paces (greater than D) did the effect of fatigue produce a deterioration in inspection efficiency.

The effects of human wear and lack of concentration are less pronounced if the inspection is not continuous, such as in *batch inspection* where a relatively small product lot is inspected after the production process or in *patrol inspection* where a quality control officer makes a round surveillance during the production process. *Continuous inspection* is more appropriate to automated procedures. The choice between human or machine inspection is mainly a consequence of the size and repetitiveness of the production line.

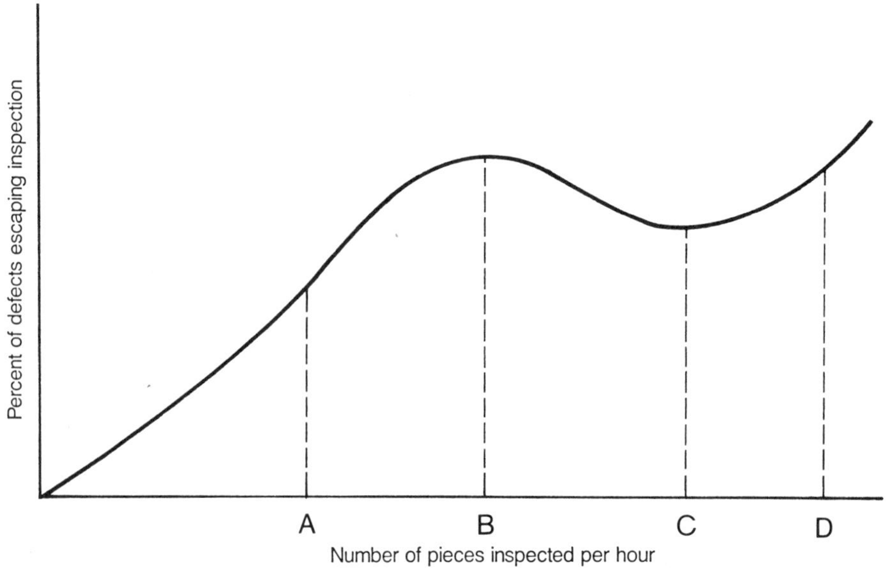

Figure 1-1 The curve shows, for one particular inspector, the relationship between number of pieces inspected per hour and number of defects escaping detection.

In a small shop, the production manager also acts as quality inspector, and for each job the quality criteria depend on the requirements of the specific customer. Usually, human inspection is sufficiently flexible to accommodate the variable requirements of such a production line. At the other extreme are, for instance, the huge automobile plants where thousands of people work on three shifts. Automated inspection is much more suitable to such processes in order to assure reliable and constant quality criteria for all the products at the different production stages. Capital investment and employee retraining capabilities are also much more favorable in such large production units (Batchelor et al., 1985, Section 2.3).

Most inspection procedures are neither totally manual nor completely automated. The human inspector often relies on gaging instruments or on defect enhancing image processing systems to facilitate decision-making activites. On the other hand, highly automated inspection systems often require operator intervention to judge whether a product which does not meet specifications can be fixed or recycled or whether an alarm signal along the production line indicates a problem sufficiently serious to suspend the process. A trend toward increased automation in the quality control and process monitoring fields is, however, clearly discernible in the modern industrial plants.

1.3 Economic Factors

Inspection tasks constitute a sizeable fraction of production costs. It has recently been estimated (Miller, 1985) that inspection activities are performed by 10% of the work force. Optical instrumentation represents a major component of the industrial inspection hardware. A recent report predicts an annual growth as high as 25% in the U.S. market for electro-optical inspection systems, rising from $900 million in 1984 to $3.7 billion by 1990 (*Sensor Review*, July 1986, p. 120).

Quality assurance activities are either compulsory or voluntary. Examples of the first are verification and testing procedures imposed by government regulations on health and safety or export quality controls. Legal requirements may result in substantial inspection costs for safety sensitive products. Quality control as a whole accounts for as much as 60% of the cost of an airplane (Papadakis, 1983).

Except for compulsory procedures, all other inspection activities are based on economic considerations. Product quality is not an objective per se, but must be justified in terms of an improvement in the present or projected profit position of the company.

1.3.1 *Quality and Income*

Product quality may affect income because of its effects on sales and prices as well as the company's reputation in the market.

Quality and Share of the Market

Once a product is actively on the market, it attains a "share of the market," i.e., a proportion of all sales by all manufacturers for that type of product. The measurement and analysis of consumer preference patterns and the implementation of the appropriate product features to suit such patterns are essential tools available to the company to improve its market share position.

The willingness of the customer to respond to product quality strongly depends on the buyer's capability and motivation to assess the product. As far as industrial products are concerned, such an assessment is performed more on technological performance than on sensory qualities. Product quality features such as fitness, integrity, functionality and durability are high on the list for industrial customers, provided they are made aware of such characteristics.

Products sold to end users in the consumer market are screened on different terms. The consumer is often more influenced by marketing skills including attractiveness of the package and appeal of prior advertisement or the manufacturer's reputation than the intrinsic quality of the product. Figure 1-2 illustrates such behavior, showing the results of a survey on 41 products sold on supermarket shelves (Juran et al., 1979). The horizontal scale shows consumer preference over the leading competitor product as determined by a statistically sound preference testing poll. The vertical scale shows the market share versus the leading competitor, considering the two as constituting 100%. In contrast to the vacant areas on the horizontal scale of consumer preference, the vertical scale of market share has data along the entire spectrum. One product (marked A on Figure 1-2) lies squarely on the 50% consumer preference line, implying that the product quality is very similar to the competitor. Yet this product has only 10% of the market share, compared to 90% for the competitor, the reason being that the competing company pioneered this particular product, thus acquiring a strong initial share of the market which is retained thereafter through good promotion.

The relation of product quality to customer preference must be based on objective marketing data rather than on manufacturer assumptions. In the

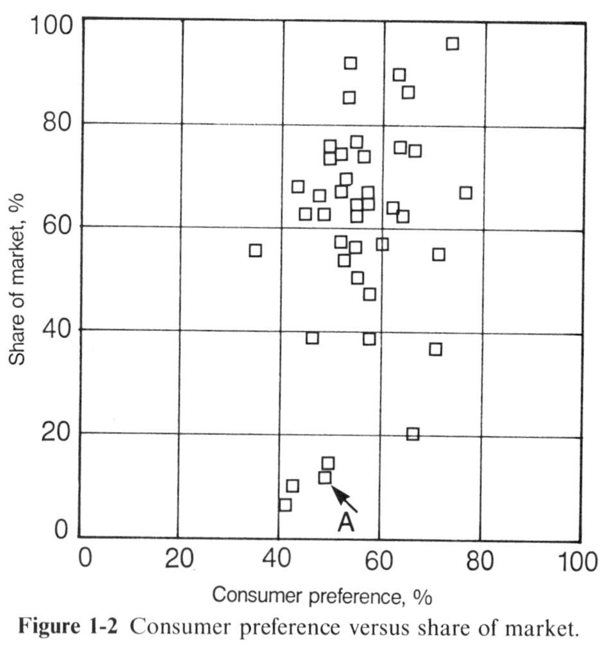

Figure 1-2 Consumer preference versus share of market.

carpet industry, the engineers devoted much effort to improving wear qualities and other technological aspects of fitness for use. However, a marketing survey established that customers' reasons for selecting carpets were primarily sensory, i.e.,

Color.	56%
Pattern	20%
Other sensory	6%
Nonsensory	18%

The determination of the quality criteria for optimum market penetration is thus strongly dependent on the customer type and product function.

Quality and Prices

While sensory product characteristics play an important role in the consumer market, the marketing of capital goods is much more dependent on life-cycle costing considerations. Under this concept, the various elements of the cost of usage of a piece of equipment are quantified in order to enable users and manufacturers to improve their decision making in ways which will optimize the user's overall cost. Figure 1-3 shows the life-cycle costing diagram for three manufacturers, A, B, and C, supplying a major piece of equipment to the user (Juran et al., 1979). The breakdown also shows how the costs of operation and maintenance differ among the three competitors.

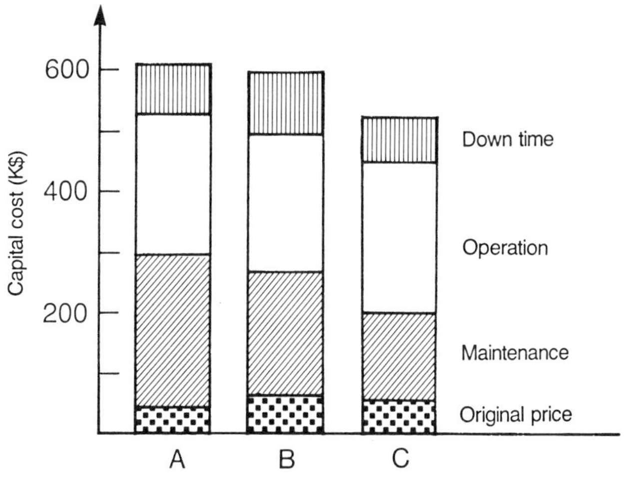

Figure 1-3 Life-cycle costing: Cumulative costs of capital goods supplied by three competitive companies.

The conclusion is that equipment C is the most convenient for the buyer, even if its original price is higher than that of the supplier A. Such an example shows that the functional quality of capital goods may justify a higher selling price based on strictly budgetary considerations.

Another example is seen in a study of life-cycle costing for a missile guidance system (Colandene, 1964). In that program several levels of reliability were considered. Estimates were prepared for the costs of achieving these levels of reliability as well as for the resulting operational and total costs. These estimates are shown in Figure 1-4. The position of the minimum in the cumulative cost curve determines the optimized amount of R&D investment for a cost-effective increase in quality and reliability.

Quality Reputation

Quality assurance may be a convenient industrial strategy in terms not only of immediate returns but also as an investment for the future. A positive quality reputation, whether for consumer products or for capital goods, is of the highest value. Even though customers may not perceive a quality increase for a specific product, they nonetheless sense the performance record of the manufacturer's products. Moreover, a manufacturer's reputation may often be transferred from one established line of its products to a new line. More generally, the reputation created by a highly visible sector of a production community may profit the whole community. The quality reputation of the Swiss watch industry has been an important asset for

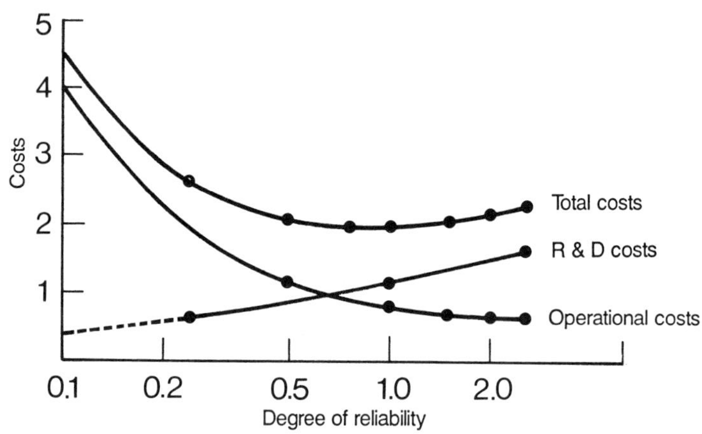

Figure 1-4 Dependence of R & D and operational costs on the level of reliability of a missile guidance system.

centuries, not only for that industry itself but reflecting on the whole national credibility. Examples of this kind have stimulated the setup of compulsory export quality controls in many countries such as Japan (Juran et al., 1979, p. 4–23).

1.3.2 *Quality and Costs*

Improved product quality assured by a more effective inspection either during or after the manufacturing process may translate into cost savings for the company. On the other hand, quality assurance has its own costs, so the optimized quality control effort must be determined for a given company in terms of a quality cost analysis (Feigenbaum, 1983; Ward, 1985).

Cost Savings Produced by Quality Control

Savings in material costs may be obtained through improvement in inspection controls. The author encountered such a situation during an interaction with a manufacturer of insulated wire. In a typical production line, electric wire with 2mm-thick polymeric coating was produced in order to assure proper cable insulation after aging effects. Without wire gaging controls, the wire diameter could vary within ± 0.5 mm because of variability in the rheological conditions within the polymer coating extruder. The extruding parameters were thus set up for a nominal wire coating thickness of 2.5 ± 0.5 mm in order to meet the minimum requirements.

The installation of an on-line wire diameter optical gaging system provided continuous control of the extruder pressure and temperature parameters, thus reducing the wire diameter variability to ± 0.1 mm. This allowed a reduction of the average coating thickness to 2.1 mm, a saving of 20% in polymer material costs.

Timely detection of faulty products by on-line inspection may also result in savings in process costs by avoiding processing of defective materials which would later be rejected in final inspection. This becomes particularly important when the faulty component is an integral part of the final assembly, causing rejection of the whole structure. A typical case was reported during a recent investigation at Pontiac Motors in order to establish a quality control program (Mundel, 1983). Among the findings was the fact that one of several sources for fasteners was supplying products, a high proportion of which proved nonconforming. The identification and correction of this source of defects changed the acceptance range on Pontiac engines from 65% to 95% on first inspection.

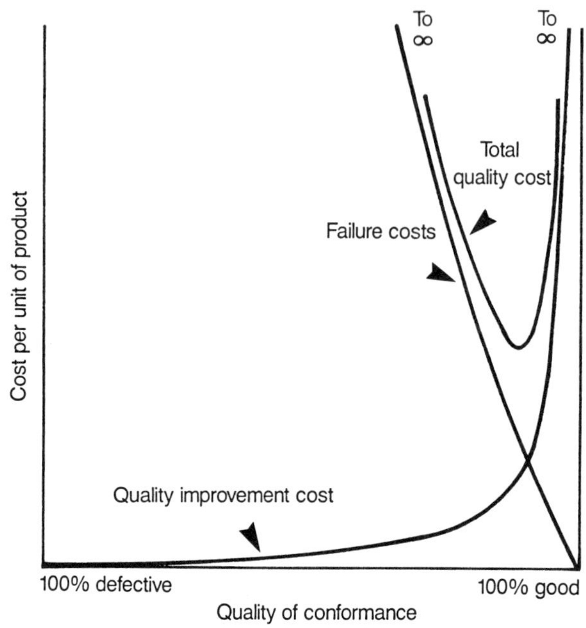

Figure 1-5 Model for optimum quality costs.

Quality Costs Analysis

The final evaluation on the optimum level of quality assurance activities within a company is usually performed through overall quality cost analysis. The concept of quality cost (Feigenbaum, 1983) corresponds to the total product cost versus the achieved product quality. Such a concept may be understood from Figure 1-5. The total quality cost is composed of failure costs, i.e., costs related to lack of quality such as scraps, rework charges, downtime, yield losses, etc. and quality improvement costs such as inspection equipment, personnel training, quality appraisal and data analysis, etc. It should be stressed that the actual cost of the equipment is often a small fraction of the total installation cost including process modifications, debugging, performance assessment and personnel training. As an example, the cost of a 3-D vision system is in the range of \$50,000 to \$90,000 and it is dropping, while total installation costs which have been made public are in the \$200,000 to \$500,000 range (Miller, 1985, p. 10). An often underestimated cost component related to quality control concerns the liquidity or working capital locked up in batches of products awaiting clearance and the cost of the space needed to store them. This factor constitutes a strong motive for the gradual evolution from batch inspection to continuous on-line inspection (Sharp, 1976).

With reference to Figure 1-5, we can see that if the product quality is very poor, the product cost is high because of the incidence of failure costs. We are in a typical situation where the company "cannot afford avoiding a quality improvement program." At the other extreme, excessive effort towards obtaining very high product quality results in skyrocketing quality improvement costs and a consequent upsurge in total quality cost. This is the "perfectionism" zone. The optimum level of product quality from a production cost point of view corresponds to the minimum of the total quality cost curve.

1.4 Statistical Techniques

Statistics is an important tool in analyzing quality. Even if 100% inspection is applied, a statistical analysis is required to evaluate the significance of product quality departure from norm, the establishment of an acceptable quality level or the choice of how immediate a feedback reaction on the process parameters must be after a given number of faulty products (Ott, 1975).

1.4.1 *The Control Chart*

One of the oldest and still widely used quality evaluation tools is the control chart. Such a chart provides in graphical form the time history of a given product parameter, such as the surface roughness of a cold-rolled sheet or the thickness of a polymeric coating so that a significant departure from normality can rapidly be assessed. An example of a chart record is shown in Figure 1-6. As the product departure from standard exceeds a predetermined level, the operator (or the central decision unit in a computer-controlled process, see, e.g., Drewery, 1985) will rapidly proceed to identify the problem and implement a corrective action to restore the standard operation. Such a method is effective because it is timely and located in the site of the operation.

The parameter recorded in the control chart may be the result of an individual measurement or an average, such as the average coating thickness over a 1 cm scanning measurement. Averaging over N samples decreases the data variability by a factor of $N^{1/2}$ (Ott, 1975, Section 1–8) and facilitates

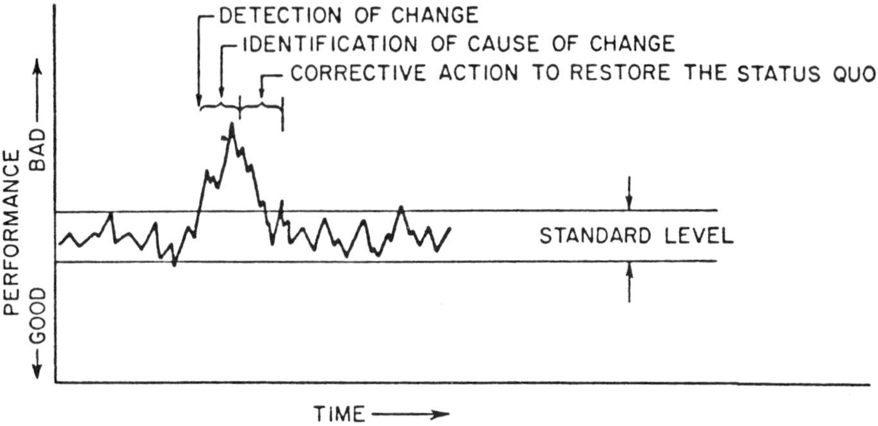

Figure 1-6 Control chart for a given quality parameter.

the interpretation of the records. However, the time or product variability of a given parameter may contain additional useful information on the process. Consequently, a pair of charts is often used for quality control, one plotting the average and the other the standard deviation or more often the range (maximum minus minimum values), in order to facilitate data collection by the operator. The average chart tells when a change has occurred in central tendency, which may be due to factors such as tool wear, a gradual increase in temperature, a new batch of materials of greater toughness or a different method used by a night shift workman. The range chart indicates when a significant gain or loss of uniformity has taken place. This may be caused by a process malfunction such as a loose tool post, erratic supply of coolant, careless chucking of stock or variability of pieces as received from a previous roughing operation.

1.4.2 *Statistical Analysis*

The example of Figure 1-6 shows the importance of determining when a given departure from standard is to be taken as significant. A wider range of accepted variability leads to lower quality standards as well as to a delayed reaction to process malfunction. At the other extreme, too narrow a variability range may require an unjustified amount of data analysis effort by the operator or generate feedback loop instabilities. One of the main purposes of statistical quality analysis is to determine which is the level of "normal" variability of a given product.

Process variations are traceable to two kinds of causes, (1) random, i.e., due solely to chance and (2) assignable, i.e., due to specific "findable" causes. Ideally, only random causes should be present in a process, because this represents the minimum possible amount of variation. A process which is operating without assignable causes of variation is said to be "in a state of statistical control," which is usually abbreviated to "in control." Frequently, a statistical analysis must be performed to analyze a frequency-distribution table such as the histogram shown in Figure 1-7 and to make comparisons with the probability distribution of a random variable in order to determine whether the parameter fluctuations are normal statistical variations.

Many engineering characteristics can be approximated by the normal distribution, which is plotted in Figure 1-8. A comparison of the frequency distribution with such a plot will provide information on whether the variable is normally distributed and what are the limits of the normal statistical fluctuations for such a variable (e.g., three times the standard deviation). A widely used method to verify the hypothesis of normal distribution employs probability paper (Ott, 1975, Section 1-7). Such diagrams are so constructed that a sample of data from a normally distributed population plots approximately as a straight line. An example of a cumulative frequency distribution from a normally distributed population is shown in Figure 1-9 (Juran et al., 1979). Departures from linearity may indicate the presence of abnormal factors. An example is shown in Figures 1-10 and 1-11 (Mundel, 1983). Figure 1-10 shows the control chart of a product showing both an inherent and a

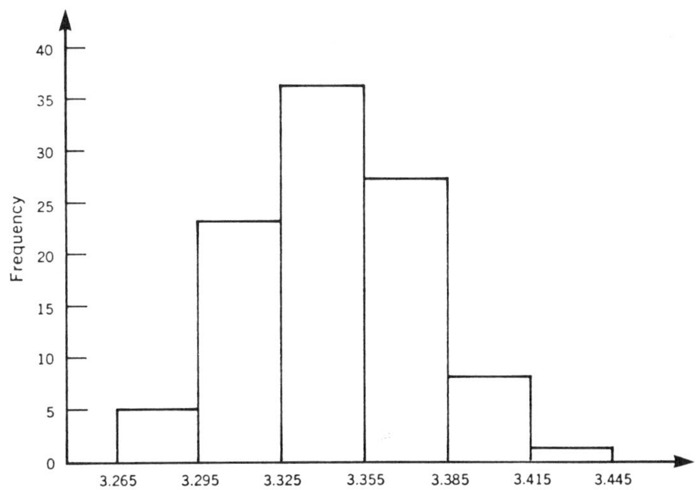

Figure 1-7 Histogram of the frequency distribution of a given parameter value.

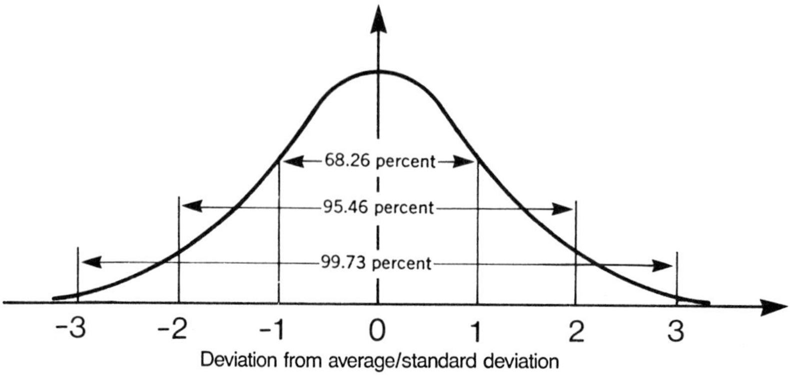

Figure 1-8 The normal probability distribution.

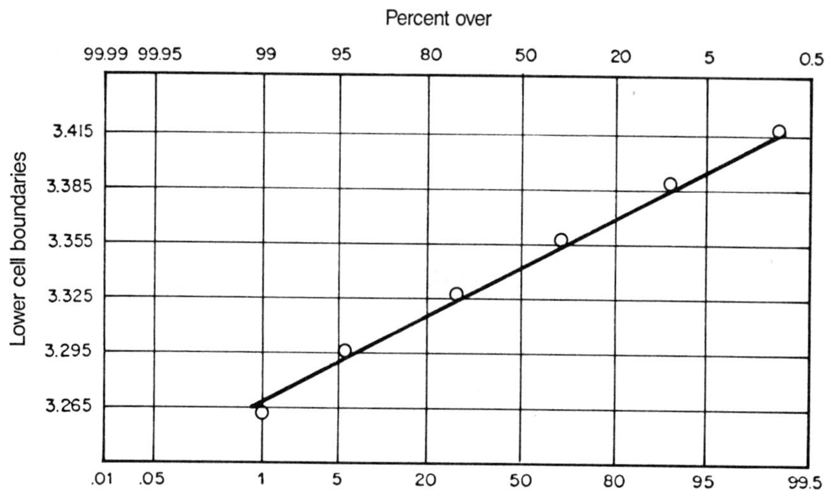

Figure 1-9 Cumulative frequency distribution from a normally distributed sample plotted on normal probability paper.

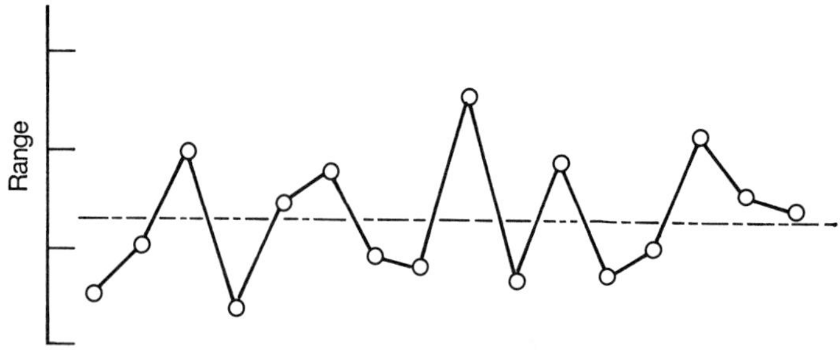

Figure 1-10 Control chart of a parameter showing a cyclical variation.

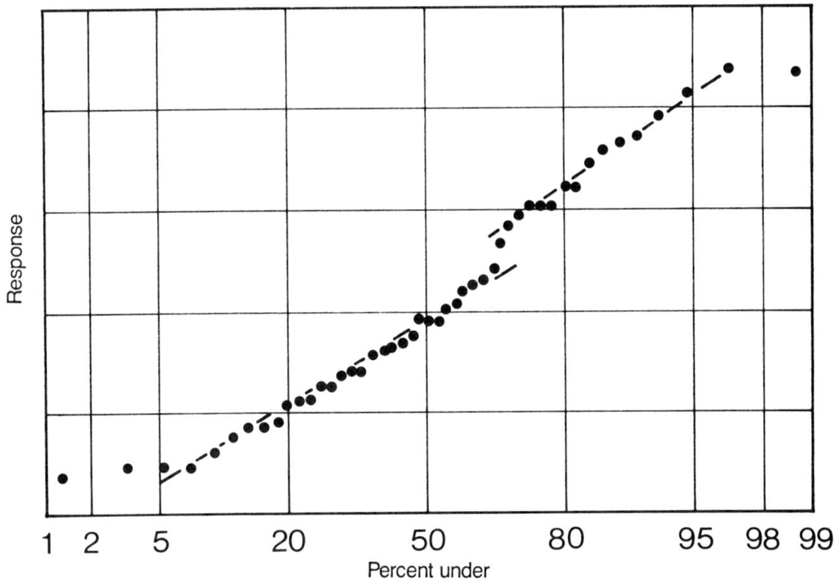

Figure 1-11 Normal probability plot for the parameter whose chart is partially reproduced in Figure 1-10. A departure from linearity is clearly visible.

cyclical variation. The normal probability plot for such a product, Figure 1-11, indicates the presence of a departure from normality for this population.

1.4.3 *Sampling*

The choice as to how much inspection effort is required in a given production environment often depends on the reliability of a suitable sampling procedure. If general quality information is required, such as in the verification of an "acceptable quality level" for the product (Mundel, 1983), a 100% inspection of the full production is not required. In this case, the general rules concerning statistical significance and lack of bias must be applied in the choice of the sampling plan (Ott, 1975, Chapt. 8; Cornish et al., 1981; Considine, 1985, p. 6–20 to 6–35). In particular, in industrial quality control the samples should be collected lot-by-lot, avoiding mixing of lots of products from different sources (different production shifts, different batches of input materials, etc). The choice of the samples within the lot must be random unless specific knowledge on the process justifies a different plan. For example, samples of the diameter of an extruded wire should be measured at random angular orientations unless we know that the wire cross section is perfectly circular. Knowledge of the product's maximum rate of change may provide important information to limit the sampling rate. An example of this may be found when sampling a liquid product. The presence of uniformity through fluidity reduces the need for random sampling as well as the required sampling rate, provided that the sampling location is chosen at a position where high agitation produces proper mixing of all the components in the liquid (Considine, 1985, p. 11–20).

The optimum sampling procedure depends on whether the product specifications are given in terms of "*attributes*" or of "*variables*". The first case refers to the specifications which are written in terms of a simple classification of product into categories of defective and nondefective (e.g., broken pieces in a shipment). In this case, the sampling plans involve comparison of the number of defectives counted in the sample to an acceptance number to determine the acceptability of the lot. Specifying variables requires a more onerous variable measurement procedure for controls, although techniques such as narrow-limit gaging are available to improve the sampling efficiency (Ott, 1975, Chapter 7). When specifications are stated in terms of continuous measurements on individual units of product (e.g., length of each of the pieces in a shipment), considerable savings in sample size may be achieved by applying a variables sampling plan directly to the measurements themselves. Sampling by variables requires a higher inspection effort, because each item must be measured rather than classified on a go or no-go basis as in sampling by attributes. However, proper unit weighting by its proximity to specifications results in an at least 30% smaller sample size for same protection (see Juran et al., 1979, Chapter 25).

Finally, a sampling program may be specified as either single sampling or double sampling. When the former is chosen, one sample of n elements per lot is taken, and the lot is either accepted or rejected depending on whether the sample proves to be acceptable or not. Obviously, unacceptable lots may provide acceptable samples and vice versa. When double sampling, a first sample of n_1 elements per lot is taken, and the lot is accepted if the sample is well above the specifications or rejected if the sample is much below the specifications. In the intermediate cases, a second sample of n_2 elements is taken before making the decision. It is normally found that for an equal probability of misjudgement, double sampling requires a sample size n_1 much smaller than the size n of single sampling, while the additional sample of size n_2 is required in a relatively small number of cases. This leads to a total inspection cost for equal output quality which is usually smaller in the case of double sampling.

1.5 Inspection Requirements in the Materials Processing and Manufacturing Industries

The installation of inspection instrumentation has followed a rapidly increasing trend in recent decades. As an example, Figure 1-12 shows the number of on-line process analyzers for chemical quality control and process monitoring which have been installed on successive chemical plants of the same kind by a British petrochemical complex (Cornish et al., 1981).

In the transformation industry, inspection needs are created by the requirements of (a) raw material control, (b) process control and (c) product control. The specific parameters to be inspected are strongly dependent on the kind of products that the company supplies so that a list of inspection needs is better made by industrial sectors.

1.5.1 *Metallurgical Industry*

(a) Raw materials: The principal raw material is the ore, which usually comes from a predetermined source. Variations in the processing

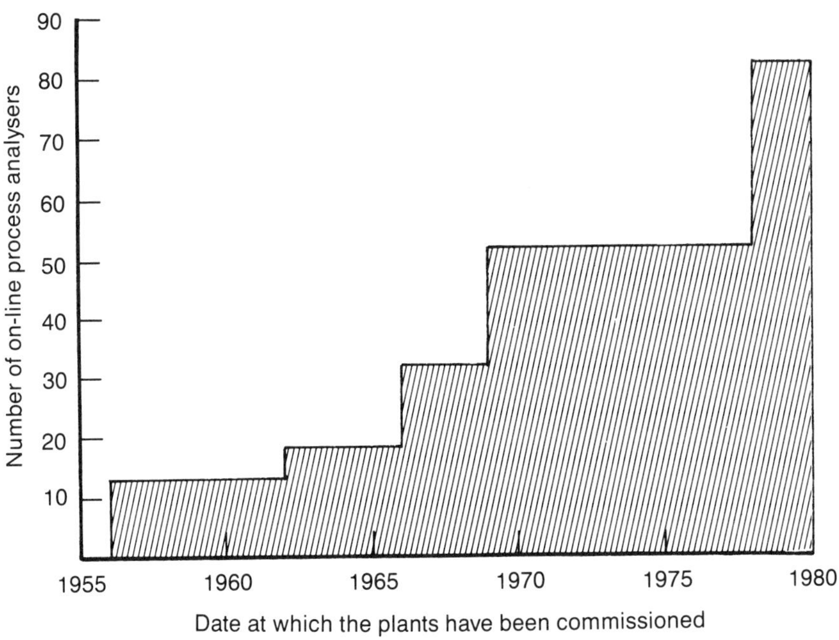

Figure 1-12 Number of on-line process analyzers installed on successive chemical plants of the same kind by a British petrochemical industry.

quality characteristics of the ore can have a profound influence upon the quality of the metal as well as the efficiency of the process. Many of the raw materials are purchased on the basis of commercial specification, and it is only on rare occasions that a plant is in an inventory or purchasing position to reject a shipment that fails to conform. Chemical composition, X-ray crystallography and morphological (e.g., particle size and uniformity) information is required to choose the process that best conforms to the quality of the ore available.

(b) Process control: Molten metal received from smelting is usually refined or modified by the addition of alloying ingredients before casting. Its composition must be determined at this stage, usually by sampling and chemical analysis. Subsequent processing operations include casting, forging and milling to produce slabs, blooms or billets, which are subsequently reduced to rods, plates, or sheets by rolling or drawing. Such products may be sold at this stage or after further processing, such as forming or coating. The optimization of process parameters such as furnace temperature or roll gap and speed requires on-line inspection of the metal temperature, mechanical and metallurgical properties, or sheet thickness.

(c) Product control: The final product must be inspected for geometrical dimensions, flatness and profile, surface cleanliness and brightness, presence of defects such as inclusions, cracks or slivers, as well as tested for mechanical composition and mechanical properties.

1.5.2 *Metal Fabricating and Automotive Industry*

(a) Raw materials: Controls similar to those performed on the finished product of the metallurgical industries are required. Examples are chemical composition testing for metals, physical and dimensional properties of parts, and integrity of anti-corrosion coatings.

(b) Process control: Metallic objects are die formed, forged and heat treated, cut and drilled, finished and grinded. Process parameters to inspect are die integrity and temperature, workpiece alignment, tool wear and proper positioning during cutting and drilling, workpiece surface and subsurface temperature during heat treating and tool orientation during grinding. The massive introduction of computer controlled equipment has spurred an immediate need for adaptive machine vision systems in this field.

(c) Product control: Robotic machining, welding, assembling or spray-coating and painting are being heavily introduced in the automotive industry, generating strong requirements for vision based, robot guiding systems and on-process surface quality inspection. Product characteristics such as the geometrical dimensions of the die-formed and cut or drilled parts, degree of surface finish, hole shape and orientation, and presence of hairline cracking must be verified after machining. Specialized products such as nuts, bolts, springs, gears, etc. must be inspected for geometrical characteristics and mechanical integrity. Consumer oriented industries such as the automotive industry require inspection of sensory quality such as painted body color and smoothness, as well as safety related tests such as brake tests and exhaust emission analysis.

1.5.3 *Pulp and Paper*

(a) Raw materials: Pulpwood must be inspected for volume and dry weight, age of logs and bark contents. Purchased wood pulp must be tested for consistency, freeness and brightness, and shives contents.

(b) Process control: The paper making process comprises two steps, the pulp mill and the paper mill. Pulping may be obtained mechanically, i.e., through grinding, or chemically, i.e., by solvent digesting and bleaching. Pulp inspection for freeness, average fiber length and degree of fibrillation may provide the required information to control the grinding or the chemical treating process. As to the paper milling operation, parameters such as amount of additives, drawing speed, vacuum suction or calender pressure may be adjusted by an in-process knowledge of the paper web basis weight, moisture, formation, density, gloss and color.

(c) Product control: Paper opacity, roughness and printability, and presence of flaws such as spots, tears, rope marks or broken edges are examples of quality requirements for the finished paper product.

1.5.4 *Chemical Process Industries*

(a) Raw materials: Most raw materials for chemical processes are purchased on the basis of chemical analysis, with standard specifications typical. The methods and philosophy of sampling and testing depend upon the nature of the material (bulk solids, bulk liquids, type of packaging, etc.).

(b) Process control: Chemical processes may comprise widely different techniques, each of which presents particular inspection and process control needs. Examples are the evaluation of the degree of sedimentation for solvent extraction, refractive index monitoring for the control of the hydrogenation process, determination of the caustic concentration, digester temperature and pressure, properties of the black liquor in the causticization process and the optical monitoring of flow, dispersion, turbidity and liquid level within the vessel (Raab, 1984).

(c) Product control: Relevant properties are the percentage of impurities, completeness of the chemical transformations and uniformity of the final product.

1.5.5 *Plastics Forming*

(a) Raw materials: Mixtures of polymers and additives, usually in pellet form, are heated and shaped in molds or die-extruded to produce a wide variety of objects. These materials are commonly bought from major chemical suppliers whose quality controls are well established. As a consequence, noncritical molders avoid the expense of laboratory confirmations and limit themselves to identity checking. In the case of degradable materials, such as uncured thermosetting polymers, inspection of the degree of precure is often performed before forming.

(b) Process control: Examples of inspection operations useful for process control are blown-film thickness gaging to control the rate and concentricity of blow extrusion, surface roughness measurement to monitor the extrusion of compounded materials, molded part temperature control to determine the speed of the molding operation, as well as sensing of temperature, pressure and mixing efficiency within the extruder.

(c) Product control: Examples are the presence of flashes or thin walls in molded products as well as film transparency, tear strength, permeability, thickness and uniformity for blown extruded films.

1.5.6 *Electronic Components*

(a) Raw materials: The electronic components have undergone a miniaturization process resulting in a packaging density of tens of thousands of components per chip in the modern VLSI devices. As a consequence, the cost of the raw materials is a small fraction of the cost of the device. Nevertheless, the standards for integrity, purity and absence of contamination for the semiconductor substrates and thin film materials are particularly high.

(b) Process control: Examples of in-process controls are the monitoring of materials during thin film electroplating or etching, temperature of the joined components during soldering or cleanliness inspection prior to bonding.

(c) Product control: Such controls are of the highest importance for micro-circuits for two reasons. First, the large number of components present on the chip make the overall probability of a circuit malfunction relatively high. Second, because of the above mentioned low material costs compared to the design and process costs, there is a tendency to fabricate such components in large batches, a large portion of these being rejected after inspection. The rejection rates are sometimes of the order of 95%. Post-process inspection thus represents an important activity for such companies. Examples of such procedures are visual observation of chip and contact integrity by optical, acoustic or scanning electron microscopy, automated pattern verification of multi-layer boards before lamination, automated circuit testing by a computer driven array of contact pins or by resident photoconductive bridges, or thermal inertia testing of the solder joints.

Chapter 2

Overview of Nonoptical Inspection Techniques

When the optical scientist is faced with an industrial inspection problem, the first and perhaps the most crucial decision to be made is whether a suitable inspection technique exists to solve such a problem and, if several approaches are possible, to choose the best one. Such a decision must take into account not only physical feasibility but also economic considerations. Few problems remain irresolvable if one can afford unlimited development efforts, but the question is whether the planned technique can provide the required information at an affordable cost. The acceptable budget level is determined by the benefits to be gained minus the life-cycle cost of the new equipment, including the added expenses incurred to adapt the production process to the new tool.

If an optical inspection technique appears to be both physically and economically suitable, a further study must be made to verify whether an alternative optical or nonoptical technique exists which would be easier to implement, or more reliable, or which would require less maintenance during operation. Otherwise, a competitor could step in with a cheaper technique just after our company has undertaken the expensive process of developing the instrument and preparing the market for it. The comparison with alternative techniques should be performed, at the latest, after the first feasibility study, in order to avoid unnecessary development efforts and also because during product development the scientist is too focused on the chosen technique to make an objective comparison with other approaches.

An overview of nonoptical industrial inspection techniques is given in this chapter. This review should provide optics specialists with a basic knowledge of the competitive techniques and with a feeling for the range of their possible application. This should enable them to determine whether an alternative technique has enough potential to justify further inquiry with vendors supplying this kind of instrumentation.

2.1 Mechanical Techniques

2.1.1 *Mechanical Gaging*

If access to the inspected workpiece is unrestricted and if measurement time is not at a premium, mechanical contact techniques may be used to provide dimensional information. Manually set vernier calipers, micrometers and fixed gages are widely used in industry. The accuracy of such gages is usually of the order of 10 μm. An accuracy of the order of 1 μm is provided by linear variable differential transformer (LVDT) devices (see, e.g., Considine, 1985, p. 9–17). The combination of a micrometer and a piezoelectric actuator (as available, e.g., from companies such as Burleigh or Tropel) provides a relative sensitivity of a fraction of 1 μm, although the absolute accuracy is limited by mechanical mismatch and deflection of the mechanical assembly, contact surface smoothness and cleanliness, temperature variations, applied pressure and gage wear.

When the workpiece is too large or awkwardly shaped to be measured by a micrometer or caliper, a test set can be obtained using sliding clamps. Different kinds of assemblies can be used to measure parts such as the outside diameter of a piston (Kennedy and Andrews, 1977, Chapter 8). The absolute value of the diameter is obtained by comparison with gage block standards. A list of commercial equipment of this kind can be found in the directory, *Inspection, Gaging and Testing*, 1984, Section 4 for manual systems and Section 3 for automated coordinate measuring systems, which is published by the Society of Manufacturing Engineers, Dearborn, MI.

One of the limitations of contact gages for on-line inspection is their reduced response speed. An example is given in Figure 2-1, showing the results of a comparison of an LVDT contact probe with an electro-optical triangulation sensor for monitoring thin metal sheet displacement in our institute's steel-galvanizing pilot line (Lamontagne et al., 1987). Although

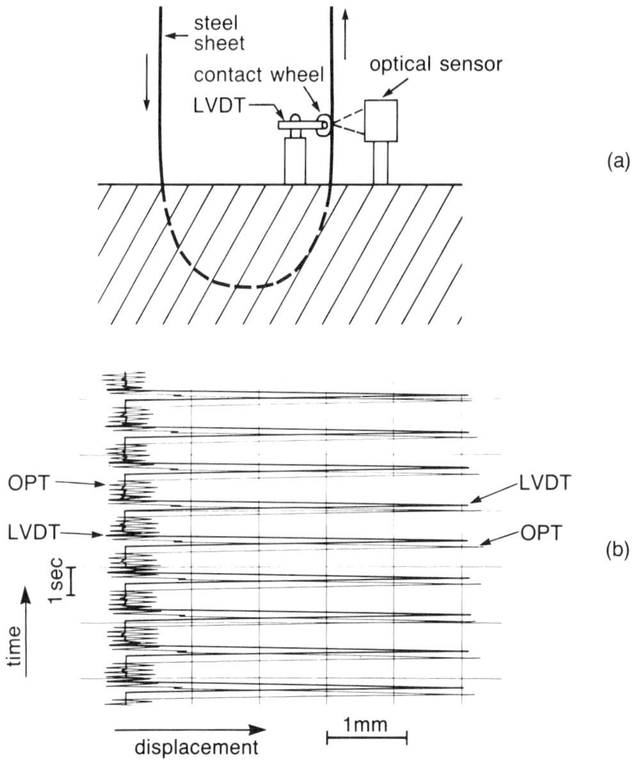

Figure 2-1 Comparison of the performance of a contact device and of an optical sensor to probe sheet displacement under periodical solicitation in the IMRI steel-galvanizing pilot line: (a) experimental setup, (b) chart-recorder outputs from the two sensors (the traces are purposely shifted laterally for better viewing).

both instruments proved to have an adequate absolute resolution for monitoring the sheet displacement under periodic solicitation, the LVDT device could not follow the small-amplitude oscillations at a frequency of nearly 5 Hz due to mechanical inertia.

For the routine measurement of the internal diameter of cylinders, holes and bases, important errors are introduced by caliper tilting or misalignment or by the presence of foreign particles. To avoid such problems, air gages of the kind shown in Figure 2-2 are often used for such measurements. An air pressure source is connected by a pressure tube to a measuring plug which is introduced in the cylinder. The pressure in the tube, which is measured by a manometer, depends on the average gap between the plug and the cylinder surface. Typical air spacings of 50 μm are used to obtain a measurement

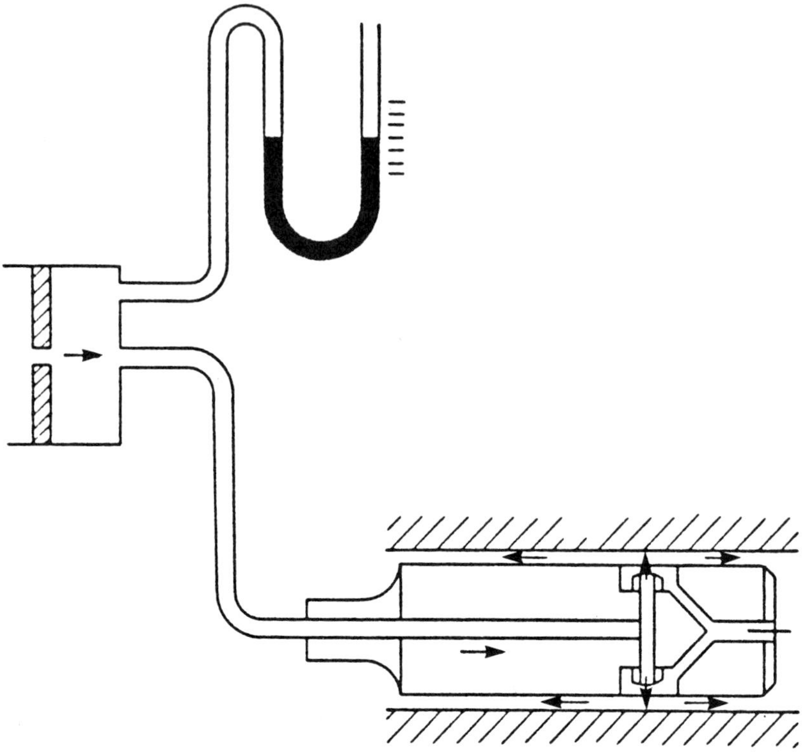

Figure 2.2 Air gage for the inside diameter of a cylinder.

accuracy of the order of 10 μm. Different plugs are used for different bore diameters. Routine gaging by this method is convenient because it requires no precise tool adjustment from the operator, while powder particles on the cylinder surface are blown away by the ejected air. Variable-width air gases are also used to perform measurements on moving surfaces, such as for real-time monitoring the thickness of a cold-rolled metallic sheet. Figure 2-3 shows an example of such an industrial system (Kennedy and Andrews, 1977). Automatic correction of bore geometry during the machining cycle has recently been reported using similar tools (Pepper, 1985).

2.1.2 *Tactile Sensors*

Tactile transducers correspond to the human ability to recognize objects by the sense of touch. The development of such sensors (Pugh, 1986) has been spurred by the requirements of automated and robotic manufacturing. In contrast with optical proximity sensors (Catros and Espiau, 1980; see also

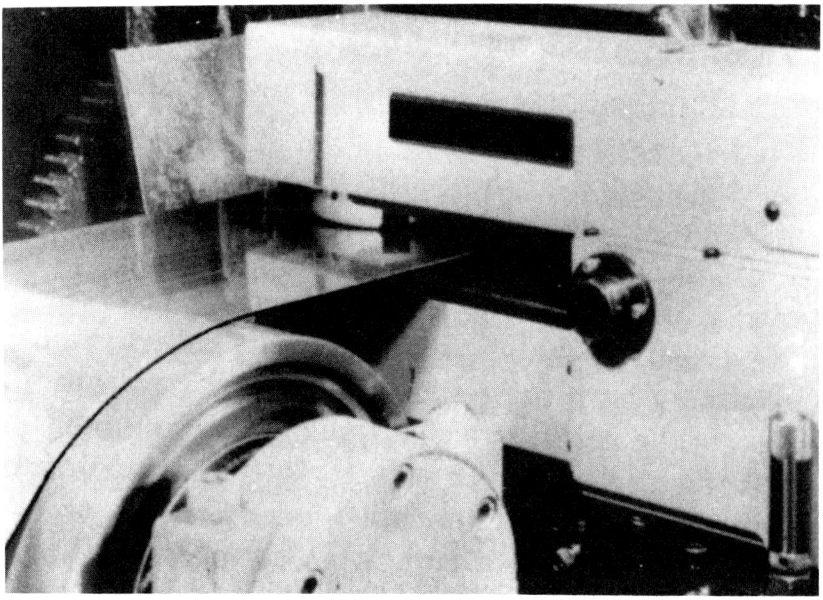

Figure 2-3 Air-cushion sheet thickness measurement at rolling mill.

Inspection, Gaging and Testing, available from the Society of Manufacturing Engineers, 1984, p. 490 for fiber optic proximity sensors), such sensors require actual contact with the inspected surface, the detected signal depending on interface pressure. A number of pressure sensitive techniques have been proposed for use in tactile sensors (Jones and Kvasnik, 1985; Allan, 1985). Thin pressure sensitive films, such as PVF_2 or magnetoresistive skins (Vranish, 1984) are under intensive development for this purpose.

Figure 2-4 shows a typical bolt-joining operation to be performed by an industrial robot. The reaction force sensed by the stress transducer during the insertion of the bolt in the hole is shown in Figure 2-5 (Nervin and Whitney, 1978). After an initial peak due to an eventual minor mismatch of the bolt with respect to the hole, the reaction force decreases and then linearly increases proportionally to the contact surface as the bolt penetrates into the hole. An eventual misalignment of the bolt may produce a sudden increase of the reaction force with risk of jamming, requiring a correction of the tool orientation by the gripper.

Such sensors can be used to finely tune the position of a robot arm, which can be positioned with an accuracy not higher than 1 or 2 mm on standard robotic systems because of part-to-part looseness and deflections of the arm

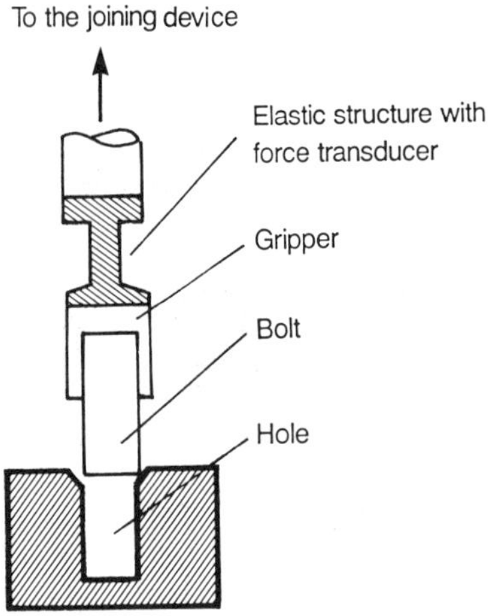

Figure 2-4 Principle of a joining device with tactile sensor.

structure. An example of such an implementation (Schweizer, 1977) is shown in Figure 2-6. The bolt gripper is driven by an xyz assembly. The z-motor moves the bolt toward the hole until a light contact with the surface is established, as detected by the pressure pickup transducer. At this point the x- and y- motors scan the surface, looking for the position of minimum axial pressure which corresponds to the position of the hole. At this position the z-motor drives the bolt into the hole, while inductive transducers measure an eventual angular misalignment which is corrected by the x- and y- motors.

Tactile sensors have also been proposed for object recognition through a

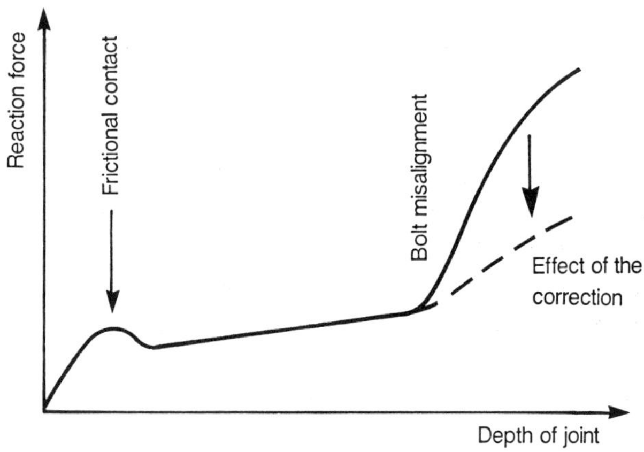

Figure 2-5 Basic curve for the reaction forces during a joining operation.

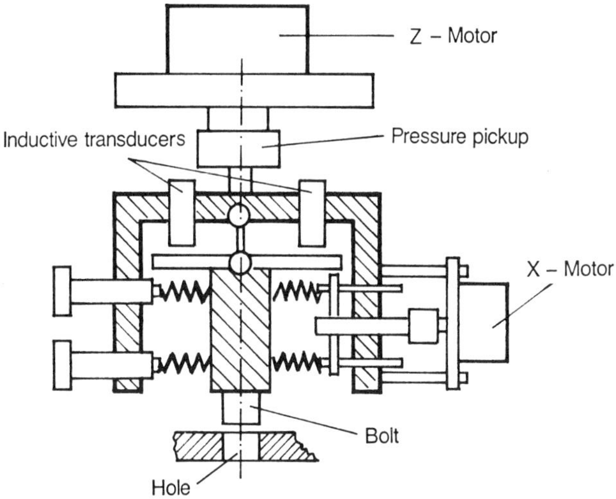

Figure 2-6 Tactile system for bolt alignment during joining.

multiple contact array. An illustration of the basic idea is shown in Figure 2-7 (Bretschi, 1981). In this case an object is recognized from the signals sent by a matrix of moving pins adapting themselves to the object morphology. As Figure 2-7 shows, such a matrix could be integrated into a gripper for object recognition during robot handling. Obviously, such functions may also be performed by optical or ultrasonic pattern recognition or proximity sensors.

2.2 Ultrasonic Inspection

After vision, hearing is the most important human sense for obtaining environmental information. Unlike vision, hearing does not provide us with

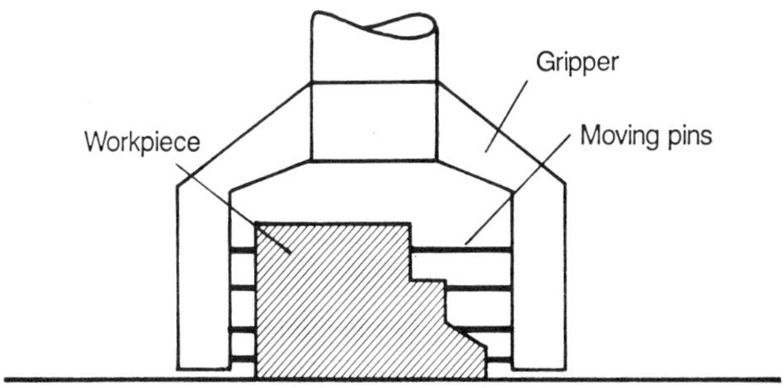

Figure 2-7 Diagram of a tactile sensor for recognition tasks.

a geometrical image of the world. The physical reason is related to the value of the sonic wavelength which is of the order of one meter as compared to one micrometer for the optical wavelength. As the pixel size of an image cannot be smaller than a wavelength, our hearing organs should extend over hundreds of meters in order to obtain an acoustic image of our environment! Similar considerations can be made concerning the difficulty of obtaining pictorial information by ultrasonic devices. Ultrasonic frequencies in the GHz range would be required in order to reduce the acoustic wavelength to the optical size. However, the penetration distance of GHz ultrasound in air or water is extremely small because the ultrasonic attenuation increases sharply with frequency (Kinsler and Frey, 1962, Chapter 9).

Nevertheless, nonpictorial acoustic information is very valuable to human beings. We can communicate acoustically without the need to angle our hearing organs in the direction of the sound source, we can find a gap under a wall panel by tapping on it or we can detect a malfunction in our car by listening to an unusual engine noise. Bats can orient themselves in total darkness by a skilled utilization of their sonar system. It is therefore not surprising that many sensors for industrial product inspection and robot guiding are under development using ultrasonic devices. This section presents an overview of such applications.

2.2.1 *Active Air Propagation Techniques*

Although optical triangulation techniques are available for depth sensing, optical distance measurements tend to be rather inaccurate if the sensing device must be kept compact. Airborne acoustic pulse-echo techniques are thus widespread for range finding applications, particularly for robot guiding (Marsh et al., 1984; Pugh, 1986).

The basic principle is similar to the underwater active sonar method. A short ultrasonic burst is emitted by a piezoelectric transducer, and the echo reflected from a surface intercepted by the ultrasonic beam is detected either by the same transducer or by a receiver situated nearby the transmitter. The time delay between the transmitted and the received pulse provides an evaluation of the distance between the transducer and the surface. Alternatively, a chirped-frequency continuous output may be emitted by the transmitter and the delayed response is simultaneously detected by the receiver, a configuration which has found application as an orientation aid for the blind (Kay, 1964; Kay and Crecraft, 1985). Ultrasonic range finding modules such as the 200 KHz device produced by Massa Products, Hingham, MA have a range of 5 to 60 cm, while 2 MHz devices have been used with a 2 to 20 cm range and 0.4 mm accuracy in a laboratory environment (Fox et al., 1983). If the reflecting surface is large and normal to the beam and if the transducer diameter is much larger than the acoustic wavelength, the angular divergence can be kept small and the maximum range is mainly limited by the air attenuation, which is of the order of 0.1 and 10 dB/cm, respectively, at 200 KHz and 2 MHz (Kinsler and Frey, 1962, p. 232).

The relation between range and target cross section is given by Gadient (1983) in a study for the applicability to robotic sensing of the Polaroid acoustic range finder used for automatically focusing some Polaroid cameras (Deustman, 1980). The far-field angular pattern of such a device has a 12° wide central lobe within the −10dB points. The lateral resolution is thus of the order of 10 cm at 60 cm range. Such a relatively high angular divergence, which is diffraction limited and thus cannot be reduced using a focused (concave shaped) transducer, is a consequence of the relatively large ultrasonic wavelength (3 mm for a 100 KHz acoustic wave in air as compared to 1 μm for the typical optical wavelength).

The large value of the acoustic wavelength makes ultrasonic range finders relatively insensitive to the roughness of the inspected surface. Optically rough surfaces appear very smooth to the incident ultrasonic beam, resulting in strongly directional specular reflections if the surface area is large. Omnidirectional reflections will, however, be produced by objects smaller than the acoustic wavelength. Specular reflections result in strong echo signals only if the reflecting surface is normal to the incident beam.

Solid surfaces are extremely good acoustic reflectors, so that ultrasonic range finders are quite unaffected by the color or cleanliness of the isonified surface. The acoustic reflectivity of an interface between two media 1 and 2 is at normal incidence (Kinsler and Frey, 1962, p. 131)

$$I_r/I_i = (\rho_2 c_2 - \rho_1 c_1)^2/(\rho_2 c_2 + \rho_1 c_1)^2, \qquad (2\text{-}1)$$

where I_r and I_i are, respectively, the sound intensities of the reflected and of the incident plane wave beams, ρ_1 and ρ_2 are the densities of the two media, and c_1 and c_2 are the sound velocities in the two media. Air-to-solid interfaces thus have typical acoustic reflectivities of 0.9999. Such considerations explain the technical difficulties which are encountered in the manufacture of efficient ultrasonic transducers operating in air. Such devices, which are usually made of piezoelectric plates with metallized opposite surfaces used as electrodes, produce a very large acoustic field within the plate, but the acoustic energy coupled to the air is extremely low. This, in turn, produces strong "ringing" of the transducer (prolongated vibrational oscillation of the plate) which is superposed to the faint acoustic reflections. Insertion losses of 52 dB were obtained with a 2 MHz transducer by coating the plate surface with a quarter-wavelength low impedence material (Fox et al., 1983). Better air coupling coefficients can be obtained at lower frequencies using induction or capacitance driven membrane transducers (Bretschi, 1981, Chapter 5).

By scanning an acoustic range finder over an object, a three-dimensional image of the object surface can be obtained. An example of such an implementation is shown in Figure 2-8 (Wang and Will, 1978). A 270 KHz receiver-transmitter system was mounted on a manipulator system. Both transducers have a conical acceptance angle of nearly 7°. An object located at the intersection of the two cones will give an echo to the receiver. By measuring the time of flight of the ultrasonic pulse at each position over the object, a three-dimensional image of the object is obtained. The main problems faced by this technique are a limited spatial resolution, ambiguities generated by spurious reflections when the object configuration is complex, as well as the long period of time required for a complete scan.

A faster method for three-dimensional object recognition by comparison with a previously recorded ultrasonic signature has recently been reported (Buckley, 1985). The part to be inspected is insonified by a continuous ultrasonic wave and the near-field scattered radiation is sampled and compared to a data base of possible signatures for recognition, in a way similar to the optical scattering approach (Swift and Dewhurst, 1983; Dewhurst and Swift, 1985). An array of 8 or 16 ultrasonic transducers sample the scattered field, and the system can typically distinguish between 15 different object signatures or part orientations. This approach is particularly useful to find small defects in repetitive parts in the assembly line. An instrument of this kind is manufactured by the company Cochlea of San José, CA. To obtain an acceptable resolution, the sensor must be within two wavelengths from the object, which means a standoff distance of a few centimeters at a frequency of 40 KHz. If the piece is repetitively positioned at a specific distance from the sensor, defects as small as 25 μm in size can be detected over smooth objects. Defects in convoluted objects are harder to find and may require special sensor geometries, such as a miniaturized probe for introduction inside a hole to check if it has a thread.

2.2.2 *Passive Airborne Techniques*

Passive acoustic techniques rely, as in the passive sonar systems, on the detection of acoustic waves emitted by the inspected object, i.e., they do not

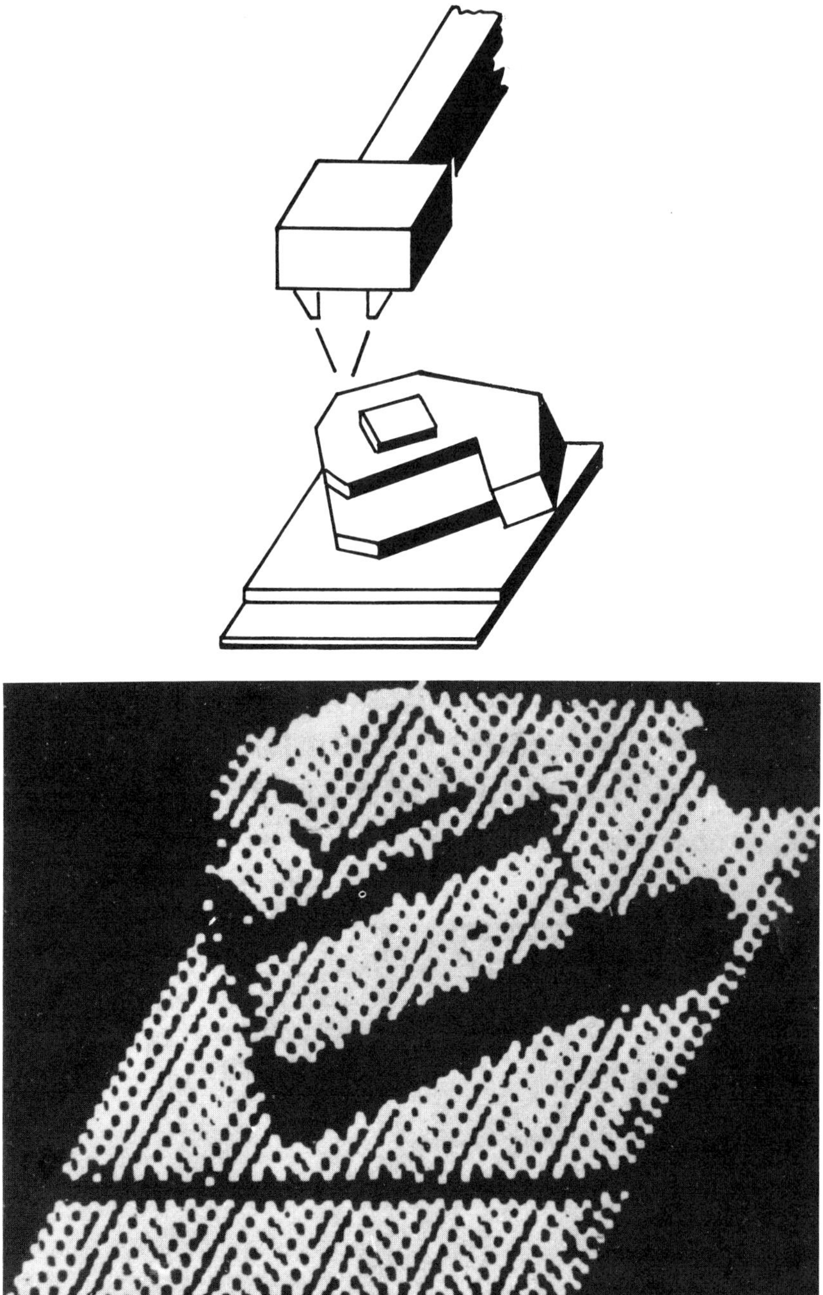

Figure 2-8 (top): diagram of an ultrasonic transmitter-receiver range finder mounted on a scanning system; (bottom): three-dimensional image of the scanned object.

require an acoustic transmitter. Some examples of industrial applications of such techniques are given in this section.

Steel Converter Monitoring

In the steel converter, the accompanying elements in the crude iron (carbon, sulphur, phosphorus) are reduced by injection of air or oxygen into the smelt. In the past, the chief engineer could monitor the rate of the oxidation reaction by optically observing the flame ejected from the mouth of the converter and by listening to the corresponding sound. The sound intensity varies during the blowing period for two reasons: first, the volume of the gas-filled inner cavity is reduced by the slag accumulation; second, more and more liquid slag surrounds the air inlet, thus muffling the sound produced by the emerging gas.

Modern steel converters are equipped with an exhaust hood for waste gas cleaning. Optical inspection of the flame is obstructed by such a hood, which also affects the free propagation of the sound. An oxidation monitoring device using an acoustic pipe has thus been developed for process control (Bretchi, 1980, Section 10.2.2). A diagram of the system is shown in Figure 2-9. The sound is picked up near the mouth of the converter by a 5 to 15 m long steel pipe of 3 to 5 cm in diameter. The microphone and signal processing electronics are thus protected against heat, flying sparks, slag eruptions and soiling. The high frequency (hissing) components of the signal which do not

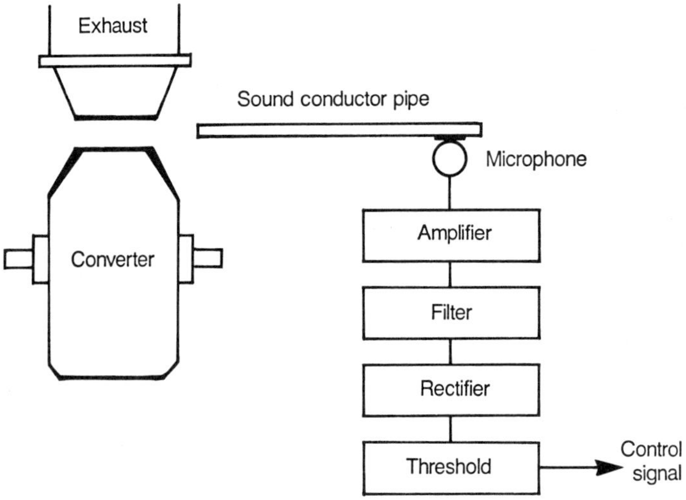

Figure 2-9 Diagram of an acoustic steel converter monitoring system.

contain useful information are filtered out and the level of the signal in the 150 to 1,000 Hz range is monitored for process control. A drop in the signal level, corresponding to the formation of a deep slag layer, is a warning to stop the gas supply in order to avoid excessive reduction or slag boil-over.

Acoustic Monitoring of Electric Arc Processes

Electric arcs are used in a variety of material processing activities such as in arc welding or arc furnaces. A precise knowledge of the electric current and of the potential difference across the arc is required to continuously monitor such processes. The voltmeter value for the voltage, which depends mainly on the arc length, is often affected by noise produced by induction from arc-current fluctuations and variable ohmic drops in the arc electrodes. An acoustic monitoring technique was thus developed to overcome such problems (Drouet and Nadeau, 1982).

As shown in the insert of Figure 2-10, the acoustic power emitted by the arc is measured by a microphone situated at 85 mm from the arc axis. Figure 2-10 shows the relation between the arc voltage V from the voltmeter and the value KV^* which is equal to the integrated acoustic power signal divided by the current. By varying from 3 mm to 23 mm the arc length and thus the arc voltage, a linear relation is found between the actual voltage and the acoustically evaluated voltage level V^*. Such a relation can be used to determine the arc voltage level, or equivalently the arc length, from an independent measurement of the acoustic noise from the arc.

2.2.3 *Active Solid Propagation Techniques*

The most attractive feature of ultrasound is its capability to propagate within opaque materials. Unlike electromagnetic waves, mechanical waves are not affected by the electrical conductivity of the material so that ultrasonic waves in the MHz range can penetrate to depths of tens of centimeters inside metals. Different configurations are used to inspect solid materials for flaws or physical characteristics.

(a) *Point-by-point manual scan*

A typical configuration is shown in Figure 2-11. A welded plate with a crack near the root of the weldment is being inspected ultrasonically. An angle transducer (piezoelectric transducer mounted on a wedge) is used to transmit an ultrasonic pulse at a given angle, as shown in the figure.

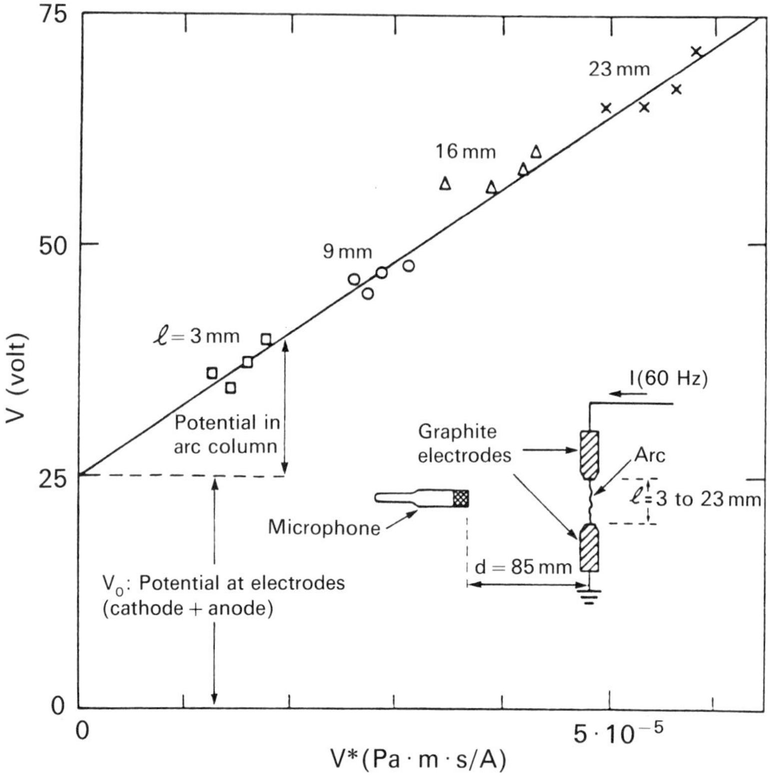

Figure 2-10 Relation between the voltage across an electric arc and the normalized acoustic power produced by the arc.

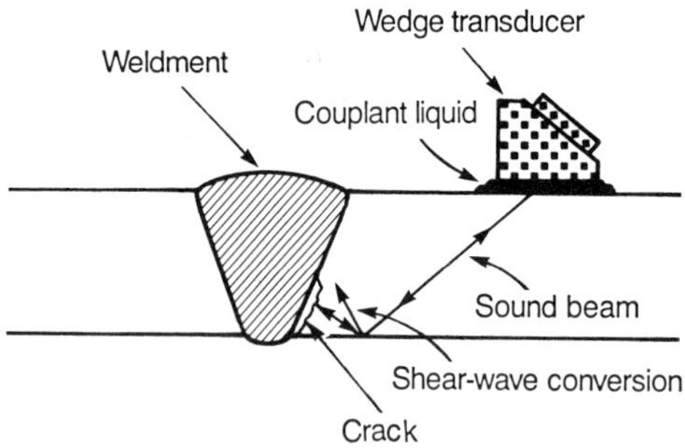

Figure 2-11 Pulse-echo ultrasonic testing to find cracks in a welded plate.

The ultrasonic beam is reflected at the bottom of the plate, so that it strikes the crack surface at a nearly normal incidence for best reflectivity. Because of the very high reflection coefficient of acoustic waves at the air-solid interface, even very narrow air gaps (down to a few micrometers in thickness) provide very strong pulse-echoes when they are normal to the incident ultrasonic beam. This explains why ultrasonic techniques are so widely used in nondestructive testing. Because the crack root acts as a stress concentrator producing crack propagation at an accelerated rate, timely detection of cracks in structures such as aircraft or nuclear reactors may prevent catastrophic failures.

As indicated in Figure 2-11, longitudinal-to-transverse mode conversion may occur at the bottom reflection (acoustic waves are longitudinal or transversal depending on whether the vibration direction is parallel or perpendicular to the direction of propagation of the ultrasonic wave). As the transverse wave velocity is smaller than the velocity of the longitudinal wave, the transverse wave is reflected at a sharper angle. This may generate spurious echoes, making signal interpretation rather complicated in certain cases. It should also be noted that a layer of coupling liquid (usually water or oil) must be present at the surface-transducer interface if good coupling of the ultrasonic beam in and out of the material is to be obtained. Liquid application and transducer proper positioning must be assured at every measurement, making manual scans of long weldments or large surfaces particularly time consuming.

(b) *Immersion techniques*

In order to avoid manual scanning over wide surfaces, immersion techniques are used when possible. A typical configuration is shown in Figure 2-12. A search unit, ultrasonic transducer mounted on a manipulator having three to five degrees of freedom (two angular degrees of freedom are often necessary if complex surfaces are to be followed), scans the workpiece immersed in a water-filled tank. This assures good ultrasonic coupling from the transducer to the workpiece without requiring a manual application of the coupling liquid. A typical signal obtained at a given transducer position is shown in Figure 2-13 (Boyer, 1976). The first echo corresponds to the reflection at the interface between the water and the top surface of the plate. A back reflection is also obtained from the interface between the water and the bottom surface of the workpiece. Additional echoes will be present between the front and

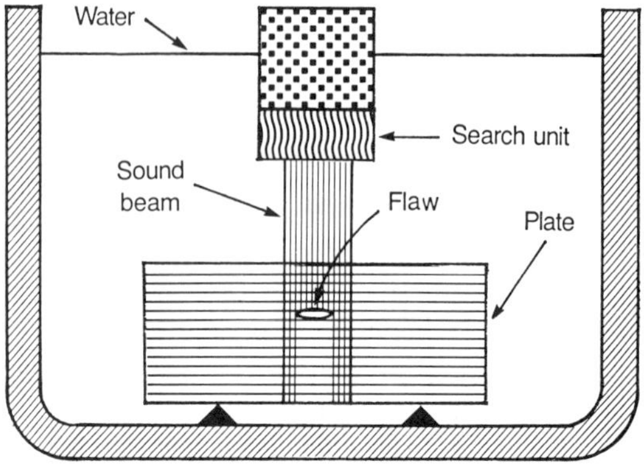

Figure 2-12 Straight beam immersion technique for the systematic search of internal flaws in a metallic plate.

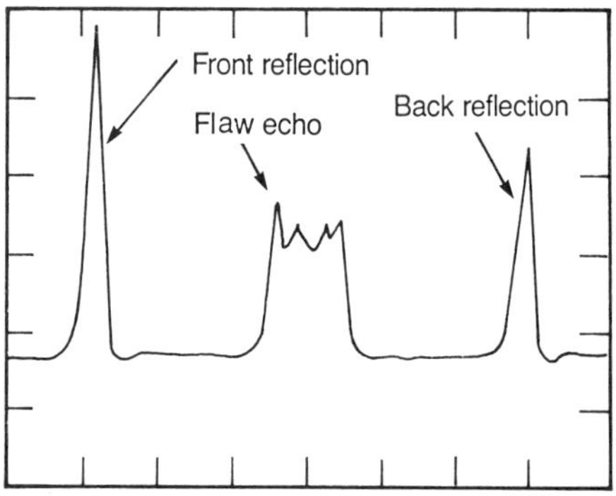

Figure 2-13 Echoes detected by the transducer shown in Figure 2-12.

the back reflection if an internal flaw is encountered along the path of the ultrasonic beam, as shown in Figure 2-12. The time delay between the front reflection and the echo indicates the depth of the flaw below the surface. The reflected signal at each X-Y position is put into memory, and the data are normally displayed in a C-scan image, showing the presence of defects within a given depth range on an X-Y matrix.

The same basic apparatus can be used to obtain a profile of the plate thickness. In this case, the time delay between the front and the back reflections provides a measurement of the plate thickness, if the sound velocity in the medium is known. The value of such a velocity may be affected by metal composition, stress or temperature. Accuracies better than 1% in the thickness measurement of plane parallel plates can be obtained by pulse superposition techniques (Boyer, 1976; Miller, 1984).

Conversely, if the plate thickness is known, the time delay between front and back reflection can be used to measure the sound velocity in the medium, while the relative amplitude of multiple reflections contains information about the ultrasonic attenuation. Sound velocity and attenuation data can be related to material properties such as chemical and structural composition (see, e.g., Piché, 1984) or the presence of residual stress.

If the dimensions of the flaws to be detected are very small, focused transducers can be used for a finer scan. The basic principle is shown in Figure 2-14, where focusing of the ultrasonic beam is obtained by bonding an ultrasonic lens (plane-concave plate) to the surface of the transducer. The converging ultrasonic beam is refracted at the workpiece surface resulting in a large acoustic beam convergence in the material. The angle of refraction follows Snell's law, the sound velocity in water being from two to five times smaller than in the most common solid materials. The depth of the focal point inside the plate may be varied by moving the transducer longitudinally.

(c) *Squirter-based techniques*

Immersion techniques for automated ultrasonic inspection are not adequate for very large structures, such as aircraft sections or a continuous strip, or very light materials, such as air-filled honeycomb laminates. In this case, squirter-type techniques using water columns continuously projected against the surface can be used (see Figure 2-15). Materials at bulk temperatures exceeding 100°C can also be inspected by this

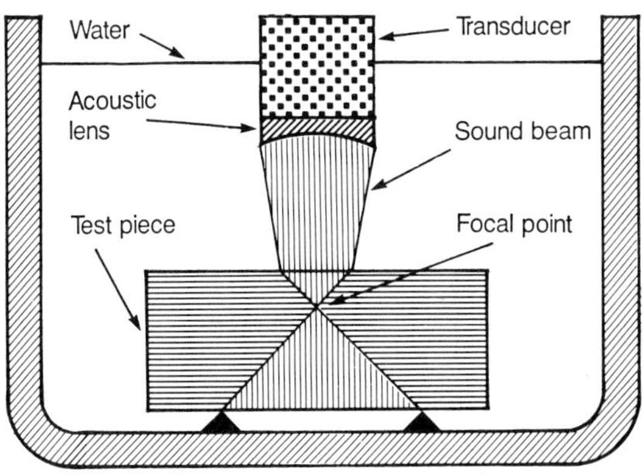

Figure 2-14 Immersion ultrasonic inspection by a focused transducer.

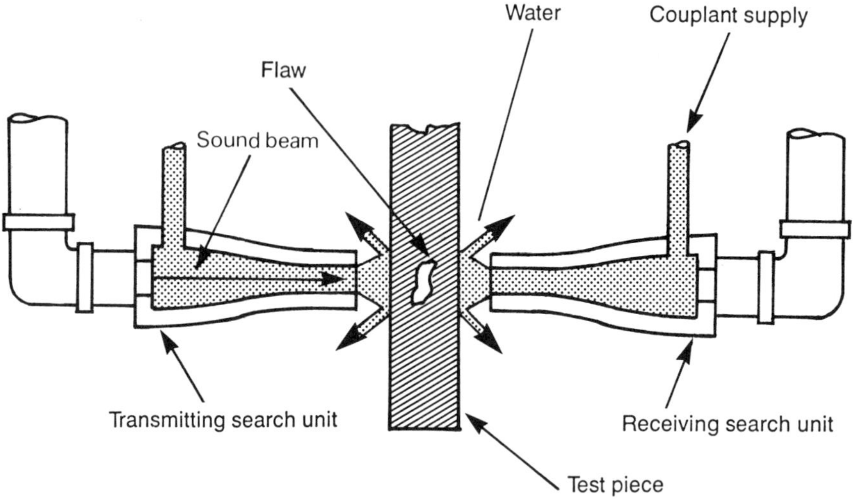

Figure 2-15 Squirter-type transmission ultrasonic inspection.

technique, taking advantage of the local surface cooling effect of the water jet. Both immersion and squirter-type systems may work in reflection, as in Figure 2-12, or in transmission, as in Figure 2-15. In the latter case, the receiving transducer detects an attenuated signal when an extended flaw is intercepted by the ultrasonic beam within the material. Transmission systems cannot provide flaw depth information. They are particularly used for complex materials, such as layered graphite-epoxy composites whose internal discontinuites produce many spurious reflections making pulse-echo recognition particularly difficult.

An example of a squirter based automated inspection system is shown in Figure 2-16. Two robot arms containing the transmitting and the receiving ultrasonic units are synchronously scanned across the inspected part, which may be as large as an aircraft wing. Alternatively, the workpiece may be moved across the two static ultrasonic units. Transmission systems generally cannot be used to inspect thick assembled structures.

2.2.4 *Passive Structure Borne Techniques*

(a) *Acoustic emission*

When a structure is subject to a mechanical load, sudden strain-release events may occur through crack growth, localized plastic deformation or phase transformations within the material. Such events, which start taking

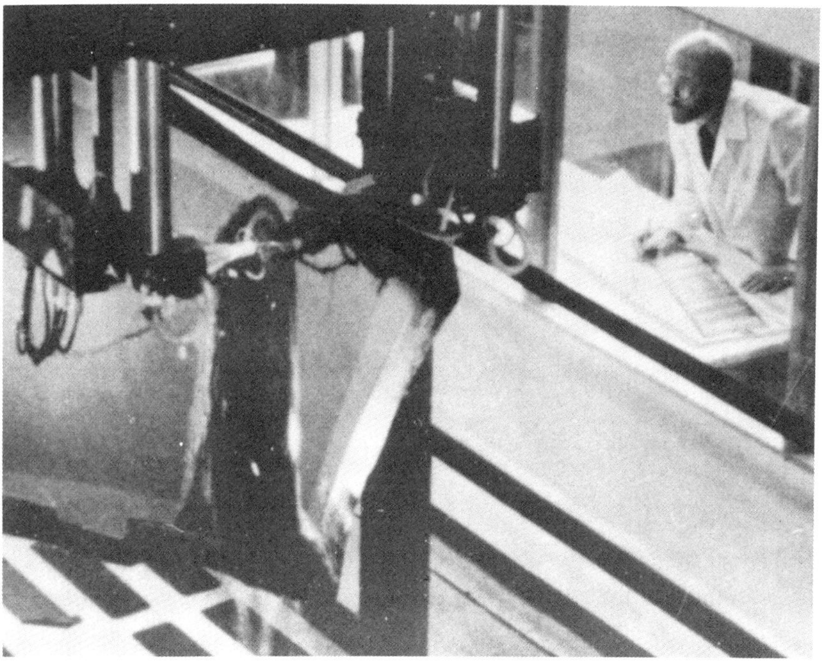

Figure 2-16 Industrial squirter-based ultrasonic inspection system. (Courtesy of McDonnel Douglas.)

place long before the failure of the structure, produce the emission of structure borne acoustic waves, mainly in the 50 KHz to 1 MHz region, which can be detected by acoustic transducers bonded to the structure.

A typical acoustic emission plot from a pressure vessel tested to rupture is shown in Figure 2-17 (Boyer, 1976). The vertical logarithmic scale gives the number of acoustic emission events recorded per second as the hydrostatic pressure within the vessel is increased. Significant acoustic emission bursts are observed for applied loads well below the critical load leading to vessel rupture, while the count rate starts increasing steadily for pressures larger than 80% of the critical load. Acoustic emission monitoring has thus been investigated as a method for detecting incipient failure on loaded structures during service. The use of three or more transducers at different positions over the structure makes it possible to locate the position of the acoustically-active flaw through a phase comparison of the signals detected by the different detectors. Implementation difficulties of such a technique are mainly related to the usually strong

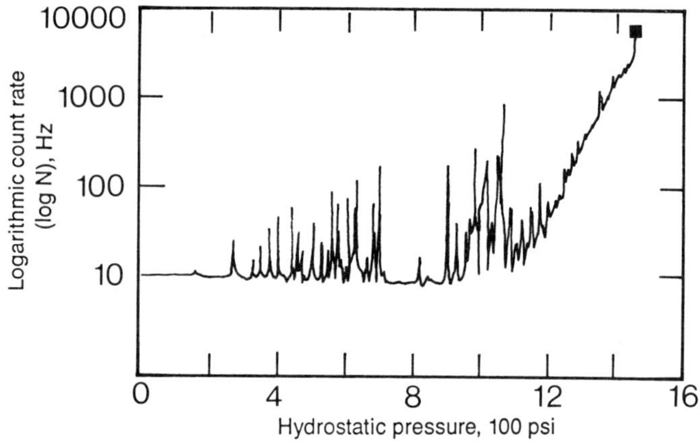

Figure 2-17 Acoustic emission plot for a pressure vessel tested to rupture.

background acoustic noise as well as to the deformation and multiple reflections of the acoustic signal as it propagates through the structure.

(b) *Resonance monitoring techniques*

The integrity of a given structure may also be verified by the observation of the normal vibration modes of the structure (Cawley, 1985; Lange, 1985). An example of the power spectrum recorded from a structure with or without internal delamination defects is shown in Figure 2-18 (Beuter and Weiss, 1985). The test object was a plate consisting of two layers of different materials, set into vibration by tapping with a metal hammer. The spectrum shows a typical line structure, which is roughly defined by the geometry and acoustical properties of the metallic layers. In case of insufficient bonding between the two layers, the lines in the spectrum are shifted to different frequencies. By identifying such characteristic lines, one can define a classification algorithm to differentiate between intact and badly bonded specimens. Implementation problems for complex structures arise from spectrum differences related to different clamping geometries and to different tapping positions or techniques.

A successful implementation of a similar method has been realized for the on-line inspection of grinder balls (Nadeau and Bussière, 1984). Such hardened-steel balls, typically 5 to 10 cm in diameter, often develop

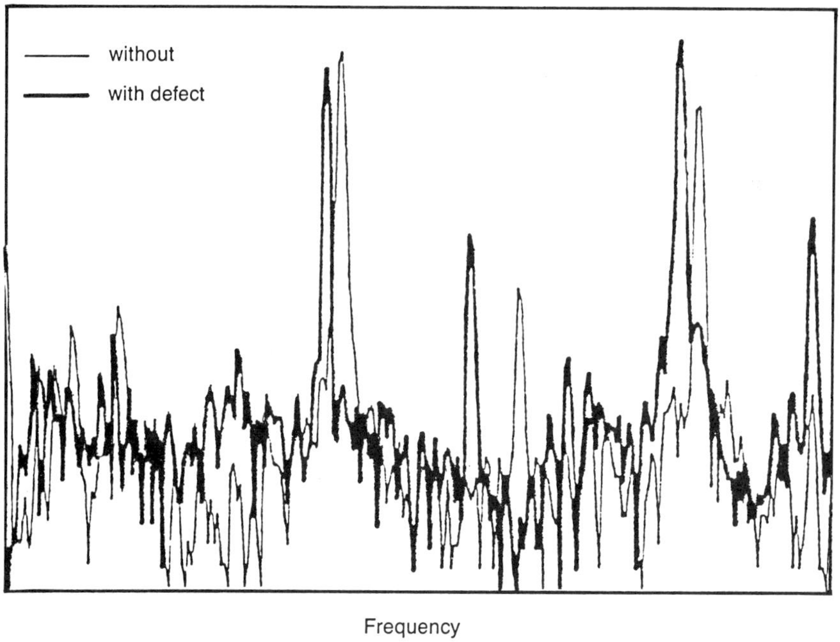

Figure 2-18 Power spectra of a layered structure with or without delamination defects.

deep-running cracks during the forging and quenching process. Testing is done on a statistical basis by manual inspection. Cracks can be detected visually or else acoustically, by knocking together two balls and listening to the characteristic high-pitched sound produced by the cracked balls. Examples of acoustic waveforms produced by the impact between two good balls and between a good and a cracked ball are shown in Figure 2-19. It can be seen that, apart from the initial low frequency (lower than 3 KHz) pulse which is common to both waveforms, a characteristic 5 to 20 KHz vibration is emitted by the cracked ball. Such a vibration range corresponds to the resonance modes of a partly-sectioned sphere.

An automated ball sorting system was built based on this principle, as shown schematically in Figure 2-20. The balls are dropped on a fixed, 5 cm-size steel cube, and the microphone-detected sound, after bandpass filtering the 10 to 20 KHz frequency band and integration, is used to activate a solenoid operated sorting device. Field test results were sufficienctly encouraging to consider installing a 100% inspection device based on this technique on the production line.

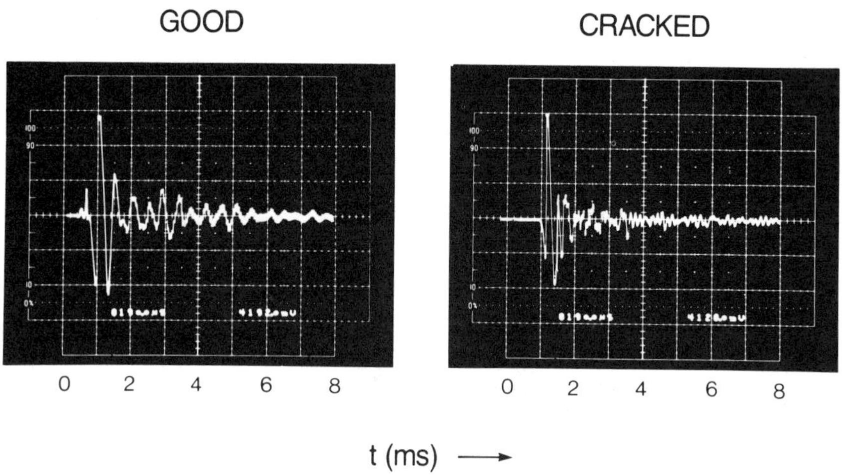

Figure 2-19 Acoustic waveforms produced by the impact of a good ball (left) and a cracked ball (right).

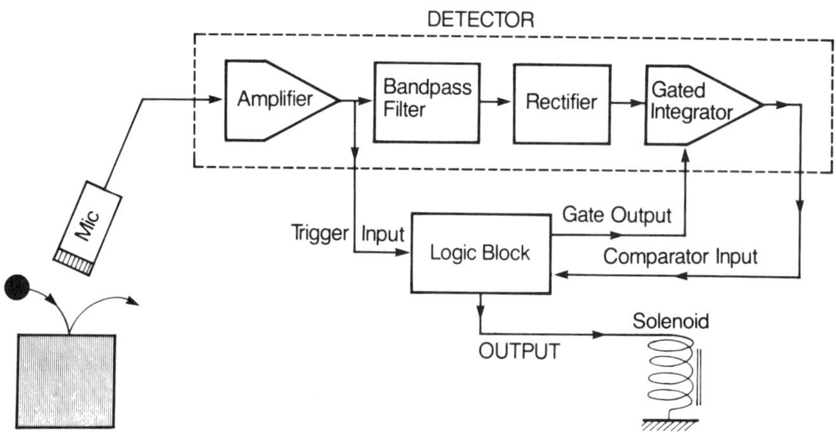

Figure 2-20 Schematic diagram of the acoustic ball-sorting system.

2.3 Radiographic Inspection

Radiography uses high frequency electromagnetic radiation (X-rays and gamma rays) or particulate beams (mainly beta rays and neutron radiation) to inspect a wide range of materials. Radiographic techniques are widespread in industry because they require no contact with the inspected workpiece, making them suitable for the inspection of moving products or sheets, and because they can penetrate through substantial thicknesses of metallic materials.

2.3.1 *Radiographic Imaging*

As X-rays are not refracted nor specularly reflected by materials, neither lenses nor reflective imaging devices can be used to produce an image of the irradiated object. Imaging methods must therefore rely on shadow projection techniques. A diagram of an X-ray imaging system is shown in Figure 2-21. A radiation source projects a radiation cone toward the object to be inspected, which is backed by a recording plate such as an X-ray film

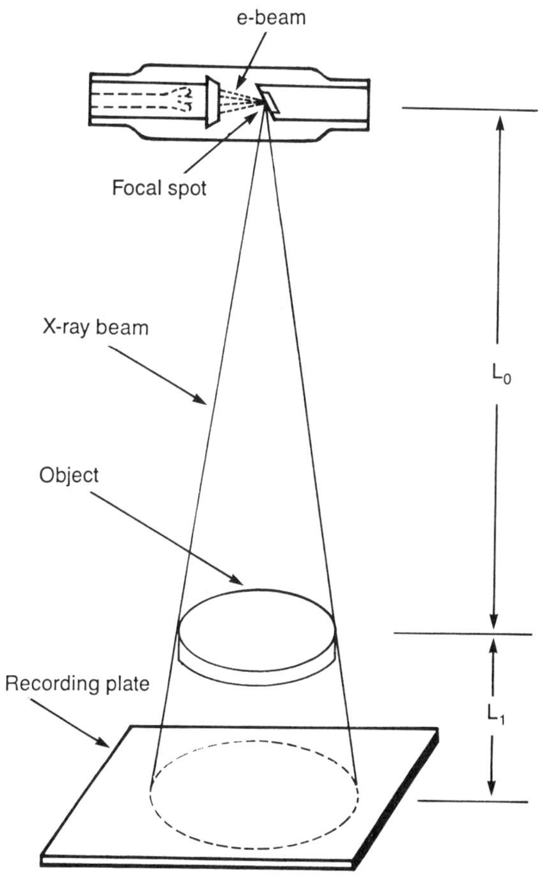

Figure 2-21 Schematic diagram of an X-ray imaging system.

or a fluorescent screen. Variations in the object thickness, density or composition produce changes in the recorded shadow image of the object on the plate.

Although compact and inexpensive isotope radiation sources are available (see next section), X-ray tubes of the kind shown in Figure 2-21 are usually required in order to obtain a sufficiently high radiation intensity. Such sources are evacuated tubes containing an electron beam gun focused on a small spot on the target, usually a tungsten layer in contact with a cooled copper substrate. Acceleration voltages up to 500 KV at 10 mA can be reached, making target cooling efficiency the main limitation to the available X-ray output power. The bombarded target area emits both bremsstrahlung and characteristic X-rays over a cone angle of the order of 30°. An increase in the acceleration voltage leads to an increase in the average photon energy, while an increase in the current increases the X-ray beam intensity.

The recording medium is either an X-ray film or a fluorescent screen. X-ray films are more sensitive and have a better resolving power than the fluorescent screens, however, they require development like a normal photographic film before they can be used. Fluorescent screens can be observed in real time, but they are limited in practice to the inspection of relatively small objects which can be inserted in a closed enclosure. Closed enclosures are used both because of safety considerations, given the relatively high X-ray intensities which are required, and because fluorescent screens are relatively bulky and lack flexibility, while X-ray film can be cut to size and wrapped around the workpiece to be inspected.

As can be appreciated from Figure 2-21, the maximum sharpness of the recorded image is limited by geometrical considerations. Best sharpness is obtained by limiting the focal spot diameter within the limits imposed by target overheating as well as by using large L_0/L_1 ratios (limited by the thickness of the object and of the filtering screens). Typical focal spot diameters of a few millimeters result in geometric lack of sharpness on an X-ray film in the 0.1 to 1 mm range. Microfocus devices with thin objects can provide resolutions close to 10 μm. Image sharpness is lost with thicker objects because of X-ray scattering by the object edges as well as because of the requirement for a longer exposure time. Longer-exposed images are more affected by vibrations and by background scattering. Finally, the spatial resolution of fluorescent screens is often in the 1 mm range, limiting the overall image sharpness to a similar value if such screens are used rather than the X-ray films.

The industrial applications of radiographic imaging are mainly in the nondestructive testing area, i.e., for the detection of hidden internal flaws which can lead to a premature failure of the piece. Defects such as voids and porosity in weldments, castings and forgings, or crushed-core defects in adhesively-bonded honeycomb laminates are readily detected by radiography. The required X-ray beam intensity and exposure time are strongly dependent on the atomic number of the inspected material as well as on the photon energy of the radiation, as Table 2-1 shows. Relatively large X-ray sources such as the one shown in Figure 2-22 are required to inspect heavy metallic walls. For the inspection of large tubular workpieces such as shown in this figure, the X-ray film may be wrapped around the internal wall of the piece. Because of the relatively high purchasing and operating costs of X-ray imaging equipment, routine inspection using such techniques is mainly limited to the testing of safety sensitive materials, such as nuclear reactors and aircraft structures, or to large volume applications where economic reasons justify their use, such as for inspecting pipeline weldments.

2.3.2 *Thickness Gaging*

Nuclear radiation gaging is a widespread technique for industrial production control. A main advantage of this approach over the ultrasonic multiple-

Table 2-1 X-Ray Narrow Beam Attenuation Coefficients of Selected Elements

		Attenuation coefficient (cm^{-1})			
Element	Atomic number	20 KeV	100 KeV	1 MeV	10 MeV
Beryllium	4	0.40	0.242	0.103	0.0295
Carbon	6	0.95	0.340	0.141	0.0433
Aluminum	13	9.40	0.459	0.166	0.0621
Silicon	14	10.6	0.428	0.149	0.0576
Titanium	22	73.5	1.24	0.266	0.122
Chromium	24	151	2.29	0.426	0.202
Iron	26	207	2.93	0.471	0.233
Nickel	28	294	3.96	0.543	0.279
Copper	29	308	4.10	0.524	0.272
Zinc	30	272	3.56	0.423	0.223
Silver	47	156	15.3	0.621	0.404
Tin	50	128	12.0	0.419	0.283
Tungsten	74	1020	81.5	1.22	0.894
Lead	82	818	62.0	0.798	0.570

Source: McMaster, 1959, Chapter 27

Figure 2-22 A 400 KV, 10 mA industrial X-ray setup for the irradiation of a thick-walled tubular piece. (Courtesy of Balteau S.A. and NDT Technologies, Montréal.)

reflection thickness gaging method is its noncontact nature, which makes it suitable for the inspection of rapidly moving sheets. An advantage of nuclear techniques over optical triangulation methods is that nuclear techniques may not require access from both side of the material.

(a) *Transmission configuration*

In the transmission configuration, a radiation source is placed on one side of the sheet while a detector is positioned on the opposite side to measure the attenuation of the through-transmitted beam. An example of an industrial setup based on this principle is shown in Figure 2-23 (McMaster, 1959). This technology has undergone no major evolution in recent years.

The radiation source may be either an X-ray tube or an isotope source.

Figure 2-23 Transmission radioisotope gage measuring rubber coating fabric.

If an X-ray tube is used, the acceleration voltage may be adjusted to produce the optimum average photon energy, depending on the sheet material type and thickness. Maximum signal-to-noise ratio is obtained when the transmitted beam has an intensity roughly equal to one-half of the incident intensity. The attenuation coefficients of Table 2-1 may thus be used to set the optimum X-ray average photon energy, which is determined by the accelerating voltage. As an example, a photon energy of 20 KeV is best for thickness gaging a 1 mm-thick aluminum sheet, while a photon energy of 100 KeV is best suited for the inspection of a 5 mm-thick steel sheet. Low energy photons are also useful to decrease the sensor inaccuracy generated by the presence of oil films on the metallic foil. As Table 2-1 shows, the X-ray attenuation in the low energy region is strongly dependent on the atomic number of the inspected material (in this region, X-ray absorption is due mainly to the photoelectric effect, increasing approximately as the fifth power of the atomic number). The relative sensitivity to low atomic number substances such as oils is thus reduced in the energy range below 50 KeV.

Different devices are available for measurement of nuclear radiation for thickness gaging. Among earliest detectors are ionization chambers where a gas which is ionized by the incident nuclear radiation is situated between two electrodes, providing an output current proportional to the radiation intensity. A variation of such a scheme is the Geiger–Müller counter, more appropriate for very low count rates. Better reliability is offered by solid-state devices where the detecting material is either a photoconductor, giving an electric current proportional to the number of electrons released by the incident radiation, or a luminescent material whose optical activity is proportional to the radiation intensity.

Scintillation counters are composed of a luminescent material backed by a photomultiplier tube to provide greater sensitivity at the expense of linearity and stability.

Because of the sensitivity variations caused by the environment (such as the temperature coefficient of phosphors), as well as by aging effects on the source and detector, calibration requires particular care if good accuracy is to be obtained. Automated recalibration techniques which are sometimes used include mounting a standard plate or wedge on a rotating device for periodical checks or comparing the detected luminescence signal with the signal obtained from a fixed light beam reference.

(b) *Backscattering Configuration*

In many instances, such as in the inspection of pipe wall thickness, access to both sides of the material is not allowed. Backscattering nuclear gages have been developed for such applications.

In a backscattering configuration, the detector is situated nearby the source to detect the intensity of the radiation scattered back by the irradiated material. The level of the backscattered intensity increases with the thickness of the sheet, approaching a saturation level when the thickness is large with respect to the half-penetration depth of the radiation. An example of an industrial device set up to measure the wall thickness of an extruded plastics pipe is shown in Figure 2-24.

Although X-ray tubes can be used, and are used, for such devices, the high cost and limited reliability of these sources can be avoided by using small radioisotope sources whenever the required source intensity is limited. Radioisotopes (either natural or artificial unstable nuclei which decay emitting nuclear radiation) in the form of small pellets are readily available as sources of gamma ray or beta ray radiation. Photon energies

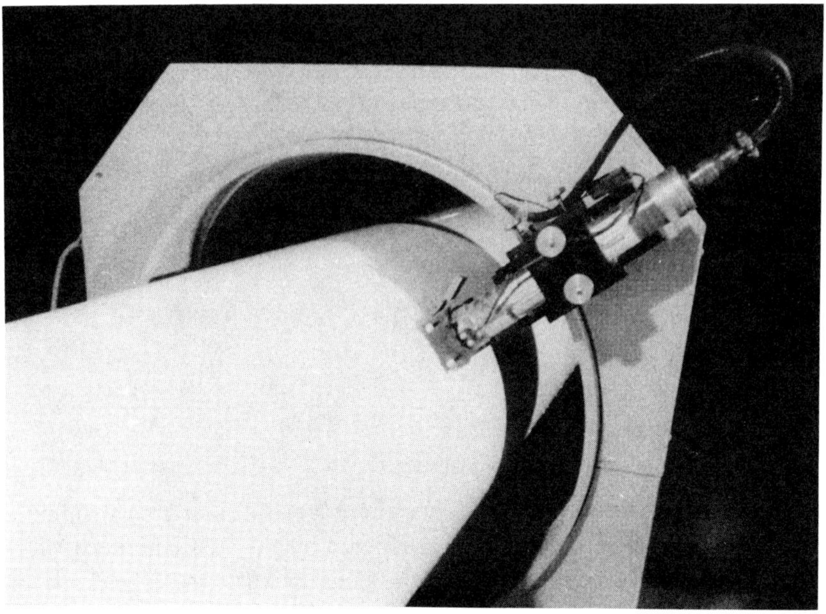

Figure 2-24 Backscattering nuclear gage set-up for measuring the thickness of a PVC pipe. (Courtesy of NDC Systems.)

of the most common sources go from the 84 KeV of the gamma rays emitted by the Thulium-170 to the 3.5 MeV of the beta ray source Ruthenium-106. Half-lives may be as short as 74 days for the Iridium-192 or 33 years for the widely used Cesium-137 source.

Beta rays have a lower penetration depth than X-rays or gamma rays so that beta rays are used mainly to inspect thin sheets or coatings. As the absorption of monoenergetic electrons is far from following an exponential law, attenuation coefficients of the kind listed in Table 2-1 cannot be given for beta ray radiation. Nevertheless, approximate thickness ranges for a typical beta ray gage are given in Table 2-2. As we can see, such ranges are not particularly dependent on the atomic number of the inspected material, being roughly inversely proportional to the material mass density.

The relatively short penetration depth of beta rays makes such sources best suited for the measurement of very thin coatings, while thicker coatings are monitored by backscattering gamma ray or X-ray sources (Baier, 1981). The thickness of the coating on a semi-infinite (much larger than the ray penetration depth) substrate is determined by measuring the

Table 2-2 Approximate Range of Thickness Measured by Strontium-90 Beta Ray Gages

Material	Density (g/cm^3)	Thickness (mm)
Aluminum	2.7	0.05–2
Asbestos	2.5	0.05–2
Copper and brass	8.9	0.015–0.5
Glass	2.7	0.05–2
Paper	1.0	0.12–5
Plastics (celluloid)	1.4	0.1–3.5
Rubber	1.0	0.12–5
Steel	7.8	0.015–0.5

Source: McMaster, 1959, p. 18–16.

backscattering levels obtained from a bare substrate and from an infinitely thick coating. The radiation backscattered by a coated sample goes from the first to the second level as the coating thickness increases from zero to infinity. Calibration curves are determined in this way. If the substrate is not infinitely thick and if its thickness does not vary during the process, the coating thickness can be determined by subtracting from the detected signal the level corresponding to backscattering by the substrate.

The backscattered signal level depends on the distance between the instrument window and the inspected surface. Such a distance must thus be kept rigorously fixed to avoid signal fluctuations or decalibration produced by sample displacements. For this reason, some companies which have in the past produced noncontact beta backscattering equipment have now discontinued production (P. Raffelsberger, Fischer Technologies Inc., private communication) in favor of more reliable surface contact types.

(c) *Safety Considerations*

Nuclear radiation is harmful to living organisms. Effects on the human body are cumulative, so that maximum allowable exposure levels for nuclear equipment operators are expressed in number of radiation units per year or even per lifetime.

Nuclear radiation sources are well shielded with lead plates in directions other than toward the inspected surface. However, some scattering of the useful beam is unavoidable. Radiation density in the vicinity of the source is monitored by radiation gages and plotted in the form of iso-dose curves for the information of the personnel. Radiation warning signs are usually displayed at the minimum distance at which a person may occupy a position continuously during a 40-hour week. Special precautions such as the use of remote-handling tools must be taken when setting in place or replacing radioactive sources.

2.4 Electromagnetic Techniques

2.4.1 *Direct Current Methods*

Resistivity measurements are often required to characterize resistive materials or electrolytic liquids or to test insulating layers (McMaster, 1959 Chapter 5). The value of the electrical resistivity may be taken as a measure of the degree of purity or of the chemical composition of a metal (Omini et al., 1981; Cohen and West, 1983). Applications to resistance weld quality control have recently been reported (Cohen and West, 1983a) and a commercial ResTest™ instrument is available from Western Electric. Moisture meters such as the one shown in Figure 2-25 may also be based on a resistivity

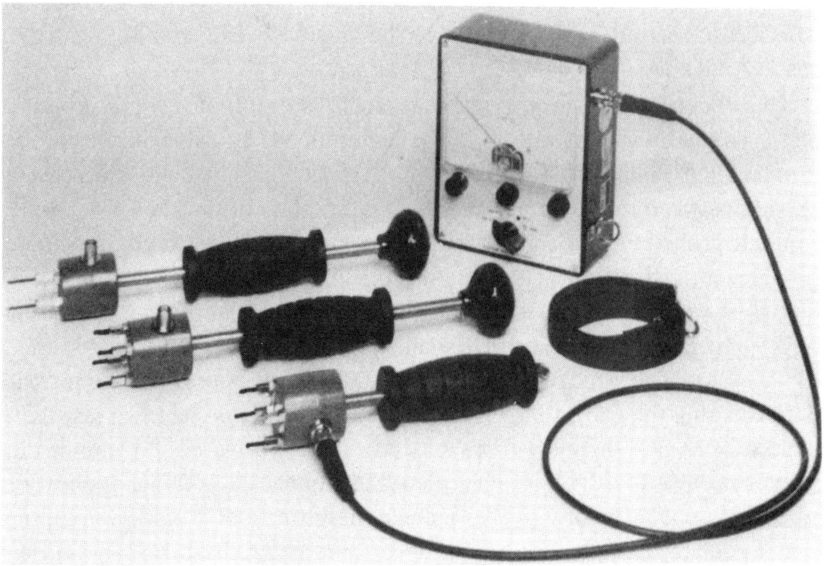

Figure 2-25 Electrical resistivity moisture meter for lumber and wood products with different needle-electrode heads. (Courtesy of Data Tech.)

measurement. The obtainable accuracy in the characterization of fairly conductive samples or coatings is mainly limited by the quality of the contacts and by the minimum value of the testing instrument internal resistance. Balanced circuits using four-point probes have been used to measure surface crack depths (Buchanan and Thurston, 1956). More recent improvements of such a technique using an AC field have been reported (Dover and Collins, 1980).

Thermoelectric gages are based on a different principle. Two probes of a known material are brought in contact with the material to be characterized. One of the contacts is heated by a hot air jet to a thermocouple monitored temperature, while the other junction remains at room temperature. The resulting thermoelectric potential difference is used to identify the unknown material (McMaster, 1959, Chapter 5). Similar techniques can be used to measure the thickness of conductive coatings (Stinebring, 1972).

Pinholes in insulating coatings on a conductive substrate can be detected by applying a high voltage between the substrate and a probe electrode in contact with the coated side of the sample. Solid electrodes such as rolls or clusters of metallic chains may be used to induce an electric spark through the air gap over the pinhole. Alternatively, a liquid contact can be made on the coating surface (see, e.g., the publication *Inspection, Gaging and Testing*, available from the Soc. Manuf. Eng., 1984, p. 456). The same publication describes in Section 7 a number of electromagnetic instruments for measuring coating thickness.

The thickness of dielectric coatings, such as paints or ceramic coatings on a conductive substrate, can be determined by capacitance probes (Birchenough and Marshall, 1972; Latter, 1983). A voltage difference is applied between the substrate and the second plate of the capacitor formed by a flat probe put into contact with the coating. The measured capacitance provides an estimation of the coating thickness if its dielectric constant is known. A variation of this scheme for the inspection of nonmagnetic coatings on ferromagnetic substrates uses a small magnet clamped to the coated surface (Baier, 1981). The force required to lift the magnet decreases with an increase of the coating thickness (Berger, 1975). Measurements of electrical resistivity, capacitance and dissipation factor have been proposed for monitoring polymers during cure (Chottiner et al., 1982; Tajima, 1982). Other applications of capacitance proximity probes include monitoring of vibration amplitude and frequency, roughness evaluation of a moving sheet, and eccentricity sensing. The accuracy may reach 0.01% of the probe-to-surface distance (see the above mentioned SME publication, pp. 188 and 205).

2.4.2 *Eddy Current Inspection*

It was mentioned in the previous section that direct-contact electrical resistivity measurements are limited by the quality of the contacts. Eddy current methods can make such measurements without the need for contact with the inspected piece. The basic principle consists in AC driving an induction coil situated in the immediate vicinity of the workpiece. As in induction heating, this alternating current causes eddy currents to flow in the piece as a result of electromagnetic induction. The level of the induced currents, which is several orders of magnitude smaller than in induction heating, is measured through the variation in electrical impedance they produce on the exciting coil or in a secondary coil.

If a flat workpiece is inspected by a flat coil, the distance variations between the coil and the surface produce spurious variations of the detected signal (lift-off effect). Such an effect may actually be used to measure the thickness of an insulating coating or a conductive substrate if the electrical properties of the substrate are known. In general, careful probe positioning as near as possible to the surface is required to avoid reading errors caused by lift-off. The positioning accuracy is relaxed when the coil can be made to encircle the workpiece, such as in the pipe inspection configuration shown in Figure 2-26. Also shown in this figure is the perturbation to the induced current flow which is introduced by a crack perpendicular to the surface. Such a

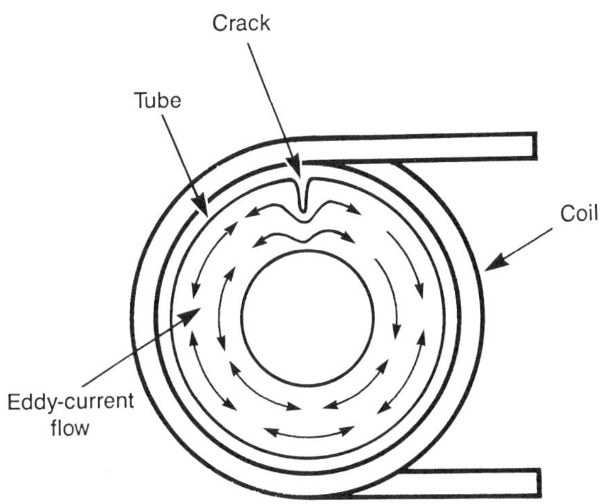

Figure 2-26 Cross section of a surface-cracked pipe inspected by an eddy current induction coil.

perturbation generates a strong variation in the signal detected by the pickup coil, making such devices very sensitive to the detection of surface cracks. The depth of material which can be probed in this way depends on the penetration depth of the eddy currents. Such a "skin depth" decreases with the excitation frequency, typical values going from 3 mm for aluminum at 1 KHz to 10 μm for steel at 1 MHz (Boyer, 1976, p. 80).

Typical applications of eddy current devices are the detection of surface or near-surface defects such as cracks, seams, voids or inclusions in a wide range of materials. Figure 2-27 shows a typical operation of an advanced eddy current probe with frequency sweeping and lift-off compensation capabilities. Other applications are the determination of chemical composition and structural characteristics of rods and pipes by conductivity measurement, the evaluation of the depth of case hardening, the analysis of alloy composition, structure and internal stress in ferromagnetic materials by hysteresis-loop tests, and the on-line inspection of hot-rolled billets (McMasters, 1959, Chapter 42).

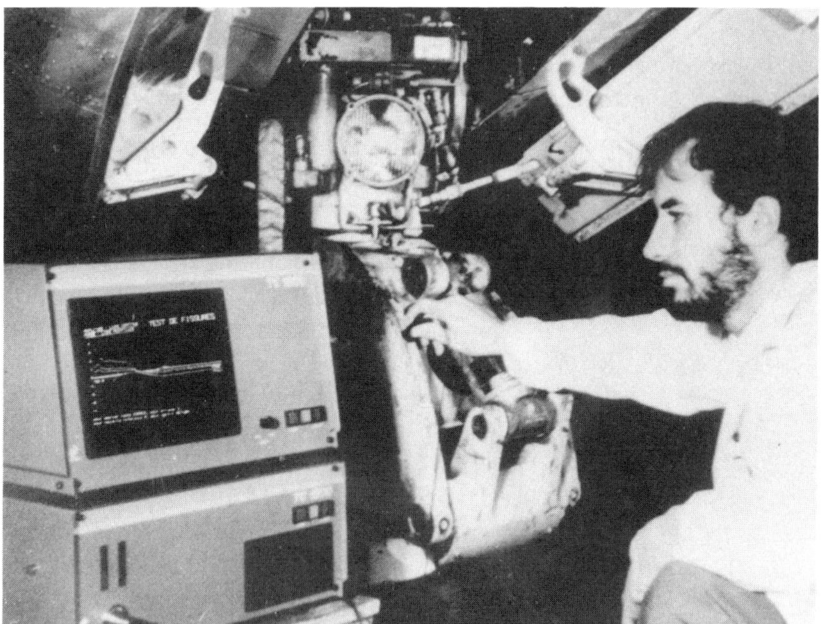

Figure 2-27 Industrial eddy current equipment used to inspect an aeronautical structure. (Courtesy of Tecrad, Québec)

2.4.3 *Magnetic Particle Inspection*

Magnetic particle inspection is a method for visualizing surface and subsurface discontinuities in ferromagnetic materials. Magnetic particles are spread, or applied after dispersion in a liquid slurry, on the surface of the piece during or after magnetization by a strong magnet or intense (in the KA range) electric currents. The presence of a surface opening or near-subsurface crack perpendicular to the field lines perturbs the magnetic field, creating a leakage field out of the piece (see Figure 2-28). The magnetic particles are gathered and held by the leakage field around the faulty areas, thus providing a visual indication of the presence of a defect.

Iron, nickel and cobalt alloys are routinely inspected by this technique. An example of the magnetic particle visualization of a surface crack in a forged propeller component is shown in Figure 2-29 (McMaster, 1959, Chapter 32; Betz, 1973). Such a crack was less than 0.5 mm deep and almost invisible to the naked eye. Another example is shown in Figure 2-30 (Boyer, 1976, p. 72). A stainless steel honeycomb panel formed by a honeycomb core brazed to top and bottom metallic skins was inspected by a magnetic printing method. After application of a plastic coating and magnetic particles on the top skin, the piece is submitted to an alternating magnetic field produced by a coil. The vibratory effect on the magnetic particles causes them to stain the plastic coating in the regions where they have been attracted by changes in magnetic permeability. As Figure 2-30 shows, the core pattern otherwise invisible to the eye is made visible through the metallic skin. Areas where the core cells are not bonded to the facing sheet can be observed as a fading out of the pattern above such areas.

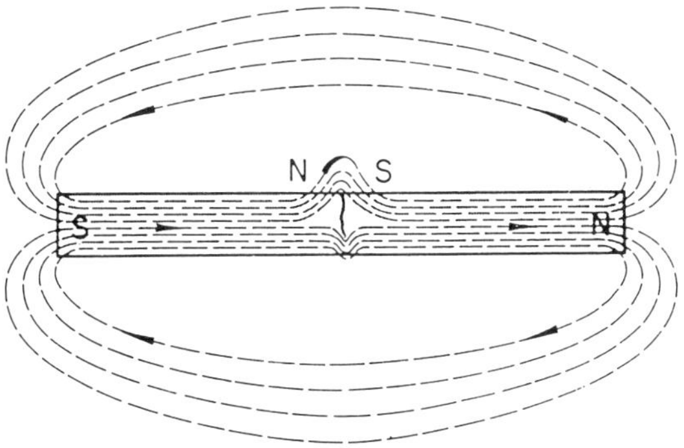

Figure 2-28 Leakage field produced by a crack on a bar magnet.

Figure 2-29 Magnetic particle indication of a forging lap in a propeller shaft lock nut shown by the white arrow.

2.4.4 *Microwave Inspection*

Microwave technology, which was mainly spurred by the development of the radar during World War II, has long been considered an attractive tool for materials inspection. Similarly to radiography and in contrast to ultrasound, microwaves do not require contact with the inspected piece and are less dangerous to human health than nuclear radiation. The basic limitation of microwaves is their inability to penetrate electrically conductive media. For the inspection of nonmetallic materials, microwaves have a greater penetration depth than optical beams, at the expense of spatial resolution and of an increased difficulty to focus, scan or image the beam with components of reasonable size.

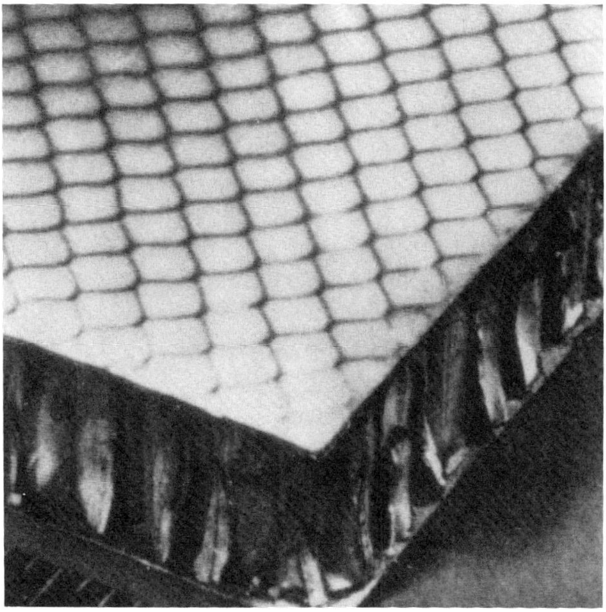

Figure 2-30 Magnetic particle indication of the core pattern on the top skin of a stainless steel brazed honeycomb panel.

The basic transmission configuration for the inspection of a nonconductive material is shown in Figure 2-31. A microwave generator feeds both a transmitting antenna and a phase-sensitive detector, which superposes this signal with the signal detected after transmission through the material, producing a phase-locked output which is not affected by incoherent environmental noise. Discontinuities within the material producing a variation of the scattered or absorbed microwave power can be seen when scanning the beam through the piece. Swept frequency methods, where a frequency band of typically one octave is covered for each position, can be used to put in evidence frequency dependent absorption. Voids, inclusions or porosity can be detected only if their size is of the order of magnitude of the wavelength, typically in the mm or cm range.

Single antenna double-transmission methods are possible by backing the material with a reflective metallic plate (Bahr, 1982). Truly reflective methods are also possible for depth ranging by using the delayed chirped frequency method mentioned in Section 2.2.1. A thickness-gaging configuration for the inspection of metallic sheets is shown in Figure 2-32. A standing wave pattern is produced within a double arm waveguide by the waves reflected from the

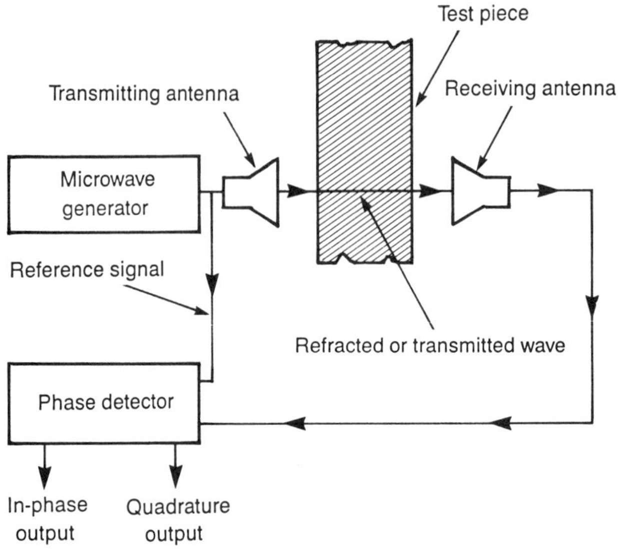

Figure 2-31 Diagram of the basic components of a transmission microwave inspection system.

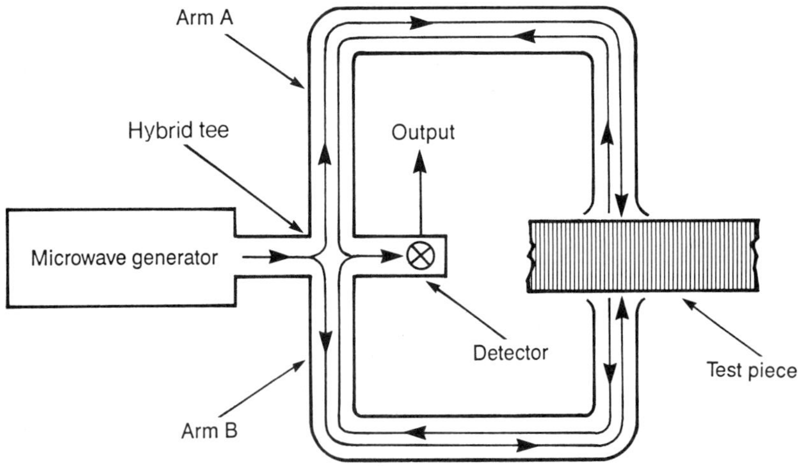

Figure 2-32 Diagram of a microwave reflectometer for measuring the thickness of a metallic sheet.

two surfaces of the sheet. The thickness of the sheet can be determined from the value of the generator frequency which is required to produce a destructive interference between the reflected waves (Boyer, 1976, p. 249). Single-side thickness measurement of metal-backed dielectric layers such as 50 μm-thick polyethylene coatings on aluminum foil or 1 cm-thick fiberglass reinforced plastics sheets have also been reported (Botsco, 1968).

Moisture content in polyethylene powder has also been determined by microwave through-transmission at 9.4 GHz (Botsco, 1968). The same technique is also used for moisture analysis in paper sheets (Bosisio and Giroux, 1970; Chu and Balls, 1976; Anderson, 1980). An analysis of the effect of temperature on the accuracy of microwave moisture measurements has recently been published (Aggarwal and Johnston, 1985). Other microwave applications that have been reported include monitoring of polymerization, oxidation, esterification and vulcanization of dielectric materials, as well as the measurement of material anisotropy in fiber resin composites (Boyer, 1976, p. 253).

2.5 Sensors for Industrial Process Control

The previous sections of this chapter are mainly devoted to sensing techniques for industrial quality control and defect detection in manufactured products. This section reviews sensors which are used to monitor the operating parameters of industrial processes (Hunter, 1978, Lenk, 1980; Considine, 1985; Clevett, 1986). Because of the vast variety of devices which are used in such an instrumentation field, the sensors will be classified with respect to the measured parameter rather than with respect to their operating physical mechanism.

2.5.1 *Pressure Sensors*

Sensors for pressure measurement in a gas or liquid are mostly based on either a liquid column mechanism or on the deformation of an elastic element (Benedict, 1969; Chesmond, 1984; Bailey, 1984; Considine, 1985).

Some typical liquid column sensors with mechanical display pointers are shown in Figure 2-33. The difference between the pressure to be measured, p_2, and a reference pressure, p_1, which may be the atmospheric pressure, causes a displacement in the liquid column level, as shown in Figure 2-33a. Such a level is related to the pressure value by a graduated scale. This type of pressure sensor is called a manometer.

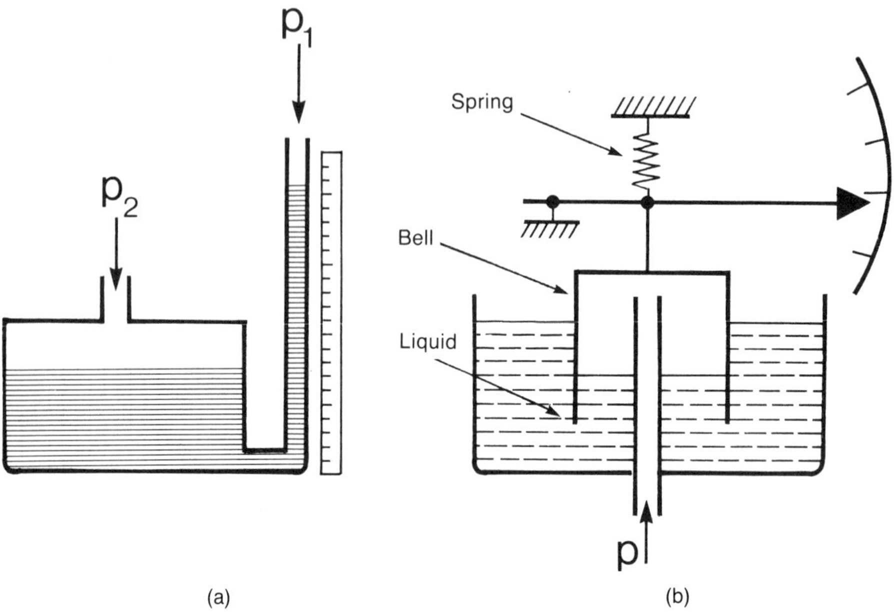

Figure 2-33 Liquid column mechanical pressure sensor of the (a) manometer type, and (b) bell type.

Figure 2-33b shows a bell type liquid-based pressure sensor. This device utilizes a bell-shaped container immersed in a sealing liquid. The pressure to be measured is applied to the inside of the bell, the motion of which is opposed by a restricting spring. The level of the bell top is proportional to the applied pressure.

Liquid column instruments are rather cumbersome and limited in pressure range. Mechanical sensors using an elastic element which deforms under pressure are generally more practical and can be used to monitor higher pressure levels. Two examples of such sensors are given in Figure 2-34.

The sensor on the left of this figure is of the bourdon type. In the original patent of E. Bourdon in 1852, the bourdon tube was described as a curved or twisted tube whose transverse section differed from a circular form. In principle, a tube of this kind closed at one end, if bent or distorted, has the property of changing its shape with internal pressure variations. An internal pressure increase causes the cross section to become more circular and the shape to straighten, resulting in motion of the closed end of the tube provided its open end is held rigid. This motion is translated to a pressure indication by a pointer. Devices of this kind using stainless steel or Inconel tubes may be used to monitor pressures of several thousands of kg/cm^2.

Another kind of sensor using an elastic transducer is shown in Figure 2-34b. The transducer is in this case a bellows element which is a thin-walled metal tube with deeply convoluted sidewalls permitting axial expansion and contraction. In the configuration of Figure 2-34b, the bellows contraction is translated by a shaft to a display of the applied pressure differential. These devices are rugged, sensitive and can be used to measure pressures higher than 100 kg/cm^2.

Another elastic element which is widely used for pressure measurements is the metallic diaphragm. This is a flexible disk, usually with concentric corrugations, which is used to convert pressure to displacement and also to serve as a fluid barrier in seal assemblies. Metallic capsules, which are made of two diaphragms joined at their periphery by soldering or welding, are also used when sealing is not required. The operating pressure range of such devices is normally below 100 kg/cm^2 and much smaller for rubber diaphragms. Figure 2-35 shows a typical device with different readout electronic transducers. This kind of transducer is gradually replacing the mechanical readout systems shown in Figures 2-33 and 2-34. Figure 2-35 shows three examples of electronic transducers using (a) a capacitor, (b) a variable reluctance magnetic transducer and (c) a linear variable differential transformer.

Other sensors, such as strain gages, optical interferometers or piezoelectric transducers can be used for the readout mechanism. Strain gages, of which two examples are shown in Figure 2-36, are widely used transducers which are bonded to the part whose strain is to be measured. They are composed of resistive elements whose length and cross section change with the application of longitudinal strain, giving a variable resistance signal. The design shown on the right can be used to monitor the amount of deflection in a circular diaphragm. Optical interferometers or other displacement sensing optical devices (see Sections 4.1.3 and 4.1.4) have also been used to monitor the central displacement of the diaphragm. The sensitivity may reach 1 Pa of pressure differential (Considine, 1985, p. 3-59), but such devices are still relatively cumbersome. Finally, piezoelectric transducers (see Section 2.2) are convenient for sensing pressure variation because they give an electric signal without need of any mechanical transducer, are small in size and very rugged. Because of low frequency leak electric currents, they are normally limited to dynamic pressure sensing at frequencies higher than 1 Hz (Considine, 1985, p. 3-50), so that they cannot be used for static pressure measurement.

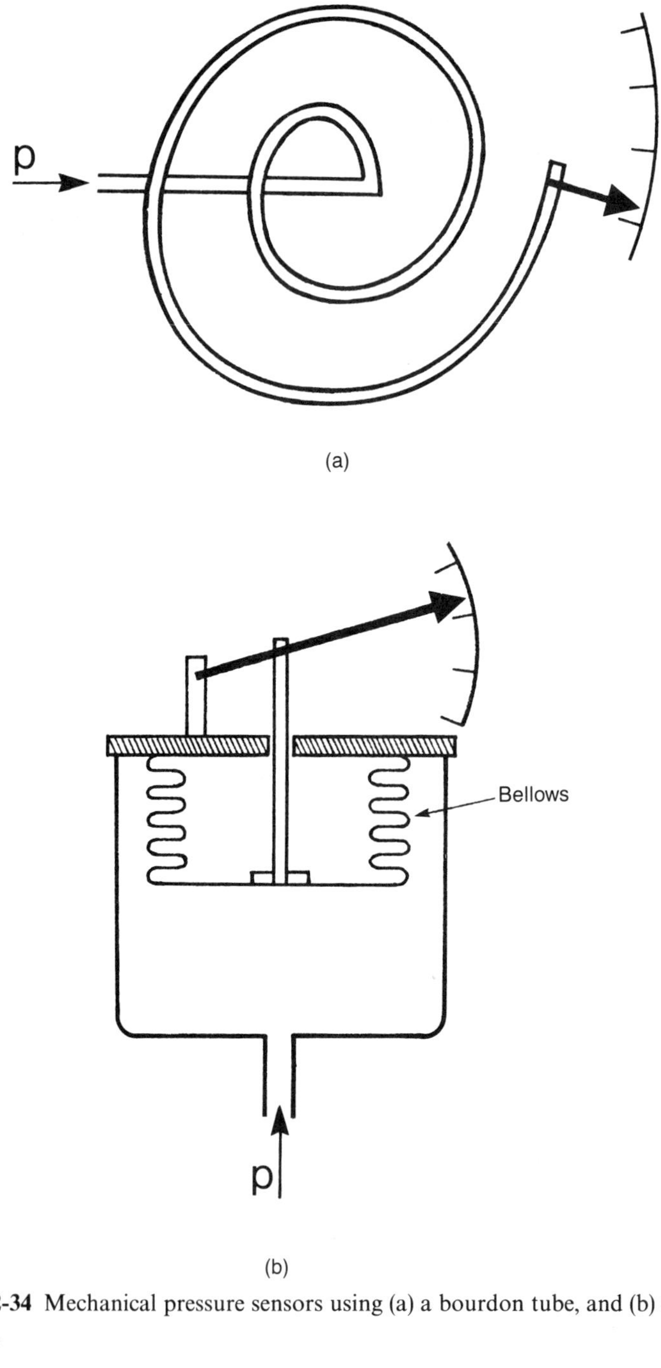

Figure 2-34 Mechanical pressure sensors using (a) a bourdon tube, and (b) a bellows element.

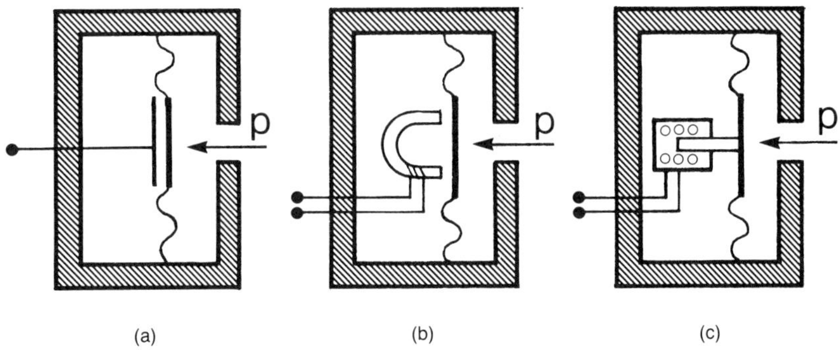

Figure 2-35 Diaphragm-based mechanical pressure sensors with (a) capacitive, (b) magnetic and (c) LVDT transducers.

2.5.2 *Fluid Sensors*

This section reviews sensing systems for fluid status including fluid flow, liquid level and rheological properties.

Fluid flow measurement

Some typical fluid flow sensors are shown in Figure 2-37 (Lenk, 1980; Miller, 1983; Considine, 1985). The device on the left monitors the pressure buildup before a contraction of the tube. The pressure building obstruction in the tube may be in the form of an orifice, a nozzle or a tapered Venturi tube section as shown in Figure 2-37a. This last design is the most convenient, though more expensive, to minimize the overall resistance to a free fluid flow. A variation of this approach is the elbow differential sensor, where two pressure sensors are installed on the inner and on the outer portions of an elbow along the tube. A centrifugal force proportional to the fluid speed produces a higher pressure on the outer port. This approach is quite convenient, since many installations already have elbows so that no additional obstruction must be inserted on the tube, but the sensitivity is low, particularly for gas flows.

The center diagram shown in Figure 2-37 shows a turbine rotor flowmeter. The rotation rate of the turbine is proportional to the fluid flow rate. Such a rate may be transmitted to an external tachometer through a shaft or be picked up by a magnetic proximity sensor. The turbine approach is accurate but rather expensive, while shaft wear problems must be accounted for in long-term industrial installations. A variation of this method is the vortex shedding approach (Knudsen and Sovik, 1984) where a sharp object is inserted

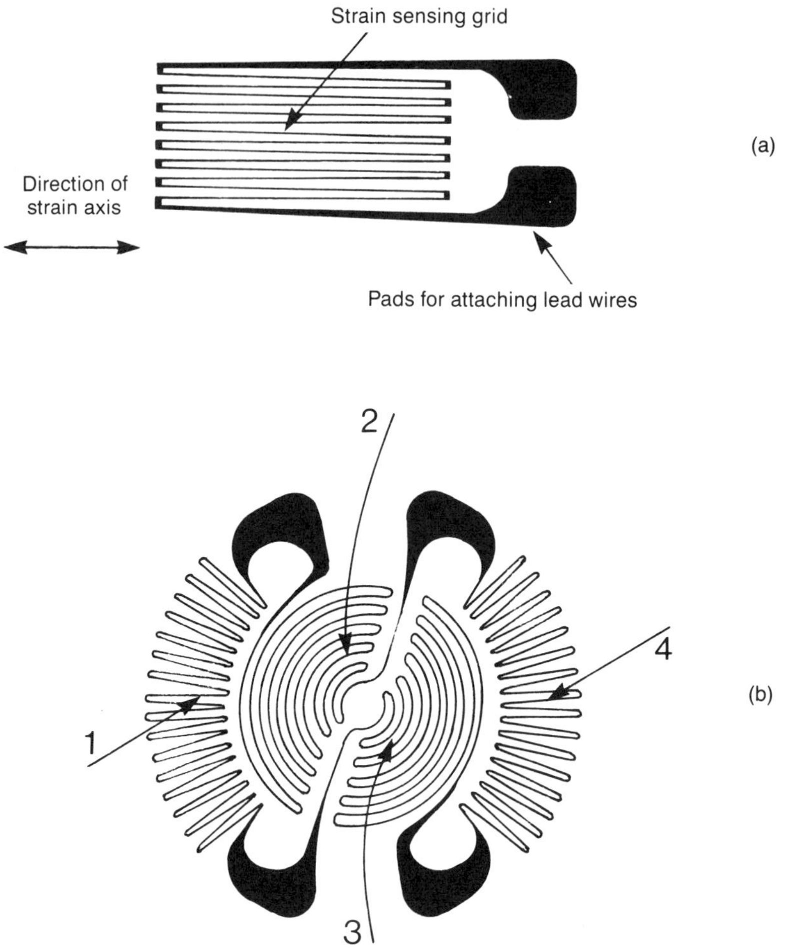

Figure 2-36 (a) Typical bonded foil strain gage for unidirectional strain; (b) pattern used to measure strain in a diaphragm: Elements (1) and (4) are subjected to compressive radial strain, while elements (2) and (3) are subjected to tensile tangential strain.

in the stream to produce a vortex whose pressure-fluctuation frequency is sensed by a piezoelectric transducer. No moving parts are required in this case, so that high fluid flows can be monitored, but the pressure loss caused by the shedder element is important.

The third flowmeter design shown in Figure 2-37 makes use of a floating plug in a vertical, tapered tube section. Gravity brings the conical plug down

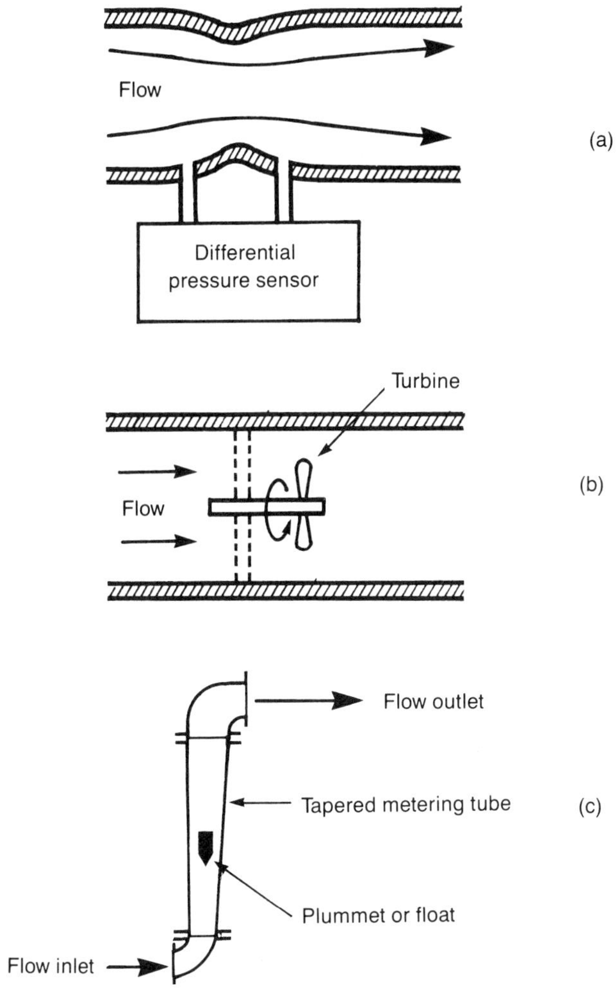

Figure 2-37 Some typical liquid flow measuring sensors using (a) a Venturi tube, (b) a turbine rotor and (c) a floating plug.

to the narrower part of the tube as the flow is reduced. The plug position can be monitored by optical or electromagnetic transducers. This design is rugged and inexpensive, but the linear range is limited and the pressure drop induced by the plug is significant.

Nonmechanical flow sensing devices have also been developed where no obstacle to fluid flow is introduced entailing, however, higher cost. One approach is the magnetic flowmeter (Thompson, 1984) where a magnetic field is applied transverse to the fluid flow in a nonmagnetic tube. The flow of a conducting fluid across such a field produces by electromagnetic induction a transverse electric field which is picked up by two electrodes. Another method is based on ultrasound. An ultrasonic wave is transmitted by a transducer along the flow direction, and its time of flight to a second transducer is dependent on the speed of the moving medium. A different technique monitors the frequency of the ultrasonic beam reflected by inhomogeneities in the fluid, measuring the Doppler shift which is dependent on the fluid flow. Similar techniques using optical beams will be reviewed in Chapter 7. Other approaches are based on the thermal dissipation rate from a heated element in contact with the flowing fluid or else on the flow time of a radioactive tracer between two sensing points (Considine, 1985, p. 4–70). Finally, mass flow rather than stream velocity can be measured by a dynamic torque-monitoring method (Blumenthal, 1984).

Level Sensors

Monitoring liquid level is an important practical requirement in many industrial processes using vessels such as distillation columns, reboilers or mixing tanks. A given optimum level may be required in vessels such as evaporators, where too low a level may uncover the heating surface and too high a level will require greater input as the head pressure increases. In general, a prompt and reliable level control allows use of smaller vessels, thus reducing capital expenses.

Several methods are effectively used to measure liquid level, the most straightforward being the measurement of hydrostatic pressure on the bottom of the vessel. This approach can be used when the specific gravity of the liquid does not change, as it could by effect of variations in composition or temperature and when the pressure imposed on the surface of the liquid in a closed vessel is known.

Another widely used approach to level monitoring makes use of buoyancy methods. A ball-float mechanism may act on a feeding valve through a shaft to keep the level constant in the vessel. Open-loop systems where the liquid communicates with an external liquid column gage for visual inspection are also quite common.

Other techniques are based on point contact with a suspended sensor, such as an electrically charged electrode which discharges upon contact with the grounded liquid. Double-point sensing allows the liquid to be maintained between two predetermined levels. Alternatively, the level can be measured by a capacitive probe partially immersed in the liquid. The capacitance of the probe changes with the liquid level. Other level sensing methods include ultrasonic pulse-echo or side-to-side transmittance across the vessel, as well as nuclear devices monitoring the attenuation of nuclear radiation by the mass of liquid in the vessel.

Viscosity and Consistency Monitoring

Newtonian materials, such as water or most mineral oils, are characterized by the viscosity coefficient which is the ratio between the applied shear stress and the shear rate flow (Sherman, 1970). Other materials, such as paper pulp or printers' ink, exhibit plastic behavior with a flow rate increasing more than proportionally to the applied stress. Such behavior may be enhanced by the strong dependence of the viscosity coefficient of such materials on temperature, a phenomenon which can be used to control the amount of ink flow in printers (Cielo and Westwood, 1978). Dilatant materials, such as quicksand or peanut butter, have the opposite property of reducing the flow rate increase with increasing shear stress. The apparent viscosity of non-Newtonian materials is called consistency. Because of the strong, although highly nonlinear, dependence of consistency on the amount of diluted material in a suspension, the term consistency is often used to designate the concentration of solute. In the pulp and paper industry, for example, consistency is expressed in terms of a weight percentage of dry fibers in the pulp (Lavigne, 1979).

Viscosity is normally measured by sensing the drag torque of a stationary element in a rotating cup. Other approaches involve measuring the flow rate through an orifice or monitoring the ultrasonic velocity in the medium (Considine, 1985, p. 11–10). Continuous viscosity measurement and control is required in many industrial processes such as those involving polymerization, where viscosity is an indication of the end point of the reaction, or those involving starch, where viscosity is a measure of the enzyme actions taking place.

Consistency is an important parameter in many industrial processes. Flowability, which is strictly related to consistency, is important as a material characterization parameter for control of a process such as wood pulp refining for papermaking. Consistency of the slurry is in this case related to the average length of the fibers or to the amount of lumps or knots of fibers, and must be known in real time in order to determine when pulp treating should be terminated. Flowability may also be of importance in itself, as in food sterilization processes. Slight changes in viscosity may produce unacceptable variations in the time of contact with the continuous agitating cookers, resulting in food degradation.

Consistency, like viscosity, is normally measured by torque sensing in rotating arms or by measuring the flow time through apertures. Laboratory analyses may be periodically performed on conveniently chosen samples. Appropriate sampling techniques (Cornish et al., 1981) include avoiding slow stream recesses where sedimentation may occur. Sampling at a point immediately after a centrifugal pump is good practice.

Real-time consistency monitoring is often combined with slurry dilution for closed-loop consistency control. An example of such a control is shown in Figure 2-38. The amount of solvent addition through the dilution valve is regulated to assure a constant value of the continuously monitored consistency (diluting is usually more practical than increasing the consistency). The blade-type sensor shown in Figure 2-38 is a typical, rugged device for on-line consistency measurement. Such a parameter is evaluated from the torque exerted on a blade by the flowing material. The sensor is so designed that it is relatively unaffected by changes in the flow rate. Sensitivities of $\pm 0.02\%$ over a consistency range of 1.75% to 6% are typical (Considine, 1985, p.11–21).

2.5.3 *Temperature Sensors*

The mercury thermometer is probably the best known temperature sensor for domestic use. A similar instrument which is more appropriate for industrial

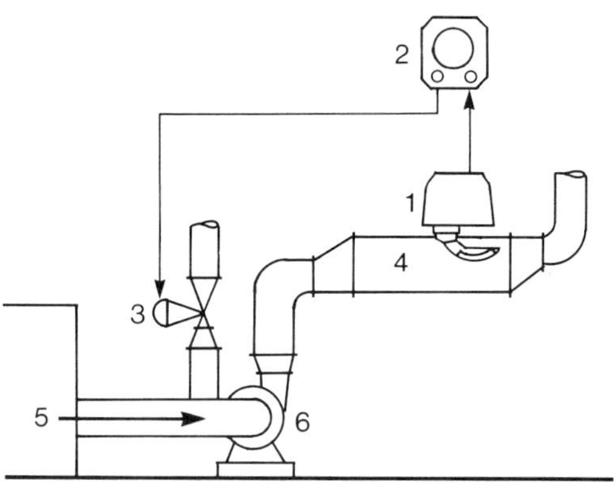

Figure 2-38 Typical installation of a blade-type consistency controller: (1) consistency sensor, (2) controller, (3) dilution valve, (4) stilling pipe, (5) stock chest and (6) pump.

use is the fluid pressure thermometer where the mercury bulb is attached to a bourdon tube of the kind shown in Figure 2-34a. Temperature is indicated by changes in pressure, rather than in volume, of the enclosed liquid, which produce a deformation of the spiral tip. Another spiral-shaped device is the bimetallic thermometer, where the mercury-filled bourdon tube is replaced by a bimetallic spiral. Movement of the spiral tip is produced by the different thermal expansion coefficients of the two dissimilar metals which are bonded together to form the bimetallic strip.

The most widely used temperature sensors in industrial controls are the thermoelectric devices, comprising thermocouple and thermoresistive systems. Thermocouples are electric circuits made of different metals, such as copper-constantan, chromel-alumel or tungsten-rhenium alloys, producing a thermoelectric current when one of the junctions is raised by contact with the probed object at a temperature different from the temperature of the other junction. Maximum temperatures may reach 400°C for the copper-constantan welded wire thermocouples and up to 2500°C for the tungsten-rhenium devices if held in an inert atmosphere. Miniature thermocouples may have response times as small as 1 ms if good thermal contact to the heating source is assured. Several thermocouples may be connected in series to reach higher sensitivity, while an uncertainty of one part in 10^3 of the absolute temperature is typical for these sensors (Considine, 1985, p. 2–15).

Thermoresistive devices include conductive-type and semiconductor systems. The first category uses electrically-insulated metal wires, such as a platinum wire, whose electric resistance increases with temperature. Such devices have better stability and accuracy properties than thermocouples, but their response time is normally higher than 0.5 seconds because of the electrical insulation thermal inertia. Thin film platinum elements have recently become available, and they may reach substantially higher response speeds. The other type of thermoresistive devices are the thermistors which use a semiconductor sensing element with a substantially higher, and reversed, thermoresistive coefficient as compared to a metallic element. Thermistors thus give an output thermoelectric voltage of magnitude much higher than platinum wire devices, making the electronics simpler. However, their stability and accuracy is much lower, while their maximum operating temperature is limited to 300°C. A different approach is that of oscillating crystal sensors, where a small quartz oscillator undergoes resonance frequency variations of the order of 1 KHz/°C which are precisely monitored by heterodyne techniques (Lenk, 1980, Section 6-7).

Radiation thermometry, where the temperature of the inspected object is monitored by detecting its infrared emission with a pyrometer, is a noncontact technique and thus potentially very fast (some infrared photodiodes have a time response of 2 ns) and readily applied to incandescent-temperature measurements. Other advantages are their applicability to rapidly moving objects, such as turbine blades in a rotor, as well as remote or hard-to-access surfaces, provided that they are in line of sight with the infrared sensor. The fast response time and optical nature of such sensors make it possible to rapidly scan large surfaces: Modern infrared thermal imaging cameras may display 200×200 pixel thermal images in a 30 ms period. Their DC accuracy is, however, limited to 1% of the absolute temperature in ideal conditions, while substantially higher temperature evaluation errors are induced by variations in the infrared emissivity of the inspected surface or by reflections from surrounding hot bodies. Different kinds of infrared detectors are described in Section 3.3, while a number of industrial applications of such devices are given in Chapters 6 and 7.

2.5.4 *In-process Analyzers*

This section reviews a number of process analyzers, i.e. on-line instruments, to analyze the physicochemical material properties of the processed product in order to control the process parameters. A description of purely chemical analysis instruments, such as gas or liquid chromatographers, pH measurement devices or oxidation-reduction monitoring systems, can be found elsewhere (see, e.g., Considine, 1985, Section 6).

In-process analyzers can be classified as either *in situ* or *sampling* instruments. *In situ* analyzers are permanently installed on the production line and continuously perform an analysis of the moving or flowing product. A typical example is a pulp turbidity optical transmission sensor which is widely used to monitor through a window the composition of a liquid flowing in a pipe. A continuous operation of *in situ* analyzers does not ensure 100% analysis of the product: the turbidity sensor may be pointed only to the central part of the pipe, or else the product may be cross machine scanned so that only discrete portions of the material at each cross machine position are periodically analyzed. *Sampling* analyzers require the periodic extraction of a sample of the material for analysis either with a laboratory instrument or by an analyzer situated nearby the production line. This approach, which is still the most widely used, is required when the analysis demands substantial sample preparation or when the analytical instrument cannot operate in the environmental conditions of the production line. An example is an electron microprobe X-ray analyzer, requiring vacuum operation for electron beam bombardment.

Density and Specific Gravity Material Analysis

The composition of liquid or powder mixtures or the nature of a material undergoing chemical or structural modifications may affect its density. Examples are the degree of crystallinity of a polymer or the amount of dissolved sugar in the residual liquid at a sugar extracting refinery.

A very common instrument for manual measurement of the specific gravity of a liquid is the hand hydrometer shown in Figure 2-39a. This instrument consists of a weighted float with a small diameter stem proportioned so that more or less of the scale is submerged according to specific gravity of the fluid in which it is floating. The vertical equilibrium position of the float depends on the density of the liquid and can be read either visually or through optical or magnetic readout devices.

A system which is better suited to on-line automated density measurements is shown in Figure 2-39b. Two pipes of different lengths are immersed in the liquid to be analyzed. A weak air flow is applied to the pipes so that the air pressure in each pipe is equal to the hydrostatic pressure at the respective depths within the liquid. The pressure differential between the pipes is proportional to the liquid density. The pipes may be replaced by pressure transducers for operation in dry powder fluids.

Devices for monitoring gas densities include gravity balance sensors at the bottom of a tall gas column as well as buoyancy gas balance systems where a hollow container is balanced in air and in the gas to be monitored and the pressure difference is related to the gas density relative to air (Considine, 1985, p. 6–193).

A number of acoustic impedance devices have been developed for density measurements (Chesmond, 1984; Considine, 1985). The best known is the U-tube sensor, where a U-shaped tube is vibrated at its natural oscillation frequency while the liquid to be analyzed is flowing within the tube. Variations in the density of the liquid produce variations in the sensed resonance frequency. Variations of this technique include the use of vibrating cylinders, vibrating vanes and vibrating twin-tube sensors. Ultrasonic velocity and attenuation measurement are also being used for on-line density measurement (see, e.g., *Control Engineering*, March 1978, p. 35; see also Piché, 1984). Finally, it should be recalled that an indirect measurement of density in

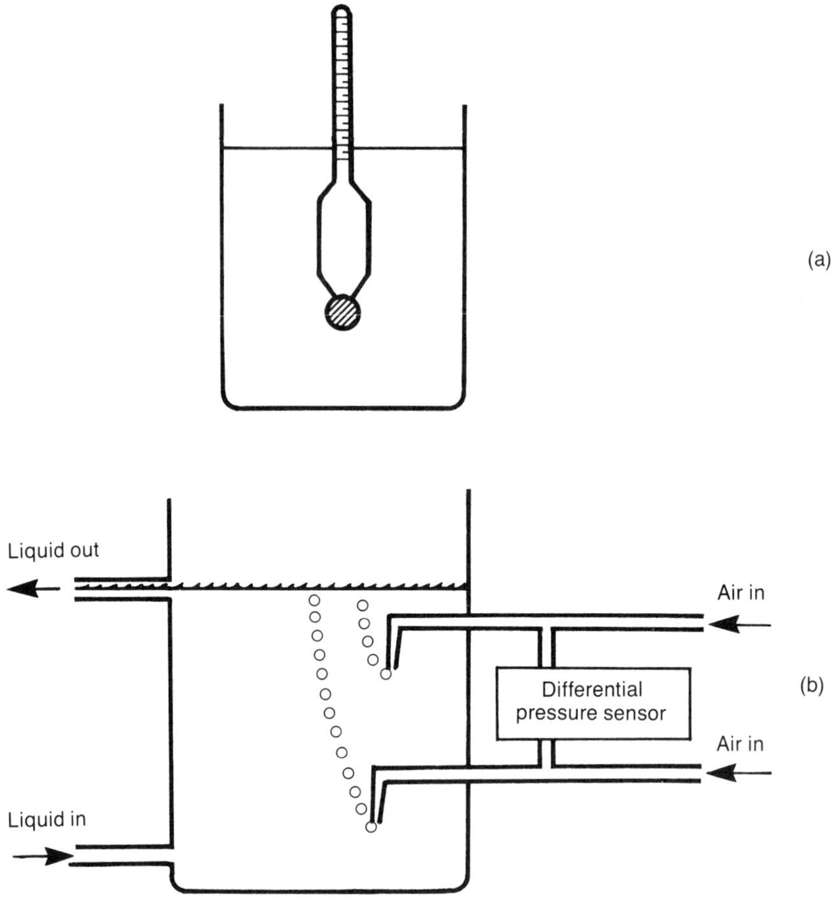

Figure 2-39 Density measurement by (a) a hand hydrometer and (b) a hydrostatic air bubbler pressure differential sensing system.

slurries may be obtained by consistency or viscosity measurements, as described before.

Humidity and Moisture Measurement

A well-known instrument for humidity measurement is the psychrometer, schematically shown in Figure 2-40a. It is composed of a wet-bulb and of a dry-bulb thermometer, typically resistive wire devices, which are inserted in a stream of air of speed in the 4 to 10 m/s range. The dry-bulb thermometer measures ambient temperature, while the wet-bulb thermometer measures

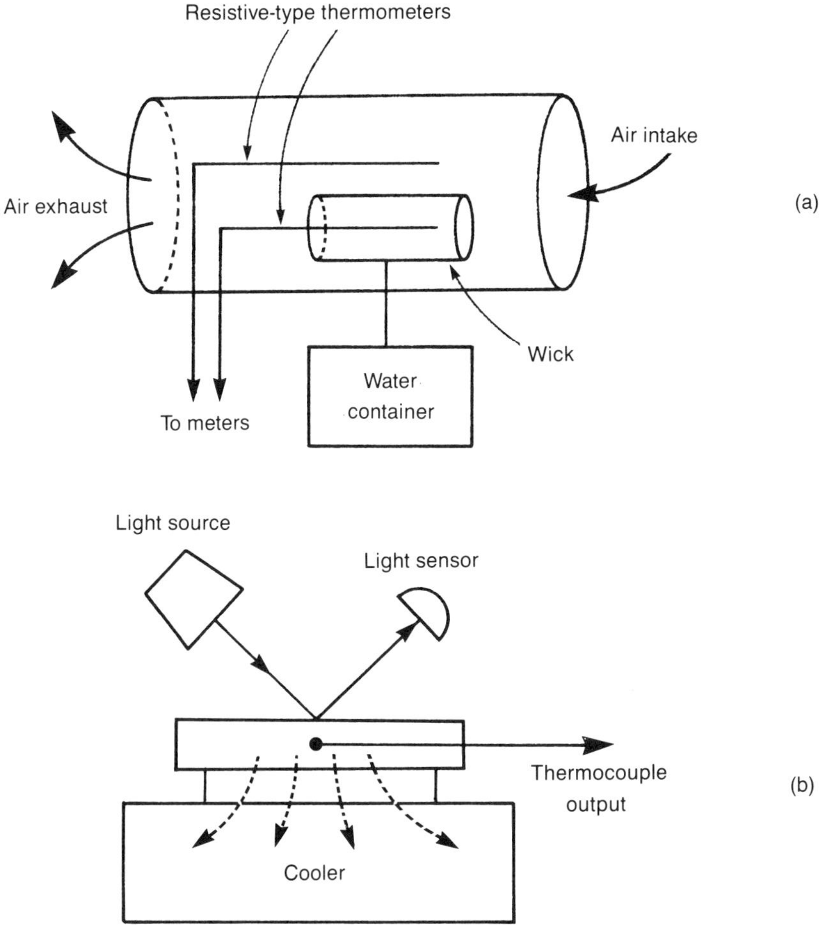

Figure 2-40 Humidity monitoring by (a) psychrometry and (b) dew point temperature sensing.

temperature reduction due to evaporative cooling. A wick or a porous ceramic sleeve saturated with water surrounds the wet-bulb thermometer, and its evaporative cooling rate depends on the air humidity. Relative humidity is determined from the two temperature readings and a reading of the barometric pressure by means of a psychrometric table (see, e.g., Considine, 1985, p. 10–5).

Another approach for humidity measurement is by dew point sensors (Figure 2-40b). A mirror in thermal contact with a thermoelectric cooler is exposed to the ambient atmosphere. When the mirror temperature is reduced below the water saturation temperature of the surrounding gas, moisture condensation forms on the mirror surface and is detected by the optical-reflectivity sensor shown in Figure 2-40b. A thermocouple in the mirror material senses the dew formation temperature, which is related to the relative humidity of the ambient gas. An alternative design uses a resistive type dew point sensor rather than the reflective mirror. Such a sensor uses a resistive element in the form of a metal conducting grid embedded in the cooled disk (Lenk, 1980, p. 79).

Other devices use resistive hygrometer sensors, whose resistance varies with relative humidity without being much affected by ambient temperature, or a hair element, whose length varies almost linearly with humidity, or else an oscillating crystal. A widely used humidity sensor is the aluminum oxide sensor. This type of device is a capacitor formed by depositing a layer of porous aluminum oxide on a conductive substrate and then coating the oxide with a thin film of gold. Water vapor penetrates the gold layer and is absorbed by the porous oxidation layer. The quantity of water absorbed determines the electric impedance of the capacity, which in turn is a measure of water vapor pressure.

While relative humidity is the fraction of water present in a gas as compared to the water amount under saturation conditions, moisture is the percentage of water contained in a solid or liquid material. Direct current moisture meters measuring the resistivity of the material between two electrodes have been reviewed in Section 2.4.1, while the use of microwaves for moisture monitoring in plastics and paper was mentioned in Section 2.4.4. Most of the humidity sensors described above, such as the aluminum oxide sensor, may also be used to measure moisture in liquids. Infrared-based moisture sensors will be reviewed in the last chapter of this book.

Conductivity Sensors

Electrical and thermal conductivity are useful parameters for the analysis of the composition or structure of many industrial materials. The electrical conductivity of electrolytes is often monitored in industrial processes to evaluate the concentration of solutes such as sodium chloride, sodium hydroxide or sulfuric acid. Conductivity measurements are also widely used to detect leaks in heat exchangers and resultant contamination of heating or cooling media in such equipment as acid coolers, condenser coils and steam coils. Any sudden increase in conductivity of the heat-exchange medium is taken to indicate a leakage of electrolyte. A break in steam condenser tubes results in an increase in condensate conductivity.

The simplest electrical conductivity sensing device is made of two electrodes immersed in the electrolyte. However, substantial reading errors are caused by electrode oxidation or fouling by precipitation of extraneous substances. A four-electrode sensor provides a conductivity measurement which is less affected by these errors. In such a device, two measuring electrodes are located between a set of current producing electrodes. The measuring electrodes are used to determine the voltage drop in the electrolyte caused by the current. The measured electrolyte electrical conductivity is strongly dependent on temperature. A temperature sensor is thus usually coupled to the conductivity-monitoring head to provide automatic temperature compensation through calibration data.

Thermal conductivity is often used in industrial processes to determine gas composition. Some gases, such as hydrogen or helium, have a thermal conductivity which is more than six times higher than the conductivity of oxygen or nitrogen. A thermal conductivity gas analysis cell is thus used to determine the composition of gas mixtures such as hydrogen in nitrogen, chlorine in air, water vapor in oxygen, etc. The cell usually consists of a heated filament in the gas to be analyzed. The equilibrium temperature, which is usually measured by monitoring the electric resistance of the heated wire, depends on the gas thermal conductivity. By proper design and geometry and by limiting the temperature rise of the heated element, it is possible to minimize reading errors due to radiation and convective flow.

X-ray and Particle Beam Analyzers

Radiographic imaging devices for the detection of hidden flaws in opaque materials and for thickness gaging were described in Section 2.3. X-ray devices are also widely used for analytical purposes. Radiographic analysis techniques include X-ray crystallography, fluorescence, selective absorption and electron microprobes (see, e.g., Clark, 1955 and Buerger, 1960).

X-ray diffraction analysis can be used to determine, e.g., the degree of crystallinity of polymeric materials. Crystalline samples give sharp angular peaks in the diffraction pattern, corresponding to the Bragg-reflection condition from the crystal layers. The position of the angular peaks is related to the lattice spacing within the crystals which are usually known for given substances. Departures of the experimental pattern from normal angular spectra may be interpreted as due to residual strain and stress. Other applications of X-ray diffraction are in the estimation of the grain size of many polycrystalline materials.

X-ray fluorescence is used to evaluate the relative composition of different elements within a given substance. When X-rays penetrate a material, they may extract electrons from the inner shells of the atoms. Such electronic positions are filled through electron transitions from outer shells, with the emission of fluorescence X-rays whose energies are characteristic of the energy levels of each atom. The fluorescence energy distribution spectrum thus provides an evaluation of the relative concentration of the different elements in the irradiated sample.

A different approach is the measurement of the absorption spectrum through a given thickness of the material. For a given element, there are abrupt changes in the absorption coefficient at certain wavelengths characteristic of the element. These abrupt changes occur at wavelengths corresponding to the energies of the K, L and M shells, which are called absorption edges of the spectrum. If an absorption measurement is made at two wavelengths just above and below the absorption edge corresponding to a given element, the ratio between such measurements provides an estimate of the relative concentration of such an element.

Still another approach in analytic radiography is the electron microprobe method where the sample surface is bombarded by a finely focused electron beam. The penetrating high-energy electrons produce re-emission of back-scattered electrons, secondary electrons and X-rays. All of these radiations may be detected and analyzed to obtain information on the elemental composition of the target. Because of the high focusing capabilities of the electron beam, a very good spatial resolution can be obtained when scanning a heterogeneous sample. Examples of applications are the analysis of phase segregation in fine-grained alloys or the inspection of semiconductors for microcircuit applications.

Finally, charged-particle beams play an important role in mass spectroscopy. In classical mass spectrometers, a stream of ions is bent by a transverse magnetic field and the measured radius of curvature is an indication of the mass-to-charge (m/e) ratio which is characteristic of the different compounds. A mass spectrometry device which is more applicable to industrial analysis is the quadrupole mass spectrometer (Considine, 1985, p. 6-57). A quadrupole mass filter is a set of four parallel rods disposed in a square matrix. A radio frequency and DC voltage of equal potential, but opposite polarity, is applied to each set of rods. By varying the absolute potential applied to the rods, it is possible to intercept all the longitudinally-oscillating ions except those having a given m/e ratio. Such instruments are increasingly used for the industrial analysis of a wide range of processed materials.

Chapter 3

Fundamentals of Optical Technology

Optics is the science of light. In common language, light is the radiation which is detected by our eyes. Since the discovery by Newton that solar radiation contains spectral components unseen by the eye, it has become common practice in science to refer to light as the ultraviolet, visible and infrared radiation. In modern technology light includes electromagnetic radiation which can be generated and detected by optical devices and manipulated by optical components such as lenses and mirrors, namely the wavelength range from approximately 100 nm to 1 mm.

This chapter is intended as an introduction to relevant optical technology for the nonoptically trained scientist and engineer. The optical scientist may also find in this review some useful information on the practical aspects of the devices which are commonly used in optical inspection.

3.1 Light Sources

Modern optical technology has been greatly stimulated by the development of the laser 25 years ago. Applications such as optical communications and information processing, optoelectronics and holography have been revived by the availability of strong, directional and coherent laser radiation. In turn, the exceptional success obtained in the development of some of these applications, such as fiber optic communications, has provided the economic incentive to develop rugged, inexpensive and reliable light sources like the modern laser diodes as well as high performance and compact optical detectors. Industrial applications such as optical inspection are gaining a lot of momentum from the proliferation of these optical components.

3.1.1 *Incoherent Sources*

Although laser light possesses superior qualities in terms of monochromaticity and directionality, incoherent sources are still widely used in industrial inspection applications where the coherence of lasers is not required. In some cases, the higher time stability and lack of speckle of incoherent sources actually make them a better choice than laser sources. In this section the most common incoherent sources used for industrial applications will be briefly described.

Radiometric and Photometric Units

As incoherent sources have typically been developed for human illumination purposes, their performance is often given in terms of visible light output. A knowledge of the photometric system of measurement is thus necessary in order to assess the specifications outlined by the vendor.

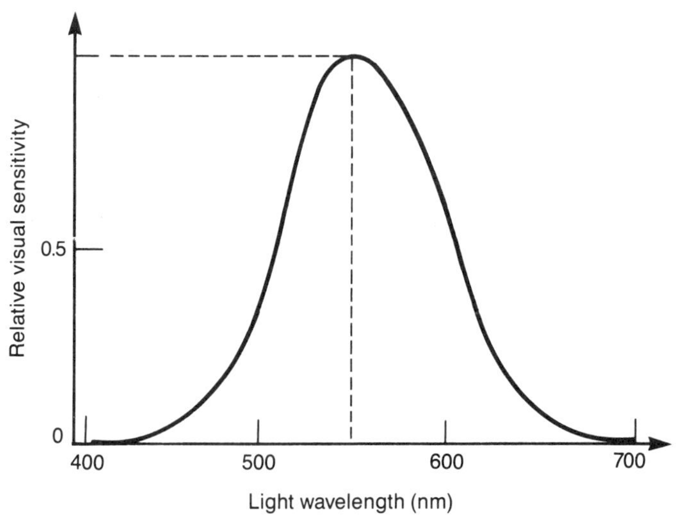

Figure 3-1 Relative sensitivity of the human eye to light of different wavelengths.

Radiometric units are given in terms of light power, the most commonly used being W and W/cm^2. Photometric units are based on the visual response, shown in Figure 3-1 (Rostagni, 1952; Teele, 1965; Hill, 1985). The luminous power is obtained by integrating the power spectral distribution of the curve after multiplication by the visual sensitivity curve and by a conversion constant of 680 lumens per watt. This value is in agreement with the photometric standard which is the luminous intensity of 1/60 cm^2 of a blackbody at the temperature of melting platinum and which is called candela or international standard candle. The nomenclature and definition of the numerous photometric units in use is given in Table 3-1 (Wolfe and Zissis, 1978, Chapter 1).

Thermal Sources

The most commonly used thermal light source is the tungsten-filament lamp, in particular the tungsten-halogen light source. These lamps have a tungsten filament mounted in a small quartz envelope containing a small amount of halogen gas. Such a gas combines with tungsten evaporated from the filament preventing its deposition on the inside of the hot quartz envelope. This provides longer lifetimes at higher operating temperatures.

Some typical lamp configurations are shown in Figure 3-2 (Carlson and Clark, 1965; Eby and Levin, 1979). The microlens-capped configuration

Table 3-1 Photometric Units

Quantity	Unit name	Unit dimensions
Luminous flux	Lumen (lm)	lm
Luminous energy	Talbot	lm s
Illuminance	Lux	lm m^{-2}
	Phot	lm cm^{-2}
	Foot-candle	lm ft^{-2}
Luminous intensity	Candela (cd)	cd = lm sr^{-1}
Luminance	Nit (candelas/m^2)	cd m^{-2}
(photometric brightness)	Stilb (candelas/cm^2)	cd cm^{-2}
	Lambert (for diffuse materials)	π^{-1} cd cm^{-2}
	Foot-lambert (for diffuse materials)	π^{-1} cd ft^{-2}
	Apostilb (for diffuse materials)	π^{-1} cd m^{-2}

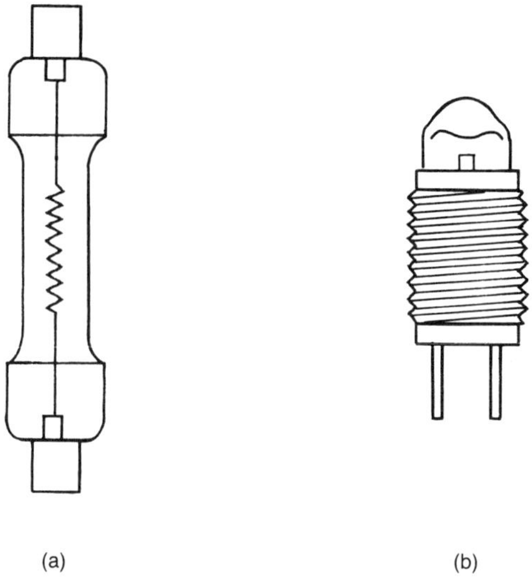

Figure 3-2 Examples of tungsten-halogen lamp configurations.

shown in (b) (by courtesy of Gilway, Woburn MA) provides a nearly 1 cm-diameter, 7000 lux illuminated area at 5 cm of distance, with a particularly compact lamp assembly. The helicoidal shape of the filament is required to minimize thermoelastic stress during operation. Such a shape introduces an angular modulation of the emitted flux because of filament shading. A typical light output angular distribution in the plane containing the axis of a linear filament lamp of the kind shown in Figure 3-2 (a) is shown in Figure 3-3.

The spectral emission curve of a quartz-halogen lamp at a typical operating temperature of 3200 K is shown in Figure 3-4. The curve is similar in shape to the blackbody emission at this temperature, modified by the emissivity curve of tungsten (going from 0.5 in the visible to 0.3 at 2 μm at such a temperature) and by the quartz transmissivity. Input electric powers may be as low as 3 watts with a 40,000 hours lifetime for the Gilway lamp shown in Figure 3-2 (b), or as high as several kilowatts for large quartz-halogen lamps with lifetimes smaller than 100 hours. The radiated power over the whole spectrum and in all the angular directions is typically 20% to 40% of the electric input; however, only a small fraction of the radiated output is in the visible range, as Figure 3-4 shows. This results in a relatively low luminous intensity, particularly at low filament temperatures, as Figure 3-5 shows. This

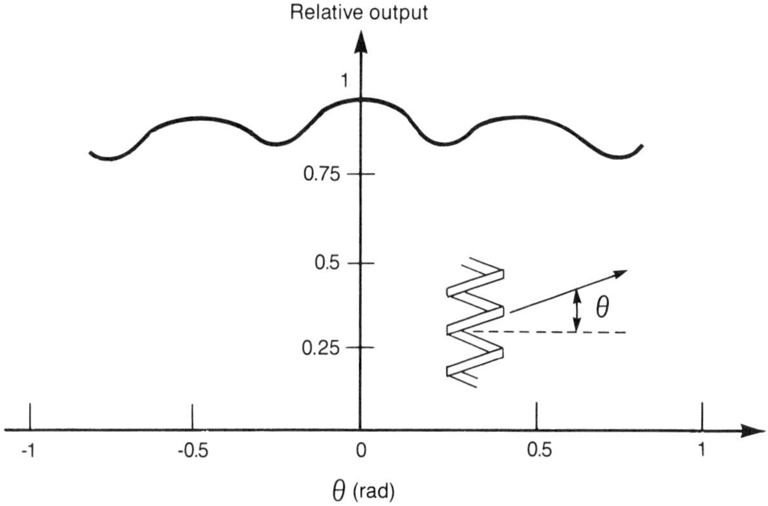

Figure 3-3 Radiometric output of a quartz-halogen linear filament lamp in the plane containing the filament axis.

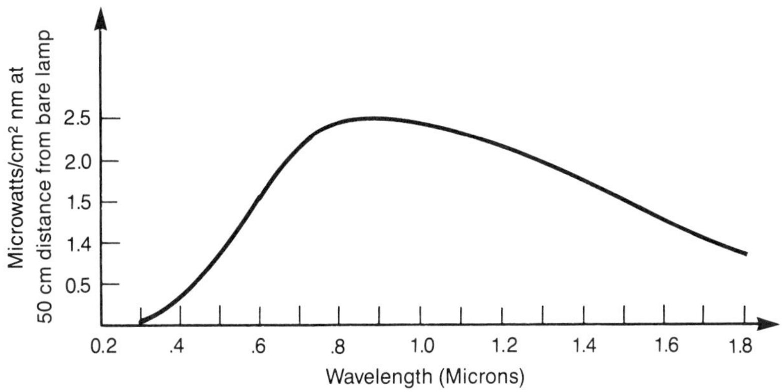

Figure 3-4 Spectral irradiance of a 100-watt quartz-halogen lamp operated at 3200 K. (Courtesy of Oriel, Stratford, Conn.)

figure also shows that an increase in the filament temperature in order to increase the luminous output is extremely expensive in terms of lifetime reduction. It should be mentioned that for optical inspection purposes, a strong visible emission is often not required. Silicon detectors have highest sensitivities in the 0.7 to 1 μm region, while good detectors in the 1.5 μm region are increasingly available.

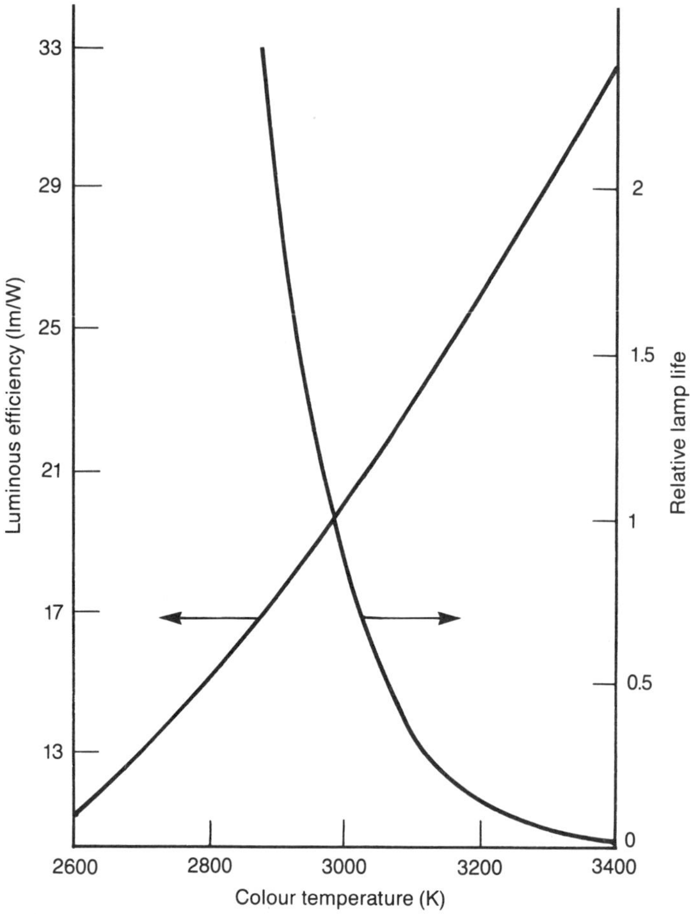

Figure 3-5 Luminous efficiency and lifetime variations with the operating filament temperature for a quartz-halogen lamp. (Courtesy of Oriel, Stratford, Conn.)

Arc Lamp Sources

High power arc sources differ from gas-discharge sources used in atomic spectroscopy and fluorescent illuminators mainly because of their high operating pressure. These lamps consist of quartz envelopes containing two electrodes whose tips are spaced from 1 mm to several centimeters depending on the required power and which are filled with a gas at 50 to 80 atmospheres at the temperature of operation. The high pressure gas is capable of carrying a large current density and of dissipating a large power in a small volume. This produces a relatively small, highly intense source of visible light as well as of ultraviolet and near-IR radiation. High-pressure mercury arc sources may also be used as far-IR incoherent sources in the wavelength range higher than 50 μm where the quartz envelope is transparent. In the range from 25 to 50 μm, thermal sources such as the Nernst glower or the Globar (air-operated semiconducting ceramic rods which are resistively heated to temperatures of the order of 1500°C) are often used in applications such as spectrometry (see, e.g., Olsen, 1975, Sections 3-2 and 3-6).

Figure 3-6 shows a typical geometrical configuration for a short-arc lamp. Figure 3-7 shows the luminance distribution in the hot-arc region. The high luminance zone near the cathode tip, having a brightness higher than 50% of the maximum brightness, has a volume smaller than 1 mm^3. The source spot size is thus much smaller than what can be obtained with a tungsten filament lamp of comparable power.

Figure 3-8 shows a typical angular light intensity distribution from an arc lamp in the axial plane. Because of the cylindrical symmetry of the lamp, much of the emitted light is lost. Some lamps are backed by a spherical mirror to double the light power collected by the collimating front lens. A more efficient arc lamp assembly is shown in Figure 3-9. An elliptical reflector

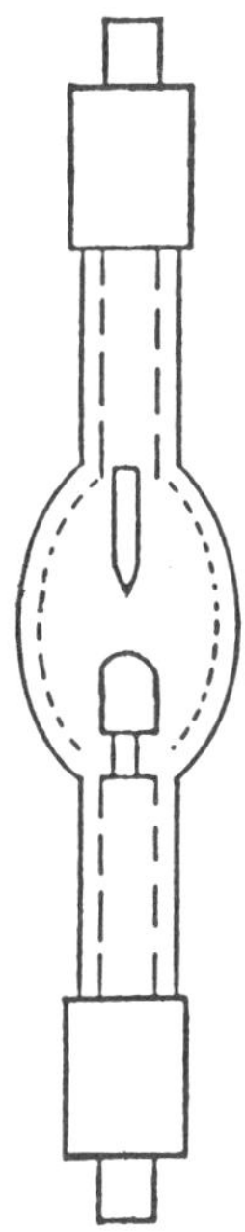

Figure 3-6 Typical configuration for an arc lamp source.

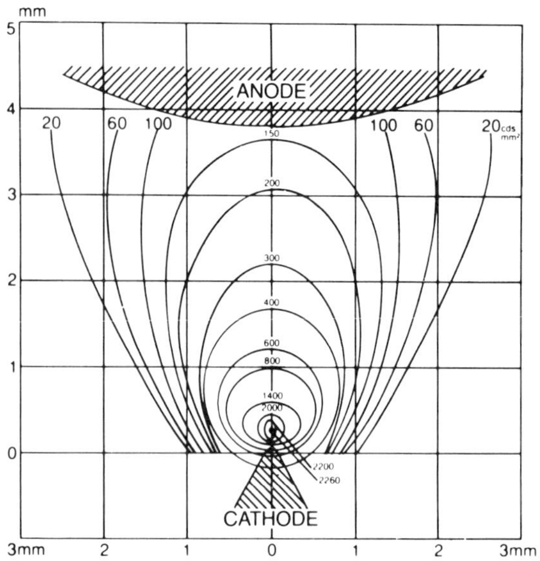

Figure 3-7 Luminance distribution in the arc of a typical 2 KW arc lamp. The iso-luminance curves are in cds/mm^2. (Courtesy of Canrad Hanovia, Newark, N.J.)

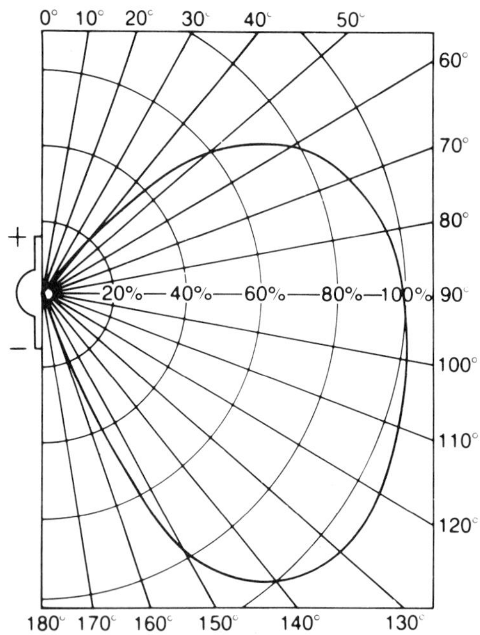

Figure 3-8 Relative light intensity distribution for a typical small-arc lamp. (Courtesy of Canrad-Hanovia, Newark, N.J.)

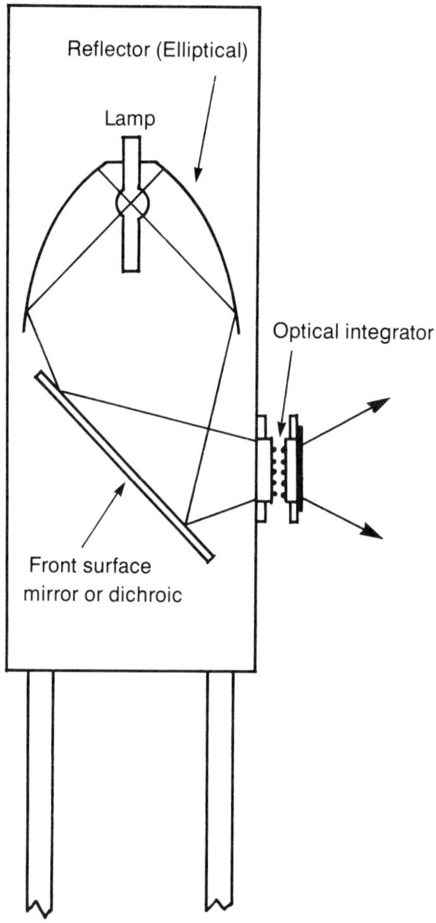

Figure 3-9 Arc lamp assembly with elliptical reflector. (Courtesy of Kratos, Ramsey, N.J.)

is used to collect most of the irradiated intensity and to produce a roughly collimated, hollow ring light beam. After deflection by a 45° mirror, the beam may be randomized by an integrator (array of small quartz lenses) to produce a roughly uniform, diverging light beam.

Arc lamp sources of input power going from less than 100 W for small-spot lamps to 300 KW for the liquid-cooled lamp produced by Vortek are available. Typical lifetimes are in the range of 100 to 2000 hours. Electrical-to-optical power conversion efficiencies are of the same order of magnitude as those of quartz-halogen lamps, however, the luminous efficiency is typically an order of magnitude higher because of the strong emission of arc sources in the visible range. The spectral distribution of the arc lamp output strongly depends on the filling gas. Figure 3-10 shows the spectra of two of the most common arc sources. Xenon lamps produce a relatively continuous spectrum approximating a 6000 K color temperature, while mercury lamps emit strong lines in the 260 to 620 nm region.

Figure 3-10 also shows that a substantial portion of the arc source emission is in the ultraviolet, particularly in the case of mercury lamps. The permissible eye and skin daily exposure limit to radiation in the 250 to 300 nm range is 3 mJ/cm^2 (Sliney and Wolbarsht, 1980, Section 10.3.4). UV-absorbing goggles and protective clothing should thus be worn when working in proximity to high power arc sources. Good ventilation should also be maintained in the working area in order to evacuate the UV-generated ozone, unless special UV-absorbing envelopes are used. Finally, these lamps operate at very high pressures and are therefore subject to explosion. They must consequently be operated inside a protective housing and must be handled with protective goggles and gloves.

In contrast with thermal sources, the time response of arc lamps is not limited by the thermal dissipation time within the filament. Short flashes of duration smaller than 1 ms can thus be obtained from arc lamps, as required in stroboscopic illumination systems. Short igniter pulses are used to generate the charge distribution required to quickly start the arc. Conversely, if the lamp is to operate continuously from the 60-Hz AC line, a ballast is required to properly operate the lamp. The lamp ballast is an electric circuit providing sufficient voltage to reignite the lamp at every half cycle after the lamp has been started by the igniter.

3.1.2 *Coherent Sources*

Lasers are sources of coherent light, both spatially and temporally. A light beam is spatially coherent if it maintains a given phase relationship across its width, as is the case for a highly collimated plane wave. Temporal coherence means that the phase of the electromagnetic field associated with the optical radiation varies in time in a predictable way, as with a perfectly sinusoidal (monochromatic) wave.

Such properties make lasers particularly suitable to many optical inspection requirements. Spatial coherence means that the laser beam can be made nearly parallel for transmission over large distances with little spread losses. Alternatively, the laser beam can be focused on a small spot for high spatial resolution or for ambient-light suppression by spatial-filtering through a pinhole. Large temporal coherence means narrow spectral bandwidth, so

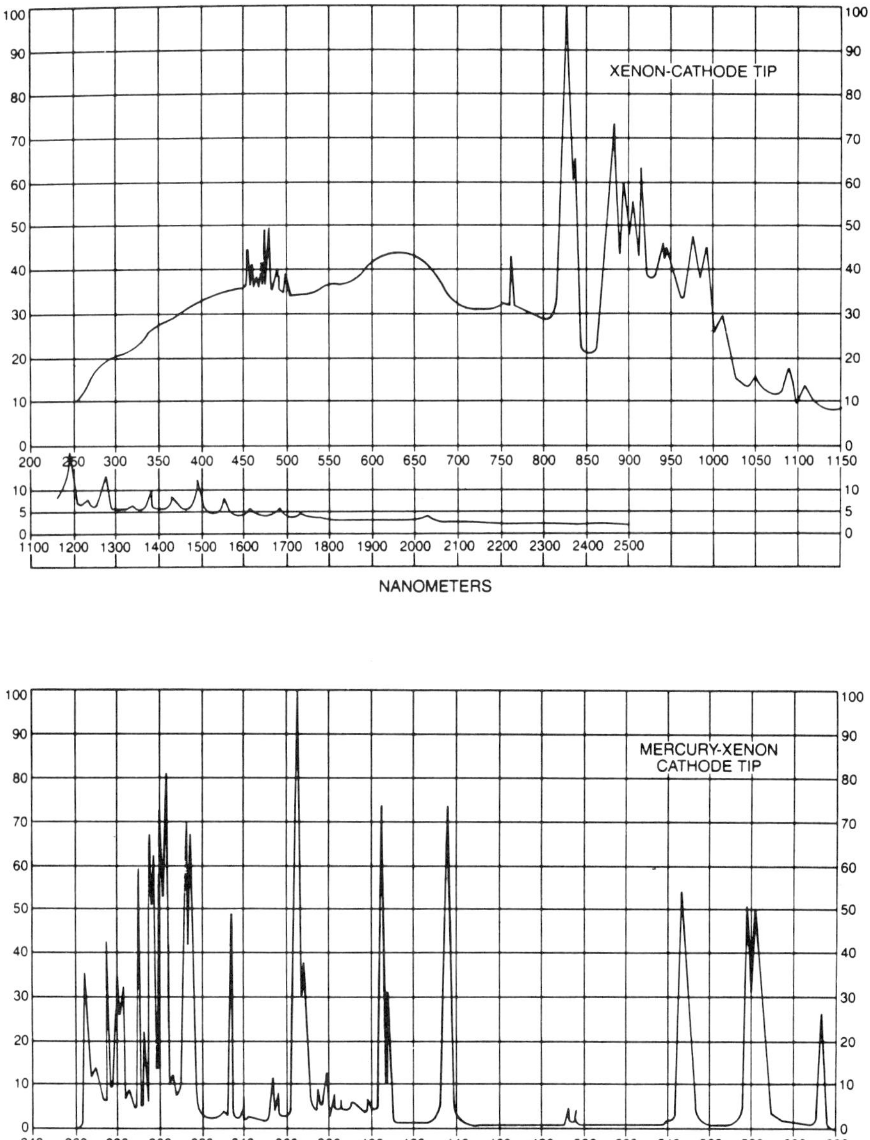

Figure 3-10 Relative spectral distribution of typical Xenon (above) and Mercury-Xenon (below) arc lamp sources. (Courtesy of Canrad-Hanovia, Newark, N.J.)

that ambient light can be further separated from the laser beam by a spectral filter, while optical components may be designed to have minimum aberrations or anti-reflection coatings at the specific laser wavelength. Techniques such as holography or interferometry could hardly be implemented with incoherent light sources.

Laser Characteristics

Several kinds of laser are available on the market. Table 3-2 outlines the most common types (for a more detailed list, publications such as the *Laser Focus Buyer's Guide* published by PennWell, Littleton, MA can be consulted). Figure 3-11 shows the schematic diagram of a gas laser. The discharge excited atoms within the gas tube are stimulated to emit a strong axial radiation which is amplified at each multiple reflection between the two mirrors. If windows at the Brewster angle are present in the cavity, only the polarization component parallel to the plane of the figure will be effectively amplified. A substantially similar scheme is valid for lasers having a solid or a liquid lasing medium, with the difference that excitation is in such cases provided by intense light radiation (optical pumping).

Because of the multiple reflections between the mirrors, only discrete values of the wavelength (corresponding to exact submultiples of a round trip between the mirrors) are effectively amplified. The result is a spectral distribution of the kind shown in Figure 3-12 for multimode lasers. The

Table 3-2 Characteristics of Some of the Most Common Laser Devices

Medium	Wavelength (μm)	Typical operation	Average output power (W)
Excimer	0.19–0.35	pulsed	1–100
Nitrogen	0.337	pulsed	0.001–0.1
He–Cd	0.325–0.442	CW	0.001–0.05
Argon	0.458–0.514	CW	0.01–15
He–Ne	0.633	CW	0.001–0.05
Ruby	0.694	pulsed	0.1–100
Dye	0.2–5	CW or pulsed	0.001–10
Semiconductor	0.6–5	CW or pulsed	0.001–1
Neodymium	1.06	CW or pulsed	0.1–1000
CO	5–7	CW or pulsed	0.2–20
CO_2	9–11.8	CW or pulsed	1–20000
Methanol	37–1217	CW or pulsed	0.001–0.1

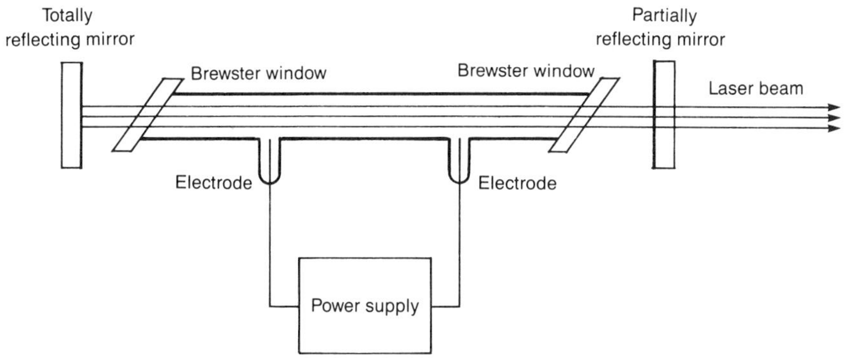

Figure 3-11 Diagram of a gas laser with electron collision pumping.

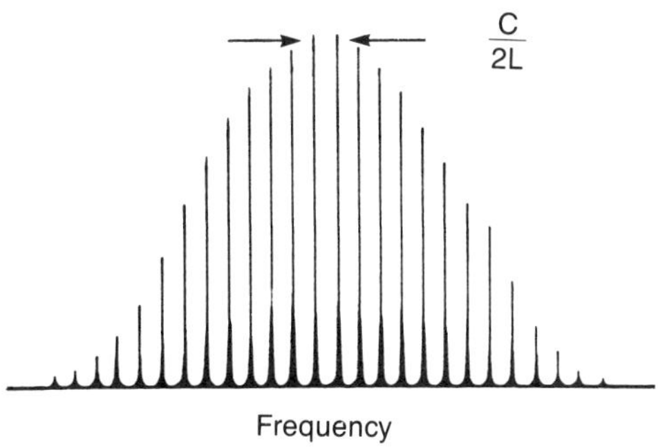

Figure 3-12 Typical intensity distribution of a multiple longitudinal mode laser.

intermode frequency separation is equal to $c/2L$, where c is the speed of light and L the optical cavity length. Even so, the overall laser bandwidth of a typical gas laser is quite narrow, in the 0.01 nm range. This is much smaller than the spectral bandwidth of an interference filter, which is typically 5 nm. However, the presence of several longitudinal modes reduces the allowed spatial range in some applications such as holography or interferometry.

In addition to the longitudinal (frequency) modes, a laser may oscillate in different transverse (spatial) modes, as shown in Figure 3-13 (Kogelnik and Li, 1966). The higher order modes may be suppressed leaving only the fundamental TEM_{00} mode by reducing the angular beam aperture within the laser cavity through a diaphragm. The laser power reduction produced by an intra-cavity diaphragm (which prevents the amplification of highly divergent beams) is smaller than the intensity loss which would be introduced by an extra-cavity diaphragm. Similarly to the longitudinal modes, the

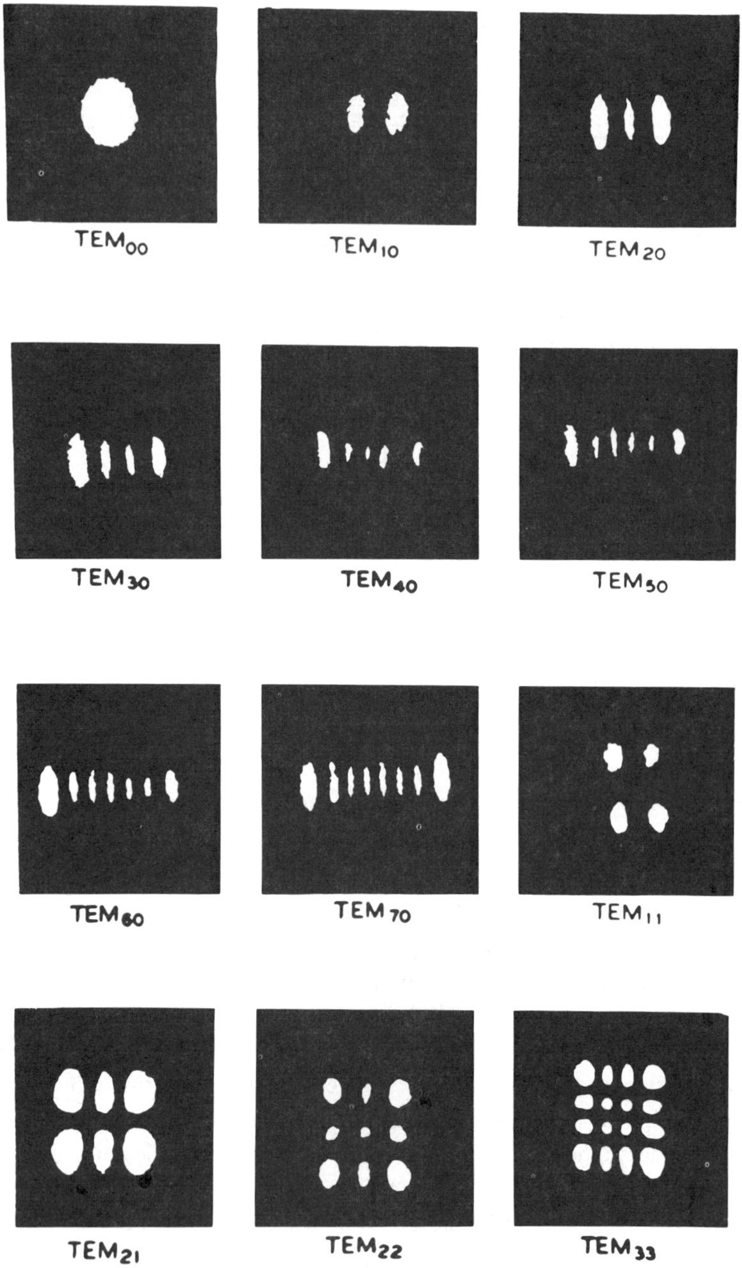

Figure 3-13 Typical transverse electromagnetic (TEM) modes in the emergent beam of a laser.

transverse modes are mutually incoherent. A multiple-spatial-mode laser beam is no longer coherent across its width, so that the beam divergence is larger than the diffraction limited divergence of the TEM_{00} mode,

$$\theta \simeq \lambda / w_0, \tag{3-1}$$

where θ is the half angular spread in radians, λ the wavelength and w_0 the beam waist (approximately equal to the beam diameter at the laser output).

The Diode Laser

A class of lasers which is playing an increasingly important role in optical inspection is that of semiconductor- or diode-lasers (see Figure 3-14 reproduced from Gorog et al., 1976). Photon emission is here induced by electron-hole recombinations produced by a forward current flow through a p-n junction. The laser radiation is multiply reflected between the cleaved faces of the semiconductor pellet along the junction layer which acts as a waveguide. The single-spatial-mode linearly polarized emerging beam, typically 0.5 μm by 3 μm in size, is highly divergent because of equation (3-1). It may either be directly injected into an optical fiber end, as is usually done in fiber optic communication links (some of these diode lasers are supplied by the manufacturer with an integral fiber pigtail for easier fiber-to-fiber

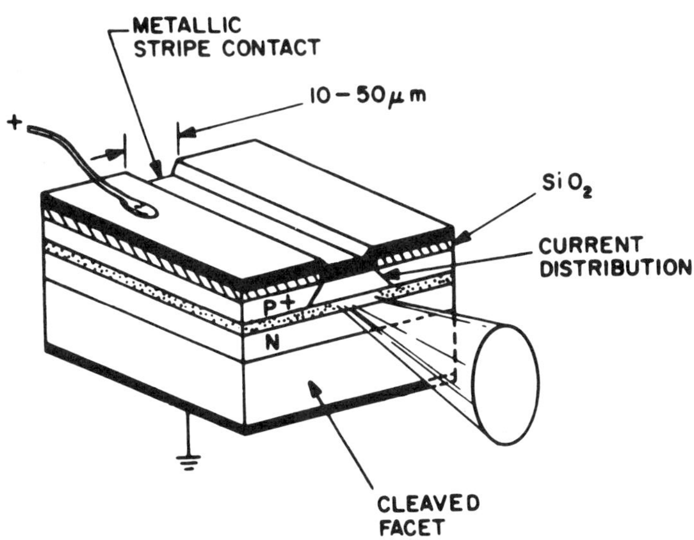

Figure 3-14 Schematic of gallium-arsenide laser with direct current (electron collision) pumping.

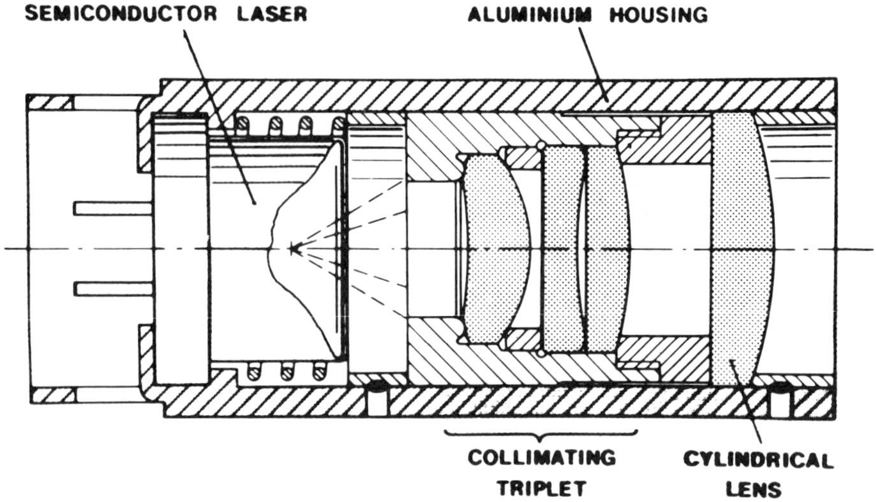

Figure 3-15 The 3.4 cm-long collimator pen housing a diode laser which was developed for video disk players. (Courtesy of Philips.)

splicing), or be transformed into a collimated beam by appropriate optics of the kind shown in Figure 3-15. The incorporation of a cylindrical lens (to compensate for the beam astigmatism because of its asymmetric shape at the output facet) produces a diffraction limited, 5 mm diameter, 2 mW laser beam, the whole 1.2 cm-diameter by 3.4 cm-length system being available at quantity prices in the $100 range.

Compared to gas lasers such as the popular He–Ne, diode lasers are rugged, compact and inexpensive coherent sources which are particularly well-suited for industrial applications. Lifetimes typically reach 100,000 hours, while their energy efficiency is in the 5 to 40 percent range. Output powers reaching 1 W of continuous emission are now available, and higher outputs should be possible in future. However, most of such lasers emit in the near-IR range, so that developmental laboratory work is often best performed with He–Ne or other visible emitting lasers.

Diode lasers have some peculiar characteristics which must be taken into consideration for proper use of such devices. A typical diagram of the light output versus applied electric current is shown in Figure 3-16. Because of the steep rise of the emitted power above threshold, current limiting devices should be included in the driving circuit to avoid inadvertent burnout of the laser output facet. The housing temperature strongly affects the current threshold. Such devices must thus be mounted in good contact with a metallic

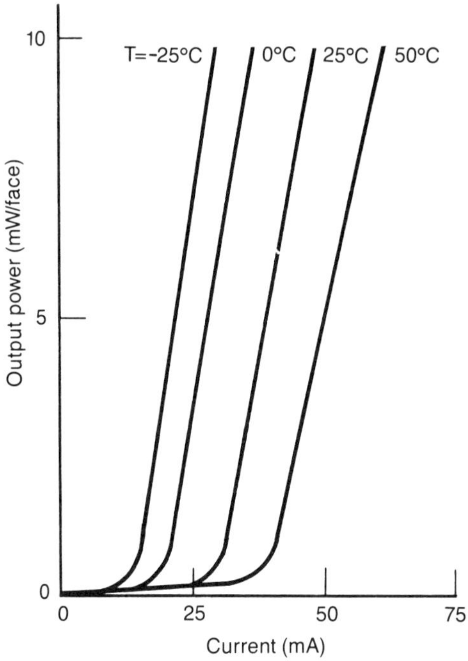

Figure 3-16 Typical light versus electric-current characteristics of a diode-laser for different heat-sink temperatures. (Courtesy of Hitachi.)

fixture for heat dissipation. Some models are packaged with a thermoelectric cooler and temperature monitoring devices to maintain a proper operating temperature. A photodiode monitoring the rear facet emission is included in some packages to stabilize the output power.

The output spectral distribution for a typical diode laser is shown in Figure 3-17. Single-mode operation is usually obtained at higher power levels, but mode hopping at a rate of nearly 0.2 nm/°C is observed if the junction temperature is not strictly controlled. Moreover, minor reflections, e.g., from the butt-coupled fiber end, tend to destabilize the laser, producing multi-mode operation even at high power levels (Miles et al., 1980). Distributed feedback cavity designs are under intense development to provide stable single-mode operation under variable conditions (Suematsu, 1985).

Another interesting development in diode laser technology is the emergence of phase-locked diode arrays. The output power of a single-stripe laser such as shown in Figure 3-14 is limited by the output-facet damaging threshold

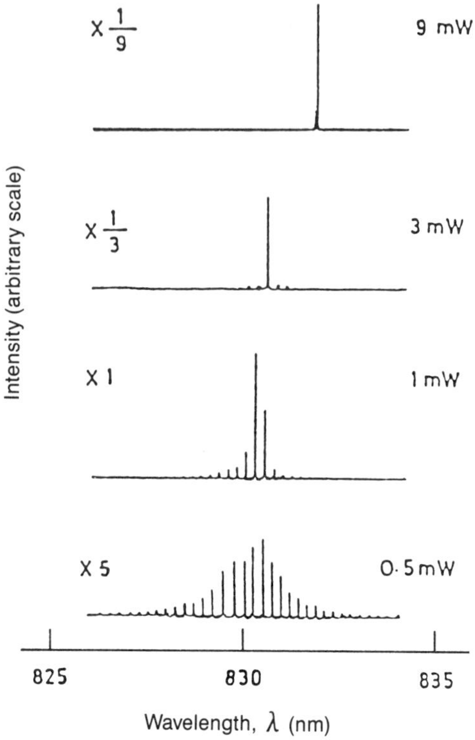

Figure 3-17 Typical emission spectra of a diode-laser at various output levels under CW operation. (Courtesy of Hitachi.)

from the highly concentrated output beam. By increasing the width of the stripe, larger outputs can be achieved but multiple transverse modes appear. If several stripes are aligned next to each other in a linear array, each being evanescent-wave coupled to the adjacent ones, high power coherent radiation is obtained (Streifer et al., 1984).

Finally, although diode lasers are much superior in terms of output power and coherence, their incoherent counterparts, namely the light emitting diodes (LEDs), may sometimes be used if a stable, low power, wide spectral bandwidth source is required. Such devices, which have been in use for a while, generate power outputs in the mW range in the near IR and lower output levels in the visible. Edge emitting superluminescent diodes operating in the 1300 to 1500 mm range are attracting renewed interest because of their good stability and reliability characteristics (Olsen, 1985; see also Section 3.5.2).

3.2 Lasers and Safety

The high parallelism of laser radiation requires special care to avoid the risk of eye damage by inadvertent exposure to the beam. Figure 3-18 shows why such precautions are necessary (Sliney and Wolbarsht, 1980). Exposure to high power infrared laser beams may produce corneal damage. This may occur more often than thought, because of the nonvisibility of such radiation and of its susceptibility to specular reflections by roughly polished surfaces. Near-UV radiation partially penetrates the eye, so that corneal surface burns are less likely in this spectral range. UV radiation may, however, induce other kinds of damage such as erythema, cataracts or photokeratitis (temporary blindness) which must be prevented by wearing appropriate goggles, as mentioned in Section 3.1.1.

The highest retinal hazard is encountered with visible and near-IR radiation, in the 400 to 1400 nm range. As Figure 3-18a shows, a parallel beam is focused on a small spot of the retina, where very strong radiation density may be reached causing irreversible damage even with relatively low

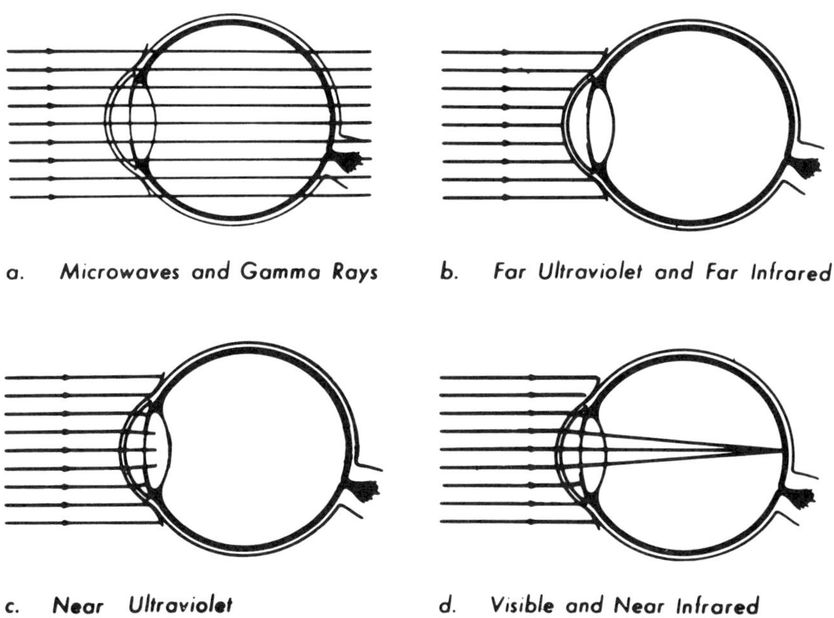

Figure 3-18a Schematic diagram of the absorption of electromagnetic radiation in the eye.

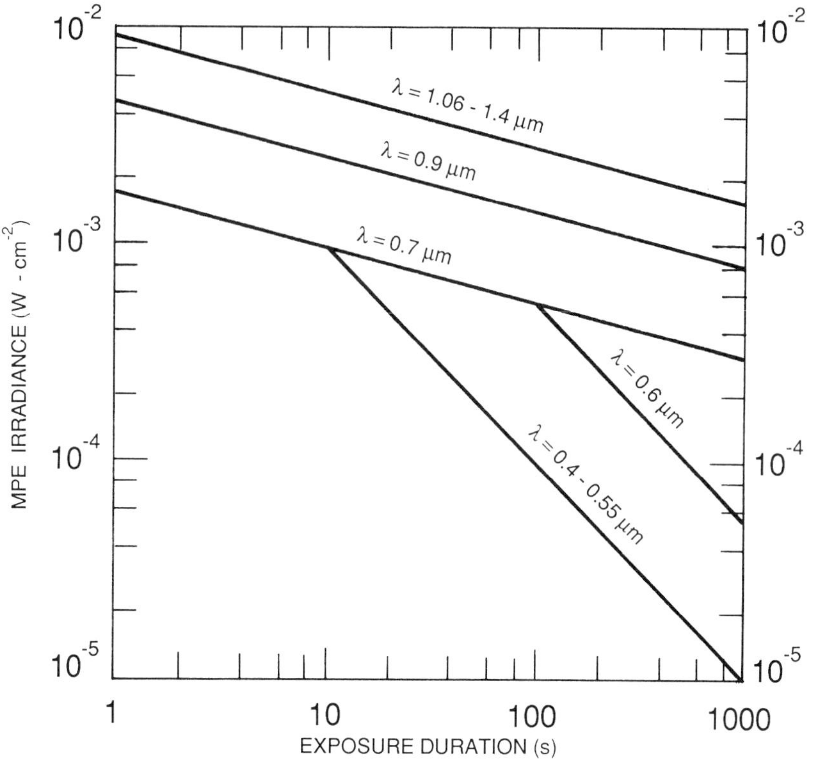

Figure 3-18b Ocular maximum permissible exposure for intrabeam viewing of a laser beam as a function of exposure duration and wavelengths.

power laser beams. Figure 3-18b shows the corresponding maximum permissible exposure as a function of the staring duration (Mallow and Chabot, 1978). Because of the limited spatial coherence of non-laser sources, safety requirements are much lower with incoherent light sources.

A number of federal or local regulations have thus been promulgated in every nation to assure that proper safety rules are followed when working with intense coherent sources. The laser manufacturer must label its product with mention of the class to which it belongs. The Bureau of Radiological Health has published guidelines for the classification of lasers into four classes, which are represented in Figure 3-19 for the CW lasers. More complex rules are specified for pulsed lasers, as illustrated in Tables 3-3 to 3-6.

The safety rules to be followed for the lasers of the different classes can be summarized as follows (Sliney and Wolbarsht, 1980).

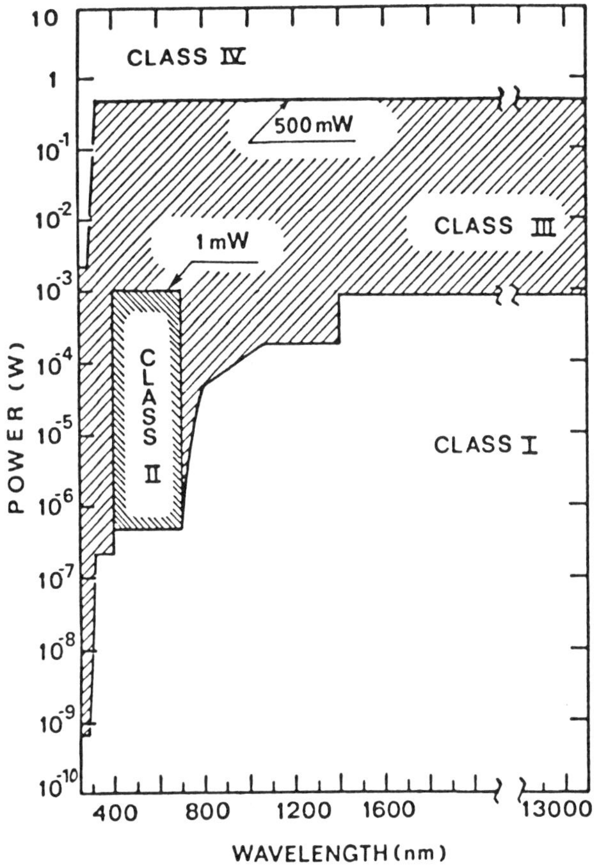

Figure 3-19 CW laser classification by the Bureau of Radiological Health in terms of safety hazards.

Class I Controls

No user safety rules are necessary.

Class II Controls

(a) Never permit a person to continuously stare into the laser source if exposure levels exceed the applicable permissible exposure level for the duration of intended staring.
(b) Never point the laser at an individual's eye unless a useful purpose exists and the exposure level and duration will not exceed the permissible limit.

Table 3-3 Class I Accessible Emission Limits for Laser Radiation

Wavelength (nanometers)	Emission duration (seconds)	Class I — Accessible emission limits
>250 but ⩽400	$\leqslant 3.0 \times 10^4$	$2.4 \times 10^{-5} k_1 k_2$ J
	$> 3.0 \times 10^4$	$8.0 \times 10^{-10} k_1 k_2 t$ J
>400 but ⩽1400	$> 1.0 \times 10^{-9}$ to 2.0×10^{-5}	$2.0 \times 10^{-7} k_1 k_2$ J
	$> 2.0 \times 10^{-5}$ to 1.0×10^1	$7.0 \times 10^{-4} k_1 k_2 t^{3/4}$ J
	$> 1.0 \times 10^1$ to 1.0×10^4	$3.9 \times 10^{-3} k_1 k_2$ J
	$> 1.0 \times 10^4$	$3.9 \times 10^{-7} k_1 k_2 t$ J
>1400 but ⩽13000	$> 1.0 \times 10^{-9}$ to 1.0×10^{-7}	$7.9 \times 10^{-5} k_1 k_2$ J
	$> 1.0 \times 10^{-7}$ to 1.0×10^1	$4.4 \times 10^{-3} k_1 k_2 t^{1/4}$ J
	$> 1.0 \times 10^1$	$7.9 \times 10^{-4} k_1 k_2 t$ J

Table 3-4 Class II Accessible Emision Limits for Laser Radiation

Wavelength (nanometers)	Emission duration (seconds)	Class II — Accessible emission limits
>400 but ⩽ 700	$> 2.5 \times 10^{-1}$	$1.0 \times 10^{-3} k_1 k_2 t$ J

Table 3-5 Class III Accessible Emission Limits for Laser Radiation

Wavelength (nanometers)	Emission Duration (seconds)	Class III — Accessible emission limits
>250 but ⩽400	$\leqslant 2.5 \times 10^{-1}$	$3.8 \times 10^{-4} k_1 k_2$ J
	$> 2.5 \times 10^{-1}$	$1.5 \times 10^{-3} k_1 k_2 t$ J.
>400 but ⩽1400	$> 1.0 \times 10^{-9}$ to 2.5×10^{-1}	$10 k_1 k_2 t^{1/3}$ J cm^{-2} to a maximum value of 10 J cm^{-2}
	$> 2.5 \times 10^{-1}$	$5.0 \times 10^{-1} t$ J
>1400 but ⩽13,000	$> 1.0 \times 10^{-9}$ to 1.0×10^1	10 J cm^{-2}
	$> 1.0 \times 10^1$	$5.0 \times 10^{-1} t$ J

Table 3-6 Selected Numerical Solutions for k_1 and k_2

Wavelength (nanometers)	k_1	k_2		
		$t \leqslant 100$ sec.	$t = 1000$ sec.	$t \geqslant 10{,}000$ sec.
250	1.0			
300	1.0			
305	3.31			
310	33.1		1.0	
315	330.0			
500	1.0			
600	1.0			
700	1.0			
750	1.25	1	5.0	50.0
800	1.56	1	10.0	100.0
850	1.95	1	10.0	100.0
900	2.44	1	10.0	100.0
950	3.05	1	10.0	100.0
1000	3.82	1	10.0	100.0
1100	5.00	1	10.0	100.0
1400	5.00	1	10.0	100.0
1500	1.00			
1600	1.00		1.0	
13.000	1.00			

Class III Controls

(a) Do not aim the laser at an individual's eye.

(b) Permit only experienced personnel to operate the laser.

(c) Enclose as much of the beam path as possible. Even a transparent enclosure will prevent individuals from placing their head or reflecting objects within the beam path. Terminations should be used at the end of the useful paths of the direct and of any secondary beams.

(d) Shutters, polarizers and optical filters should be placed at the laser exit port to reduce the beam power to the minimal useful level.

(e) Control spectators.

(f) A warning light or buzzer should indicate laser operation. This is especially needed if the beam is not visible, i.e., for infrared lasers.

(g) Do not permit laser tracking of nontarget vehicles or aircraft.

(h) Operate the laser only in a restricted area — for example, in a closed room without windows, and place a warning sign on the door.
(i) Place the laser beam path well above or well below the eye level of any sitting or standing observers whenever possible. The laser should be mounted firmly to assure that the beam travels only along its intended path.
(j) Always use proper laser eye protection if a potential eye hazard exists for the direct beam or a specular reflection.
(k) A key switch should be installed to minimize tampering by unauthorized individuals.
(l) The beam or its specular reflection should never be directly viewed with optical instruments such as binoculars or telescopes without sufficient protective filters.
(m) Remove all unnecessary mirrorlike surfaces from within the vicinity of the laser beam path.

Class IV Controls

(a) Fortunately, these high power lasers are seldom used outside of research laboratories and restricted industrial environments where personnel access is carefully controlled.
(b) These lasers should only be operated within a localized enclosure, or in a controlled workplace, or where the beam is directed into outer space. If a complete local enclosure is not possible, laser operation indoors should be in a light-tight room with interlocked entrances to assure that the laser cannot emit while a door is open.
(c) Eye protection is needed for all individuals working within the controlled area. If the laser beam irradiance is sufficient to be a serious skin or fire hazard, a suitable shielding should be used between the laser beam and any personnel.
(d) Remote firing with video monitoring or other remote (safe) viewing techniques should be chosen when feasible.
(e) Outdoor high-power laser devices such as satellite laser transmission systems and laser radar (LIDAR) should have positive stops on the azimuth and elevation traverse to assure that the beam cannot intercept occupied areas or nontarget aircraft.
(f) Beam shutters, beam polarizers and beam filters should always be used to limit use to authorized personnel only. The flash lamps in optical pump systems should be shielded to eliminate any direct viewing.
(g) Backstops should be diffusely reflecting-fire resistant target materials

where feasible. Safety enclosures should be used around microwelding and microdrilling workpieces to contain hazardous reflections from the work area. Microscopic viewing systems used to study the workpiece should ensure against hazardous levels of reflection of laser irradiation back through the optics.

It is important to note that the classification guidelines given in Tables 3-3 to 3-6 concern the "accessible emission limits", not necessarily the laser emission level. A tightly enclosed laser system may be of Class I even if it contains a laser of higher class. A definition of accessibility has been given in these terms (Sliney and Wolbarsht, 1980, p. 310):

> Laser radiation capable of meeting a part of the human body to which it is hazardous, either as emitted from an aperture, by redirection from within a housing to the outside by the use of a reflector or optical guide or by the insertion of part of the body through an enclosure port into the interior of the laser system, or by a failure mode of the laser.

The interpretation of the concept of accessibility is sometimes a matter of some controversy.

3.3 Optical Detectors

Light detection is a critical component of optical inspection. This section is an introduction to the characteristics of the optical detectors which are most often used in this technology. It should be kept in mind that the detector capabilities may be further restricted by the characteristics of other components in the optical inspection system, such as the spectral transmissivity of lenses and windows or the speed limitations of the amplifying electronics.

There are two classes of detection mechanisms, namely thermal effect and photon effect devices. The second type is much more often used in optical inspection, mainly in the photoemissive, photoconductive and photovoltaic forms which will be briefly described below. A description of less common detection mechanisms such as the photon-drag, the photo-electromagnetic, the photo-parametric and the quantum-amplifier methods may be found in more specialized publications (Kingston, 1978, Chapter 8; Duley, 1976, Section 3.4.5; Ross, 1966, Chapter 5). This section is devoted to real-time detectors only. Multiple-step light detection mechanisms such as photographic film recording are briefly described in Section 3.7.3.

3.3.1 *Thermal Detectors*

The responsive element of thermal detectors is sensitive to changes in temperature brought about by variations of incident radiation. Transduction processes include the bolometric, thermovoltaic, thermopneumatic and pyroelectric effects.

Bolometric detectors are based on the change in the electrical conductivity of the sensing element when heated by absorption of the optical radiation. The metallic or semiconductor sensing thermistor is usually mounted in a sensitive bridge circuit. Typical sensitivities of 10^2–10^3 V/W in the signal frequency range of 1 Hz to 1 KHz are obtained from commercial devices.

If the sensing element is the junction of a thermocouple or a series of thermocouples (thermopile), an electric signal is produced by the thermovoltaic effect upon heating by the incident radiation. Responsivities of the order of 10 V/W are usually obtained with time constants of the order of 10 ms, which correspond to the thermal propagation time through the radiation absorbing material on the sensing element.

Thermopneumatic detectors such as Golay cells sense the distortion of a membrane induced by air heating in the irradiated chamber. Responsivities of 10^4 V/W are reached, but these devices have a limited bandwidth and are very sensitive to vibrations.

Pyroelectric sensors are made of materials such as barium titanate, having a permanent electrical polarization that is a strong function of temperature. Transient heating produces a variation of the surface charges, inducing an electric current in the detector circuit. As Figure 3-20 shows, the responsivity level and signal bandwidth of such devices are strongly dependent on the value of the load resistance.

3.3.2 *Photoemissive Detectors*

In these devices, an electron flux is produced through the photoelectric effect when the cathode is exposed to optical radiation. In photomultipliers, the electrons are accelerated and multiplied by secondary emission through a series of dynodes as shown in Figure 3-21. Multiplication factors of 10^5 to 10^7 are obtained with typically ten stages in commercial devices. Photomultipliers are mainly used for the detection of faint light intensities in the μW range or lower. Higher radiation levels may produce damage to the last dynodes.

The spectral sensitivity of such devices is determined by the photocathode response curve of which some examples are shown in Figure 3-22 (Ross,

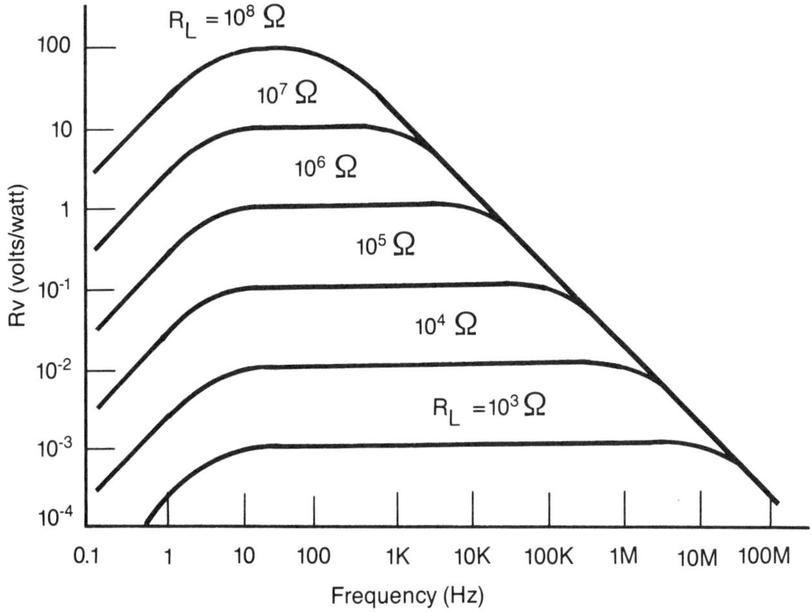

Figure 3-20 Responsivity versus signal frequency for a typical pyroelectric detector with different load resistors. (Courtesy of Molectron Inc.)

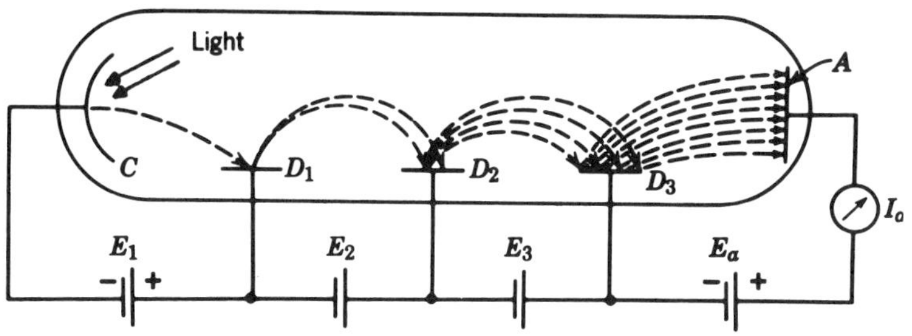

Figure 3-21 Photomultiplier schematic diagram. C: cathode; A: anode; D_i: dynodes.

1966). The infrared sensitivity is limited by the minimum photon energy which is required to eject an electron. The detector time constant is typically 10 ns, corresponding to the time dispersion in the amplifying stages. Linear output to within 3% is normally achieved through a dynamic range of 60 dB. It should be stressed that photomultipliers are rather complex devices which may give deceptive results if not properly used. Absolute readings should not be trusted before some tens of minutes have elapsed from the switch-on time;

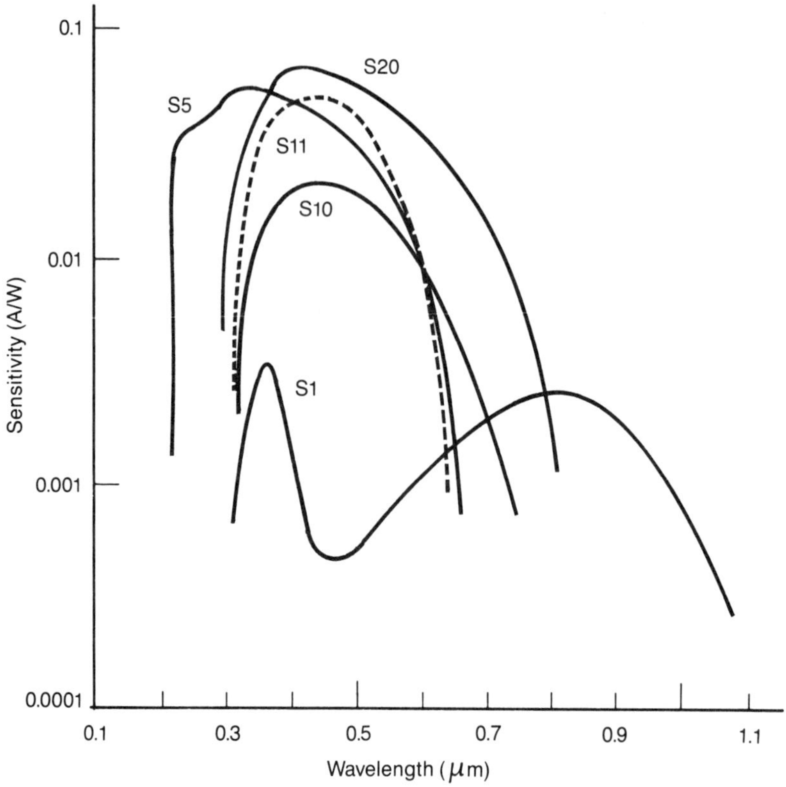

Figure 3-22 Typical photocathode response curves.

use of a ground glass in front of the cathode window is recommended when detecting structured light in order to avoid spatial response nonuniformity across the cathode aperture.

3.3.3 *Photoconductive and Photovoltaic Devices*

This category of solid state detectors has become very popular because of their ruggedness, compactness, speed and reliability. The semiconductor sensors generate electron-hole charges upon absorption of the incident photons. In photoconductive detectors, the released carriers increase the conductivity of the target, producing a variation in the polarizing current. Photovoltaic devices operate as photodiodes with a p-n junction which separates the photon generated electron-hole carriers without need for a polarizing voltage. A back-bias voltage may, however, be applied to reduce the carrier migration time and thus increase the detector response speed.

If the back-bias voltage is increased to several hundreds of volts, carrier acceleration becomes large enough to produce secondary emission on collision with the lattice. Avalanche photodiodes using such a mechanism are available, the best known being the silicon devices having multiplication factors of the order of 100. Electronic multiplication is convenient when the signal-to-noise ratio or the signal bandwidth are amplifier-limited. Hybrid detectors having sensor and amplifier integrated in the same device are often sufficiently sensitive while being more rugged and reliable than avalanche photodiodes.

The spectral detectivities (defined in Section 3.5.3) of some of the most common detectors are shown in Figure 3-23. Semiconductor detectors are characterized by a sharp upper wavelength limit which corresponds to the width of their forbidden energy gap. Long-wavelength detectors must normally be cooled at cryogenic temperatures to avoid energy level depletion by thermal excitation of the impurities and background radiation from the

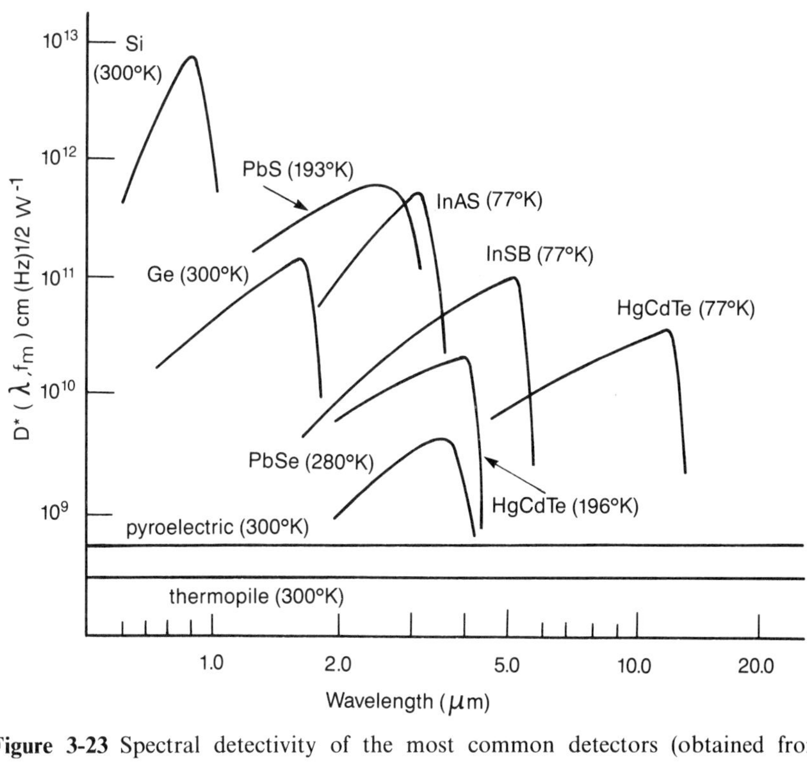

Figure 3-23 Spectral detectivity of the most common detectors (obtained from specifications of commercial devices manufactured by firms such as Judson, Sanders, UDT and Barnes).

device itself. Such detectors are usually mounted on Dewar flasks which are filled every four to eight hours with liquid nitrogen at 77 K. Joule–Thompson cooling devices using a high-pressure gas jet may be used to avoid liquid nitrogen supply but require a cumbersome gas supply (Wolfe and Zissis, 1978, Section 15.4). Thermoelectric coolers are more practical but are limited to cooling temperatures larger than 190 K. Thermal detectors have a flat wavelength response and do not require cooling, but their detectivity is much lower as Figure 3-23 shows. Time constants of infrared detectors may be as small as 10^{-11} s for ultrafast Si detectors or may extend over several ms for PbS detectors.

3.4 Image Sensing Devices

When the inspected object is to be scanned in one or two dimensions, two choices are available: mechanical scanning by rotating mirrors or similar devices, which will be described in Section 3.8, and electronic scanning through the use of discrete or continuous detector arrays. The latter approach has the advantage of not requiring any moving parts. Moreover, each element of the array is pointed to the scene for most of the observation period (the readout time usually being a small fraction of the exposure time), while a single detector mechanically scanned over N pixels records each image element during a fraction $1/N$ of the observation time. Detector arrays require, however, more powerful broad sources for scene illumination as compared with a lower power directional laser beam for flying-spot scanners, while response uniformity between adjacent pixels is much better with the approach using a scanned single sensor.

3.4.1 *Discrete Detector Arrays*

Analog Position Sensors

The simplest detector array is the two-field, or four-quadrant, position sensing photodiode. Figure 3-24 shows schematically such a device. The sensing area of a photodiode is divided into four parts, each giving a separate output. A light beam is pointed toward the center of the field. The difference between the two top and the two bottom fields is a measure of the average vertical position of the beam, while the difference between the left and the right elements gives the horizontal position. Beam position measurements may be required in applications such as air turbulence monitoring, measuring

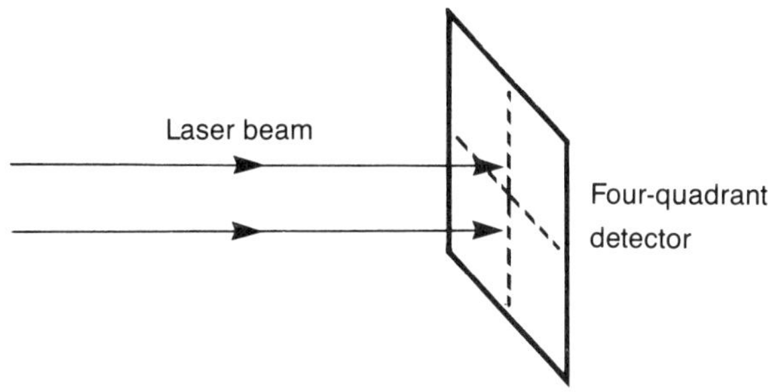

Figure 3-24 Beam position sensing four-quadrant detector.

angular displacements of a specularly reflecting surface, beam pointing and tracking, triangulation range finding, vibration monitoring, etc.

Other types of position sensors have a continuous photoconductive layer between an electrode and a resistive layer. Electric contact between the electrode and the resistive layer is thus obtained at the position of the incident beam. Such a position is evaluated from the value of the electric current in each of the two parts of the resistive layer (Kelly, 1976; Bamba et al., 1984). The continuous sensors are mainly useful when the position of a very narrow incident beam must be determined, while the two-field or four-quadrant sensors accept wider beams because they effectively determine the center of gravity of the beam intensity distribution. Typical position resolutions are in the 10 μm range. Absolute position measurement may, however, be affected by spurious light sources close to the target because of the light integration performed by such devices.

Detector Array Devices

Detector arrays with several tens of sensors of various shapes are commercially available. Examples are the 48-element thermoelectrically cooled PbSe or HgCdTe linear arrays which are used in some thermal imaging devices, or the radial array detector which is manufactured by Recognition Systems in the United States. Circular arrays are also available for inspection of parts such as circular gears (Pryor and North, 1985, p. 175). If several hundreds of sensors, each with a parallel output, are lined up in the same device, the large number of signal connections to be made to such arrays becomes a major practical problem. Self-scanned solid state arrays have been developed containing the scanning circuit integrated in the same sensor chip. Most such devices employ silicon sensing elements which are sensitive in the 400 to 1100 nm spectral range.

A first type of self-scanned linear array is the photodiode array, of which a schematic diagram is shown in Figure 3-25. Up to 4096 photodiode elements may be integrated in a silicon chip at typical center-to-center distances of 12 to 25 km. Each photodiode acts both as sensing element and as storage medium for the electric charge liberated by the incident radiation during the exposure period. A digital shift register, integrated close to the detector array in the same chip, switches an array of metal-oxide-semiconductor (MOS) transistors which connect each photodiode sequentially with a common video line. The charge accumulated in each photodiode is thus monitored in turn by the video line, which reestablishes the diode initial voltage level. Element scanning rates of several MHz are currently achieved, resulting in array scanning rates in the KHz range.

Linear or two-dimensional charge coupled device (CCD) arrays are scanned in a different way. The charge packets accumulated in the sensing elements are shifted from one element to the next until the end of the array where they are read through the output register. As they require no switching array in close proximity with the sensors on the device surface, CCDs are much more suitable than scanned photodiodes for closely packed two-dimensional arrays. The basic charge shifting mechanism in a CCD is shown in Figure 3-26. A series of adjacent surface electrodes are driven by an alternated periodical waveform with phase shifts ϕ_1, ϕ_2 and ϕ_3. The surface potential profile below the electrodes is arranged in the form of a number of potential wells which collect the signal charge packets and move them through the array as the phase of each electrode is shifted. Using transparent electrodes, the elements of the CCD register could theoretically be used as sensing elements. However, separate sensing elements, such as photodiodes or MOS sensors, are normally used in modern devices to avoid image smear during readout. At the end of the exposure period, the charge developed in the sensing elements is loaded in parallel into the shielded CCD shift register and read out during

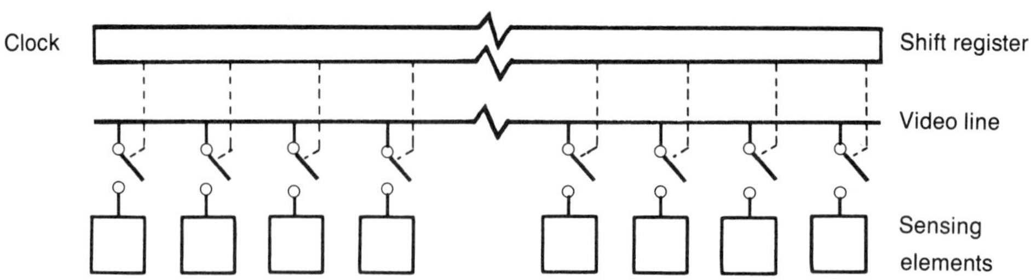

Figure 3-25 Diagram of a scanned photodiode array containing the sensing elements as well as a digital shift register integrated in the same silicon chip.

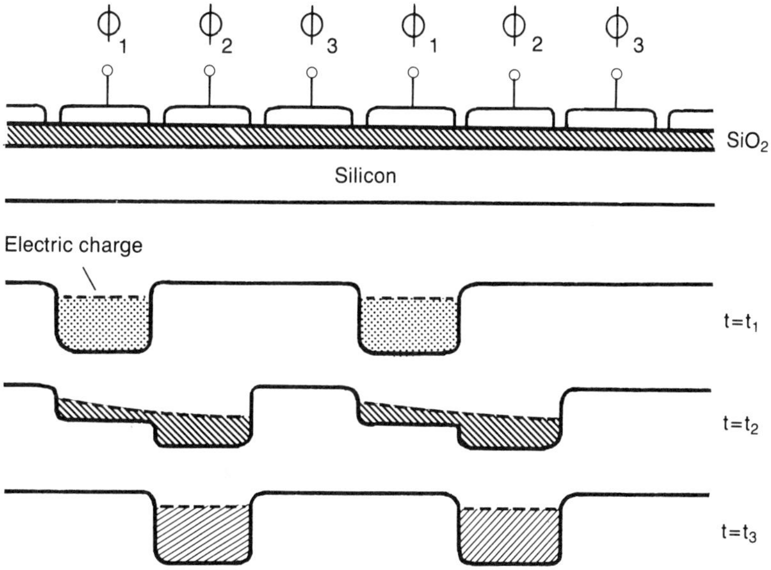

Figure 3-26 Schematic diagram of a CCD register and surface potential at sequential times t_1, t_2 and t_3.

the next exposure period. The charge may alternatively be injected into the substrate in charge injection devices (CID) to gain space in two-dimensional arrays (Burke and Michon, 1976).

Detector Array Performances

Even though technological advances in detector arrays are rapidly making mechanically scanned detectors obsolete, detector arrays still have a number of limitations compared to single-spot or scanned sensors. An overview of such limitations is given here. A more detailed analysis of the performance of specific photodiode-array or CCD-array devices can be found in specialized publications (Hopwood, 1980; Talmi and Simpson, 1980; Purll, 1983; Hodor et al., 1984).

The first limitation is speed. Line scan rates for large arrays are in the KHz range, while image scan rates of fast two-dimensional cameras are typically 30 to 300 frames per second, to be compared to the 10 GHz bandwidth of high-speed silicon photodiode single sensors. If limited spatial resolution but high speed is required, consideration should be given to the possibility of using a number of single detectors with parallel outputs.

A second limitation is resolution. Although up to 4096 elements are available in a linear array, cross-talk between adjacent elements (Hodor et al., 1984) and the aberrations of wide-spectral-bandwidth optics usually limit the resolution to 1000 elements per line, to be compared with 10,000 elements per line for high quality laser scanners. Cross-talk in silicon devices is usually larger at longer wavelengths, so that best spatial resolution with such devices is obtained by filtering out the red and IR incident radiation.

Detector arrays are severely limited in terms of dynamic range. Although good quality, single sensor silicon detectors have linear dynamic ranges of 100 dB or more, the dynamic range of ordinary detector arrays is 10^3 to 10^4. One reason for such a limitation is the saturation of charge buildup during the exposure time, which limits the linear range to less than 70% of the saturation level. In single detectors the output current is continuously evacuated, thus avoiding the saturation problem. Another limitation to the dynamic range is related to the presence of thermally-excited dark current which adds to the photo-current and may actually saturate the sensors at low scanning rates when imaging with a low light level.

Larger dynamic ranges are theoretically possible if the detector array is cooled, although the required thermoelectric cooling power is higher than for single sensor devices. However, the light-measurement accuracy is limited by the response nonuniformity across the array (Hodor et al., 1984). Figure 3-27 shows the sensitivity of the different elements of a 256-element array in different spectral regions (Purll, 1983). An element-to-element nonuniformity of 1% or more at higher wavelenghts is normally present in detector arrays. Such variations could be corrected through multiplication by a memory-stored normalization vector. However, different normalization vectors would be required for different wavelength ranges, as Figure 3-27 shows, as well as for different operating temperatures or output levels. All these limitations are obviously of little consequence for high-contrast imaging applications.

Finally, all detector arrays are susceptible to blooming when very large light intensity variations are present across the imaged scene. When the element which is subject to strong illumination saturates, its charge will spill over the adjacent elements, thus blanking out part of the image. CCD arrays are particularly susceptible to blooming, while scanned photodiode arrays and CID devices have higher blooming thresholds.

It should be mentioned, at this point, that CCD technology is rapidly evolving, so that many of the limitations described above may be overcome in the near future. The present trend is toward dramatic reductions in dark currents and toward an increased spatial resolution for such devices.

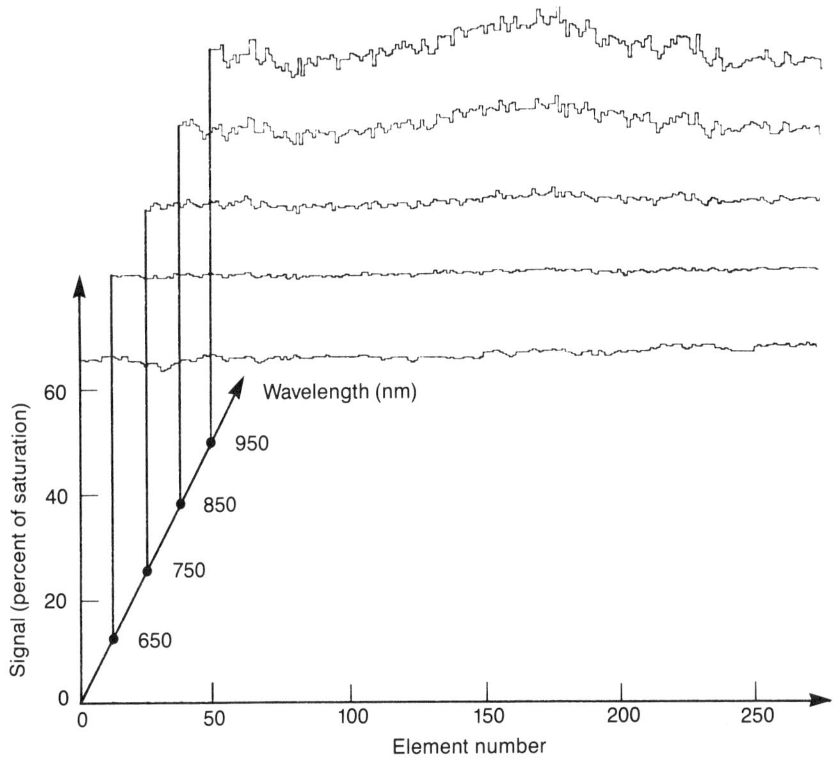

Figure 3-27 Response nonuniformity along a 256-element detector array under illuminating light of different wavelengths.

2048 × 2048-element imagers and 5000-element single-line arrays have been recently announced (see *Photonics Spectra*, Jan. 1987, p. 87).

3.4.2 *Television Tubes*

Solid state two-dimensional arrays are being increasingly adopted in imaging applications because of their compactness, speed and ruggedness. Nevertheless, the older vidicon-type imaging tubes are still widely used because of their low price and easy availability.

A schematic view of a vidicon device is shown in Figure 3-28. The luminous image is focused by the front optics on the retina, which is composed of an outer transparent electrode layer and an inner photoconductive multilayer. The inner surface, or target, is scanned by an electron beam which is accelerated by a mesh held typically at 300 V above the cathode potential.

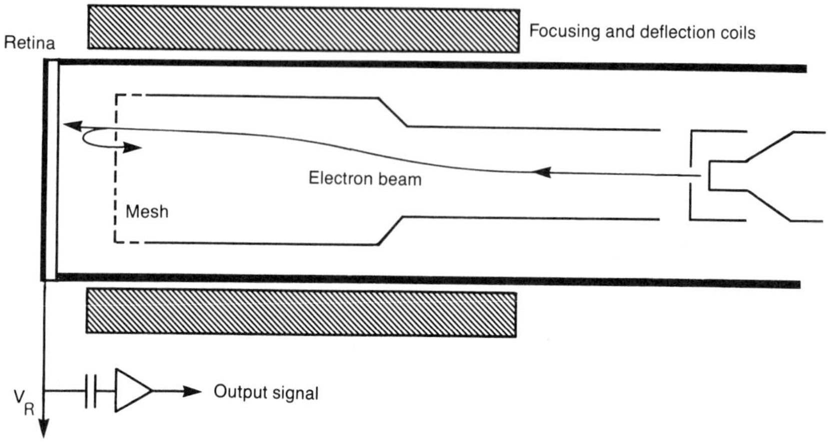

Figure 3-28 Schematic diagram of a vidicon device.

The e-beam is subsequently decelerated between the mesh and the retina, whose electrode potential, V_R, is typically 10 V above the cathode reference. If no image irradiation is falling on the retina, the charge deposited by the electron beam on the target raises its inner surface potential slightly above the cathode potential, and the electron beam is accelerated back to the mesh or to the cathode. However, if the incident photon flux increases the photoconductor conductivity, part of such charges flow to the outer electrode during a frame period. At the next scan, the discharged element may thus absorb a given amount of surface charge, producing a displacement current on the electrode which constitutes the output signal.

Photoconductive materials such as antimony trisulfide or lead oxide (in the latter case the tube is usually called Plumbicon, which is a trademark of Philips) provide a good response in the visible, while silicon targets are used for a higher sensitivity in the 400 to 1000 nm range. Pyroelectric vidicons are commercially available for thermal imaging in the 3 to 15 μm area, although their sensitivity is lower than the sensitivity of liquid-nitrogen cooled, mechanically scanned IR cameras. As pyroelectric detectors are only sensitive to input variations, a chopper must be used to blank out alternate frames (Wolfe and Zissis, 1978, Section 13.5.3).

The response speed of camera tubes is generally limited to 30 frames per second by the image lag on the retina. Photoconductors require a finite time to return to the original highly-resistive state after the light has been cut off. Silicon-diode targets are exempt from such a phenomenon, but still exhibit some lag due to the dynamic impedence of the scanning electron beam and to the target storage capacitance (Wolfe and Zissis, 1978, Section 13.3.1). Solid state CCD cameras are inherently faster, scanning rates of several hundreds of frames per second being possible with such devices. The spatial resolution of solid state cameras is also higher than the resolution of most camera tubes, typically 300 × 300 pixels.

The dynamic range of camera tubes is limited to less than 10^3 by dark currents and by the limited amount of charge which can be supplied by the electron beam during the short dwell-time on a single pixel. Signal linearity is poor in comparison with solid state cameras. Long-term retina saturation effects result in semipermanent image retention when a vidicon tube is exposed to a fixed scene for a long period of time, while silicon diode cameras have higher resistance to such a degradation. Typical lifetimes of vidicon cameras are in the 10^4 hours range, while the lifetime of solid state cameras is practically unlimited.

3.5 Detection and Noise

The detectability of optical signals is determined by the amount of noise which is encountered in the detection process. This section reviews the sources of noise which are most often found in optical inspection, while the next section will describe some practical noise reduction techniques. Only time-domain noise is considered here, while image-processing techniques to reduce spatial noise in two-dimensional images are described in Section 5.2.

3.5.1 *Environmental Noise*

The first category of noise in the detected signal, which is most important in industrial inspection, is noise produced by sources other than the illumination and detection devices, i.e., by the inspection environment. Environmental noise encountered in optical inspection may be classified as either optical or nonoptical noise.

Sources of optical noise

Optical noise is here defined as the random fluctuations of the detected signal which are produced by variations of the detected light intensity. Such variations may be produced by intensity fluctuations of light sources which are present in the detection environment, such as ambient light, flame flickering in furnaces, arc irradiation during robotic arc welding, etc. Spurious light may reach the detector either directly or after diffuse reflection from the inspected surface. Ambient illumination gives rise to strong optical noise components at twice the line frequency of 60 Hz in North America and at higher harmonics. Operation at such frequencies should be avoided if possible.

Optical noise may also originate from variations in the surrounding medium. Not infrequent examples in a production line are the presence of vapor turbulence or of moving reflective surfaces in the background. Ambient vibration, which is particularly strong at low frequencies, may be a major source of noise in interferometric inspection or in high resolution beam positioning. Slight movement of a rough surface illuminated by a coherent light beam may produce considerable speckle noise, as will be described in Section 3.10.

Nonoptical noise

A major source of environmental nonoptical noise is electromagnetic interference introduced on the detector, wiring or amplifying electronics from external electric sources. According to a European study (Kopsala and Paivarinne, 1984) over 99% of process control problems are related to interference in interface circuitries. Examples are electronic noise induced by the electric power line at 60 Hz and harmonics, the presence of radio waves from broadcasting stations, ignition noise from nearby engines or spark gaps, and power switches in heavy machinery. In general, long signal-carrying wires before amplification are the most vulnerable components to interference noise.

Other sources of nonoptical noise include ionizing radiation, which may produce substantial carrier generation in photon-effect detectors in the proximity of nuclear reactors or nearby radiographic inspection devices, and transduction mechanisms such as structure vibration transmitted to coaxial cables as a result of dimensional, and thus capacitive, changes in the cable.

3.5.2 *Light-Source Noise*

Noise in the detected signal may be produced by amplitude or phase fluctuations of the light source used to illuminate the inspected scene. Incoherent thermal sources have the lowest emission noise. If they are driven by a well stabilized DC current source, thermal sources usually exhibit shot-noise limited behavior. Such a limit is determined by the random arrival of the discrete photons in the detected light beam, as will be described in the next section. Arc discharge sources have higher noise levels because of arc instability, particularly in high pressure devices.

Laser sources have noise levels which are often higher than the shot-noise limit by several tens of DB in the low frequency spectrum. Figure 3-29 (reproduced from Cole, 1980) shows the noise levels versus the observation frequency of some commercial gas lasers for a constant detected intensity of 220 μW. Figure 3-30 shows a comparison of noise from a diode laser and from a single-mode gas laser (Dandrige et al., 1980). Emision noise characteristics of lasers are rather complex and often not fully understood. The following discussion will outline some of these characteristics.

Sources of Laser Noise

There are a number of noise sources in excess of shot noise in laser devices. Because laser radiation is produced by the coherent amplification of spontaneously emitted photons, one basic source of noise in lasers is related to spontaneous emission. This noise component is, however, overshadowed by other amplitude noise components in normal laser operation. Similar considerations are valid for the phase noise component, which would lead

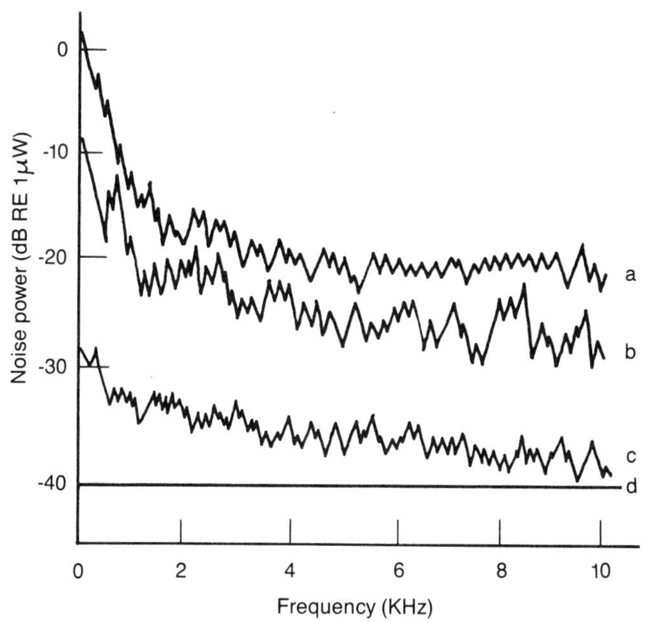

Figure 3-29 Noise power spectrum of some commercial gas lasers: (a) Argon laser, Spectra Physics 165-B; (b) He-Ne laser, multimode, Spectra Physics 124-B; (c) He-Ne single-mode laser, Tropel 100; and (d) shot-noise quantum limit in a 30 Hz bandwidth.

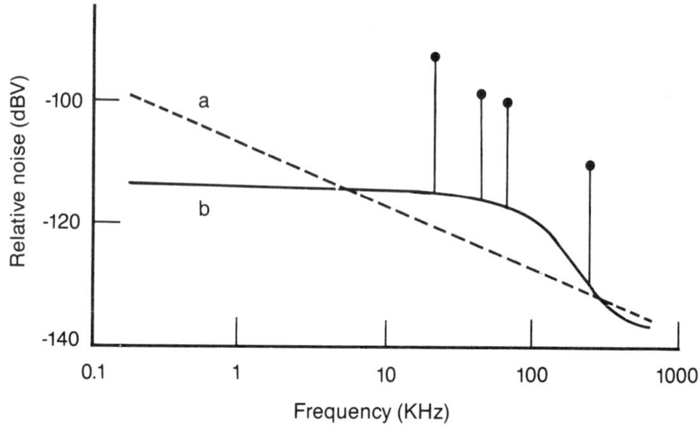

Figure 3-30 Relative noise spectra of: (a) a Hitachi HLP 1400 diode laser and (b) a Tropel 100 He-Ne laser.

in theory to laser frequency bandwidths as small as 0.01 Hz in gas lasers if only the spontaneous emission noise component were present (Mooradian, 1985).

A more important source of noise in DC-operated gas lasers is plasma noise. This is introduced by fluctuations of current density in the gas discharge and represents a macroscopic fluctuation in gain of the laser. Because this kind of noise is generated by acoustic instabilities of the plasma medium, its magnitude and frequency spectrum is related to a number of parameters such as gas pressure, discharge length and diameter, and current level. The magnitude of the plasma noise increases rapidly above a current threshold whose value is lowered by an increase in either the discharge length, diameter or gas pressure (Bloom, 1965). The spectrum of plasma noise in typical He–Ne lasers does not extend above 100 KHz (Prescott and Van der Ziel, 1966) and is the main cause for the "shoulder" which appears at such a frequency in the noise spectral distribution of commercial He–Ne lasers as shown by curve b in Figure 3-30.

Another major source of noise in the 0.1 to 10 MHz frequency range is given by relaxation oscillations. The basic physical mechanism of such oscillations is an interplay between the laser amplification and the atomic inversion (Yariv, 1976). This results in randomly occurring, either damped or undamped oscillations of the laser output power at some specific frequencies which produce "spikes" in the noise spectrum of the kind shown in curve b of Figure 3-30.

Mode interference noise

An important source of laser noise is caused by interference between the emission modes. If two or more longitudinal modes are present (see Figures 3-12 and 3-17), amplitude fluctuations at the beat frequency in the 10^8–10^9 Hz range will occur. Indeed, if the phase between the independently oscillating modes is synchronized by a modulator or a saturable absorber cell, the laser will exhibit a series of periodic spikes at such frequencies (mode-locked laser operation). The frequency difference between adjacent laser modes varies because of nonlinearities in the laser medium. As a consequence, fluctuations at the difference between beat frequencies are also observed and constitute a major source of noise down to the KHz range. The exact location of such frequencies is very sensitive to changes in the resonator length which may be produced by ambient vibrations and sometimes to local magnetic fields at the plasma tube. Consequently, the pattern of such fluctuations tends to be highly unstable. The effect of mode-interference noise appears very clearly when the spectrum of a multimode and a single-mode gas laser are compared (curves b and d of Figure 3-29). Such a noise source is also responsible for the strong increase in noise from a diode laser operated at near threshold conditions as shown in Figure 3-31a (Dandrige, 1982). Because typical diode lasers change from multimode to monomode operation as the pumping current is increased above threshold (see Figure 3-17), the noise is significantly higher in the region just above threshold when the laser is operating on several longitudinal modes. The noise curve refers to a Mitsubishi diode laser at a frequency of 1 KHz.

Similar behavior is shown in Figure 3-31b (Gotoh et al., 1982). In this case a Hitachi single-mode diode laser was operated at gradually rising temperature, producing mode hopping (see Section 3.1.2) from one mode to the next at given temperature levels. The broken curve A shows the output wavelength when the temperature is increased. Curve B shows the noise output in a 30 KHz bandwidth. The correspondence between the occurrence of strong noise peaks and the overlapping of two adjacent modes is clearly evident.

Noise data shown in Figures 3-29 and 3-31 were obtained by directly probing the output from the laser at a relatively large distance from the device. If additional optical components, such as the input end of a fiber pigtail, are present along the beam path, additional degradation in the noise characteristics is produced by optical feedback. Partial reflections of the laser beam back into the laser cavity tend to destabilize the laser, producing multimode operation from nominally single-mode lasers. Substantial broad-

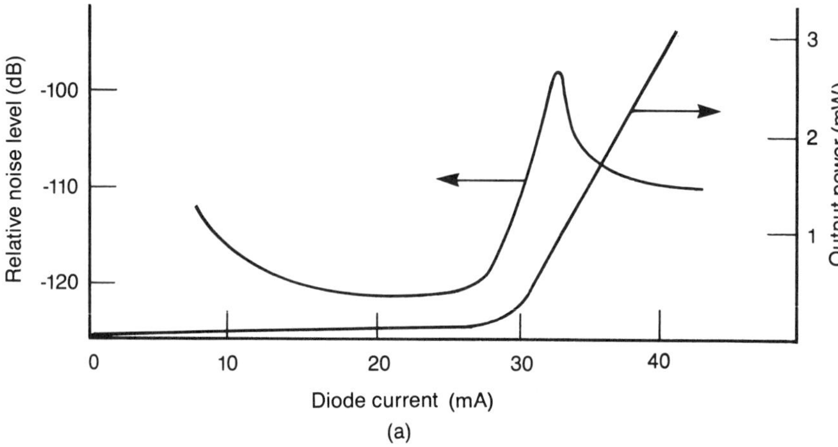

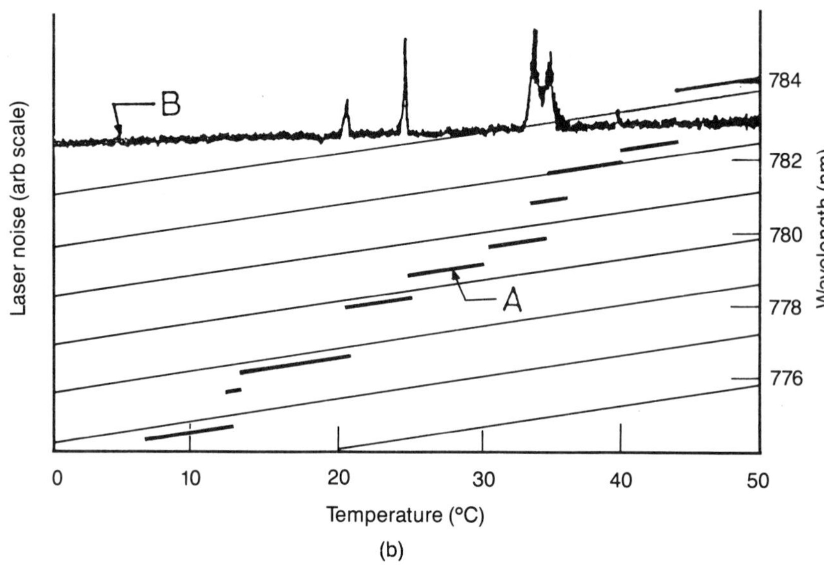

Figure 3-31 Evidence of mode-interference noise in diode lasers: (a) increase of the noise level in the multimode operation region just above threshold; (b) appearance of noise peaks when two longitudinal modes overlap at given temperatures.

ening of the laser bandwidth is produced by feedback levels of 10^{-4} in diode lasers (Miles et al., 1980). This results in a strong increase of the amplitude noise as well as in a substantial reduction of the coherence length. Optical reflections may be substantially reduced by using devices such as optical isolators (a sandwich of a polarizer followed by a 45°-oriented quarter-wave plate) or by liquid coupling the output fiber end in a fiber-coupled laser.

Reflections from the input end of a fiber pigtail are, however, quite difficult to avoid. Distributed feedback single-mode diode lasers might bring a solution to such noise problems.

Additional mode interference noise may be generated on a truly single-mode laser coupled into a multimode fiber. Interference between the different spatial modes within the fiber (see Section 3.9) produces large signal excursions as the fiber is touched or heated. A solution to both the laser mode interference and the fiber mode interference problems appears to be possible using large-bandwidth diode lasers coupled into a single-mode anti-reflection coated fiber end. A mode free, wide-line emission may be obtained from a diode laser by anti-reflection coating the output facet (super-luminescent diodes: see Olsen, 1985). The noise properties of such lasers are attractive (Dandridge, 1982), although the output power from such devices is presently limited to 100 μW, subtantially lower than the normal diode laser power of several mW.

3.5.3 *Electronic Noise*

This section will describe the main sources of noise present in the signal after the light beam has been sensed by the optical detector (Levinstein, 1965; Wolfe and Zissis, 1978, Chapter 11; Kingston, 1978). It is assumed that normal care has been taken to avoid noise sources external to the detection and amplification devices, such as noise produced by an imperfectly stabilized power supply for the polarization of the detector.

Shot Noise

Shot, or quantum, noise is generated by the random arrival and absorption of the discrete photons in the detected light beam. Shot noise is the ultimate limit to light detection accuracy.

Consider a steady flux of light incident on the detector producing $N = i\tau/e$ photoelectrons during the detector half-rise time τ, where i is the detector current and $e = 1.59.10^{-19}$ coulomb is the electronic charge. N is an average value which is subject to a root mean square fluctuation of $N^{1/2}$ according to Poisson statistics. The shot-noise current in amperes is thus given by

$$i_{SN} = N^{1/2}e/\tau = (2ieB)^{1/2}, \tag{3-2}$$

where $B = 1/2\tau$ is the detector bandwidth. For low light levels, the detector current i is mainly given by the dark current. For high light levels, i is equal to the product of the incident light flux and the detector quantum efficiency in amperes per watt. In this case, shot noise is proportional to the square root of the incident light power. By attenuating the incident light beam, one can thus recognize whether the main noise component is due to shot noise or to light source noise (proportional to the input power) or else to thermal electronic noise (independent of the input light power).

Johnson Noise

Johnson, or thermal electronic noise, is caused by random motion of thermally agitated electrons in resistive materials. It can be shown from thermodynamic considerations (see, e.g., Van der Ziel, 1970) that the rms electric current fluctuation due to thermal noise is given by

$$i_T = \left(4kT\frac{B}{R}\right)^{1/2}, \tag{3-3}$$

where $k = 1.38.10^{-23}\ J/K$ is Boltzmann's constant, T is the device temperature in degrees Kelvin and R is the load resistance. The voltage rms noise, equal to Ri_T, is thus proportional to the square root of the load resistance.

Johnson noise may be reduced by cooling the detector. It should be mentioned that thermal fluctuation noise may originate both in the source resistance and in the resistors of the amplifier. Hybrid detector packages including cooled sensor and preamplifier on the same device are particularly convenient for thermal noise minimization. Signal amplification by electron multiplication using photomultipliers or avalanche photodiodes is another approach for avoiding thermal noise from the amplifier.

Flicker Noise

Both shot and Johnson noise are "white noise", i.e., they do not depend on the observation frequency. Consequently, one can reduce the amount of noise by reducing the detection bandwidth, matching it to the expected signal bandwidth. Flicker noise, however, is proportional to $1/f^n$, where f is the signal frequency and n is typically comprised between 0.9 and 1.35. For a DC signal, no significant improvement can be obtained by increasing the time constant if flicker noise is the predominant noise source.

Sources of flicker noise are fluctuations in the emission from spots on a vacuum tube cathode and trapping of charge carriers near the sensing surface in semiconductor detectors (Pratt, 1969, Chapter 8; Van der Ziel, 1970). The low frequency spectral distribution from a typical solid state detector is shown in Figure 3-32 (Lin and Giallorenzi, 1979). We can see that flicker noise

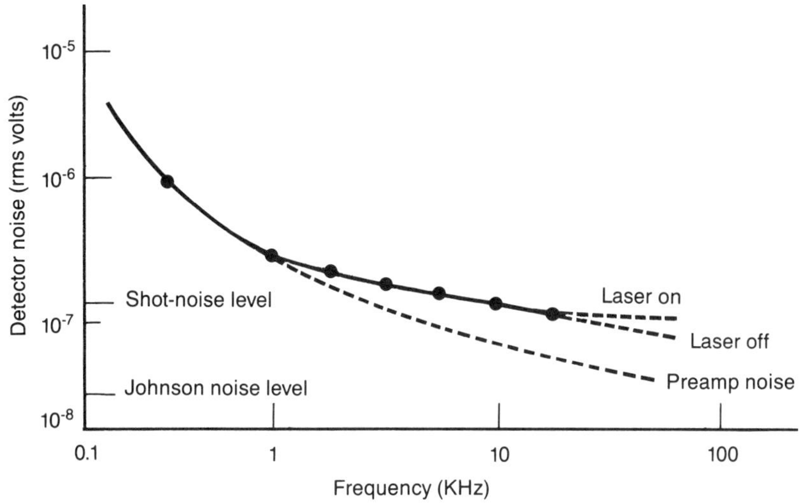

Figure 3-32 Low frequency noise spectrum from an InAs solid-state detector.

exceeds the shot-noise level at frequencies smaller than 10 KHz. Photomultiplier noise was reported in the same work to exceed shot noise only at frequencies lower than 0.1 Hz.

Normalized Spectral Detectivity

Optical detectors are often characterized by the value of their normalized spectral detectivity D^* (see Figure 3-23). To define such a parameter, we must first consider the concept of noise equivalent power (NEP), which is the input light power (in watts) required to produce a signal-to-noise ratio equal to 1. As the detector sensitivity is in general a function of the light wavelength, while noise from the detector is frequency dependent, NEP must be specified for a given wavelength, signal frequency and bandwidth. Moreover, the value of NEP is dependent on the detector area A_d.A normalized spectral detectivity parameter has thus been defined in cm $\mathrm{Hz}^{1/2}\ W^{-1}$ as

$$D^* = \frac{(A_d B)^{1/2}}{\mathrm{NEP}}. \tag{3-4}$$

As can be seen in Figure 3-23, D^* is a function of the light wavelength λ, as well as of the light-modulation frequency f_m, which is often 1 KHz but may differ for different detectors or for different manufacturers. The manufacturer specified D^* value thus refers to the inverse of the noise equivalent power, normally at the wavelength of maximum sensitivity, of a 1 cm^2-area detector of bandwidth equal to 1 Hz, with an incident light beam modulated at the optimum noise frequency.

3.6 Noise Reduction Techniques

3.6.1 *Environmental Noise Reduction*

Optical noise from spurious ambient light must first be minimized by proper shielding of the optical detection system. If a laser or a collimated narrow band source are used for illumination, spectral and spatial filtering techniques can be resorted to as shown in Figure 3-33. A spectral filter, typically an interference filter with a 1 nm to 10 nm half-bandwidth, can be used to suppress ambient light whose wavelength is not in close proximity to the laser wavelength. The interference filter should be situated in a position where the detected beam is nearly collimated to avoid angle-related detuning. Indeed, the peak transmission wavelength λ_θ of an interference filter for an incident light beam at an angle θ to the normal is given by the following expression (see, e.g., Dufour and Cielo, 1984)

$$\lambda_\theta = \lambda\left(\frac{1 - \sin^2\theta}{n^2}\right)^{1/2}, \tag{3-5}$$

where n is the refractive index of the spacing layer. If the filter is positioned on a converging or diverging beam of angular aperture 0.1 rad, a shift in the peak transmittance wavelength of the order of 3 nm would be observed for the peripheral rays of a near-IR beam.

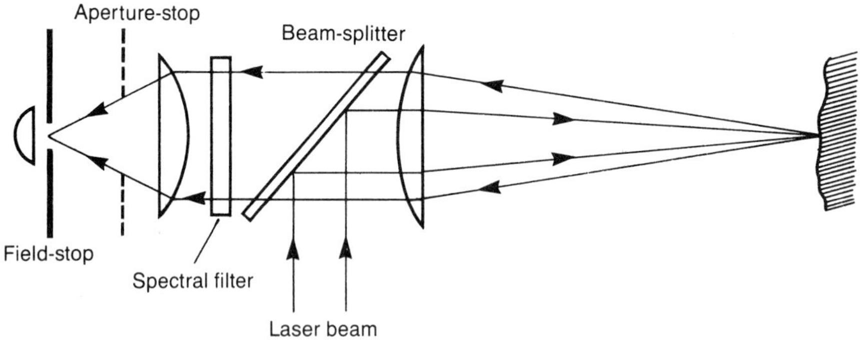

Figure 3-33 Environmental noise reduction with laser illumination.

Further suppression of ambient light scattered by the surface close to the laser illuminated spot may be obtained by spatial filtering through a pinhole or field-stop, as shown in Figure 3-33. The pinhole diameter, or the sensitive area of the detector when possible, is matched to the image of the projected laser spot on the surface. A compromise must be reached between noise reduction and depth of field requirements when a configuration of the kind of Figure 3-33 is used in applications such as flying spot scanning of a given surface. In this case a somewhat larger pinhole diameter may be required to permit a certain amount of beam defocusing across the scanned surface.

If a cryogenically-cooled detector is used in a configuration of the kind of Figure 3-33, e.g., in conjunction with a pulsed CO_2 laser for range-finding applications, or else in a thermal imaging configuration without laser illumination, it is advantageous to include any field- or aperture-stop within the cooled Dewar container. This will avoid direct exposure of the detector to thermal radiation from a room-temperature diaphragm.

Electromagnetic intereference (EMI) noise can first be reduced by following sound experimental practice including proper grounding and shielding procedures (Baxendall, 1968). Long unshielded wires should be avoided, and preamplifiers should be situated as close as possible to the detecting device. Batteries are to be preferred to rectified power supplies as stable polarizing sources for the detector. To assure full EMI shielding across the input aperture of the optics, the input lens or window may be coated with a transparent conductive coating. Typical performances of such coatings for different frequencies of the interfering noise are shown in Figure 3-34. Such conductive coatings may also be Joule-heated for defogging or deicing applications.

Signal transmission through optical fibers should be preferred to electric wire transmission in regions where strong EMI noise is present. Scene scanning may be performed by a scanner situated in front of the input end of the fiber or else by imaging the scene on the input end of a linear or two-dimensional fiber-optic light image guide.

3.6.2 *Light Source Noise Compensation*

Different common mode noise rejection techniques can be used to compensate for light source noise. A first possibility is illustrated in Figure 3-35a, representing a surface inspection configuration of the kind shown in Figure 3-33. The instantaneous intensity of the light source beam is continuously probed by detector 1. Changes in the reflected beam intensity monitored by detector 2 which are due to light intensity fluctuations may be eliminated by

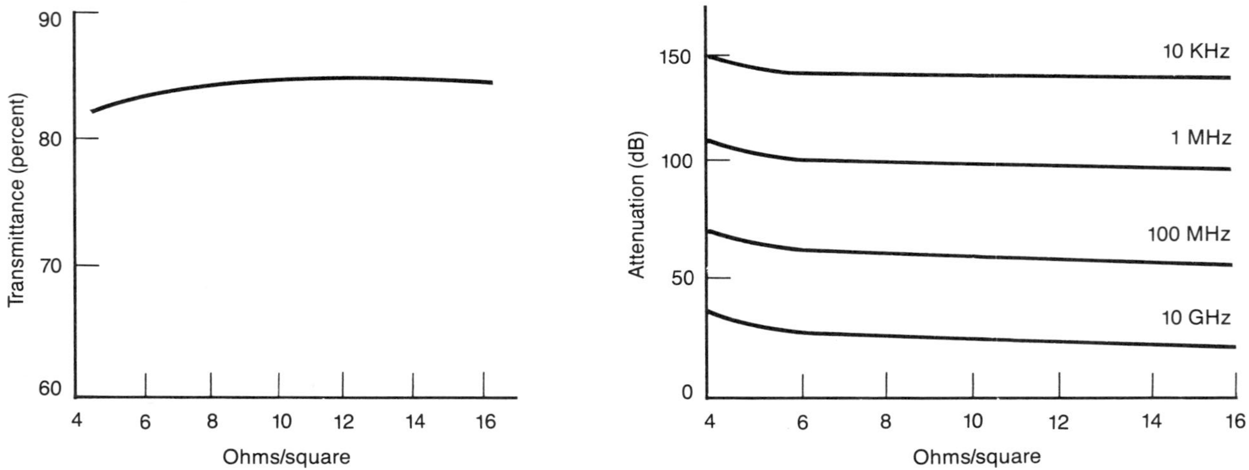

Figure 3-34 Typical optical transmissivity in the visible range (left) and EM-noise attenuation (right) of a transparent conductive shield (Courtesy of Optical & Conductive Coatings, Inc.)

taking the ratio of the signals obtained from detectors 2 and 1. Only variations of the inspected surface reflectivity will thus affect the ratio signal. If the signal is too fast for real-time division, the two signals may be separately recorded and their ratio calculated subsequently.

A slightly different approach is shown in Figure 3-35b. The diagram shows a fiber optic interferometer scheme (Dandrige and Tveten, 1981; a similar concept may be found in Palmer, 1984) where the two in-quadrature light beams interfere after propagation through similar optical fibers. The fiber optic interferometer can detect phase shifts in the beam propagating through the sensing fiber. Such phase shifts have opposite effects on the intensities sensed by the two detectors, while amplitude fluctuations of the light source equally affect such intensities. The difference signal is thus unaffected by light

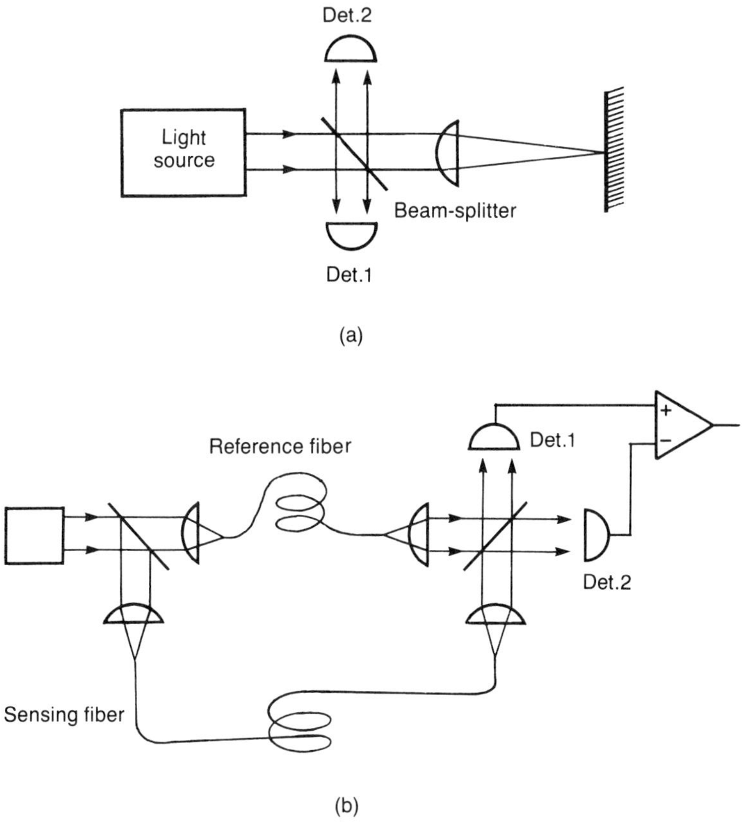

(a)

(b)

Figure 3-35 Common-mode light source noise reduction techniques: (a) signal normalization and (b) common-mode subtraction.

source noise. Note that in this case the two signals have similar amplitudes, making it possible to simply operate a difference, rather than a division, of the two signals which requires a longer time. This is gnerally not possible in a configuration of the kind shown in Figure 3-35a, where the reflected intensity from the inspected surface may fluctuate unpredictably over a wide range.

Feedback from a reference detector may alternatively be used to stabilize the laser (Dandridge and Tveten, 1983; Sollberger, 1985; Dakin and Withers, 1986). The back emission from a diode laser may be probed by a backside reference detector. Variations in the level of the back emission signal which are produced by amplitude fluctuations of the laser output may be used in a feedback loop to change the forward current applied to the diode laser junction. Similar feedback controls may be used to frequency stabilize the laser (Favre and LeGuen, 1980; Ohi, 1980).

3.6.3 *Reduction of Electronic Noise*

As can be seen from equations (2) and (3) above, both quantum and thermal noise are proportional to the square root of the detection bandwidth. It is thus convenient to reduce the system bandwidth to a frequency range not larger than the expected frequency range of the signal to minimize such noise sources. However, if the signal has a very low frequency and the detector exhibits $1/f$ noise (see Figure 3-32), no advantage is gained by low-pass filtering the detector output. A method to improve the signal-to-noise ratio (SNR) in this case makes use of synchronous detection.

A schematic diagram of a synchronous detection system is shown in Figure 3-36. A chopper, such as a rotating wheel or a tuning fork, is placed in front of the optical detector. Signal sidebands will thus appear in the spectrum of the detected signal at both sides of the chopper frequency, f_c, which is chosen high enough to be little affected by the $1/f$ noise. Mechanical choppers are

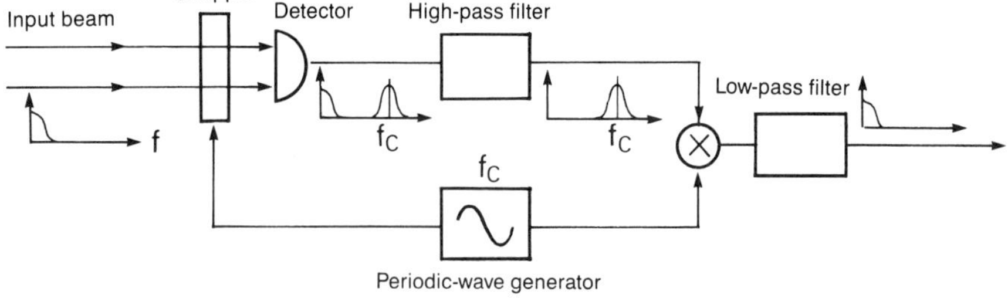

Figure 3-36 Schematic diagram of a passive synchronous detection system.

typically limited to frequencies of 10^3–10^4 Hz, while acousto-optical or electro-optical modulators can reach higher frequencies. After high-pass filtering, multiplication by the periodic wave generator signal with the proper phase and low pass filtering (these three operations are the basic functions of lock-in amplifiers), the baseband signal is restored without the low frequency noise which would be encountered in direct detection.

The technique just described is called passive synchronous detection, to distinguish it from the active technique where a chopped or modulated light beam is projected on the object to be inspected. The active synchronous detection apparatus is very similar to the one shown in Figure 3-36, with the only difference that the chopper is placed in front of the light source or that the periodic wave modulator is directly used to intensity modulate the light source such as a diode laser. Active synchronous detection is convenient not only because of the lower electronic noise at the modulation frequency but also as an additional protection against ambient light noise.

If a coherent light source is used to illuminate the inspected scene, heterodyne detection methods of the kind shown in Figure 3-37 may be applied. (Ross, 1966, Chapter 4; Kingston, 1978, Chapter 3).Part of the illumination laser beam, of optical frequency f_L, is shifted to a frequency f_S and superposed through a beam-splitter to the beam reflected from the inspected scene. The optical frequency shifter may be a Bragg cell (see Section 3.8) operating at a typical f_L–f_S frequency of 50 MHz. The optical detector output is proportional to the square of the superposed beam amplitude

$$I = [A_L \cos(\omega_L t + \phi) + A_S \cos \omega_S t]^2, \tag{3-6}$$

where A_L is the backscattered beam amplitude, A_S the amplitude of the frequency-shifted local beam, $\omega_L = 2\pi f_L$, $\omega_S = 2\pi f_S$ and ϕ is the phase difference between the two beams. The detected signal will thus contain a

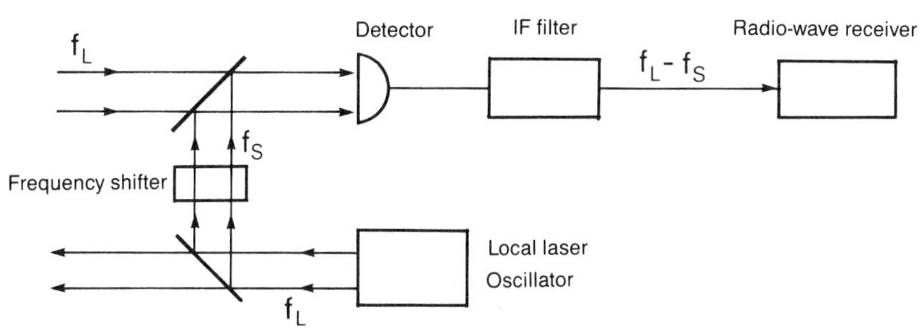

Figure 3-37 Active heterodyne detection system.

component

$$I_{IF} = A_L A_S \cos[(\omega_L - \omega_S)t + \phi] \quad (3\text{-}7)$$

at the intermediate frequency (IF) of 50 MHz which is selected by the electronic IF filter and contains the required information on the reflected optical beam. A radio-wave receiver tuned at 50 MHz will demodulate such a signal providing an output proportional to the amplitude A_L or the phase ϕ of the reflected probing beam.

The result of the heterodyne system is to uplift the detection frequency to a convenient IF frequency as well as to eliminate all ambient light noise whose optical frequency differs from f_L by more than the IF filter bandwidth. Such a sharp optical filtering capability, much narrower than is normally possible by an interference optical filter, is realized at the expense of strong beam collimation requirements for the detected beam. Indeed, the simple expression (3–6) is valid only if the two superposed beams are perfectly aligned, spatially and temporally coherent, and of same polarization. Otherwise additional loss of signal will occur (Ross, 1966, Chapter 4; Pratt, 1969, Section 10.3)

The heterodyne technique is particularly useful for the detection of very low level backscattered beams, as may be required in laser range finding or laser-Doppler velocimetry. In this case the amplitude of the IF term (3–7) may be increased by increasing the level of the local oscillator amplitude A_S. If a stable local laser oscillator is available, the detected signal level may in this way be raised well above the thermal noise level, in which case the signal-to-noise ratio will be limited only by the local-oscillator related shot noise. Variations of the heterodyne technique are passive heterodyning, such as used in laser communication receivers containing a local oscillator independent of the laser transmitter, and the homodyne approach, where the two superposed beams have the same optical frequency (Ross, 1966; Pratt, 1969).

3.7 Optical Elements

One of the advantages of using optical technology as compared to ultrasonic, nuclear or nonoptical electromagnetic radiation is the relative ease with which optical beams can be collimated, focused or scanned across the inspected scene. This section overviews beam-shaping optical components, while scanning techniques will be described in Section 3-8.

As can be seen from typical optical system diagrams such as those shown in Figures 3-33, 3-35 and 3-37, the most common optical components are mirrors, windows, beam-splitters and lenses. Mirrors are usually front surface metal coated, usually with a thin protective layer to avoid oxidation. Dielectric mirrors are available with multiple surface layers to obtain very high reflectivities (larger than 99%) at a given wavelength and angle of incidence, or to selectively reflect in a certain spectral range and transmit in another range (dichroic mirrors). Similar interference coatings may be used to reduce unwanted reflections from the surfaces of windows or lenses, always for a given wavelength and angle of incidence.

Beam-splitters may simply be transparent plates, i.e., windows, with a partially metallized front surface. Partial reflections on the back surface of a 45°-oriented beam-splitting plate may give rise to laterally displaced spurious beams ("ghost images"). To avoid this, the back surface may be anti-reflection coated for the center wavelength of the incident beam, or a pellicle beam-splitter (a partially metallized plastic membrane) may be used if the beam quality is not critical. Another possibility is to use a cube beam-splitter of the kind shown in Figure 3-38.

Cube beam-splitters are sometimes manufactured with anisotropic materials to obtain polarization dependent reflecting properties. Figure 3-38 shows a component of this kind having large transmissivity for the light component polarized in a direction parallel to the plane of the figure and large reflectivity for the perpendicular component, at a given wavelength and angle of incidence. This figure shows a typical configuration where the polarizing beam-splitter is used in conjunction with a polarized laser beam and a $\lambda/4$ (quarter-wavelength) retardation plate to obtain a more efficient beam- splitting operation. A $\lambda/4$ retardation plate uses a birefringent material to introduce a quarter-wavelength phase retardation between the two polarization components of the incident light beam. The polarization of the beam transmitted twice across the plate will thus be rotated by 90° so that the

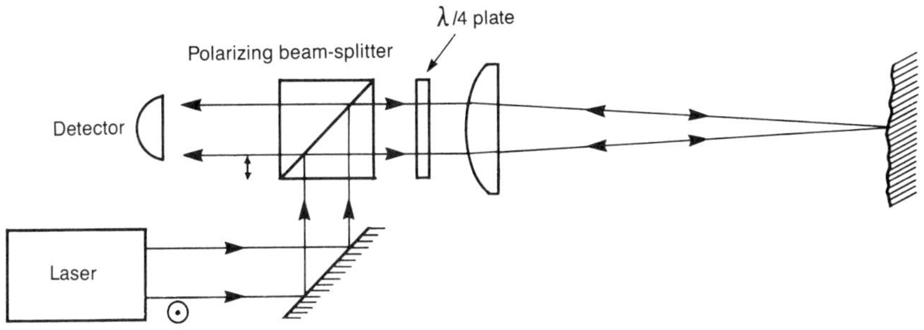

Figure 3-38 Polarized light optics for efficient beam-handling.

backscattered beam will be totally transmitted through the beam-splitter without reflection losses.

3.7.1 *Imaging Systems*

Imaging optics may be of the dioptric (using refractive elements) or catoptric (using reflecting components) type. Figure 3-39 shows examples of both kinds. Reflecting optics has the advantages of having a very wide spectral range in the visible and infrared and of being free from chromatic aberration, qualities which are particularly convenient for astronomical observation. Refractive optics is easier to align and is usually preferred when working with a narrow-band source such as a laser. This section will give a short introduction to the basic lens-system concepts which are most significant for optical inspection applications. Similar concepts are valid for reflecting components.

A thin lens is characterized by its focal length f, distance of its center plane to the focal point of convergence for axial rays and F-number, ratio of the focal length and the open diameter of the lens. The value of the focal length is important to determine the location of the object and image planes, while the F-number determines the amount of light reaching the image, which is inversely proportional to the exposure time in photography. In a thin lens, the focal length is related to the radii of curvature R_1 and R_2 of the front and back lens surfaces (the radius being negative for concave surfaces) by the expression

$$\frac{1}{f} = \Delta n\left(\frac{1}{R_1} + \frac{1}{R_2}\right), \qquad (3\text{-}8)$$

where Δn is the difference between the refractive index of the lens and that of the surrounding medium.

Figure 3-40 shows the familiar ray-tracing diagram to determine the position and the size of the image when the focal points and the object are given. The distances of the object, o, and of the image, i, from the lens are

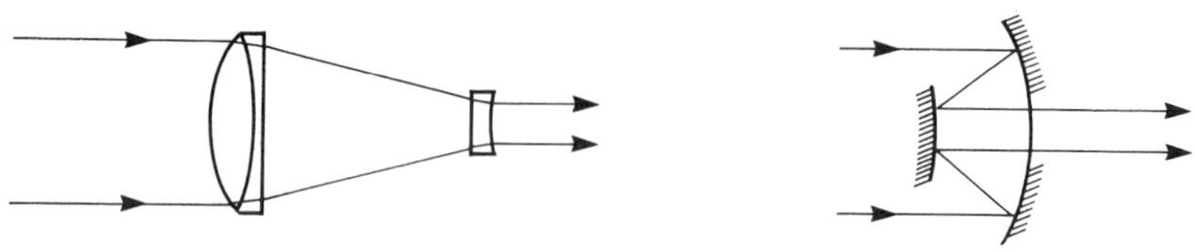

Figure 3-39 Examples of a refractive (galilean) and of a reflective (cassegrain) telescope.

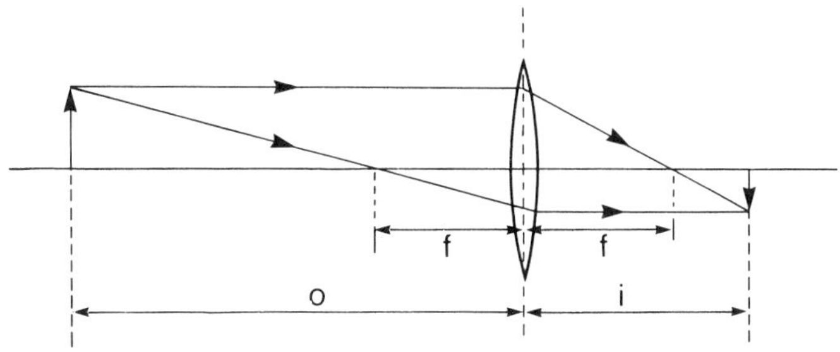

Figure 3-40 Ray tracing with a thin lens.

related by the expression

$$\frac{1}{o}+\frac{1}{i}=\frac{1}{f}, \tag{3-9}$$

where the values of o and i are negative for virtual objects or images, while f is negative for diverging lenses (Meyer–Arendt, 1972, Chapter 1.3). The ratio i/o corresponds to the lateral magnification of the image diameter with respect to the object diameter.

A similar expression is valid for thick or composite lenses, provided that the respective distances are taken from the principal planes of the lens system. The principal planes are two geometrical planes where refraction may be assumed to take place for the purpose of ray tracing, as Figure 3-41 shows. These two planes are superposed to the center plane in a thin lens. The manufacturer usually specifies for thick lenses the value of the focal length as well as the distances of the two focal points from the front and the back surfaces of the lens system, from which the positions of the principal planes may be determined. Principal points are defined as the intersection of the principal planes with the optical axis. Another useful concept is that of nodal points, which are the points through which rays can be traced with no refraction taking place (Kingslake, 1965; Meyer–Arendt, 1972, Chapter 1.4).

Microscope objectives are usually specified in terms of their power and numerical aperture, e.g., 20X and 0.54 (Benford, 1965). The power is defined so that the product of the objective and of the eyepiece powers in a microscope gives the image magnification wth respect to the image size when viewed at the standard observation distance for the unaided eye of 250 mm. If the objective is designed for a standard tube length of 160 mm, its focal length can thus be estimated as 160 mm divided by its power, while the focal length of an eyepiece is obtained as the ratio of 250 mm and its power. The numerical

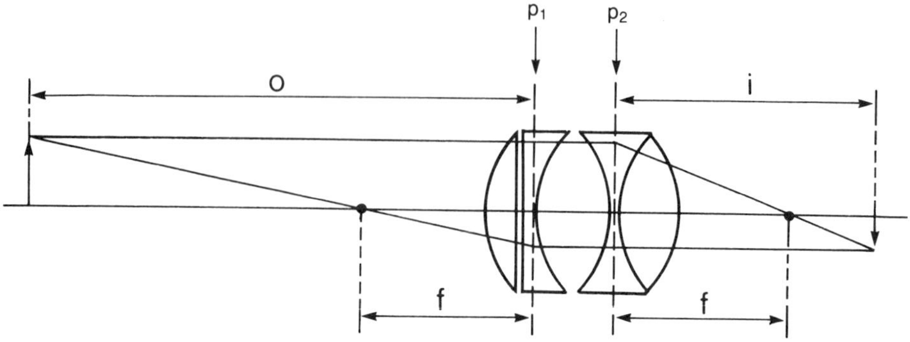

Figure 3-41 Ray tracing with a thick lens; p_1 and p_2 are, respectively, the object and image principal planes.

aperture of an objective is defined as

$$\text{N.A.} = n \sin \theta, \tag{3-10}$$

where n is the refractive index of the medium in front of the instrument (usually air) and θ is the half-angle of the cone of light entering the objective.

Depth of Field

A question of considerable importance in optical inspection is the depth of field of a focused beam (Benford, 1965; Holik, 1986). With reference to Figure 3-42, the depth of field may be defined as the distance, d, along which the diameter, w, of a focused light beam does not change appreciably. The light beam may be issued from a point source or, as in Figure 3-42, it may be the light beam collected by a narrow-aperture detector. Large values of the depth of field are usually desirable in industrial inspection, where the surface toward which the beam is pointed may be irregularly shaped or may move in time.

The exact value of spot unsharpness which may be accepted varies for every specific application, but as a general rule we may say that the displacement $d/2$ from the position of maximum sharpness is negligible provided that the geometrical spot magnification, $(d/2) \tan \theta$, is smaller than a certain fraction of the half beam-waist $w/2$ or

$$d \tan \theta = Kw \tag{3-11}$$

where K is a constant. As a rule of thumb, and for small values of the angular

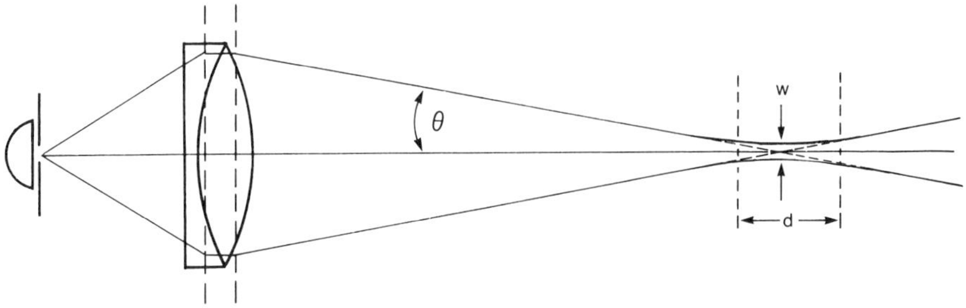

Figure 3-42 Depth of field d for a spot-focused light beam.

aperture θ, we may say that the amount of blur is negligible until the geometrical spot section is equal to the beam waist, i.e.,

$$d = w/\theta \tag{3-12}$$

where θ is expressed in radians. The value of w, in the case illustrated in Figure 3-42, is typically equal to the image of the detector sensing area, increased by the aberrations of the lens system. For perfect optics and a very small detector aperture (or for a point source in the case of a projected beam), the beam waist diameter is given by the diffraction limit

$$w \simeq \lambda/\theta, \tag{3-13}$$

where λ is the light wavelength, resulting in a depth of field

$$d \simeq \lambda/\theta^2 \tag{3-14}$$

through which the central energy of the focused spot changes by not more than 20% (Hopkins and Buzawa, 1976).

Aberrations

Although the ultimate limit to the minimum spot size of a focused light beam is given by the diffraction limit, equation (3-13), the aberrations of the focusing lenses usually set a higher limit to the resolution of the optical system. The main types of aberrations are summarized below (Conrady, 1960; Smith, 1966; Wolfe and Zissis, 1978).

(a) *Spherical aberration.* In a lens with spherical surfaces, light rays close to the optical axis and far from the axis come to focus at different distances, as Figure 3-43a shows. The result is an increase of the focused spot diameter as the aperture of the lens is increased.

(b) *Coma and astigmatism.* If a light beam passes obliquely through a lens, the peripheral rays come to focus to one side of the central ray position. This produces a comet-shaped spot aberration known as coma. In addition, the focal distance for the tangential and sagittal rays (see Figure 3-43b) is different from the focal distance of the rays lying out of this plane. The result is that two line-focuses at different positions along the beam (astigmatism) are obtained.

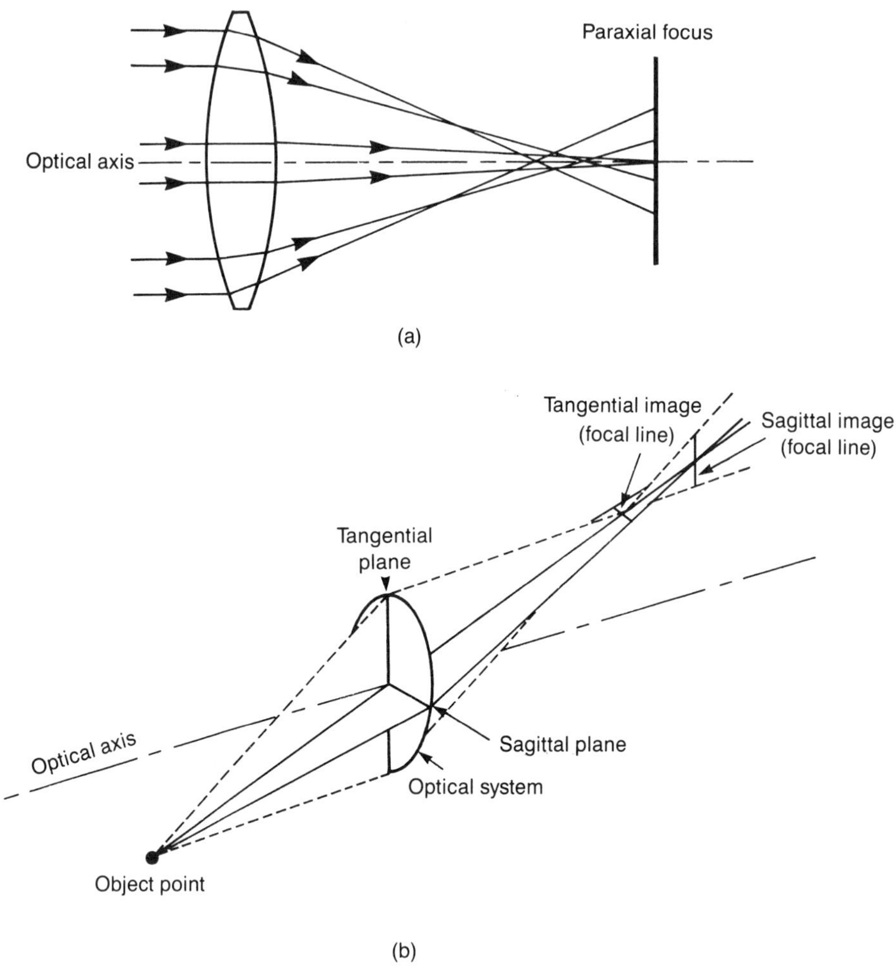

Figure 3-43 Ray tracing for a single spherical lens showing the effects of (a) spherical aberration and (b) astigmatism.

(c) *Chromatic aberration.* The refractive index of optical materials is generally different at different wavelengths (dispersion; see next section). As equation (3-8) shows, the focal length varies accordingly. Consequently, if a beam of wide spectral bandwidth is incident on the lens, a focal spot of different diameter is obtained for each color component (longitudinal chromatic aberration; see Figure 3-44a), while the image size depends on the wavelength (transverse chromatic aberration).

(d) *Field curvature.* As Figure 3-44b shows, the focal surface is generally not flat. If an image is projected on a flat surface, part of the image will be out of focus. This problem can be only partially removed by adopting a curved image plane (e.g., a curved film holder in a

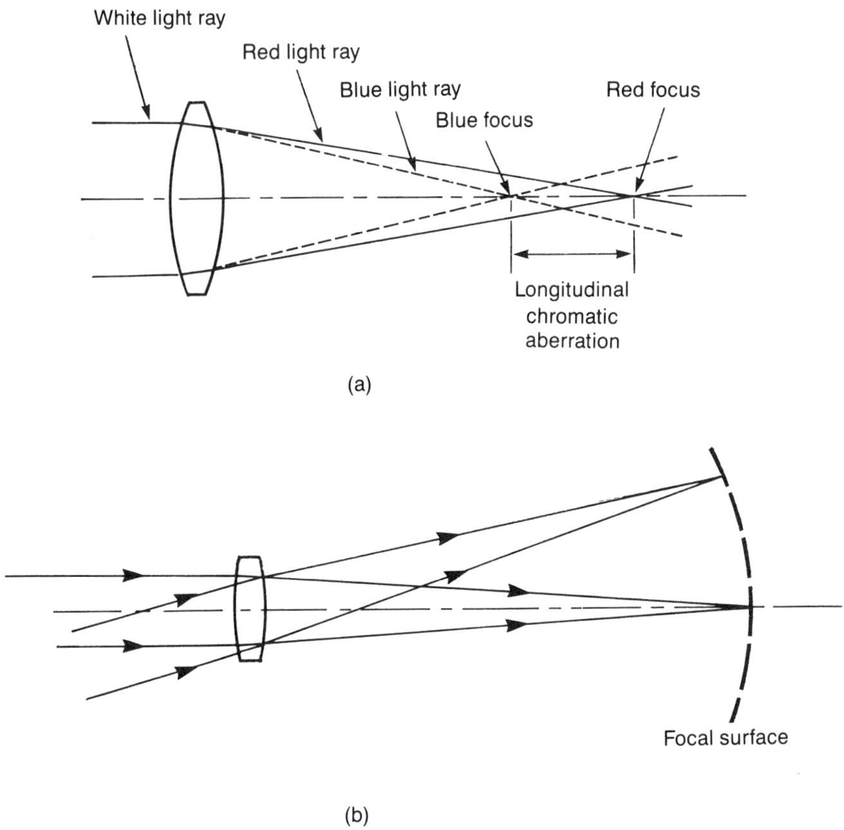

Figure 3-44 Illustration of the effect of (a) longitudinal chromatic aberration and (b) field curvature.

photographic camera) because the focal surface is in general different for the rays lying in the plane of the figure and out of this plane.

(e) *Distortion.* This kind of aberration modifies the shape of the image which, even if well on focus over its full width, is distorted with respect to the object. This is caused by an increase ("pincushion" distortion) or a decrease ("barrel" distortion) of the lateral magnification away from the optical axis (see Figure 3-45).

Except for the chromatic aberration, the main aberration components for a spherical lens are of the third order, i.e., they are proportional to the third power of either the lens aperture or the angle of the incident ray with respect to the optical axis. Consequently, the amount of aberration is very small for the rays close to the axis (paraxial rays), but it increases very rapidly as the lens aperture or the angle of incidence increase.

Lens Design for Minimum Aberration

Because of the nonlinearity of third-order aberrations, even a simple spherical lens may provide a diffraction-limited spot resolution when operating at large F-numbers, reduced image field and with monochromatic illumination. Such conditions may be found when focusing a laser beam parallel to the axis and covering only the central portion of the lens, as in Figure 3-38. Particular care will be taken in such a case to align the lens perpendicularly to the incident beam, e.g., by observing the multiple-reflection pattern, to avoid coma or astigmatism.

It will be observed in Figure 3-38, as well as in other configurations such

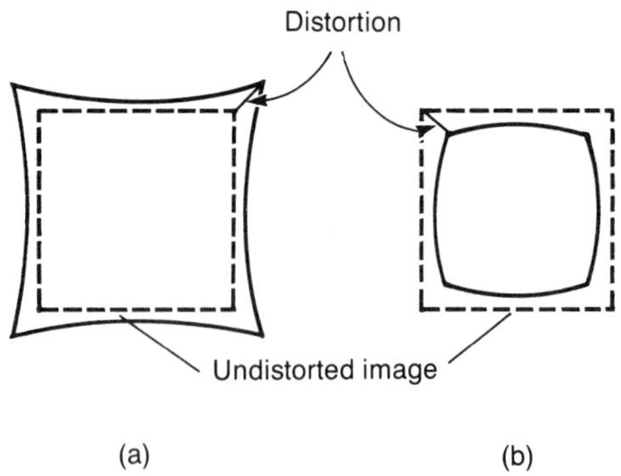

Figure 3-45 Distorted images for a square-shaped object: (a) positive or pincushion distortion; (b) negative or barrel distortion.

as Figures 3-33 or 3-35, that the best choice for focusing a parallel beam is a plano-convex lens (for optical materials with $n \simeq 1.5$ such as fused-silica; for materials such as germanium with $n \simeq 4$, a meniscus-shaped lens gives the best performance). This is a direct consequence of the nonlinearity of the spherical aberration: a plano-convex configuration assures nearly equal ray deviations at both lens surfaces, thus minimizing the total aberration. Care must be taken to orient such a lens properly, i.e., with the flat side toward the focused spot. If the lens were inversed, the entire ray deviation would take place at the curved lens surface, producing a very large third-order aberration. Similar considerations indicate that the best lens configuration for a 1 : 1 object-image magnification is a symmetric biconvex design.

Reducing the lens aperture is, however, often not desirable because of signal amplitude requirements. In such cases, the lens must be designed to minimize the aberration for nonparaxial rays. One approach to minimize spherical aberration in wide-aperture lenses is to resort to an aspheric design (Shannon, 1980). Aspherical lenses are available whose surfaces have a radius of curvature which is larger near the lens center than far from the center. Low spherical aberration lenses of this kind with F-number smaller than 1 are available at moderate prices. Such lenses are mainly used as condensers, their optical grade being usually not adequate for precise beam-focusing applications. Aspheric surfaces are difficult to produce with a precisely-controlled curvature profile and with optical quality surface smoothness, because heavy surface polishing normally produces a spherical profile.

Aberrations are usually minimized by proper design of lens combinations. An example is a lens doublet such as the one shown in Figure 3-42. By combining a positive and a negative lens with opposite aberrations, a spherical aberration-free doublet can be obtained for a given wavelength of operation and for axial focusing. Chromatic aberration may also be reduced to a certain extent with a doublet using two materials of different dispersion (achromatic doublet or "achromat").

Lens design for aberration minimization over the full image field is much more demanding, so that compromises must be made on the image sharpness when a wide-field lens is required. Examples of camera objectives designed for specific purposes are shown in Figure 3-46. The Cooke triplet is designed to correct for spherical and chromatic aberration while minimizing coma and astigmatism. A microscope objective provides minimum aberration with high numerical aperture over a reduced field; it is widely used to finely focus a laser beam. The telephoto objective is a close-up system with limited field of view, while the reverse telephoto design provides an extended field of view.

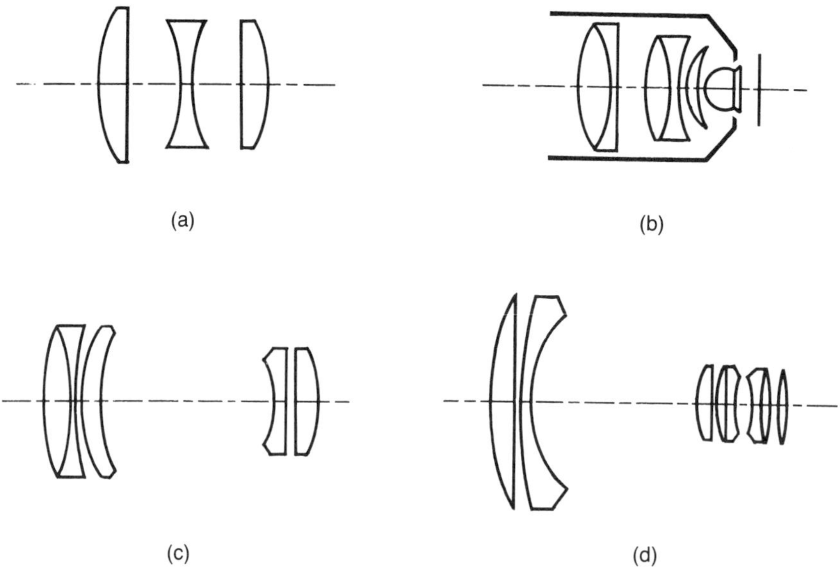

Figure 3-46 Examples of compound objectives: (a) Cooke anastigmat; (b) 80X microscope objective, (c) telephoto and (d) reversed telephoto.

Lens Evaluation

It should be clear from the preceding discussion that the quality of a lens is strictly dependent on the required optical configuration. A camera lens which is intended to produce a focal plane image of a distant object will generally perform poorly in a 1:1 imaging configuration. A lens which is diffraction limited when operating with an axial laser beam may be totally inadequate for wide-field imaging or when operating at different wavelengths.

Such considerations must be kept in mind when a lens is assessed on the basis of the manufacturer supplied lens specifications, of which the best known is the modulated transfer function (MTF). The MTF is the modulus of the optical transfer function, which is the Fourier transform of the point-spread function (the point-spread function is the light distribution in the image plane obtained with a point source in the object plane). The MTF gives the spectral response of the optical system to an input object having different spatial frequencies. It may be derived by recording the contrast obtained in the image plane when line patterns of given numbers of lines/mm are inserted in the object plane.

The MTF curve supplied by the manufacturer is usually obtained under optimum conditions, i.e., at the best aperture setting, with monochromatic illumination and near the axial focal point. A much lower resolution may be obtained when such conditions are not satisfied, or when the object-image magnification is not close to the magnification for which the lens was designed. Finally, it should be remembered that the transfer function of the full system may be further limited by some additional factors, such as the dimensions of the detector sensitive area, the quality of other optical components such as dirty windows or damaged mirror surfaces, or the presence of air turbulence along the line of sight.

3.7.2 *Optical Materials*

Materials for optical components must be chosen in terms of the spectral range of operation, resistance to the industrial inspection environment and easy maintenance.

Reflective components having front surface metallic layers made of gold, silver, aluminum or copper have a good reflectivity in the infrared, as Figure 3-47 shows (Ross, 1966). Silver has a good reflectivity throughout the visible as well, while aluminum has good reflectivity down to the near ultraviolet. Most base metallic surfaces cannot be used in an industrial environment without being coated with a thin overlayer for protection against oxidation, which particularly affects silver and copper, or for good resistance to scratches in the case of soft metals such as gold. If the spectral bandwidth of operation

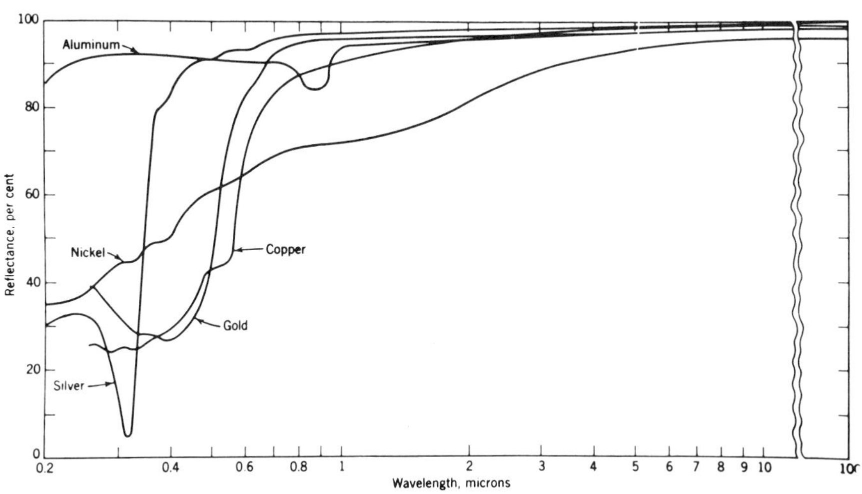

Figure 3-47 Reflectivity of the most common metals.

is relatively narrow, protective coatings of controlled thickness, such as a $\lambda/2$ SiO layer on aluminum, may be chosen for surface protection while enhancing reflectivity in the spectral region of interest.

Refractive elements such as lenses or windows are available in a wide range of materials. The primary consideration for the choice of the material is good transparency in the spectral region of operation. Figure 3-48 shows the transmissivity of different materials over the optical spectrum (uncoated 3 mm-thick windows, adapted from a number of optical manufacturer data sheets). For optical components without anti-reflection coatings, the maximum transmissivity in a spectral range where negligible material absorption takes place is determined by the Fresnel reflection losses at normal incidence (Born and Wolf, 1970, Section 1.5).

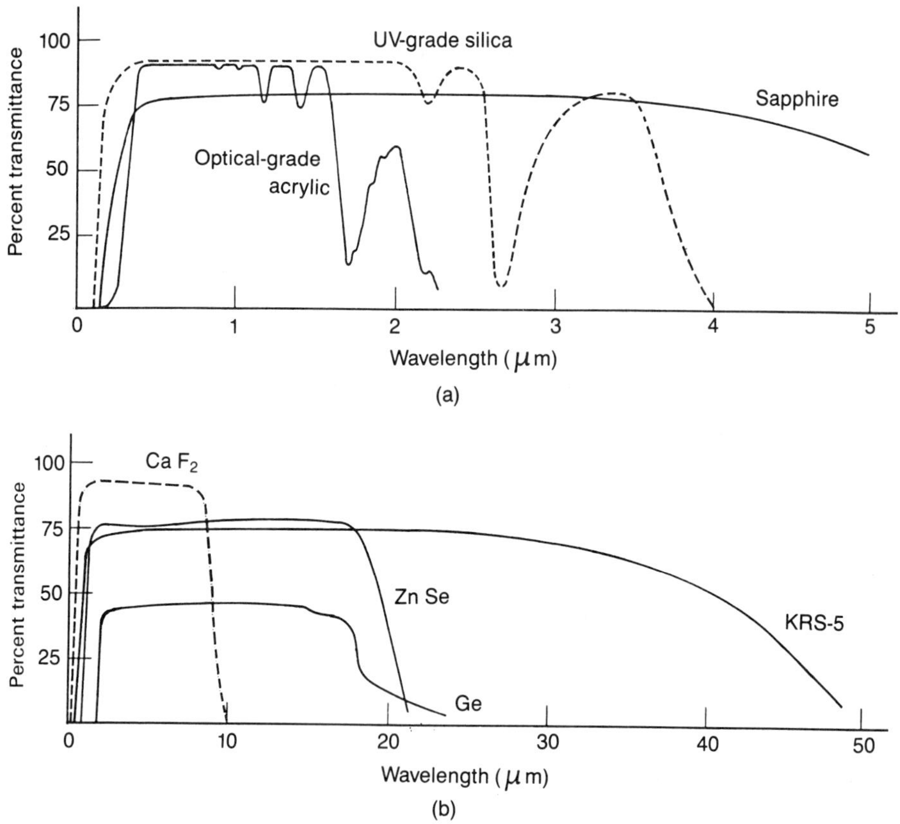

Figure 3-48 Spectral transmissivity of common optical materials used: (a) in the near-visible; (b) in the infrared range.

$$R = \frac{(n-1)^2}{(n+1)^2} \tag{3-15}$$

at each air-bounded surface, where n is the refractive index. Materials such as fused silica or calcium fluoride, which have $n \approx 1.5$ (see Table 3-7) may be used without coating with reflection losses of nearly 8% per lens or window. An uncoated germanium component, with $n = 4$, has a transmissivity of nearly 40%, so that anti-reflection coatings designed for the wavelength of operation are usually required for such a material.

Table 3-7 shows a number of relevant properties for optical materials to be used in an industrial environment. Water solubility determines whether the material considered can be used in contact with water or in a water vapor rich atmosphere without substantial degradation. Hardness (measured by Knoop indentation) is important because it is related to the resistance of the optical component to scratches and to the possibility of cleaning the piece by sheet wiping. Sapphire is a particularly hard material which is recommended when the inspection system must be installed in hostile environments. High melting temperatures are required for high-temperature applications such as furnace inspection, while a high value of the thermal conductivity determines whether the optical component can be efficiently cooled by forced air or water circulation along the rim.

Dispersion, i.e., the variation of refractive index with wavelength, is an important parameter in optical design because it determines the extent of

Table 3-7 Properties of Optical Materials

Material	Average refractive index	Water solubility (g/100 g of water)	Knoop hardness (Kg/mm^2)	Melting temperature (K)	Thermal conductivity (W/cm K)
Fused silica	1.45	insoluble	500	1750	0.01
Sapphire (Al_2O_3)	1.7	insoluble	1370	2300	0.25
Acrylic	1.5	insoluble	10	400	0.003
CaF_2	1.4	0.0017	90	1630	0.04
Ge	4.0	insoluble	200	1210	0.5
ZnSe	2.45	insoluble	137	1700	0.1
KRS-5 (Thallium) bromoiodide)	2.4	0.05	40	687	0.006

Table 3-8 Refractive Index for the Ordinary Ray of Synthetic Sapphire

$\lambda(\mu m)$	n	$\lambda(\mu m)$	n	$\lambda(\mu m)$	n
0.3	1.814	1.5	1.747	3.5	1.695
0.5	1.775	2.0	1.738	4.0	1.675
0.7	1.763	2.5	1.727	4.5	1.651
1.0	1.756	3.0	1.712	5.0	1.622

chromatic aberration. Another effect of dispersion is the variation of the optical parameters, such as the focal length, across the spectrum. Infrared transmitting materials such as sapphire which are also transparent in the visible range can be easily aligned by using a visible light beam for pointing. However, if an infrared beam in the 4 to 5 μm range must be focused with a sapphire lens, consideration must be given to the fact that the refractive index varies by as much as 0.12 between the visible and the infrared regions (see Table 3-8). This result in a variation of nearly 20% in the focal length, as can be inferred from equation (3-8). Such a shift in the focal point position must be taken into account when pointing the system with a visible beam.

3.7.3 *Holographic Optical Elements*

In recent years, two new approaches have been developed for the production of optical elements using modern optical technology methods. One such approach, which was spurred by advances in fiber optics, uses graded-index technology and will be described in Section 3-9. The other approach, which is briefly described in this section, relies on holographic techniques to produce optical elements.

To introduce simply the concept of a holographic lens, consider first the intersection of two coherent light beams, Figure 3-49 (Abramson, 1981, p. 6).

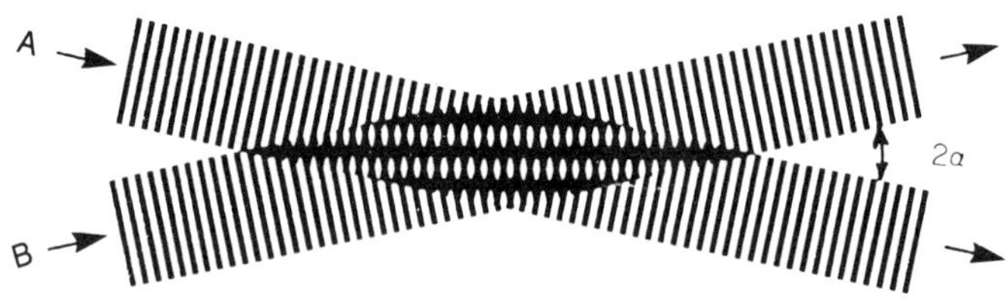

Figure 3-49 Interference between two coherent light beams A and B. The longitudinal fringes in the area of interaction correspond to the planes of equal phase difference between the two waves.

The stripes across each beam correspond to successive wavefronts. The Moiré longitudinal fringes in the intersection region visualize the planes where the two beams are in phase (white fringes) or out of phase (dark fringes). The spacing d between such planes is given in air or vacuum by the expression (Abramson, 1981)

$$\frac{\lambda}{d} = 2 \sin \alpha, \tag{3-16}$$

where λ is the wavelength and 2α is the angle between the beams. Larger angles give closer fringes.

Consider now two beams, one collimated and one focused, intersecting over the high-resolution photographic emulsion of a holographic plate, Figure 3-50a. The light intensity distribution across the line O′O″ is shown in the graph (not on scale). Maximum intensity is obtained along the planes where the two beams are in phase. The photographic emulsion is strongly exposed along such planes, while minimum exposure takes place in the out-of-phase planes. After development, the transmissivity (or the refractive index, or the thickness, depending on the kind of emulsion used) of the emulsion along the O′O″ line varies in a way similar to such an intensity distribution, with a fringe spacing which is larger near O′ and smaller near O″, because of equation (3-16). Figure 3-50b shows the reconstruction phase. The developed holographic plate acts as a variable-period diffraction grating on the reconstruction beam, producing two diffracted beams (corresponding to the +1 and −1 diffraction orders, see Meyer-Arendt, 1972, Chapter 2.7) on each side of the undeviated transmitted beam. Because the grating period varies, the diffraction angle is smaller in O′ than in O″, resulting in a converging and a diverging diffracted beam as shown in Figure 3-50b. The holographic plate thus acts as either a converging or a diverging lens on the incident collimated beam.

The presence of three different output beams for each input beam makes a holographic lens of the kind just described not very practical. A much better diffraction efficiency can be obtained if the thicknes of the emulsion layer on the holographic plate is much larger than λ, as is often the case in practice. As shown in Figure 3-51, the planes of equal exposure corresponding to the Moiré fringes of Figure 3-49 act as venetian blinds reflecting most of the incident light beam toward the first diffraction order. Diffraction efficiencies (fraction of light diffracted into the first order) of the order of 70% are commonly obtained with thick holograms (Chang and Winick, 1981; Margarinos and Coleman, 1985). We can see from Figure 3-51 that the diffracted beam will have a strong amplitude only if the rays reflected by

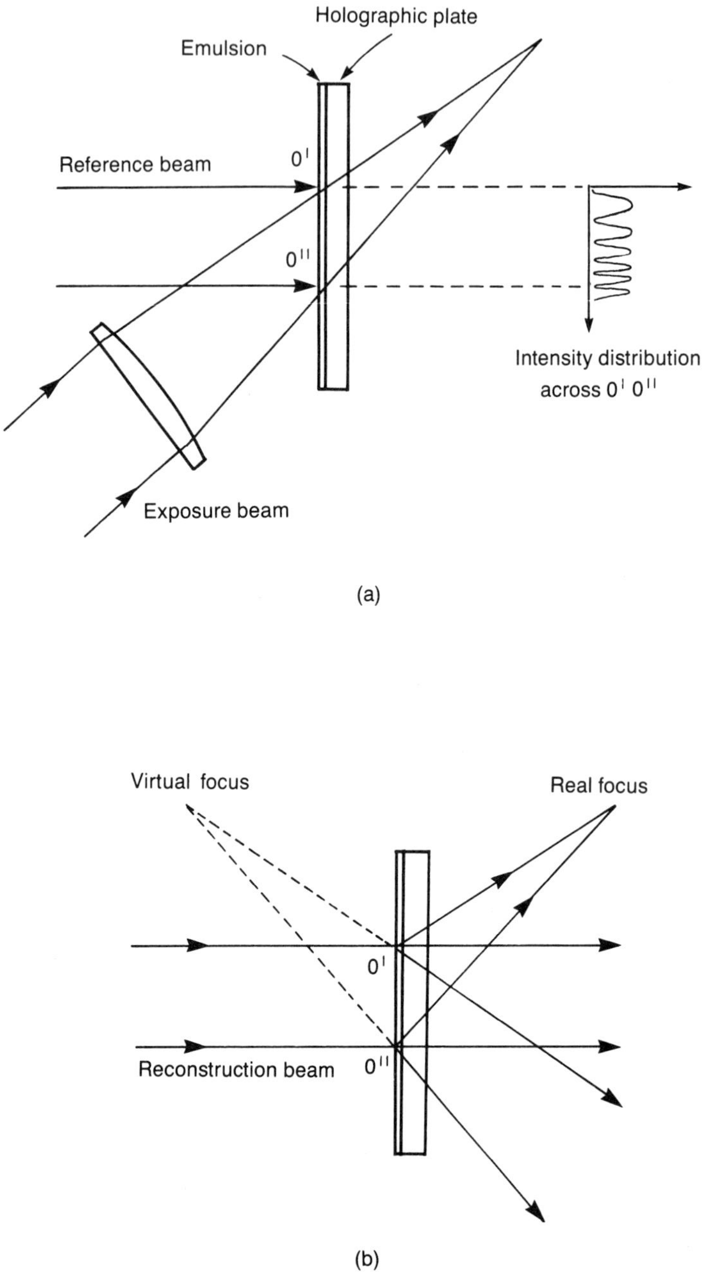

Figure 3-50 Production of a thin holographic lens: (a) exposure of the holographic plate; (b) reconstructed beams after plate development.

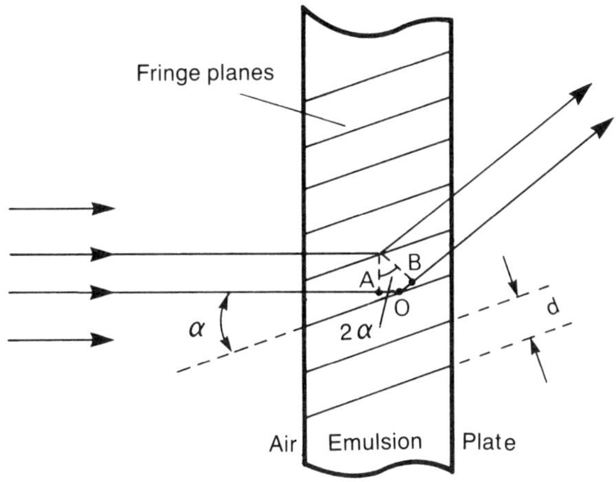

Figure 3-51 Reconstruction of a single diffracted beam with a thick emulsion hologram.

consecutive planes are in phase, i.e., if the optical path difference AOB between such rays is equal to a multiple of the wavelength (Bragg condition)

$$2nd \sin \alpha = m\lambda, \tag{3-17}$$

where n is the refractive index of the emulsion and m is an integer. A holographic optical element made of a thick emulsion hologram will thus only operate effectively at a given angle of incidence and for a given wavelength.

Holographic optical elements have some unique characteristics when compared to refractive or reflective optical components. In contrast to conventional lenses and spherical mirrors, holographic lenses do not introduce spherical aberration on an axial incident beam, provided the beams used to record the hologram are of good quality. Alternatively, the aberrations produced by fixed extraneous components along the beam path can be corrected by using the aberrated beam to expose the holographic optical element (Kim et al., 1985). Coma and astigmatism are, however, present for nonaxial angles of incidence (Close, 1979), while strong dispersion is encountered with wide spectral bandwidth light. Consequently, holographic optical elements are most interesting when working with laser beams close to the optical axis. Moreover, most hologram recording media require a short wavelength for exposure so that holographic elements are difficult to implement for infrared beams.

One of the most attractive features of holographic elements, apart from their low weight and wide aperture, is the possibility of mass production by contact reproduction. Relief holograms, where the recording medium is made of a photoresist whose thickness is modulated during development, are particularly easy to duplicate. After the original is made, the surface profile is transferred to an epoxy master, from which thousands of duplicate elements can be embossed (Brody, 1982). Thick holograms can be reproduced by putting the master hologram and an unexposed plate in close contact and illuminating with the laser beam. The undiffracted and diffracted waves from the master interfere, and the resulting interference pattern is recorded in the copy plate. Computer generation of holograhic elements is another powerful tool for the design and production of such components.

To summarize, the holographic approach shows promise for the realization of low cost optical elements with high aperture and low weight, particularly for short-wavelength laser beam focusing applications. Such a potential has not yet completely materialized, but this technology could soon reach its full development.

3.8 Scanning Techniques

As mentioned in Section 3.4, surface scanning may be realized either electronically, using one- or two-dimensional detector arrays, or mechanically, using beam scanning devices as decribed in this section. Figures 3-52 and 3-53 show two examples of mechanical scanning configurations which

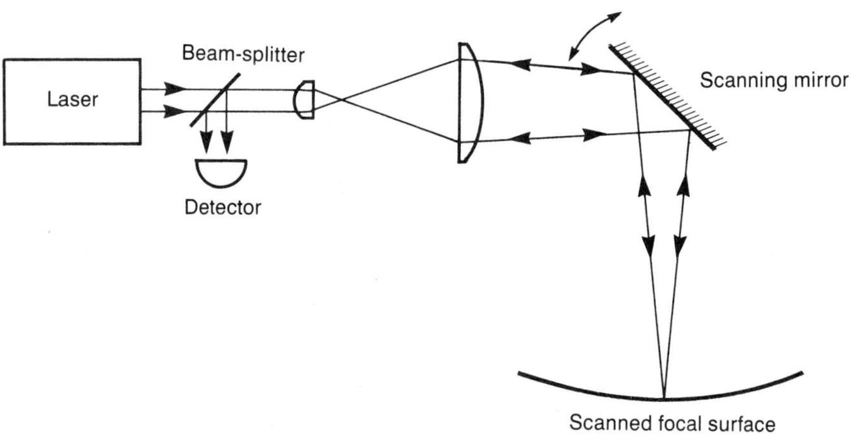

Figure 3-52 Laser flying-spot scanning and detection of back-scattered light. Post-objective scanning configuration.

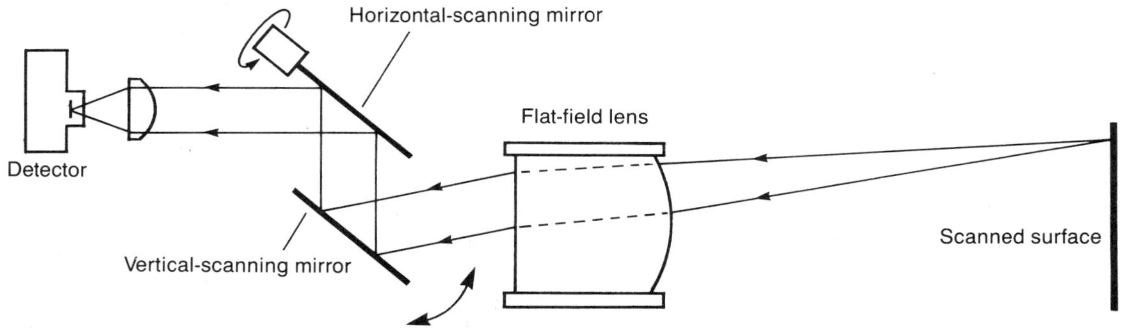

Figure 3-53 Two-axis scanning of a flat surface. Preobjective scanning configuration.

are typical of optical inspection applications. These two configurations differ under a number of aspects.

Figure 3-52 shows a laser beam scanning scheme which is often used, e.g., to scan a line across a moving sheet of a material whose surface must be inspected for defects. The laser beam is expanded to increase the angular aperture of the focused beam and thus to obtain a smaller focused spot on the inspected surface, in agreement with equation (3-13) in Section 3.7.1. A rotating mirror sweeps the focused spot over the scanned surface, while at the same time collecting the back-scattered light beam which is reflected by the beam-splitter toward the optical detector. A polarizing beam-splitter and $\lambda/4$ retardation plate in a configuration similar to that of Figure 3-38 might also be used to avoid reflection losses at the beam-splitter.

Figure 3-53 shows a passive configuration where no light beam is projected toward the surface. In this case, the scanning devices are used to point the detector across the inspected surface, to collect light either reflected or emitted (in the case of a thermal-radiation monitoring device) by the surface. A configuration of this kind may be preferred to the use of an electronically scanned array when the detector is bulky and expensive, as in the case of liquid-nitrogen cooled thermal detectors.

Two-axis scanning is obtained in the case of Figure 3-53 by using two rotating mirrors, the upper one rotating rapidly around an axis contained in the plane of the figure for a line-by-line horizontal scan of the surface, while the lower one provides the vertical scan. The two rotating mechanisms are well synchronized electronically.

Optics for Beam Scanning

The configurations shown in Figures 3-52 and 3-53 differ in another important aspect, the position of the focusing optics with respect to the scanning devices. In Figure 3-52 the focusing lens is situated before the scanning mirror. Such a lens needs to be well corrected only for the axial focusing position and for a single wavelength if used to scan a laser beam, as Figure 3-52 shows. Aberrations can thus be effectively minimized with inexpensive optics. The problem with this configuration is that the focal surface is curved, its center of curvature being near the axis of the rotating mirror. As most inspected surfaces are nearly plane, this limits the angular range over which the scanned spot is well focused.

Furthermore, the translation speed of the flying spot over a flat surface is slightly higher on the edges than on the central position of the scanned path for a constant mirror rotation rate. This is a source of errors when a precise metrology of the scanned surface is required.

Figure 3-53 illustrates a preobjective scanning configuration where the scanning devices are situated before the focusing lens. The lens can thus be designed to have a nearly flat focal surface (flat-field lens) for a given scanner-to-lens distance. Moreover, the negative ("barrel," see Figure 3-45) field distortion of the lens may be deliberately under-corrected to provide a uniform translation speed of the flying spot (Fθ lenses, see Hopkins and Buzawa, 1976). Lens systems much more expensive than in the case of Figure 3-52 are obviously required to satisfy such requirements while at the same time minimizing the aberrations for different angles of incidence and eventually for a wide wavelength range. Compromises must usually be made on either the flatness of the field, the scanning speed uniformity or the spatial resolution over the inspected surface. Such compromises are easier to reach if the lens system is designed for operation at a single wavelength, as is the case for laser beam scanning. As an example, Figure 3-54 shows the performances of a flat-field, Fθ lens providing a 34 μm-diameter focused beam over a 510 mm line-scanning length at a 804 mm focal distance with a $\lambda = 363.8\ \mu$m wavelength (Mikami et al., 1983).

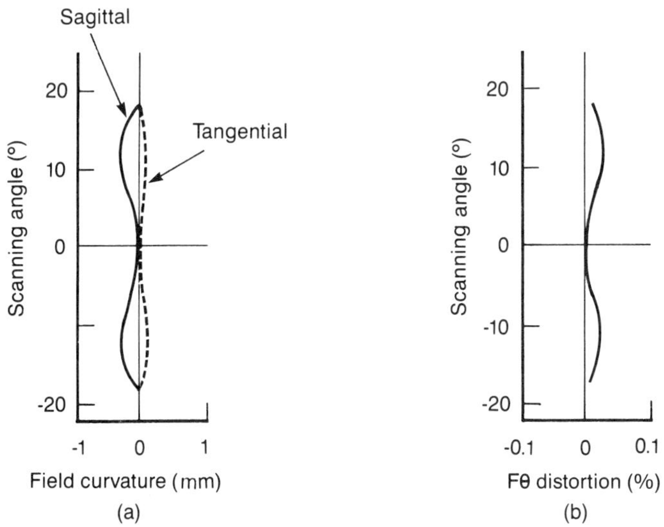

Figure 3-54 Performance of a flat-field, $F\theta$ laser line-scanning lens system. (a) departure from flatness of the focal surface, at a 804 mm focal distance, for the sagittal and tangential rays (see figure 3-43b); (b) percent departure from linearity in the spot position along the scanned line.

In some cases, such as in laser scanning for on-line rod diameter gaging as shown in Figure 3-55, depth-of-field requirements are relatively large because of unavoidable positional errors along the longitudinal axis (this may be produced, for example, by low frequency oscillations of a wire whose diameter is monitored at the exit of the extruding machine). For such an application, the wire positioning accuracy may be relaxed by signal interpolation if the center of the laser spot intersected by the rod does not change as the rod moves in the axial direction, i.e., if the axis of the scanned beam is parallel to the axis of the optical system. This can be achieved by using a telecentric scan lens arrangement as shown in Figure 3-55 (Hopkins and Buzawa, 1976). A telecentric configuration is obtained when the entrance or exit pupil (i.e., the cross section of a bundle which is common to all rays, see Meyer-Arendt, 1972, Chapter 1.5) of the lens system, here the scanning mirror, is situated in its focal plane. Similar arrangements may be employed for precise dimensioning using an imaging camera (see Batchelor et al., 1985, Section 5.6). Telecentric lens systems must have a large diameter, larger than the length of the scanned line. They are thus only convenient when relatively short paths must be scanned.

Signal-to-noise considerations

As mentioned in Section 3.5.3, electronic noise in a detector is determined at a given wavelength and signal frequency by the value of its detectivity D^*. The signal-to-noise ratio (SNR) is thus obtained as the ratio of the detected optical power W_d and the noise-equivalent-power, i.e., from equation (3-4) in Section 3.5.3.

$$\mathrm{SNR} = D^* W_d (A_d B)^{-1/2}, \tag{3-18}$$

where the signal bandwidth for a scanned detector is given by (Wolfe and Zissis, 1978, Section 10.2)

$$B = \frac{1}{t_d} = \frac{N(1 + r_0)}{t_f - t'_f} \tag{3-19}$$

where t_d is the dwell time, i.e., the time period required to scan over a resolution element (or pixel) on the scanned surface, N is the number of pixels per frame, r_0 is the line overlap ratio, t_f is the time required to scan a full frame and t'_f is the portion of t_f during which no useful information is generated, such as the dead time between successive line scans in a two-dimensional scanning operation.

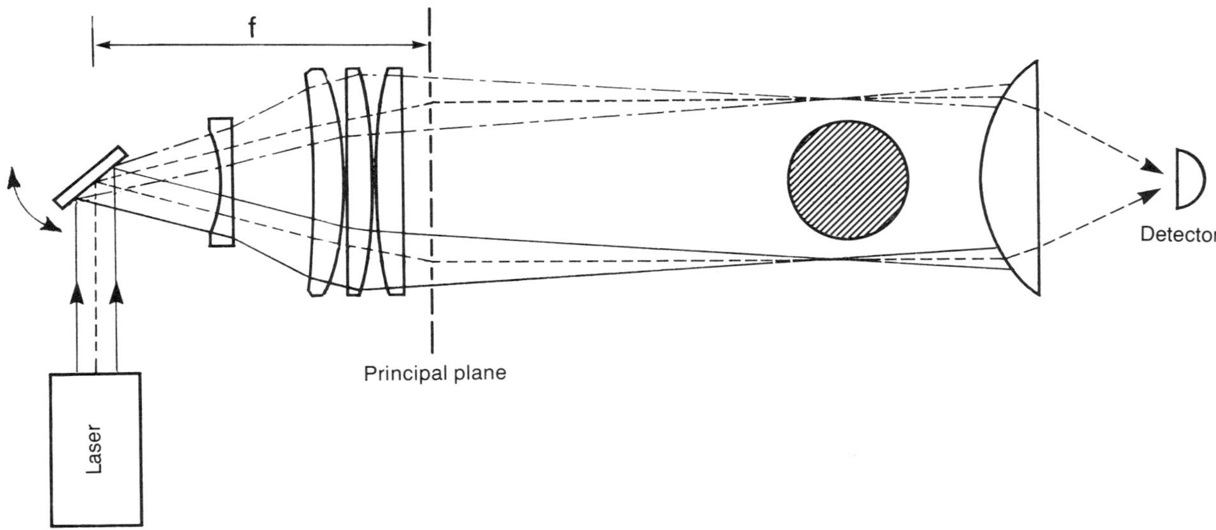

Figure 3-55 Telecentric scan lens arrangement for precise dimensioning over a relatively long longitudinal range.

If the degree of overlapping as well as the percentage of dead time in a scanned system are comparable to those produced by element-to-element cross talk and readout time in an electronically-scanned camera (see Section 3.4), the SNR may appear from equations (3-18) and (3-19) to be lower in the scanned system by a factor of $N^{1/2}$. In reality, this is usually not the case for a number of reasons.

First, the signal power level, W_d in equation (3-18), is much higher in a laser-scanning system of the kind shown in Figure 3-52, where the inspected spot is always strongly illuminated by the focused laser beam, than in an inspection system using a wide-field camera under incoherent illumination. As an example, consider a configuration where a CCD camera is used to inspect a $1\,\text{m}^2$ surface at a 5 m distance, with illumination from a 1 KW halogen lamp. If we take into account the light-conversion efficiency, angular-spread losses, as well as the limited spectral range of the camera sensitivity as compared to the wide emission bandwidth of the lamp (see Section 3.1), only 20W will typically be available for the illumination of the 400×500 pixels of the image, i.e., $100\,\mu\text{W/pixel}$. This is 100 times smaller than the 10 mW from a single diode laser focused on the inspected pixel, even if the electrical input energy of the incoherent source is 10,000 times larger!

Second, the electronic noise referred to in equation (3-18) is often overshadowed by environmental and light-source noise, as explained in Section 3.5. Ambient light noise is much easier to avoid in a laser-scanned system using spectral or spatial filtering techniques (see Section 3.6.1) than in the case of incoherent lamp illumination. Variations of light illumination across the inspected surface caused by the spatial nonuniformity of the light emission from the incoherent lamp are typically some orders of value larger than the laser intensity fluctuations. Electronic noise reduction by heterodyne detection is only possible with coherent illumination (see Section 3.6.3), while synchronous detection is inconvenient with camera devices whose modulation speed is limited by the frame grabbing time of the order of 10 ms in two-dimensional cameras.

Third, pixel-to-pixel response nonuniformity and limited dynamic range problems encountered with staring arrays (see Section 3.4.1) are avoided with scanning devices. Moreover, the single detector used in the scanning system is usually of higher quality, and its detectivity can be improved by cryogenically cooling and cold shielding techniques. This explains why infrared thermal-imaging cameras using two-dimensional scanning systems of the kind shown in Figure 3-53 provide a better image quality than pyroelectric vidicons or one-dimensionally scanned linear arrays, even though none of the coherent-illumination advantages mentioned above are present in this case.

An argument against the quality of laser scanned surface inspection, however, is the presence of speckle noise which appears when a coherent source is focused on a small spot. This point will be discussed in Section 3.10. Another point which must be considered when using a powerful focused laser beam is eye safety, particularly for visible and near-IR lasers (see Section 3.2). Finally, the lack of any moving part is a definite advantage of electronically scanned cameras, particularly in an industrial environment.

Scanning Devices

Mechanical scanning devices using either oscillating mirrors, rotating mirrors, polygons, pyramids or prisms (Marshall, 1980) are shown in Figure 3-56. Figure 3-57 shows two typical scanners available on the market.

Oscillating mirrors, usually mounted on a galvanometer shaft, are relatively inexpensive but typically limited to response times of the order of tens of milliseconds for discrete displacements, while their positioning reliability is limited by suspension balancing problems (Beiser, 1985). Resonant oscillation at frequencies up to 5 KHz is possible with small mass mirrors, but in this case the scanning speed varies sinusoidally.

Polygon and pyramid rotating mirrors can achieve much higher scanning rates when continuous line scanning is required. The polygon scanner shown in Figure 3-56b can reach rotation rates of 18,000 RPM, resulting in 3,600 line scans per second with a 12-facet polygon. In laser-scanning configurations of the kind shown in Figure 3-52, diffraction limited spots are easily obtained by matching the optics to the single laser wavelength in the axial direction, providing a spot diameter given by equation (3-13) in Section 3.7.1. In a typical optical inspection configuration with a beam diameter of 1 cm at the polygon facet, an operating distance of 50 cm and a 12-side polygon giving a scanned angular range of 60°, the spot diameter obtained with a $\lambda = 0.5\ \mu m$ laser is of the order of 50 μm, giving 10,000 resolution points across the curved 50 cm-long scanned line. As a comparison, it will be recalled from Section 3.4 that the resolution of a linear detector array is limited by lens aberrations and element-to-element cross talk to typical values of 1000 elements per line.

The beam deviation angle is equal to twice the element rotation angle for the oscillating-mirror and the rotating-polygon scanner, while it is equal to the rotation angle in the case of the rotating pyramid, Figure 3-56c. The

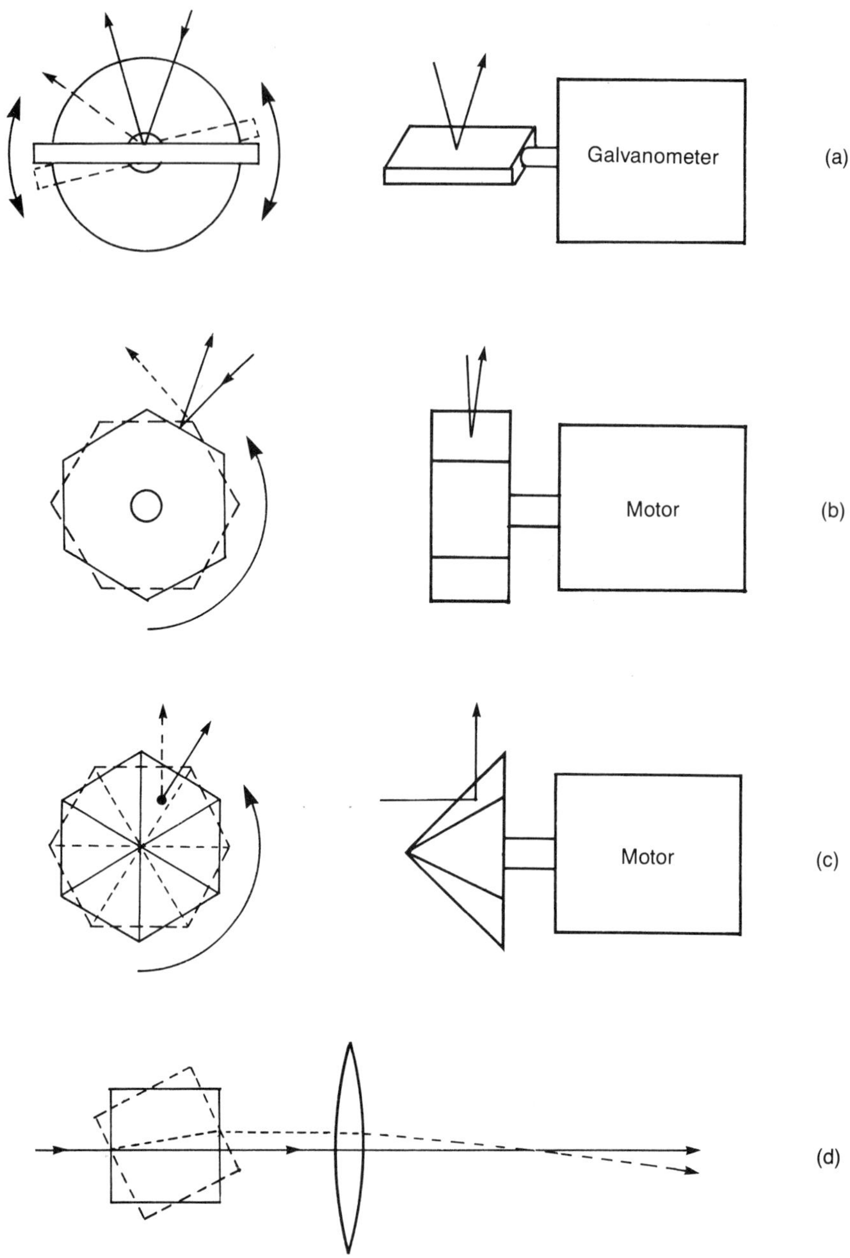

Figure 3-56 Some mechanical scanning devices: (a) oscillating mirror, (b) rotating polygon, (c) rotating pyramid and (d) rotating prism. Broken lines refer to the device and ray positions after a small rotation.

latter element thus requires twice the rotation speed to achieve the scanning rate and scanning field which are obtainable with the polygon scanner. The rotating-pyramid scanner is mainly used when the optical configuration requires a scan direction perpendicular to the plane of the incident and reflected beams. Both the rotating-polygon and the rotating-pyramid scanners produce during a one-facet scan a displacement of the position at which the beam reflection takes place, as can be seen in Figure 3-56b. Such a displacement produces a spot position error of the kind shown in Figure 3-54b, which must be accounted for in the lens design. The rotating-prism scanner, Figure 3-56d, has an angular beam rotation rate and scan linearity which depend on the characteristics of the lens following the prism. Such a design, which was used, for example, in a thermal-imaging camera manufactured by Agema, results in a straight optical axis through the scanners and the lens system.

The high spatial resolution and line-to-line positioning accuracy which are obtainable with fast scanning polygons require high quality rotating elements and synchronization electronics to reach top performances. Polygons of the kind shown in Figure 3-57b are usually available with

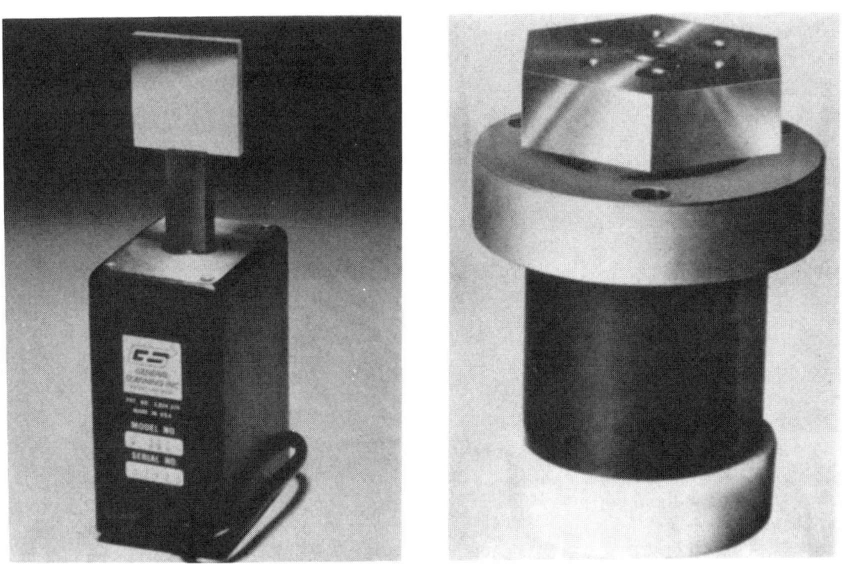

(a) (b)

Figure 3-57 Examples of (a) a galvonometer-driven mirror and (b) a rotating-hexagon scanner commercially available. (Courtesy of General Scanning and of Lincoln Laser.)

facet-to-facet and facet-to-rotational-axis (pyramidal) errors lower than 2.10^{-4} radians. This corresponds to a positional error from facet to facet larger than one resolution element if a resolution of 10,000 points per line over a 1 radian scanning field is to be obtained. Similar errors are generated by axial vibration of the polygon spindle and revolutional variations of the polygon driving motor.

Several methods can be used to correct for such errors. Pitch errors produced by pyramidal facet misalignment or from axial vibration of the polygon spindle may be corrected using cylindrical or toroidal lenses (Fleischer, 1973; Fisli, 1983; Kramer and Pizzutelli, 1985) or cylindrical mirrors (Batchelor et al., 1985, p. 316) to increase the acceptance angle of the beam reflected by the polygon facet in the plane parallel to the rotation axis. Another method is illustrated in Figure 3-58 (Mikami et al., 1983). The output from a reference facet detector sensing the position of the beam reflected by a polygon facet is fed into an acousto-optic deflector (see next section) to finely tune the angle of incidence of the laser beam on the polygon facet in order to compensate for the pitch error. Also shown in Figure 3-58 is a typical method used to compensate for jitter produced by facet-to-facet angle inaccuracy or from low frequency revolutional variations of the motor speed. An optical detector situated at the start of the scanned line gives a synchronization signal to the line recording electronics for each line scan. A

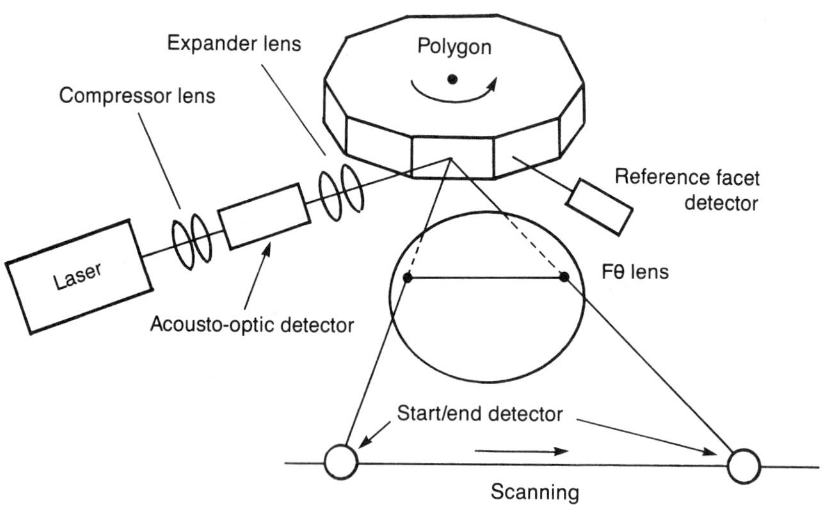

Figure 3-58 Scanning system using a reference facet detector as well as start- and end-of-line detectors for correction of the polygon pitch and jitter scanning errors.

similar detector situated at the end of the line provides a second signal from which the scanning speed can be continuously monitored so that the signal sampling rate can be varied following the low frequency variations of the motor rotation speed. Other methods which have been used to correct for such variations include simultaneous scanning of reference bar gratings to obtain a phase signal to which the sampling clock is locked (Toyen, 1976; Paul, 1983).

Acousto-Optic Deflectors

Acousto-optic and electro-optic deflectors (Korpel, 1980; Gottlieb et al., 1983) have the important advantage over mechanical scanners of not requiring any moving parts. However, the angular scanning range and number of resolvable spots which can be obtained with such devices are reduced in comparison with the mechanical systems.

The basic principle of an acousto-optic deflector is illustrated in Figure 3-59. A clear material, such as glass, or germanium for operation in the medium IR, or an anisotropic material such as a TeO_2 crystal, is insonified by a piezoelectric transducer at an ultrasonic frequency of typically 40 to

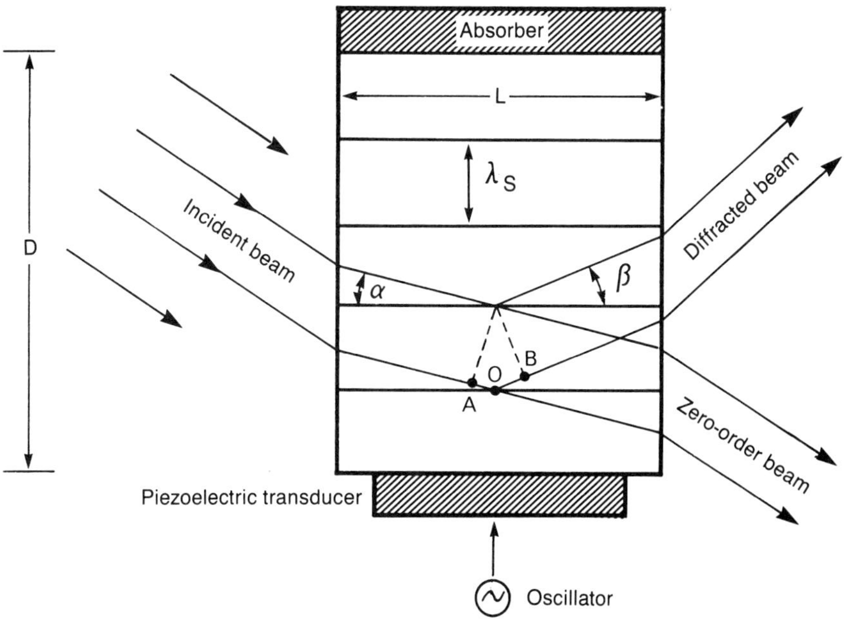

Figure 3-59 Diagram of an acousto-optic beam deflector.

150 MHz. At a given instant of time, the refractive index of the material is sinusoidally modulated with peak-to-peak distance equal to the ultrasonic wavelength λ_s as shown in Figure 3-59. A light beam incident at an angle α may be partially diffracted by such a modulated medium at an angle β. If the length L of the cell is much larger than the modulation period λ_s, the geometrical-optics reflection law implies that a diffracted beam satisfying the Bragg condition (equation (3-17) in Section 3.7.3) will exist only if $\beta = \alpha$. On the other hand, if L is smaller than λ_s, the cell behaves as a thin diffraction grating producing diffracted beams at angles β satisfying the equation (see, e.g., Meyer–Arendt, 1972, equation 2.7–2)

$$n\lambda_s (\sin \alpha + \sin \beta) = m\lambda \tag{3-20}$$

where, as usual, n is the average refractive index within the material, m is an integer and λ is the light wavelength in vacuum. In the paraxial approximation, a variation of the ultrasonic frequency, and thus of λ_s, produces a proportional variation in the beam-deviation angle $\alpha + \beta$ over a given diffraction order according to (3-20). The diffraction efficiency will, however, be quite poor in this case so that a compromise is made in the choice of the L/λ_s ratio to obtain diffraction efficiencies within the first diffraction order reaching 80% with deviation angles of $\pm 25\%$ from the central frequency deviation, as Figure 3-60 shows.

Such devices are also used as beam modulators by deviating the beam at an angle which is alternatively larger or smaller than the cut-off angle determined by the position of a knife edge. These modulators can also be driven at a constant frequency to efficiently produce a diffracted beam whose

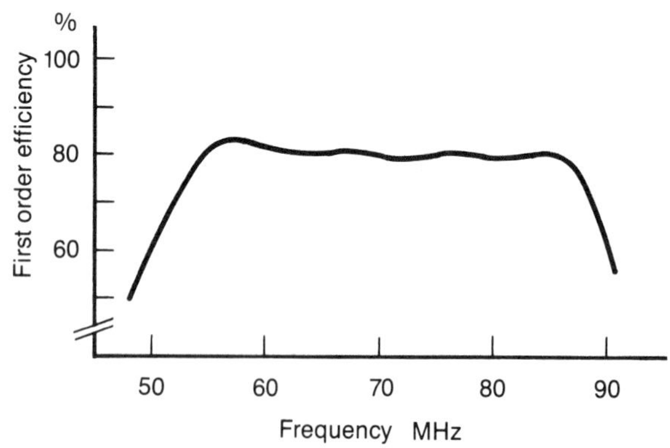

Figure 3-60 Diffraction efficiency of a typical acousto-optic deflector. (Courtesy of InterAction Corp.)

optical frequency is increased by an amount equal to the ultrasonic frequency, as can be shown by Doppler-shift or equivalently by photon-phonon interaction considerations (see, e.g., Yariv, 1976, Section 12.2). An application of such a device (Bragg cell) was described in Section 3.6.3.

One important limitation of acousto-optic deflectors is their reduced angle of deflection. For a typical acousto-optic cell with a sound velocity equal to 4.10^3 m/s and a driving frequency of 70 MHz, the acoustic wavelength λ_s is equal to 57 μm. The total deviation of the first diffraction order of a light beam of wavelength $\lambda = 0.633$ μm after transmission through the cell is equal in the paraxial approximation to $n(\alpha + \beta) = 11$ milliradians from equation (3-20). The diffracted beam is thus very close to the zero-order beam and can be separated from this by placing a knife edge at a relatively large distance from the cell or else by spatial filtering in the focal plane of a lens.

The small deviation angle obtainable with an acousto-optic device is responsible for the relatively small number of resolvable spots through its scanning range. Let us assume that the laser beam diffracted by the device described above is focused over the scanned surface by a lens of focal length $f = 1$ m. If the driving frequency can be varied from 50 to 90 MHz, the diffracted beam will span an angular range of nearly 6.5 milliradians, giving a scanned line of length equal to 6.5 mm in the focal plane of the lens. If the aperture D of the cell is equal to 5 mm, the half-angle θ of the focused beam is equal to $D/2f = 2.5$ milliradians. The beam-waist of the focused spot on the scanned surface will consequently be equal to $w = 250$ μm from equation (3-13) in Section 3.7.1, assuming that the lens is diffraction limited. If two spots are considered as resolvable when they are separated by half their beam-waist, the number of resolvable spots along the scanned line will thus be equal to 50.

Both the angular deflection range and the number of resolvable spots can be increased with the cylindrical telescope configuration shown in Figure 3-61. The incident laser beam, whose initial diameter might be equal to 2 mm, is expanded in one direction by a 25X cylindrical telescope to a beam of 2 mm × 50 mm cross section. After reconstitution of the beam initial shape by the second telescope, it can be seen by ray tracing that the angular deflection range is increased by a factor of 25, from typically 6.5 milliradians to more than 0.15 radians. Moreover, the number of resolvable points is now determined by the width $D = 50$ mm of the beam incident on the acousto-optic cell, which is ten times larger than in the example above, giving a number of resolvable spots equal to 500. This is true provided that the optics, particularly the last cylindrical lens which must operate over a 0.15 radians angular

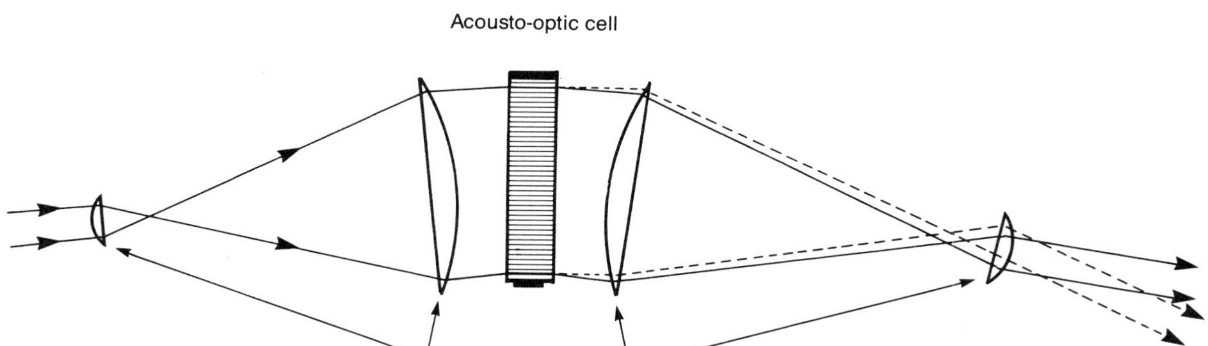

Figure 3-61 Cylindrical telescope optical configuration to obtain higher deflection range from an acousto-optic deflection.

range, is diffraction limited. Up to 2000 resolvable spots have been obtained with higher driving frequencies and lower acoustic velocity materials (Beiser, 1985).

However, the increased number of resolvable spots which can be reached by increasing the cell vertical aperture or by using lower acoustic velocity materials is obtained at the expense of the scanning speed. In the example given above, the dwell time of the focused beam on each resolvable position along the scanned line must be larger than the acoustic propagation time through the extended aperture of the cell if a diffraction limited spot diameter is to be obtained. With an acoustic velocity of 4.10^3 m/s and a vertical beam aperture of 50 mm, the dwell time for each spot is equal to 12.5 μs, resulting in a line-scanning rate of 160 lines per second with a 500-element line. This is much lower than the maximum scanning rate of 3600 line-scans per second with 10,000 resolution points per line mentioned above for the rotating polygon scanner.

To summarize, acousto-optic deflectors are convenient scanning devices requiring no moving parts, but they are limited to low scanning rates and reduced spatial resolution. Advances in this field, such as an increase in the driving frequency using acousto-optic materials which can support high frequencies without introducing excessive ultrasonic attenuation, may take place in the future.

Holographic Laser Scanners

Another unconventional technology which is making rapid progress in the scanning field makes use of holographic elements (Kramer and Pizzutelli, 1985). Earlier in this section we noted that the main limitations of rotating polygons are related to the maximum rotational speed which can be reached with massive multifacet polygons, as well as to the limited accuracy which can be obtained with a given facet size and manufacturing precision. We have also seen in Section 3.7.3 that holographic optical elements have good potential as low weight, high aperture laser deflecting and focusing components. It is thus not surprising that holographic elements are finding increasing application in scanning technology.

Two examples of holographic scanners are shown in Figure 3-62 (Brody, 1982). The left diagram shows the Xerox holographic polygon, or hologon, which has been developed for use in the laser printer produced by this company (Kramer, 1981). A number of adjacent surface-relief transmissive holograms are engraved in a rotating disk, each hologram having its grating

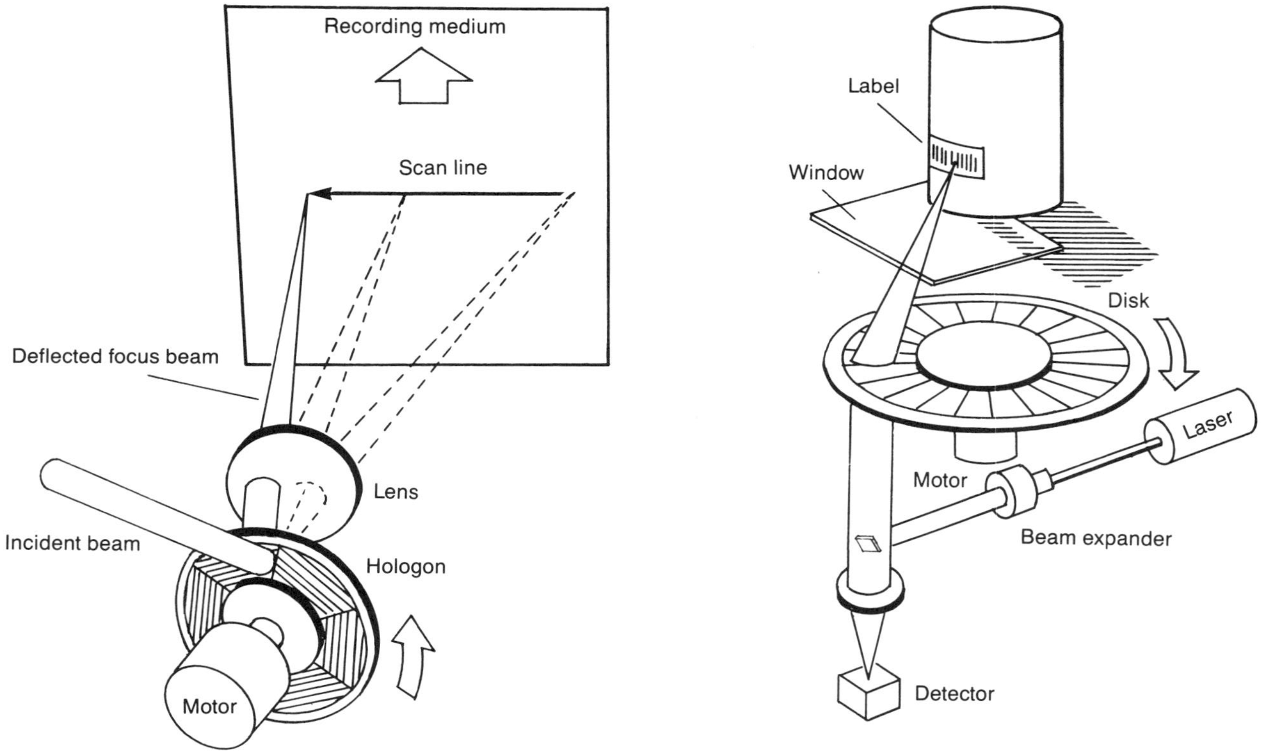

Figure 3-62 Two examples of holographic laser beam scanners: the Xerox laser printer hologon scanner (left) and the self-focusing IBM supermarket bar code reading holographic scanner (right).

lines perpendicular to the radial direction. When the disk rotates, the orientation of the grating lines within each hologram, and thus the angle at which the grating will diffract the laser beam, continuously changes. Each hologram performs a line-scanning function similar to that performed by each facet of a polygon scanner.

The Xerox hologon is designed to meet the stringent high speed and spot positioning requirements of a laser printer while keeping the costs at an accessible level. The surface relief holograms can be replicated quite inexpensively while being relatively independent of rotating wobble. This can be seen from equation (3-20): If the angle α of the incident beam is nearly equal to the angle β of the diffracted beam, a variation $\Delta\alpha$ of the angle of incidence produced by wobble results in an equal and opposite variation $\Delta\alpha$ of the diffracted beam because of (3-20). The total deflection angle $\alpha + \beta$ is thus constant, eliminating the need for expensive dynamic coupling devices to mechanically minimize wobble. Similar geometrical considerations show that for $\alpha \simeq \beta \simeq 45^\circ$ the scanned line is relatively free from bow with less than 0.0001 units of bow per unit of scan (Rallison and Lowe, 1983).

The diagram on the right of Figure 3-62 shows the IBM supermarket scanner device. The use of a holographic scanner is in this case the direct answer to the need to read the bar code on the product label with a depth of field of the order of 15 cm (Dickson and Sincerbox, 1981). For this purpose, the IBM hologram disk contains 21 volume holographic lenses performing both the scanning and the beam focusing tasks. Each of the holographic lenses has a slightly different focal length and deviation angle, so that the rotation of the disk produces a web of scan lines within a scanned volume. One of such lines will be well-focused on the product label without the need for the product to be kept at a precise distance above the scanner window.

To summarize the present status of holographic laser scanners, it should be noted that such devices, although not eliminating the need for a mechanical rotating element, may provide lighter and potentially inexpensive scanning systems which may be adapted to unusual object scanning requirements. Such elements make it possible in some cases to obtain very high scanning performance, such as the 20,000 elements per scan, 200 Mpixels per second capabilities which can be reached by the Holofacet™ scanner (Beiser, 1985). However, such performance requires the use of very high quality optics and high quality and stable holographic surfaces. One problem with holographic elements is the difficulty of producing good quality devices operating in the near-IR region due to the limited availability of holographic materials which can be photographically exposed by laser beams in this wavelength range.

3.9 Fiber Optics

The advent of low-loss optical fibers in the early 1970s was a major event in the communications field. Optical fibers are now definitely superior to competing technologies such as the coaxial cable in terms of information rate and signal attenuation. The economic impulse provided by fiber optic communications has stimulated the whole optical technology field, and particularly the development of low cost and reliable optical components such as fast detectors and small and powerful diode lasers. Such developments are providing extensive benefits to parallel technological areas such as optical inspection in the industrial environment.

Apart from such indirect effects, optical fibers are a powerful tool for many optical inspection activities. The optical signal may be acquired through a fiber optic bundle in hostile environments where high temperatures, corrosive gases or other factors would preclude the use of detectors and signal processing electronics. Transmission of the optical signal directly through fiber optics makes optical inspection less vulnerable to electromagnetic interference, which is particularly intense in the presence of heavy industrial machinery. Fiber optic probes are compact and thus convenient to use where relatively bulky optical systems would be inappropriate. Finally, optical fibers may be directly used as sensing elements, as described in Section 7.4.

Geometrical Ray Propagation in an Optical Step-Index Fiber

Figure 3-63 shows the cross section of a step-index multimode fiber (such fiber properties are defined below). These fibers are made of a cylindrical

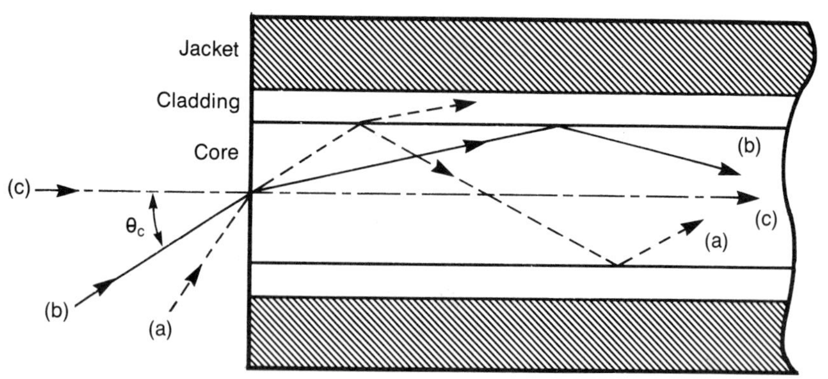

Figure 3-63 Meridional ray propagation in a multimode optical fiber.

core, typically 50 μm in diameter, surrounded by a cladding whose refractive index is lower than the index of the core. Also shown are some meridional rays, i.e., light rays passing through the fiber axis, which are incident on the input flat end of the fiber. Ray (a) is partially coupled to the cladding and thus to the light-attenuating jacket at each core-cladding reflection, resulting in a complete loss of the ray intensity after propagation through a few meters along the fiber. Rays (b) and (c) are incident at the core-cladding interface at an angle larger than the critical angle for total internal reflection and consequently propagate along the fiber with no reflection losses. A simple trigonometrical computation shows that the numerical aperture of the fiber in air is equal to (Keck, 1976)

$$\text{N.A.} = \sin \theta_c = (n_1^2 - n_2^2)^{1/2}, \tag{3-21}$$

where n_1 and n_2 are, respectively, the refractive indexes of the core and of the cladding, and the acceptance angle θ_c is the maximum angle of incidence of the meridional rays guided by the fiber, i.e., the angle of incidence of the ray which is incident at the core-cladding interface at the critical angle for total internal reflection.

The value of the angle of acceptance θ_c given by equation (3-21) is valid for meridional rays. Skew rays, i.e., rays which do not intersect the fiber axis, may be guided by the fiber even if their angle of incidence is larger than the limit given by equation (3-21) (Keck, 1976). Skew rays can propagate within the fiber along a helical-like path; they may be incident on the core-cladding interface at a nearly grazing incidence even if their angle of incidence at the input end of the fiber is larger than θ_c. Such rays undergo a larger number of reflections and cover a longer path for a given length of the fiber as compared to meridional rays, so that they normally suffer a higher attenuation.

Figure 3-63 shows two rays, (b) and (c), which are guided by the fiber but which propagate at different angles with respect to the fiber axis. Simple geometrical considerations (see, e.g., Okashi, 1982, Section 3.2.4) show that the transit time of the guided light beam through a unit length of the fiber is comprised between n_2/c and n_1/c depending on the propagation angle, where c is the speed of light in vacuum. Such a variation in the transit time of the different angular components of a transmitted light pulse causes a broadening of the pulse width detected at the end of the fiber. This phenomenon, called multimode dispersion, limits the rate of the information which can be transmitted through the fiber. Optical fibers used for long distance communications are thus chosen with a $(n_1 - n_2)/n_1$ ratio of the order of 1% to minimize multimode dispersion, at the expense of a smaller numerical aperture.

Graded-Index Fibers

The ray propagation behavior shown in Figure 3-63 is typical of step-index fibers, i.e., fibers having a constant value of the refractive index across their core, as shown in Figure 3-64a. Another important class of optical fibers is represented by graded-index fibers, whose refractive index distribution is shown in Figure 3-64b. Such a distribution may be obtained by coextrusion of different doped silica layers followed by interdiffusion of the dopant concentration. If the refractive index distribution within the fiber core is nearly parabolic, the ray propagation within the fiber for a point source at the input end is shown in Figure 3-65. The fiber acts as a series of lenses, refocusing the light beam periodically along the fiber axis. As in an imaging lens, the phase delay of all rays is the same between two points of convergence (the rays propagating near the axis cover a shorter path length than the peripheral rays, but they propagate in a denser medium, so that the optical path difference is the same for all rays). Consequently, graded-index fibers can guide higher aperture light beams than step-index fibers with little multimode dispersion.

Both step-index and graded-index fibers are, however, subject to a second form of dispersion called material dispersion, which is strongest when the light source has a wide spectral bandwidth. In this case, light beam components of different wavelengths travel at different speeds because of the dependence of the refractive index on the wavelength. The result is a time broadening of the transmitted pulses which further limits the information

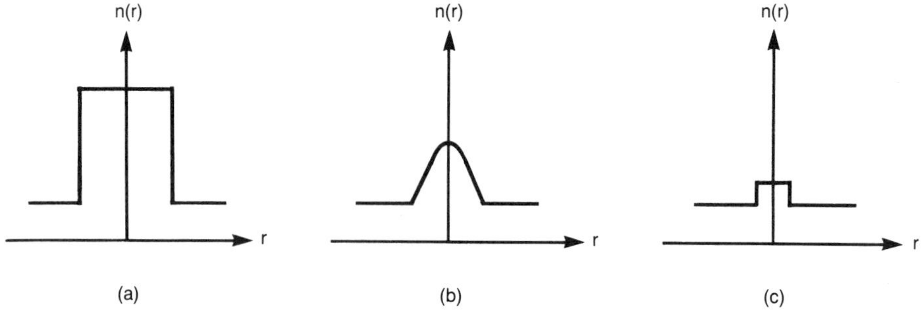

Figure 3-64 Radial refractive index profiles for: (a) a multimode step-index fiber, (b) a graded-index fiber and (c) a single-mode fiber.

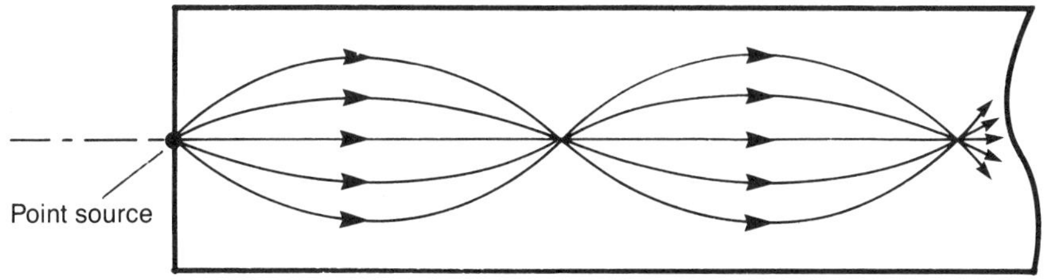

Figure 3-65 Ray propagation in a graded-index fiber.

rate which can be carried through long fiber lengths. Material dispersion is negligible for narrow bandwidth laser sources, such as the single-mode diode laser.

As it was briefly mentioned in Section 3.7.3, the graded-index fiber technology has spurred the development of a new type of optical elements, namely the graded-index (GRIN) lenses. Such lenses have focusing and imaging capabilities similar to those of spherical lenses, but their refractive power is based on their radial index distribution rather than on their shape, as is the case for classical lenses. GRIN lenses are produced by slicing a graded-index rod, which is similar to a graded-index fiber but whose diameter is typically in the 2 mm range, at convenient intervals. A large number of lenses can be obtained by a single rod so that such lenses can be mass produced at potentially low prices.

Figure 3-66 shows a typical application of such components. A quarter-pitch GRIN lens is used to collimate the output beam from a fiber (a full pitch corresponds to the length along a graded-index rod required to bring a meridional ray back to the same radial position and angular orientation). One of the nice features about such lenses is that, in contrast to classical lenses, their end surfaces may be epoxy-bonded to the fiber without affecting their focusing properties. Robust and compact (a quarter-pitch lens is typically 5 mm long) assemblies of the kind shown in Figure 3-66 can be obtained

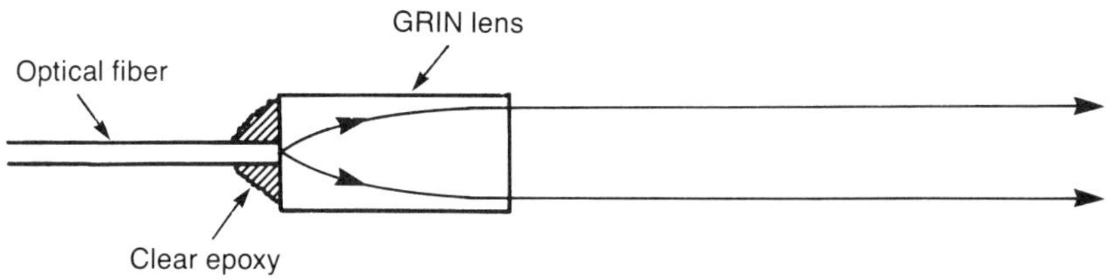

Figure 3-66 A quarter-pitch graded-index lens used as a collimator at the output of an optical fiber.

using such lenses. Other applications include their use as fiber-to-fiber splicers and four-port connectors. High quality performance, such as an on-axis resolution of 2 μm, requires, however, that one of the end faces be spherically ground to compensate for the spherical aberration (Yamamoto et al., 1982).

Waveguide Modes in Optical Fibers

Until now we have considered ray propagation properties using the laws of geometrical optics. This is adequate for relatively high aperture, large core fibers, but for thinner fibers the wave theory of electromagnetic optics must be used to describe the waveguide properties.

As within a laser resonator, light beams following different angular paths within either a graded-index (with an extended light source and skew rays) or a step-index fiber superpose giving rise to maxima and minima in the intensitiy distribution, depending on whether the different angular components superpose in phase or out of phase. Although a detailed review of the electromagnetic-mode theory in optical fibers (see, e.g., Keck, 1976 or Okoshi, 1982) is beyond the scope of this book, some qualitative insight can be obtained from the analogy between meridional-ray propagation in a step-index fiber, Figure 3-63, and a flat waveguide. Rays such as (b) and (c) in this figure will be in phase if they are delayed by 0, 1, 2, etc. wavelengths, giving rise to different possible modes of propagation. If the core diameter or the accepted aperture of the fiber are gradually reduced, the path difference between rays covering different angular paths will always be smaller than one wavelength. In this case the only mode guided by the fiber is the zero mode. This is the propagation mode where, in a simplified geometrical-optics visualization, all the rays propagate parallel to the axis. In this case, the fiber is called single mode and exhibits none of the multimode dispersion effects mentioned above. Information rates of several Gbit/s can be obtained with single-mode fibers fed by single-mode diode lasers. Moreover a single -mode fiber transmits a laser beam without any loss of its coherence properties, an essential requirement in fiber optic sensor applications.

A more formal treatment of the diffraction theory within cylindrical waveguides results in mode diagrams such as the one shown in Figure 3-67 (Keck, 1976). This plot shows the normalized propagation constant for the first modes in a step-index fiber versus the parameter given in the abscissa. Such a parameter is proportional to the product of the core radius a and the numerical aperture, as given by equation (3-21), and inversely proportional to the wavelength. Below the cutoff value of 2.4 for this parameter, only the mode HE_{11} can be guided by the fiber. Single-mode operation thus requires

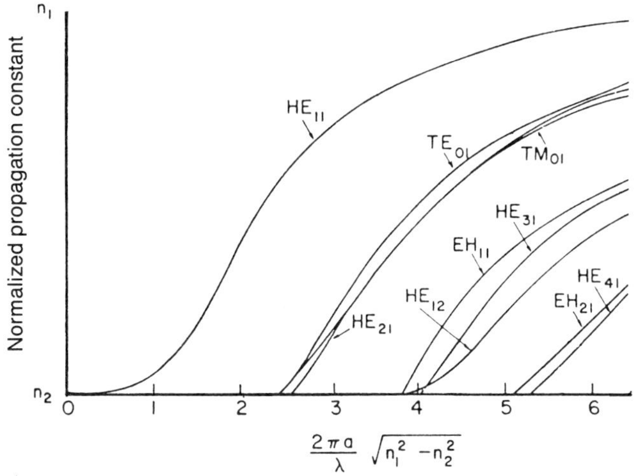

Figure 3-67 Effective refractive index of the first modes guided in a step-index fiber versus the normalized product of the core radius a and the angular aperture of the fiber.

fibers with very narrow apertures or very thin cores, as mentioned above. If $n_1 = 1.453$ and $n_2 = 1.45$ for a doped-core silica fiber and $\lambda = 0.85\ \mu$m, the cutoff condition requires a core diameter smaller than 7 μm. Single-mode fibers thus have tiny cores, as Figure 3-64c indicates, and for this reason suffer higher coupling losses than multimode fibers.

When a guided mode approaches cutoff, much of the light intensity guided by the fiber is contained in the cladding rather than the core (evanescent wave). A relatively thick cladding is thus required to avoid an excessive attenuation of the evanescent wave. This phenomenon is related to the finite penetration of the totally reflected beam into the lower index medium, which may reach several wavelengths of depth at angles of incidence approaching the critical angle (see, e.g., Cielo and Delisle, 1975a). Such a penetration, and thus the phase delay of the totally reflected beam, is different for beams with the electric field polarized either in a direction parallel or perpendicular to the core-cladding interface. This explains why the dispersion curves of oppositely polarized modes, such as the transverse elèctric (TE) and the hybrid electric-magnetic (HE) modes, are different.

Light Attenuation in Fibers

While the attentuation of the early fibers, as well as of present-day low quality plastics or glass fibers, is given in terms of dB/m, the attentuation of communication-quality fibers can now be as low as a fraction of a dB/Km.

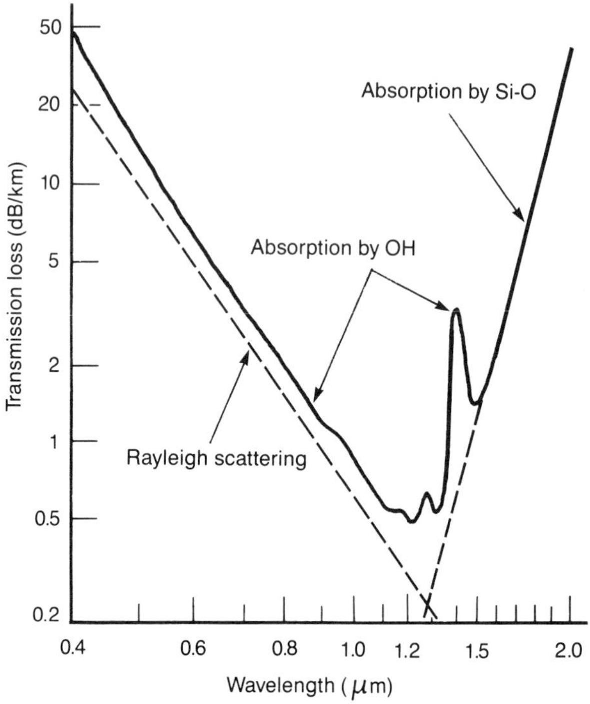

Figure 3-68 Typical transmission loss curve for a fused-silica fiber.

Figure 3-68 shows a typical plot of the light beam attenuation through a fused-silica fiber as a function of the wavelength (Okoshi, 1982). In the low wavelength region, the attenuation is mainly due to Rayleigh scattering by density fluctuations within the fiber material, which is proportional to λ^{-4}. In the spectral region of lowest attenuation, 0.8 to 1.5 μm, the transmissivity is mainly limited by residual OH impurities which are difficult to eliminate completely by dehydration. After 1.5 μm the absorption by the Si-O link in the silica molecule predominates. These characteristics of the loss curves are spurring the development of optical sources and detectors in the 1.2 to 1.5 μm region.

The attenuation diagram suggests that higher transmissivity may possibly be achieved using fiber materials which do not present strong absorption at larger wavelengths, where Rayleigh scattering is very low. Figure 3-69 shows some theoretical calculations, based upon the loss data of high purity materials, of what could be the spectral loss curve for medium-IR fibers (Miyashita and Manabe, 1982). Such low values of the projected loss in this

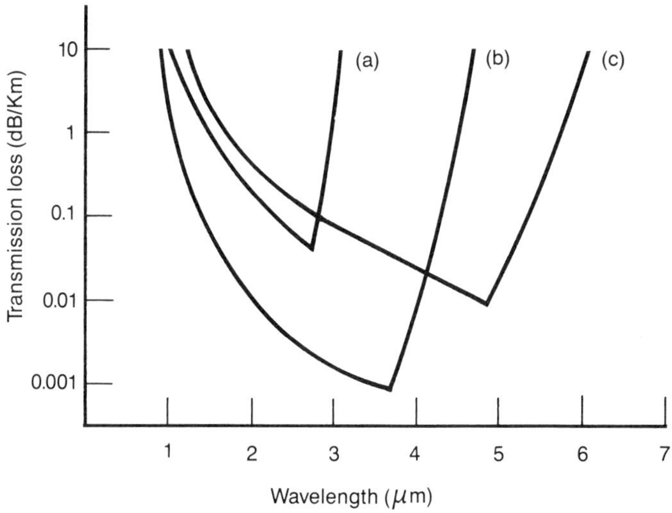

Figure 3-69 Calculated minimum loss spectra which may be achieved with (a) GeO_2 glass, (b) fluoride glass and (c) chalcogenide GeS glass optical fibers.

spectral region are strongly pushing the development of IR fibers (Sigel and Tran, 1984). However, the presently available IR fibers are still limited to a transmission loss of several tens of dB/Km, as Figure 3-70 shows (adapted from Takahashi et al., 1982; Maze et al., 1984 and Arai et al., 1985). Another possible fiber material for long wavelength transmission (Devyatykh, 1984) is polycrystalline KRS-5 (see Figure 3-48). An optical fiber with low loss at the CO_2-laser wavelength of 10.6 μm using this material was announced some years ago by the Japanese company Horiba but was subsequently withdrawn apparently because of material stability problems.

The light attenuation mechanisms described above are produced by material absorption and scattering within the fiber core. Other sources of propagation loss may be present in actual fibers. Attenuation of the evanescent wave within the cladding may be an important loss component in low price plastics-cladded fibers. Radiation loss due to bending may appear when the radius of curvature of the fiber becomes smaller than a critical value. The reason for such losses may be simply understood by ray tracing in multimode step-index fibers as shown in Figure 3-71 (Kapany, 1967): A ray which is reflected near the critical angle at the core-cladding interface in a flat waveguide may be incident at an angle smaller than the critical angle and thus coupled into the cladding if the waveguide is bent. A more general argument is the following (Miller et al., 1973): Part of the light intensity I

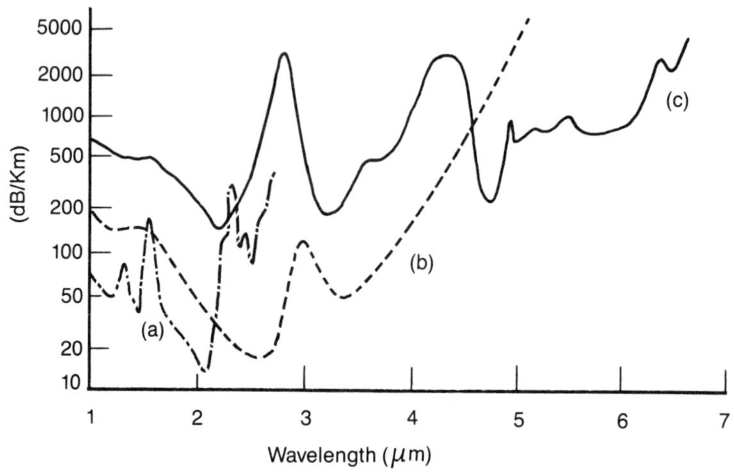

Figure 3-70 Experimental transmission loss curves for presently available medium-IR fibers made of (a) GeO_2 glass, (b) fluoride glass and (c) chalcogenide As_2S_3 glass.

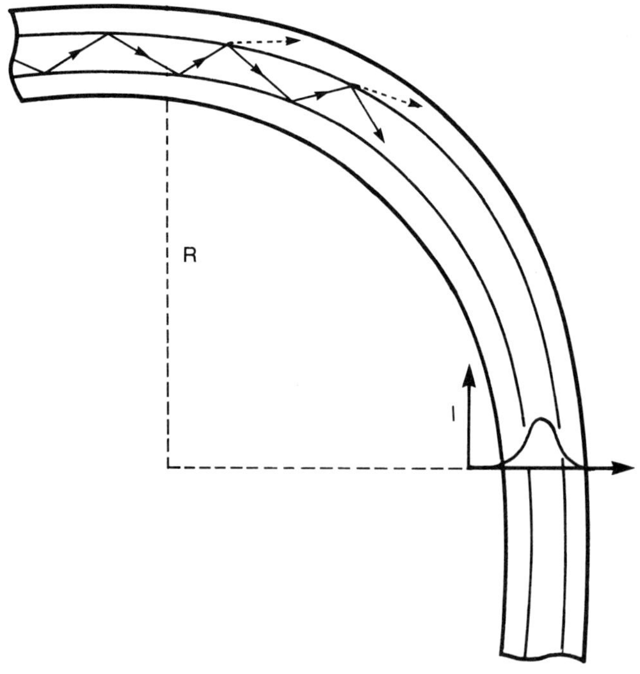

Figure 3-71 Radiation loss due to fiber curvature.

extends within the cladding, as Figure 3-71 shows; if the fiber is curved, the speed of the evanescent wave in the cladding must be larger than the speed c/n_1 along the fiber axis; if such a speed exceeds the velocity of light in the cladding c/n_2, which is physically impossible, radiation loss will appear. Simple geometry thus shows that the maximum allowable radius of curvature R of the fiber before radiation loss appears must satisfy the equation $R/R + r = n_2/n_1$, where r is the depth penetration of the evanescent wave, of the order of the cladded fiber radius for weakly coupled modes. For $r = 50\ \mu m$, we obtain $R = 5$ mm for $n_2/n_1 = 0.99$ and $R = 5$ cm for $n_2/n_1 = 0.999$. The allowed fiber bending radius is thus larger for small aperture fibers.

3.10 Speckle Under Coherent Illumination

Speckle effects are an aspect of optical technology with which the optical inspection specialist must be well acquainted. Speckle appears in a wide range of optical inspection applications when a monochromatic source such as a laser beam is used for the illumination. It may be observed across a small aperture camera image, on the light scatter pattern of a flying-spot scanner, over the output beam of a multimode fiber, etc. This section provides an introduction to the physical mechanisms which are responsible for the formation of speckle, as well as to its practical significance in terms of deterioration of the detected optical signal. Useful applications of speckle for surface inspection or three-dimensional imaging will be described in later chapters.

Intensity Distribution of Scattered Coherent Light

Figure 3-72 shows a typical configuration for the observation of the speckle pattern across a scattered light beam. A narrow laser beam illuminates a spot of diameter D on the rough surface of a reflecting object. The surface must be rough in optical terms, i.e., with random height variations whose scale is greater than a light wavelength. This encompasses most surfaces other than those which are polished to optical quality. The scattered light distribution can be observed on a viewing plane, such as a sheet of white paper, situated at a distance l from the illuminated spot. The light distribution pattern over such a plane is of the kind shown in Figure 3-73, showing highly contrasted clear and dark areas ("speckles") whose average diameter increases

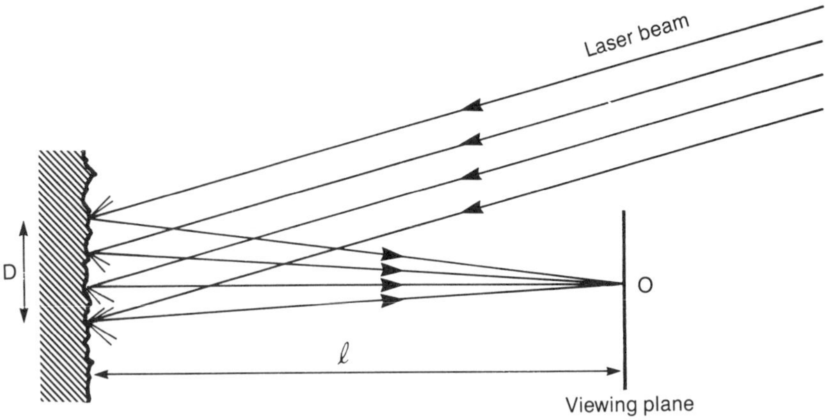

Figure 3-72 Speckle formation in the light scattered by a rough surface illuminated by a laser beam.

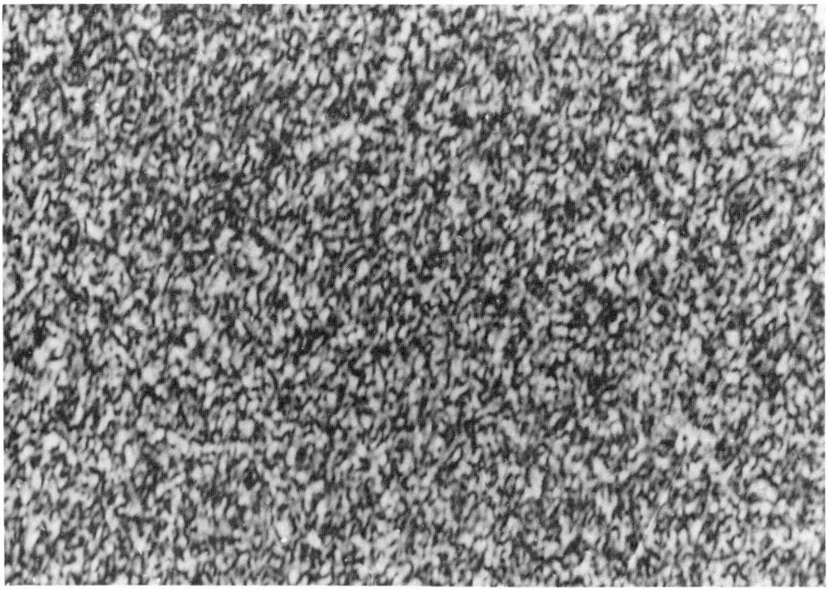

Figure 3-73 Typical laser speckle pattern.

if the ratio l/D is increased. If the illuminated surface is moved transversally by a distance larger than the spot diameter D, the new speckle pattern will be statistically similar, but totally uncorrelated, to the previously obtained pattern.

Such a phenomenon may be understood by considering the different scattered light components received at a given point O of the viewing plane, as shown in Figure 3-72. If the illuminating beam is coherent and uniquely polarized, all the scattered light rays reaching O will superpose coherently. The resulting intensity at O depends on both the amplitude and the relative phase of each component.

The superposition may be visualized by a phasor diagram of the kind shown in Figure 3-73. Each light ray reaching the point 0 can be represented by a vector whose length is proportional to the amplitude of the received light and whose angle with respect to the positive real axis corresponds to its relative phase (see, e.g., Meyer-Arendt, 1972, p. 170). The phase of each light component depends on the height of the scattering point on the rough surface, which is randomly distributed. Consequently, the angles of the component vectors in Figure 3-74 are uniformly distributed over all values between 0° and 360°. The amplitude of the resultant sum vector, which is equivalent to the distance from the starting point in the two-dimensional random walk problem (Strutt, 1919), corresponds to the amplitude of the resultant light illumination at the point O.

A statistical analysis of this problem (Goodman, 1975) yields an expression for the probability density function of the light intensity at the observation

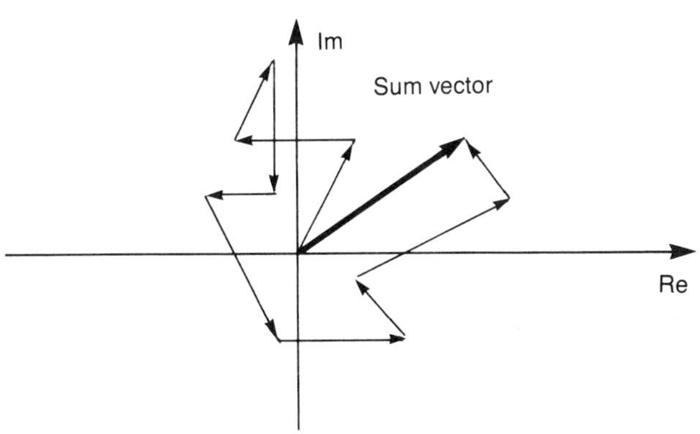

Figure 3-74 Phasor diagram showing the coherent superposition of different light components to obtain the resultant light amplitude.

point 0 of the form

$$P(I) = \exp\left(\frac{-I}{I_o}\right), \tag{3-22}$$

where $P(I)$ is proportional to the probability of obtaining a value I of the light intensity at the observation point and I_o is the average intensity at this point as could be obtained by rapidly displacing the rough surface in the transversal direction.

Curve (a) of Figure 3-75 is a plot of expression (3-22). The exponential shape of such a distribution implies that dark areas have a higher probability of occurrence, but intensities higher than twice the average intensity can still be found with significant probability. This explains why a speckle pattern has such a high contrast. A quantitative measure of the constrast of speckle is the ratio σ/I_o of the standard deviation σ and of the average value I_o of the light intensity. As the standard deviation of the right term of (3-22) is $\sigma = I_0$, the speckle contrast is equal to 1 in this case.

The speckle analysis presented above assumes that the scattered beam is coherent and uniquely polarized, as is the case if a polarized laser beam is scattered by a clean metallic surface. If the laser beam is unpolarized, or if the reflecting object (such as a sheet of paper) depolarizes the scattered light,

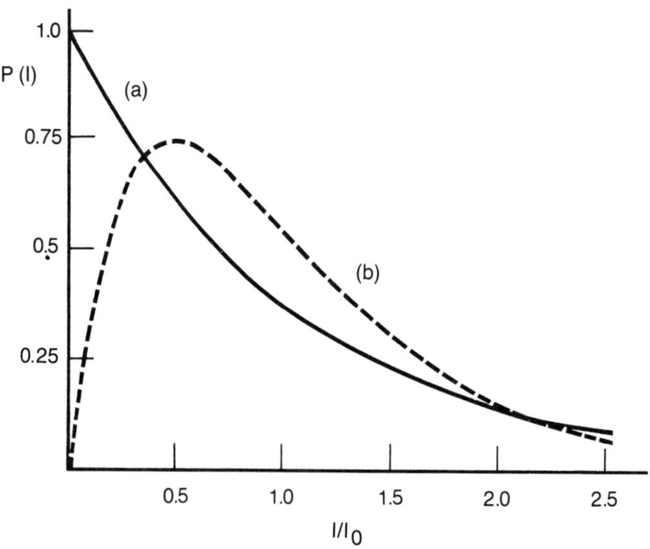

Figure 3-75 Probability distribution of the light intensity at a given point of a speckle pattern produced by (a) a polarized or (b) an unpolarized scattered beam.

the two uncorrelated polarization components of the scattered light field add incoherently. This means that the two speckle patterns corresponding to each polarization component add in intensity rather than in amplitude (see, e.g, Born and Wolf, 1970, p. 502). The result is the probability distribution given by curve (b) of Figure 3-75 (Ennos, 1975). This probability curve is less dispersed than curve (a), resulting in a slightly smaller contrast of the speckle pattern.

If the light beams scattered by the different portions of the illuminated surface in Figure 3-72 are not fully coherent, the respective speckle patterns superpose in intensity, further reducing the speckle contrast. This may occur if the beam is not spatially coherent over the whole illuminated area, or if its temporal coherence length is shorter than the path difference between the light rays reflected by the different scattering centers over the surface. In the limit of a white light beam illuminating a surface of average roughness much larger than the nearly 1 μm-long coherence length of the source, the light components scattered by the peaks and valleys of the rough surface will be totally incoherent. The resulting probability distribution curve will be strongly peaked around the average I_o value, with a small standard deviation and thus a small speckle contrast.

Size Distribution of the Speckle Grains

Speckle patterns of the kind shown in Figure 3-73 comprise a number of characteristic "grains". It was mentioned at the beginning of this section that the average grain size of the speckle pattern is proportional to the ratio l/D. Simple interferometric arguments can be used to explain the physical basis of such a relation (Vest, 1979, p. 33–36). Consider two scattering points within the illuminated surface in Figure 3-72. The two overlapping spherical waves radiating from such points produce quasi-rectilinear interference fringes on the viewing plane, as in the case of the well-known Young's double hole experiment. The spacing between successive interference maxima decreases as the distance between the two source points increases. The minimum fringe spacing is obtained when the two points are separated by the maximum distance D and is given in the paraxial approximation (see equation 3-16 in Section 3.7.3) by the expression

$$d = \lambda l/D. \tag{3-23}$$

As all the other pairs of points produce interference fringes with larger spatial periods, the superposition of all such fringe systems cannot give speckle grains of size smaller than the value d given by equation (3-23). This is confirmed by more involved computations resulting in a spatial frequency distribution across the speckle pattern of the kind shown in Figure 3-76 (Stetson, 1975). This figure shows the envelope of the two-dimensional frequency spectrum plotted versus the normalized radial frequency ν/ν_c. The cutoff frequency ν_c has a value very near to the inverse of the minimum fringe spacing given by (3-23). If we take a value of the order of $\nu_c/2$ as the average speckle spatial frequency, as well as a typical half-width grain size equal to half of the average spatial period, the expression (3-23) can be taken as the typical diameter of a speckle grain.

Speckle in Imaging Systems

The previous analysis refers to the reflective configuration shown in Figure 3-72. Similar results would be obtained in a transmissive configuration where the reflective surface is replaced by a ground glass plate back-illuminated by a laser beam of diameter D. A practical situation where such a speckle pattern may be observed in an industrial inspection application is when a laser projection device contains a window whose outer surface becomes dirty with grease or other light-scattering substances. This may occur rather

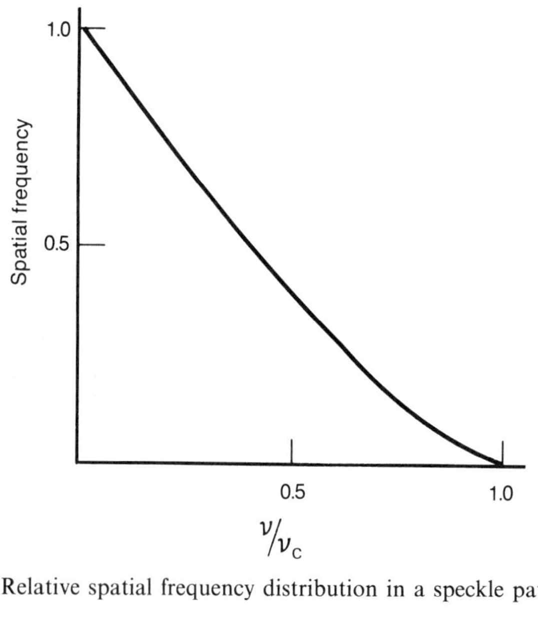

Figure 3-76 Relative spatial frequency distribution in a speckle pattern for circular apertures.

frequently when an optical inspection device is mounted on the production line.

Another configuration where a speckle pattern can be observed in transmission is shown in Figure 3-77. In this case a rough object, coherently illuminated either in reflection or in transmission, is imaged by a lens with entrance pupil of diameter D on the viewing plane. Each point O on this plane receives the light components originating from the area surrounding point P on the object surface, the surface of such area being equal to the point spread function of the lens. If the roughness amplitude of the object surface within such area is larger than the light wavelength, the scattered light beam emerging from this area will produce a random light distribution on the plane of the lens. This light distribution is equivalent to the light waveform emerging from the illuminated area and directed toward point O in Figure 3-72. The speckle size in the imaging plane of a lens is thus given by the same expression (3-23), where D and l are shown in Figure 3-77.

To gain an appreciation of the magnitudes involved, let us consider a typical camera focused at a distant object, so that $l \simeq f$, the focal length of the camera. If $\lambda = 1\ \mu\text{m}$, the speckle size (in μm) is of the order of the F-number f/D of the camera lens. Typical lens apertures go from $D = f/1.4$ to $D = f/32$, so that the speckle size goes from 1.4 μm to 32 μm depending on the camera aperture setting. We can conclude that the speckle grains in an image are extremely small except for small aperture lenses. Similar considerations apply to the speckle pattern which one may see when looking directly at a laser illuminated rough-surface object. The very small speckles that one can see in this case are produced by the lens of the eye on the retina. If one looks at the

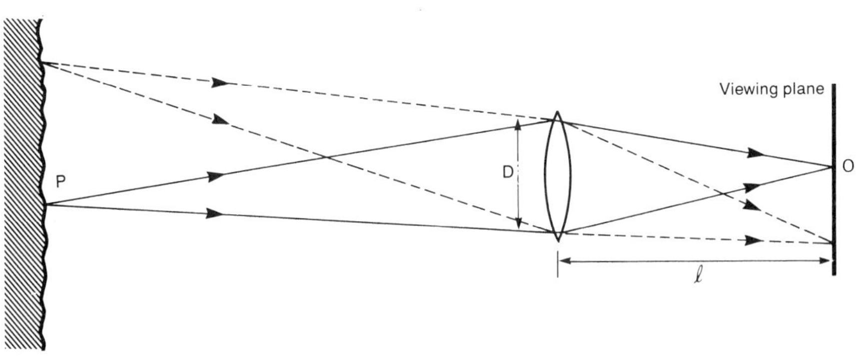

Figure 3-77 Speckle formation in the image of a coherently illuminated rough object.

same object through a small pinhole, thus reducing the entrance pupil, the speckle size appears much larger.

Speckle Noise in Optical Inspection Systems

Speckle is a major source of noise in coherently illuminated optical inspection systems. As noted in Section 3.1.1, the advantages of coherent sources must be weighed against their higher noise level when choosing between a coherent and an incoherent light source.

The speckle noise amplitude may be evaluated with reference to Figure 3-72. The configuration illustrated by this figure is typical of a flying-spot inspection system where a narrow laser beam is scanned over the object whose surface reflectivity is evaluated by sensing the scattered light intensity with an optical detector situated at point O. The laser beam is made as narrow as possible to increase the spatial resolution over the scanned surface. However, this results in greater speckle noise as explained below.

If the diameter of the detector sensitive area is equal to a_d and the average center-to-center distance between the speckle grains is equal to $2d$, the average number of grains collected by the detector is, from (3-23)

$$N \simeq \left(\frac{a_d}{2d}\right)^2 \simeq \left(\frac{a_d D}{2\lambda l}\right)^2. \tag{3-24}$$

The intensity of each speckle grain falls on a different position on the detector sensitive area so that the resultant photoelectric current is simply proportional to the sum of the intensity of each speckle grain whose probability distribution is given by (3-22). It is well known from statistics (see, e.g., Davenport, 1970, Chapter 12) that the sum of N mutually independent variables of same average I_o and standard deviation σ is a variable with average NI_o and standard deviation $\sigma N^{1/2}$. If we recall that the speckle intensity distribution, given by (3-22) in the case of a fully-coherent and polarized beam illumination, has a standard deviation $\sigma = I_o$, we obtain for the photoelectric signal an average value CNI_o and a standard deviation $CI_oN^{1/2}$ where C is a calibration constant. The standard deviation corresponds to the speckle-related noise, so that the signal-to-noise ratio related to speckle is, from (3-24)

$$\mathrm{SNR} = N^{1/2} \simeq \frac{a_d D}{2\lambda l}. \tag{3-25}$$

It can be seen that speckle noise can be reduced by increasing the diameter of the illuminated area D, at the expense of spatial resolution. Another approach for increasing the SNR for a given value of λ is to increase the viewing aperture a_d/l of the detector. This can be done by increasing the detector sensitive area or by using a large aperture collecting lens or ground glass in front of the detector.

Chapter 4

Surface Inspection

Optical inspection techniques may be broadly classified in two categories, those which provide information on the morphology of the inspected objects and those analyzing properties other than the materials shape.

Examples of the first group are the monitoring of surface finish, three-dimensional surface profiling, on-line wire diameter gaging or object-shape recognition for robot guiding. Such techniques are probably the ones which are most commonly associated with the idea of optical inspection, because they closely correspond to visual inspection activities performed by human inspectors. Optical inspection devices make it possible to obtain such visual information in hard-to-access or hostile environments, such as close to an arc welding head, inside furnaces or near hot-rolling mills using rugged optical probes such as fiber optic bundles. The main reason for the fast growth pace of this instrumentation field is related to the accelerating industrial trend toward factory automation, requiring powerful and flexible sensors as can be provided by modern optical technology.

The second category of optical inspection techniques encompasses a wide range of fields which extend much farther than human vision capabilities. Examples are the detection of overheated engine components by infrared thermography, monitoring of moisture or of the degree of curing in polymeric materials by reflection spectroscopy, material characterization by laser-induced breakdown analysis, coating evaluation by polarized ellipsometry, holographic detection of microdisplacements, pressure and temperature monitoring by fiber optic sensors, and subsurface probing by laser generation and detection of ultrasonic pulses.

Optical techniques for morphological inspection in the industrial environment are reviewed in the next few chapters of this book. The later chapters will review nonmorphological applications of modern optical inspection technology.

This chapter is devoted to optical techniques for monitoring random surface morphology features, such as roughness monitoring and the detection of scratches or other surface defects. Deterministic object shape metrology techniques will be described in the next chapter.

4.1 Random Surface Profilometry

The first two sections of this chapter deal with methods for surface roughness characterization. Surface finish is an obvious quality factor for many consumer-oriented manufactured products. Moreover, surface texture may provide relevant information on a wide range of industrial parameters as described in Table 4-1. The present availability of low-cost and reliable optical devices to perform such functions without contact is spurring the development of a number of such devices. Figure 4-1 shows an example of a commercial optical surface finish monitoring device set up for testing a mass-production product.

Surface roughness can be evaluated by two methods: (1) by recording a surface profile along a line on the sample and calculating the surface statistics from such a record or (2) by using a technique which directly produces a statistical parameter averaged over a given area of the inspected surface. This section gives an overview of methods of the first type. Techniques belonging to the second category will be described in Section 4.2.

4.1.1 *Surface Roughness Statistics*

Engineering surfaces are produced by a wide variety of processes and their structures may be highly varied and complex. Figure 4-2 shows micrographs of metallic surfaces after the operations of grinding, turning or cold rolling. Milled surfaces have a strong and regular lay pattern along the machining direction, while rolled or sand-blasted surfaces have a random isotropic texture. To quantify the average properties of surface topography, a number

Table 4-1 Applications of Roughness Monitoring to Industrial Controls

Industrial Process	Parameter of interest	References
Tool machining	Degree of tool vibration	Takeyama et al., 1976 Mitsui and Sato, 1978 Inasaki, 1985
Friction and wear	Friction evaluation	Bassani et al., 1984
Road surface evaluation	Skid characteristics	Cantor, 1979
Shot blasting	Degree of material removal	Mansour & McGoldrick, 1985
Paper making	Paper sheet printability	Lucas et al., 1978 Richter et al., 1983
Extrusion of reinforced plastics	Dispersion and orientation of charge	Favis et al., 1984
Aluminum sheet rolling	Surface quality	Fairlie et al., 1986
Cold rolling of steel sheets	Cold workability, luster, surface coating effectiveness	Konczakowski et al., 1983, Asano, 1985
Grinding, turning, calendering, rolling	Surface quality	Brodmann, 1984, Brodmann et al., 1985

of statistical functions and parameters have been developed (Vorburger and Teague, 1981). These characterize the two basic aspects of topography, the heights of the asperities (amplitude) and the longitudinal spacings beween the asperities (spatial period). Because surface profile samples are normally taken along a line, such parameters are defined in terms of a two-dimensional surface profile $y(x)$, although three-dimensional expressions can also be used.

A typical surface profile is shown in Figure 4-3. The function $y(x)$ is defined as the difference between the surface height and the mean surface level over the sampled line. The most often used roughness parameter (see American National Standards Institute B46.1-1978, *Surface Texture*, Appendix C.3, ASME, New York, 1978, as well as the ISO Standard No ISO/DIS 4287/1.2, *Surface roughness, terminology*, 1984) is the roughness average R_a defined as

$$R_a = L^{-1} \int_0^L |y(x)|\, dx, \tag{4-1}$$

where L is the sampling length. A similar parameter is the rms roughness

Figure 4-1 Optical surface roughness measurement of ball pins. (Courtesy of Rodenstock, Munich.)

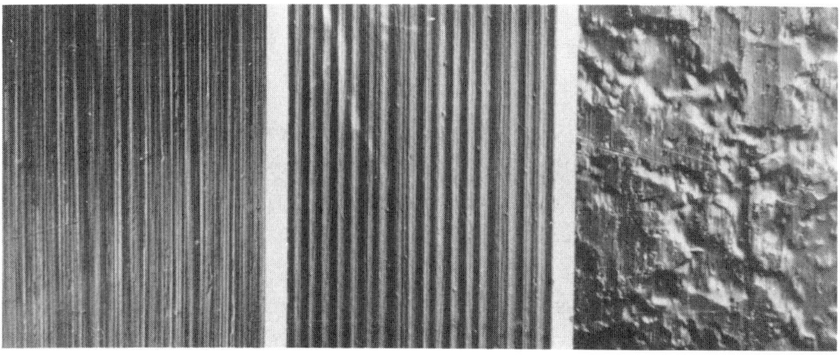

Figure 4-2 Examples of metallic surface micrographs after the operations of cold rolling (right), turning (center) and grinding (left). (Courtesy of Rodenstock, Munich.)

R_q which is defined as the standard deviation of the surface profile along the sampled line L

$$R_q = \left(L^{-1} \int_0^L y^2(x)\, dx \right)^{1/2} \tag{4-2}$$

It should be noted that Figure 4-3 shows the ideal case of a "flat" surface

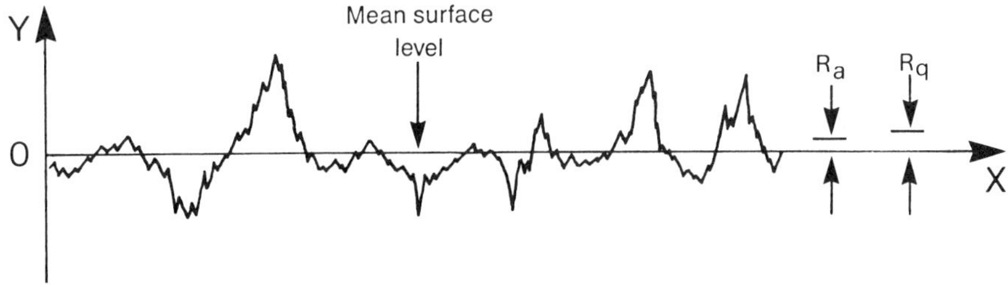

Figure 4-3 Schematic surface profile $y(x)$ corresponding to the departure from the mean surface level. The values of the two roughness parameters, R_a and R_q, are indicated.

for which the values of $y(x)$ are closely distributed around the mean surface level over the whole sampled distance. If the surface profile of a curved or tilted surface is taken, the "local" average roughness may be considerably smaller than the R_a value calculated from (4-1) as Figure 4-4 shows. To avoid such anomalies, only the local value of the mean surface level is usually considered by profiling instruments in order to obtain the actual (local) roughness. This is normally achieved by AC filtering the output of the profiling instrument. Such a procedure implies, however, that the measured R_a or R_q values depend on the chosen value of the lower cutoff frequency.

Both the R_a and R_q parameters provide a measurement of the roughness amplitude. Information on the spatial frequency f or periodicity can be obtained from the power spectral density function

$$Y^2(f) = L^{-1}\left|\int_0^L y(x)\exp(-i2\pi f)\,dx\right|^2. \tag{4-3}$$

If the surface profile has a marked periodicity, such as in the case of turned surfaces, the power spectrum will present strong components at the fundamental spatial frequency (inverse of the spatial period) and its harmonics.

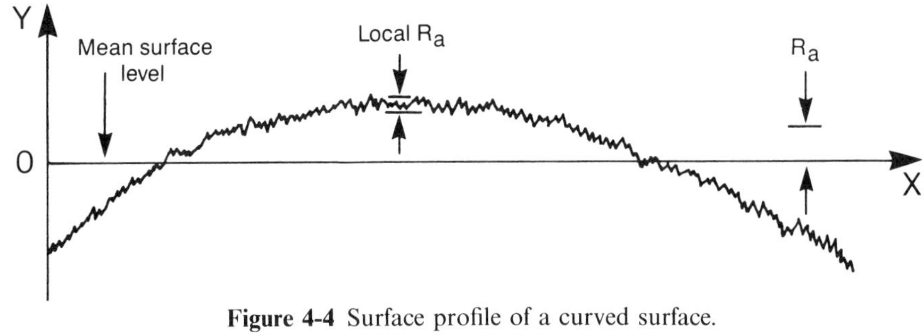

Figure 4-4 Surface profile of a curved surface.

For a nonperiodic surface the average spatial freqency

$$\frac{\int_0^\infty fY^2(f)\,df}{\int_0^\infty Y^2(f)\,df} \tag{4-4}$$

gives an order of value of the typical inverse spatial period on the inspected surface independently of the roughness amplitude.

The autocorrelation function of the surface profile

$$y_a(\tau) = L^{-1}\int_0^L y(x)y(x+\tau)\,dx \tag{4-5}$$

is a plot of how similar the surface profile is to itself when it is shifted by a distance τ. The autocorrelation function contains information similar to that provided by the power spectral density. If the surface profile is periodic, the autocorrelation function is also periodic with the same period. For a random surface profile, the autocorrelation length (length τ at which the autocorrelation function drops to a certain fraction, typically 10%, of its initial value) constitutes an estimate of the typical spatial period.

Finally, the average slope

$$\mathrm{S_a} = L^{-1}\int_0^L \left|\frac{dy}{dx}\right| dx \tag{4-6}$$

and the rms slope

$$\mathrm{S_q} = \left(L^{-1}\int_0^L \left(\frac{dy}{dx}\right)^2 dx\right)^{1/2} \tag{4-7}$$

are roughness parameters which are important mainly because of their relation with angular reflectance data. Such parameters are dependent on both the amplitude and spatial period properties of the surface. The value of these parameters heavily depends on the high spatial frequency cutoff of the measuring instrument, which is determined by the stylus radius in a mechanical profilometer or by the light wavelength in a light-scattering measurement.

4.1.2 *Tactile Profilometry*

Although other contact methods such as thermal or pneumatic techniques can be applied to surface roughness evaluation (Powell, 1969; Thomas, 1978;

Reason, 1979), stylus instruments are the standard method for precise surface profilometry (Vorburger and Teague, 1981; Bennet, 1985). The diamond stylus, whose tip radius typically measures some micrometers, is slowly trailed along the surface to be inspected while its vertical motion is sensed by an electromechanical pickup transducer. This method is not completely non-destructive, because a small groove mark is often left along the scanned line (Bennet and Dancy, 1981).

The precision which can be obtained with a mechanical profilometer is quite high, as Figure 4-5 shows (Bennet and Dancy, 1981). The two traces correspond to the $y(x)$ profile obtained along the surface of the same sample, with different speed and range settings. The profilometer has built-in software to calculate the R_a parameter using equation (4-1). In this case, the R_a-values are of the order of 1 nm, showing the excellent sensitivity of this instrument. The height resolution, which is mainly determined by the noise level of the electromagnetic pickup transducer, may reach a fraction of a nanometer. As the stylus displacement represents a convolution of the tip geometry and the surface profile, the lateral resolution is limited by the stylus

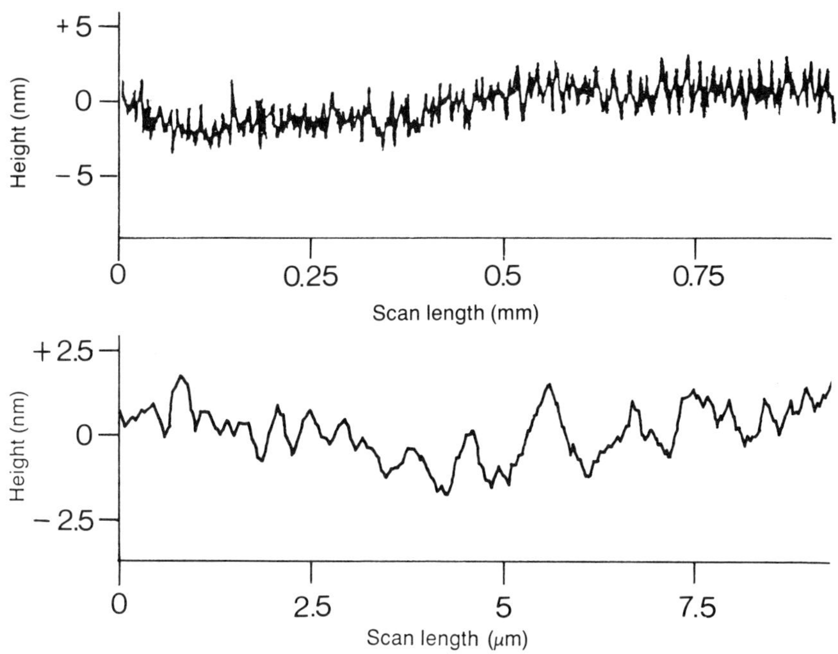

Figure 4-5 Surface profile traces obtained with a high-precision mechanical profilometer.

radius. A tip radius of 1 μm provides a lateral resolution of the order of 0.1 μm. Such performances are yet unmatched by optical profilometers (Bennet, 1985).

The limitations of mechanical profilometers for industrial inspection are their low speed and need for contact with the inspected surface. Although a stylus instrument is perfectly adequate for precise laboratory inspection of surface samples, it is certainly not applicable to on-line inspection of metal sheets during rolling or to awkwardly shaped machined components.

4.1.3 *Interferometric Optical Profilometry*

Different optical interferometers (Françon, 1966; Steel, 1967; Tolanski, 1973; Malacara, 1978) have long been used for the inspection of smooth surfaces, particularly for testing optical components. Figure 4-6 shows a classical interferometer system for the inspection of a flat surface. The tested surface and the reference flat are placed in the two arms of the interferometer so that their images are superposed on the observation plane. If the path difference along the two arms is nearly equal and the two surfaces are slightly tilted, a set of parallel fringes can be observed. Deformation of the fringe pattern can be related to surface microstructure. Micro-interferometers are

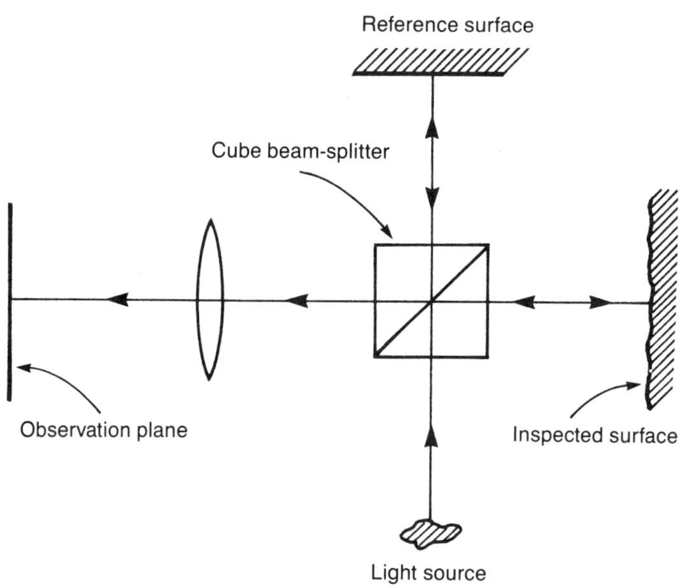

Figure 4-6 Typical interferometer setup for testing an optical flat.

available for close inspection of small areas (Benford and Rosenberger, 1967). Curved surfaces can be tested by comparison with a similarly curved reference.

Such a configuration is of limited value for the inspection of surfaces other than optical components. Visual observation of the fringe pattern results in a height resolution of the order of a quarter of a wavelength. Better resolution can, however, be obtained by fringe-scanning techniques (Bruning, 1978). As to the maximum height excursion, the vertical range is limited to a few wavelengths for nonpolished surfaces having a wide surface slope distribution. If the images of two such surfaces are superposed with a path difference of several wavelengths, the interference fringes tend to merge into one another making it impossible with conventional interferometry to track one fringe at a time (Vorburger and Teague, 1981). Long-wavelength light sources such as a CO_2 laser must be used to inspect rough surfaces with R_q-values of several micrometers (Kwon et al., 1980). Double-arm interferometers can be used for surface profilometry (Evtikhiev et al., 1977), but they are relatively sensitive to ambient vibrations, making shear interferometers such as the Nomarski interference microscope (Makosh and Solf, 1981; Eastman and Zavislan, 1983; Church and Takacs, 1985) easier to use. However, such an instrument can provide only qualitative information about the surface profile. Another problem of simple fringe-scanning interferometers is that the fringe intensity pattern cannot be unambiguously converted into a profile curve unless one can assume that the surface height profile varies monotonically along the scanned line.

A number of modified interferometric configurations have been developed in order to avoid such problems and improve the resolution performance offered by a profiling interferometer (Vorburger and Teague, 1981; Bennet, 1985; Whitehouse, 1985). An optical heterodyne profilometer (Sommargren, 1979) uses two frequency-shifted beams (see Section 3.6.3) to inspect a rotating sample independently of its surface reflectivity. One of the beams is focused on the axis of the rotating sample and acts as the reference point, while the other is focused at a short distance in order to scan a circular path around the reference point. By precisely measuring the phase of the beat signal produced by superposing the two beams, a height sensitivity of nearly 0.1 nm with a lateral resolution of 2 μm can be achieved. A substantially similar instrument using common-mode noise rejection techniques could achieve a 0.01 nm sensitivity (Huang, 1983). A low resolution version (50 nm) of the heterodyne interferometer using a long focal-length lens to focus the probing beam on the surface has also been reported (Lavan et al., 1977). A version using two concentric beams projected on the same surface to minimize mechanical instability effects has recently been described (Pantzer et al., 1986).

Another interferometric profilometer has recently been developed using a Mirau-type interferometer and a linear detector array (Wyant et al., 1984). An instrument of this kind is manufactured by Wyko Optical Inc., Tucson. The basic scheme of such an instrument is shown in Figure 4-7. An incoherent light source is imaged on the test surface through two thin plates. The second plate acts as the beam-splitter, while the first, containing a small aluminized area, acts as the reference surface. Such a compact configuration is less sensitive to ambient mechanical vibrations than the classical two-arm configuration shown in Figure 4-6 (Palmer, 1984). The beams reflected by the reference and by the test surface are superposed on a linear photodetector array. The whole focusing system is mounted on a PZT transducer to produce a slight oscillation which is recorded by the detector array to resolve the height variation ambiguity. A height resolution of 0.1 nm with a lateral

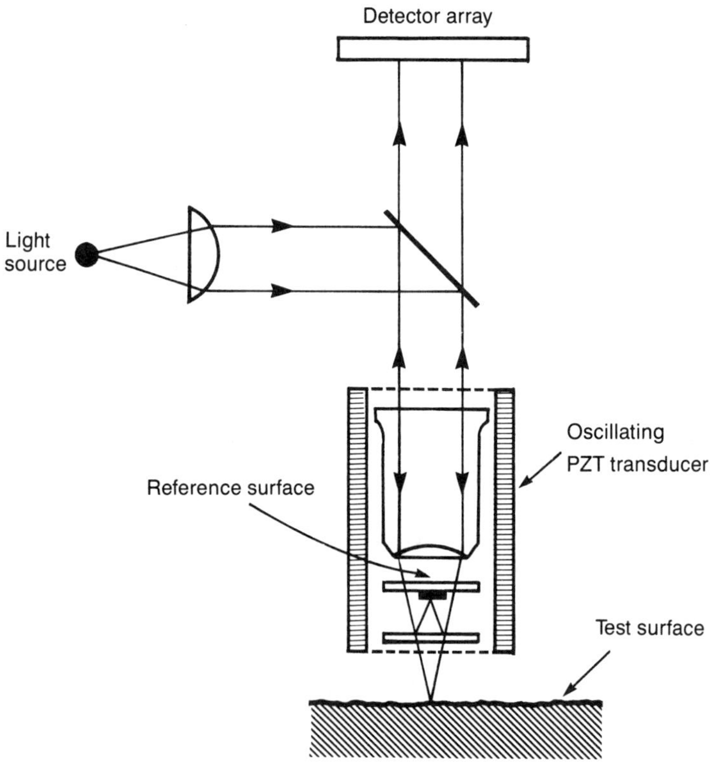

Figure 4-7 Schematic diagram of the Mirau interferometric profilometer.

resolution of a few micrometers can be reached over a 1.3 mm profile length whith the standard optical head. A maximum resolvable surface height of 5 μm can be obtained, while the maximum surface slope is 6.4°. Figure 4-8 shows a typical profile obtained for the surface of a magnetic tape. The maximum data acquisition time is 48 ms.

In conclusion, the resolution of interferometric profilometers is approaching that of mechanical profilometers, while the optical devices are faster and noncontact.

4.1.4 *Geometrical Optics Based Profilometry*

Interferometric profilometers are essentially laboratory devices which, similarly to mechanical profilometers, are typically used to inspect samples for quality control on a statistical basis. Continuous inspection in the production line is better perfomed by light-scattering techniques, which will be reviewed in Section 4.2, or by geometrical optics devices, which are described in this section.

Geometrical optics profilometers are essentially based on triangulation (Keller, 1967), a method which has been used for a long time in land surveillance and navigation (Dickson and Harkness, 1969). The basic principle of triangulation ranging is shown in Figure 4-9. A point A on the inspected surface is imaged on the image planes, p_1 and p_2, of two imaging systems solidly mounted on the same device. The plane p_1 may contain a pinhole acting as the projection light source, or may be the output facet plane of the

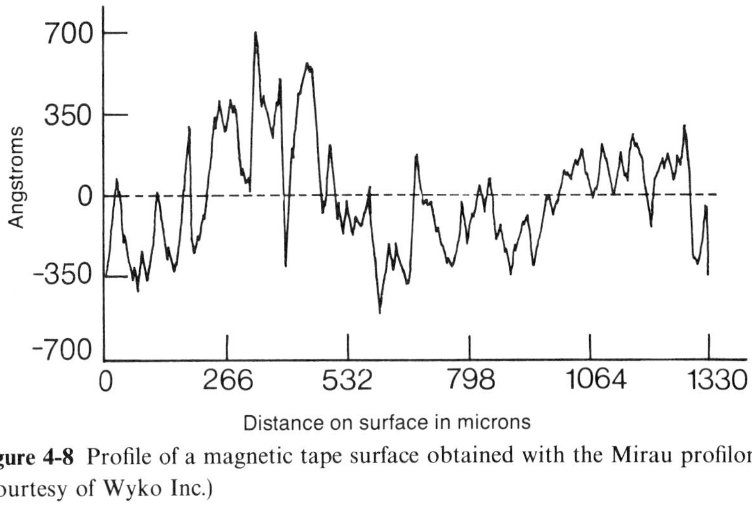

Figure 4-8 Profile of a magnetic tape surface obtained with the Mirau profilometer. (Courtesy of Wyko Inc.)

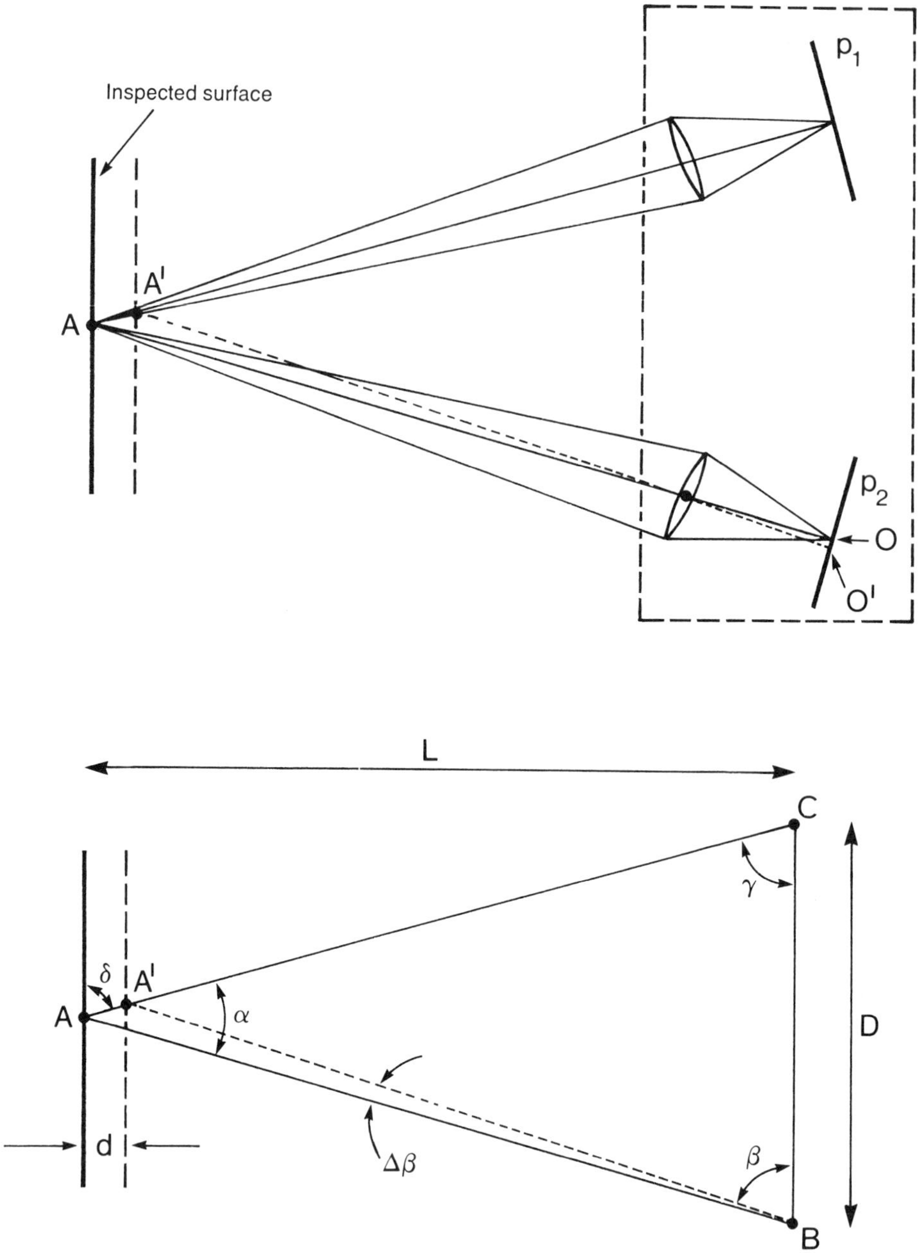

Figure 4-9 Basic principle of optical triangulation ranging. Top: physical system; bottom: geometrical equivalent.

laser diode source, or else may contain a viewing reticle in a passive rangefinding device. Plane p_2 may be a viewing reticle or the image plane of a camera. If the longitudinal position of the imaged point changes from A to A′, its image on p_2 will move from O to O′. By monitoring the position of the imaged point, one can thus determine the position of the surface along the pointing beam axis. If the inspected surface moves laterally with respect to the ranging device or vice versa, a surface profile can be obtained along the scanned direction.

The bottom diagram in Figure 4-9 shows the geometry of a triangulation system, with B and C being the object nodal points of the two lens systems (see Section 3.7). The distance D between the two imaging systems is a known parameter, as is the angle γ. The angle β is measured for a given surface position. The range AC can thus be obtained from the law of sines

$$\mathrm{AC} = \mathrm{BC}\frac{\sin \beta}{\sin \alpha} = \frac{D \sin \beta}{\sin(\pi - \beta - \gamma)}. \tag{4-8}$$

The minimum surface displacement d which can be detected is related to the minimum angle $\Delta\beta$ which can be resolved by the viewing lens. If, as is often the case, $D \ll L$ and $L \gg d$, we can write by differentiation of (4-8), assuming $\sin \beta \simeq 1$ and $\mathrm{BC} \simeq L$,

$$d \simeq \frac{D}{\alpha^2}\Delta\beta \simeq \frac{L}{\alpha}\Delta\beta \tag{4-9}$$

where α is in radians. Equation (4-9) shows that the accuracy of a triangulation measurement increases as the observation angle α increases and the working distance L decreases. As an example, if $D = 10$ cm, $\alpha = 0.1$ radians and the angular resolution is $\Delta\beta = 10^{-4}$ radians, the height resolution is $d \simeq 1$ mm at a working distance $L \simeq 1$ m. Conversely, for a microscopic set up with $D \simeq L \simeq 1$ mm and $\alpha \simeq 1$ radian, we obtain $d \simeq 0.1$ μm. An application of such a method using a microscope mounted camera to view a projected strip at an angle has recently been described for monitoring the roughness of a machined surface with a 1 μm resolution (Mitsui and Sato, 1978; Uchida et al., 1979).

Several modifications of the basic triangulation configuration shown in Figure 4-9 are possible (Whitehouse, 1985). An optical design which was used for optical profilometry of surfaces with R_a roughnesses from less than one micrometer to several tens of micrometers is shown in Figure 4-10 (Arecchi et al., 1979). A laser beam is focused on the inspected surface by a microscope objective lens of numerical aperture equal to 0.85. The reflected beam is

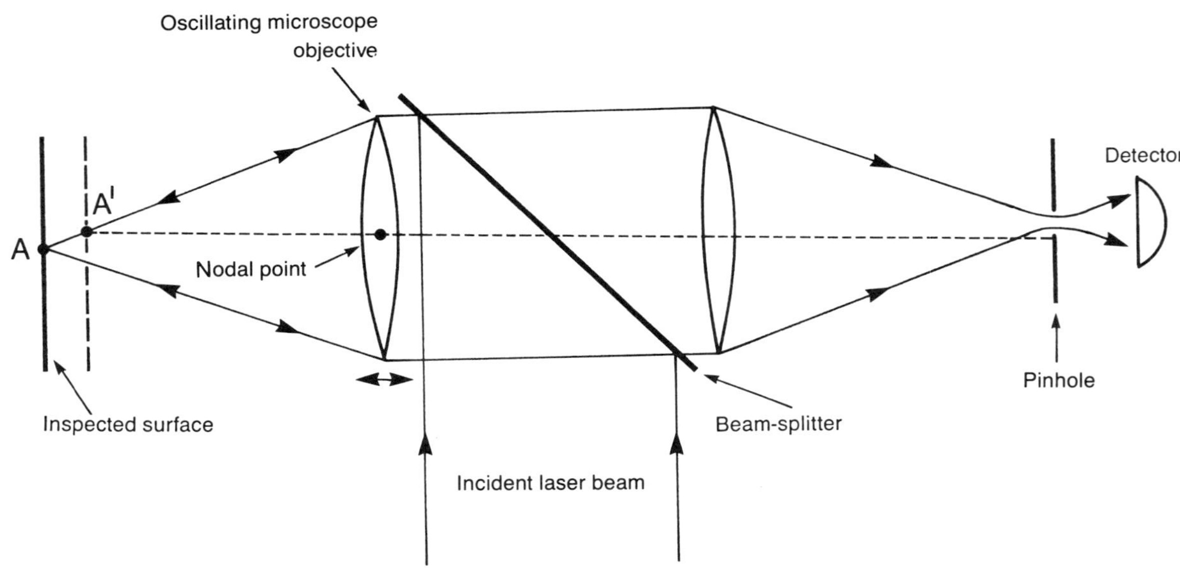

Figure 4-10 Schematic diagram of an image-plane locating optical profilometer system.

focused on a pinhole aperture by a second lens and the beam intensity transmitted through the pinhole is monitored by an optical detector. If the surface height changes, the imaged spot will no longer be well-focused on the pinhole, so that the detected intensity will decrease. By longitudinally oscillating the microscope objective with respect to the surface using a 100 μm-range, 100 Hz piezoceramic transducer, and by monitoring the position of the maximum on the detected signal, the surface height could be evaluated with a precision of 0.1 μm over a 100 μm range. The horizontal resolution is ultimately limited by the nearly 1 μm diameter of the focused spot on the surface but was here determined by the sample stepping distance of 10 μm. The maximum translation speed is of the order of 10 μm $\times$ 100 Hz = 1 mm/s.

If we compare Figure 4-10 with Figure 4-9, we realize that the lens focusing approach is essentially a triangulation method which discriminates the nearly-axial scattered beam against the light beam scattered by peripheral points such as A′ in an out-of-focus configuration. The main difference between these two configurations is that in Figure 4-9 only the peripheral beam is taken into consideration, while in Figure 4-10 the whole cone of light is detected. The central part of the cone of light corresponds to low α values of the observation angle which, as equation (4-9) shows, result in lower measurement accuracy. This part of the beam could thus be eliminated (see, e.g., Idesawa, 1985, or Blais and Rioux, 1986; see also Figure 4-44 on page 245), resulting in a sharper in-focus maximum and a lower shot noise level. Equivalently, one could say that the configuration shown in Figure 4-10 is not optimized because the light intensity transmitted through the pinhole varies as a quadratic function of the surface position. The detected signal level is thus almost constant for small surface displacements within the depth of field. Conversely, if the pinhole is moved out of the image plane, the transmitted intensity becomes a linear function of the inspected surface position. In this case, the signal level is much more sensitive to small surface displacements (Simon, 1970; Hartman et al., 1973; Elliott and Mickelson, 1979).

An approach of this kind has recently been followed by a number of investigators (Sawatari and Zipin, 1979; Fainman et al., 1982; Kleinknecht and Meier, 1983). A typical configuration is shown in Figure 4-11. The reflected beam is now monitored by two detectors, symmetrically located with respect to the image plane. The output from the two detectors is the same for a given surface position (point A in Figure 4-11). If the surface moves closer to the lens, the image plane will move closer to one detector

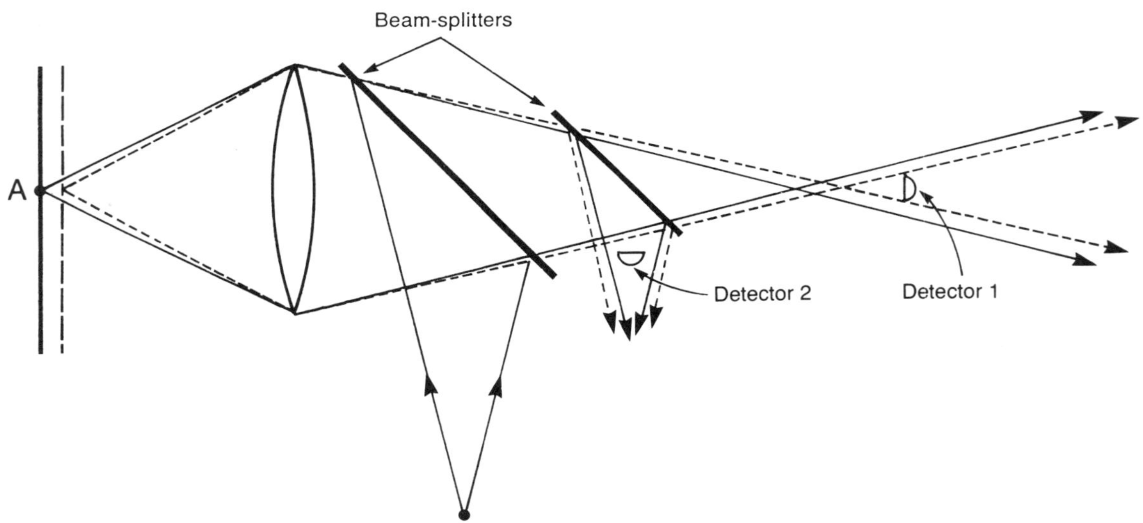

Figure 4-11 An alternative configuration for an image-plane locating system.

and farther away from the other, so that the outputs from the two detectors will no longer be equal. If the difference between the two detector outputs is taken, the signal recorded by scanning along the surface is a measure of the surface profile.

The differential approach gives a signal varying linearly with the surface position, thus providing good sensitivity. A focal point sensing technique using a substantially similar configuration could reach a precision of 10 nm using a high aperture microscope objective (Hartmann et al., 1973). Moreover, this approach is inherently faster than the technique shown in Figure 4-10, because it does not require a maximum searching oscillation of the focusing objective. It should, however, be noted that only the zero-output surface position can precisely be located by the technique shown in Figure 4-11. The absolute value of the surface displacement from such a position may not be exactly related to the signal level in a scanning device because of surface reflectivity variations or speckle noise if a coherent source is used. A mechanical, inherently slower, focus tracking loop is thus required to maintain good accuracy over a wide displacement range.

Other approaches are possible for the inspection of surface fluctuations by geometrical optics techniques. A slope monitoring method (Whitefield, 1975; Ennos and Virdee, 1983) has been described which uses autocollimation to monitor the angular orientation of the beam reflected from the inspected surface. The surface profile is obtained from the recorded surface slope by integration. Accuracies of up to 1 nm can be reached, provided, however, that the inspected surface is substantially smooth and specularly reflecting. Similar techniques using spatial filtering in the Fourier-transform plane (Goodman, 1968) by an annular diaphragm (Sawatari and Mueller, 1977), an optical-fiber bundle (McKeown et al., 1975; Locke, 1975) or a multiple-radial-element photodetector (Mead et al., 1975) have been used to map small localized distortions on relatively flat surfaces.

At the other extreme, if a surface with large and sharp fluctuations must be inspected, one can resort to oblique-incidence illumination techniques (Sawatari, 1972; Gumpertz, 1984). As can be seen from Figure 4-9, the angular displacement $\Delta\beta$ is more sensitive to the normal surface displacement d if the surface is tilted to decrease the value of the angle δ between the projected beam and the surface. However, the pointing accuracy is correspondingly reduced because the projected spot assumes an elliptical shape with a longer axis proportional to $1/\tan\delta$. Moreover, shading by sharp protuberances will mask parts of the surface. Such a configuration may nevertheless be useful for rapid visualization of large-scale surface roughness. The basic scheme is shown in Figure 4-12. If the graxing angle δ is known, one can estimate the amplitude of abnormal surface fluctuations from the extent of their shadows. Figure 4-12 shows a camera image obtained by such a technique on an extruded polymer-coated 4 mm-diameter wire with R_a roughness in the 30 μm range (Cielo et al., 1983).

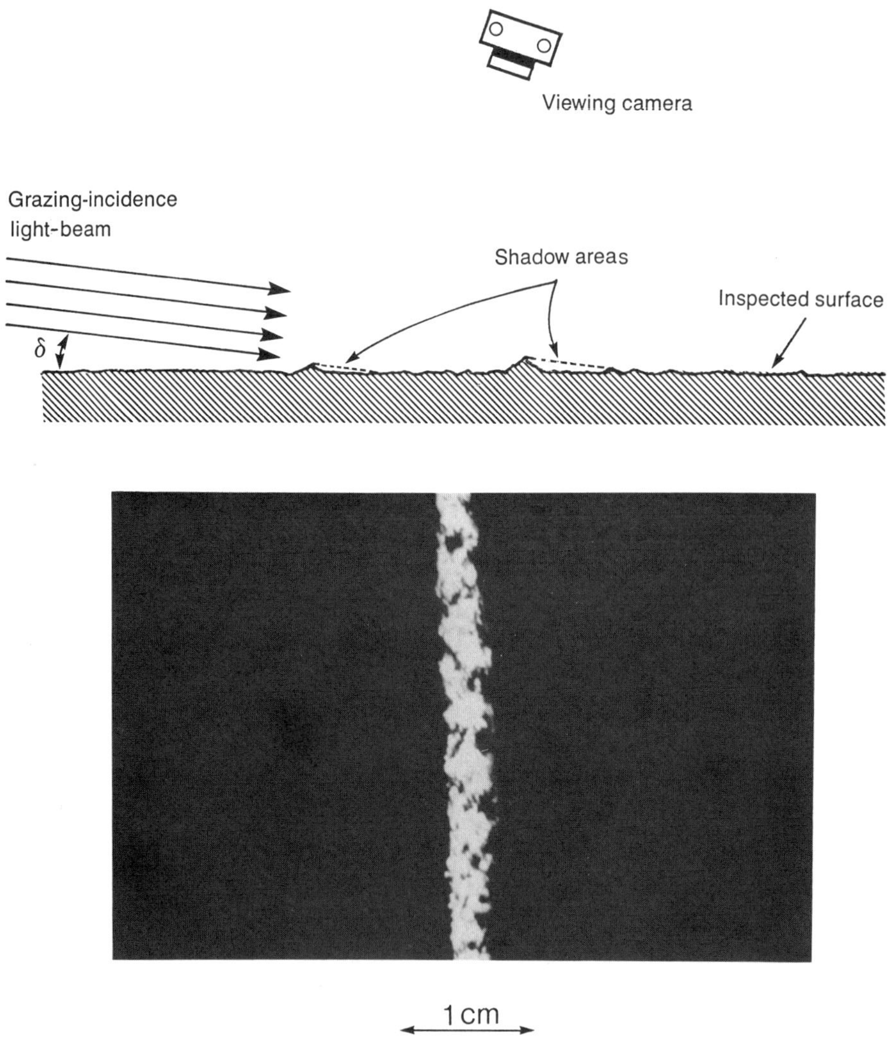

Figure 4-12 Grazing-incidence illumination for the inspection of rough surfaces; top, basic principle; bottom, visualization of surface-roughness on extruded polymer-coated wires.

4.2 Surface Monitoring by Light-Scattering Techniques

A number of mechanical and optical roughness monitoring methods have been described in the last section. As a general rule, it can be said that stylus-based profilometers are sensitive and wide-range instruments, but their use is essentially limited to laboratory use because of their need for contact, their fragility and low scanning speed. Interferometric optical profilometers are essentially limited to relatively smooth surfaces and perform best in a laboratory environment because of their sensitivity to ambient vibrations and need for a close operating distance. Geometrical optics based profilometers are mainly limited to relatively large surface roughness amplitudes in the micrometer range or higher.

In contrast to the above methods, light-scattering techniques are potentially compatible with industrial micro-roughness monitoring applications in the industrial environment because they are sensitive, can operate at a relatively large distance from the surface, are unaffected by ambient vibrations and directly provide a roughness-related evaluation averaged over a given area of the inspected surface. Statistically significant data can thus be collected at a much faster pace than with line-profilometers. As to their limitations, the light-scattering techniques cannot provide the specific surface profile, while the measured parameters are hard to quantify in terms of classical surface-roughness statistical parameters. Moreover, light-scattering methods have a limited sensitivity in the technologically important range of roughness amplitudes higher than one micrometer.

4.2.1 *Angular Spectrum of the Scattered Light*

The interaction of light with front-reflecting rough surfaces has been extensively investigated (Beckmann and Spizzichino, 1963; Elson et al., 1979). When a collimated light beam is incident on a rough surface, the angular spectrum of the scattered light has a maximum at the specular direction, as Figure 4-13 shows. The shape of this curve being strongly related to the surface topography, a lot of research work has been done to relate the angular spectrum parameters to the statistical parameters of the rough surface. Because of the wide variety of surface configurations which are found in engineering practice, a univocal relationship does not exist in general. Apart from some particular cases, the scattered field appears to depend on several statistical parameters at once, as well as on some surface properties, such as

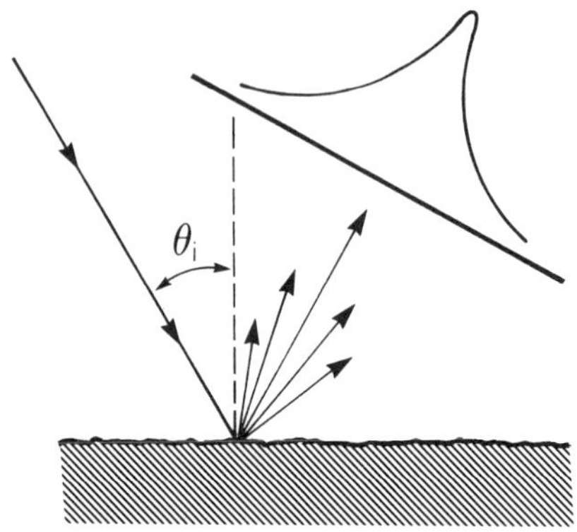

Figure 4-13 Scattering of light by a rough surface.

skewness and kurtosis (Stout et al., 1985), which cannot be quantified by the classical surface roughness parameters described in Section 4.1.1. Some experimentally observed features, such as an abnormal retroreflection along the direction of incidence (Egan and Hilgemann, 1979, Chapter 7), are not yet fully understood. Under such circumstances, many workers prefer to follow an empirical approach by which the engineering significant surface properties are directly related to measurable light-scattering parameters through experimental correlation curves (Stout et al., 1985).

Micro-roughness

The agreement between the light scattering theory and the experiments is best for relatively smooth materials for which $R_q \ll \lambda$. In this case, a sharp specular component is distinguishable in the angular spectrum of the reflected light, superposed to a much weaker and broad scattered-light distribution. The fraction of the incident light intensity which is specularly reflected has been found in many experimental investigations to be well correlated to the rms roughness R_q. This is in agreement with the diffraction theory from which the following expression has been derived for the specular reflectivity of a surface with Gaussian height distribution (Beckman and Spizzichino, 1963, Chapter 5)

$$\rho_s = \rho_o \exp\left[-\left(4\pi R_q \cos\frac{\theta_i}{\lambda}\right)^2\right], \tag{4-10}$$

where ρ_s is the reflectivity in the specular direction, ρ_o is the total reflectance of the surface, θ_i is the angle of incidence and λ is the light wavelength. The fraction ρ_s/ρ_o of the specularly reflected light is larger for smoother surfaces and for grazing angles of incidence. If the reflectivity ρ_o for the inspected material is known or has been previously measured, equation (4-10) can be used to evaluate the surface roughness R_q from a measurement of the specularly reflected intensity. Instruments based on this principle are commonly called glossmeters. An analogous, but less practical, technique consists of measuring the total amount of the light scattered away from the specular direction by using a collecting sphere (Bennet, 1978; Detrio and Miner, 1985).

Equation (4-10) has been found to be in reasonable agreement with experimental data for $R_q \cos \theta_i < \lambda/20$. For larger roughness amplitudes, the specular reflection peak in the angular spectrum tends to be hardly discernible (Vorburger et a., 1984), so that the measurement of the specularly reflected component becomes dependent on the dectector aperture. In order to extend the applicability of equation (4-10) to a higher roughness range, one can choose longer wavelengths or equivalently operate at grazing incidence.

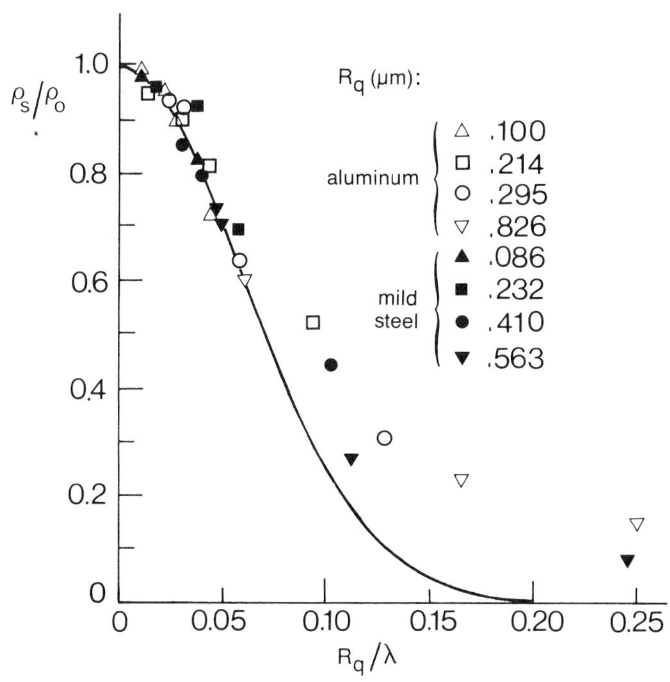

Figure 4-14 Measurements of the specular reflectance from metallic surfaces of different roughness with different light wavelengths. The solid line is obtained from equation (4-10).

Equation (4-10) was found to agree with experimental results for $R_q = 0.28\,\mu m$ and $\lambda = 633$ nm for $\theta_i > 65°$ (Hensler, 1972). Alternatively, an acceptable agreement with theory was obtained at $\theta_i = 20°$ and surface roughness up to $R_q = 0.826$ by using light of wavelength in the 2 to 15.5 μm range from a monochromator (Depew and Weir, 1971). Figure 4-14 shows the experimental data obtained with different wavelengths and for samples of different roughnesses, as well as the theoretical curve obtained from (4-10). As we can see, the agreement is good for $R_q < \lambda/10$.

Intermediate Roughness

When the product $R_q \cos\theta_i$ is larger, the angular spectrum of the diffracted beam (Beckmann and Spizzichino, 1963) becomes dependent both on the roughness amplitude R_q and on the spatial frequency distribution which is described by the power spectral density $Y^2(f)$, as Table 4-2 indicates.

The dependence on such parameters may be quite complex and depends on the specific shape of the roughness profile. A simple situation is encountered when a relatively smooth surface contains a periodic structure, such as a turned workpiece of the kind shown in the central micrograph of Figure 4-2. An example of the scattered light distribution obtained from this kind of surface is shown in Figure 4-15. This figure shows the angular spectrum of the light scattered by a diamond-turned metallic surface (Church et al., 1977). When the angular scan is performed in a plane parallel to the lay direction (curve a), the angular spectrum contains only a sharp specular reflectance peak, showing that the surface is very smooth in this direction. In

Table 4-2 Light-Scattering Regimes

Roughness amplitude	Important parameters	Physical mechanisms	Scattering techniques
$R_q \cos\theta_i \lesssim \lambda/20$	R_q	diffraction	specular reflectance monitoring
$R_q \cos\theta_i \approx \lambda/5$	R_q, $Y^2(f)$	diffraction	scattered light angular profiling
$R_q \cos\theta_i \gtrsim \lambda$	S_q	reflection	reflected light angular profiling

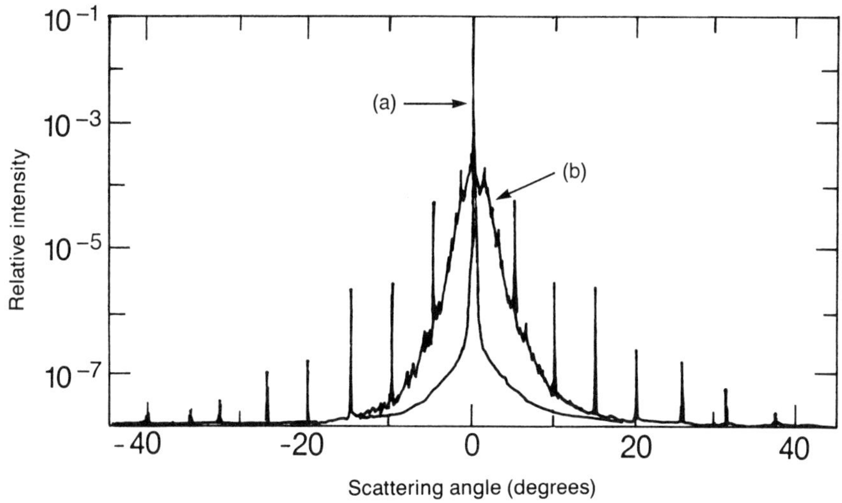

Figure 4-15 Angular spectrum of the light scattered by a diamond-turned surface in a plane (a) parallel and (b) perpendicular to the lay direction. The discrete diffraction lines produced by the periodic structure of the turned surface are clearly evident in curve (b).

the plane perpendicular to the lay direction (curve b), the specular peak is much broader, reflecting the microstructure of the engraving tool profile, and a number of discrete lines are visible. Such lines correspond to the diffraction orders of the grating-like periodical structure produced by the successive passages of the diamond tool. The same concept can be extended to random roughness by decomposing the surface profile into its Fourier spatial frequency components. If $R_q \ll \lambda$, each of these components can be considered to scatter a normally incident light beam at an angle $\theta = \sin^{-1}(\lambda f)$, where f is the spatial frequency of the sinusoidal component. The shape of the angular distribution can thus be related to the power spectral density of the surface profile (Elson and Bennett, 1979). The spread of the power spectral density can be rapidly evaluated by taking the ratio of the intensity measured 15° off-specular and the specular intensity (Corey, 1980).

Macro-roughness

When $R_q \cos \theta_i$ is equal to or larger than a light wavelength, the angular distribution of the reflected light is determined by geometrical optics rather than by diffraction. In this case, the surface may be considered as a series of facets, each making an angle θ with the mean surface plane and producing a reflected beam at an angle 2θ with respect to the direction of specular reflection (Church, 1979). An evaluation of the spread of the angular spectrum thus provides information on the S_a or S_q average-slope parameters defined in Section 4.1.1.

The average-slope monitoring approach covers the important roughness range of 1 μm to 10 μm where many industrial-grade surfaces such as cold-rolled or ground plates are found. Inspection speed is usually an important requirement in this kind of industrial inspection applications. Methods for a fast evaluation of the angular spread of the reflected light have thus been developed, such as taking the ratio of the specular reflection over the back-scattered intensity 30° off-normal (Takeyama et al., 1976), scanning a homogeneous surface with a rotating mirror (Clarke and Thomas, 1979) or using a self-scanned photodiode array (Tsukada and Yanagi, 1983; Brodmann et al., 1985).

The scanned-array approach has resulted in a particularly versatile and practical instrument which is now commercially available (see Figure 4-1). The basic configuration of such a system is shown in Figure 4-16 (Brodmann and Thum, 1984). Light from a light-emitting diode is projected on a typically

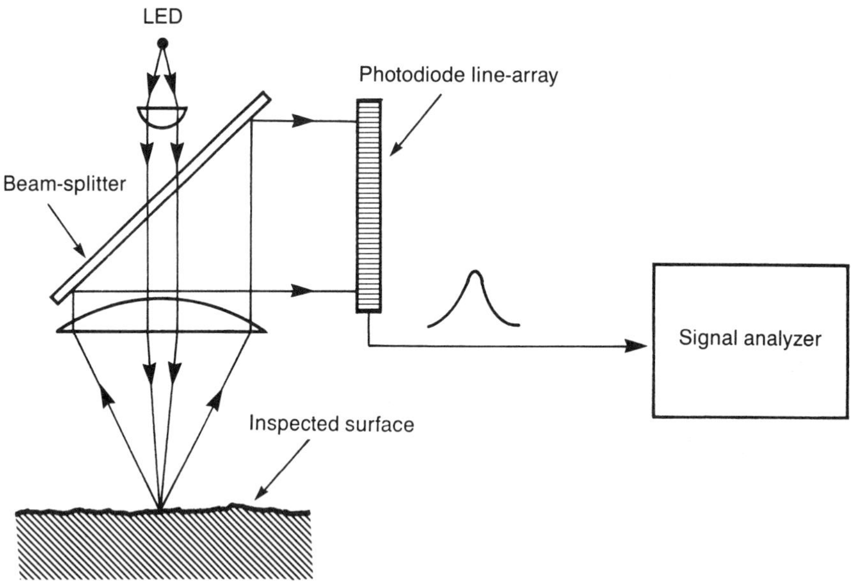

Figure 4-16 Schematic diagram of the Rodenstock surface-roughness monitoring instrument.

2 mm-diameter spot on the surface to be inspected. The scattered light is collected by a wide-aperture lens and projected on a linear photodiode array whose output gives the angular spectrum of the scattered beam. Rather than trying to evaluate the classical statistical parameters from such a spectrum, the authors have followed a more direct approach by measuring the width of the angular spectrum, which is broader for rougher surfaces, and establishing empirical correlations of such a width with the important properties of engineering surfaces. The output parameter is a normalized quadratic mean value of the scattering angles over the angular spectrum $I(\theta)$,

$$S_N = K \sum_{i=1}^{n} (\theta_i - \theta_s)^2 I(\theta_i), \tag{4-11}$$

where K is a normalizing constant, θ_i is the scattering angle probed by the ith element of the detector array and θ_s is the angle corresponding to the specular reflection. For large R_q values, it has been mentioned above that a proportional relationship holds between the slope of each portion of the illuminated surface and the angle of the light beam scattered by such a portion. It can thus be seen that the S_N parameter corresponds fo rough surfaces to the square of the S_q parameter (see equation 4-7), except for the fact that the former is averaged over the illuminated area while the latter is defined for a line path.

The Rodenstock roughness sensor is finding applications in a wide range of industrial applications, such as the inspection of machined surfaces, of the belt surface quality in a zinc-rolling production line or of the fine surface roughness of audio heads on a 15,000 parts/day measuring station (Brodmann and Thum, 1984). Although a certain correlation has been found (Brodmann et al., 1985) between the S_N and the classical R_a parameters for ground surfaces (see Figure 4-17), the authors propose the S_N parameter as an independent measure of the surface characteristics. In some cases, as shown in Figure 4-18, the S_N value appears to correlate better than the stylus-measured R_z parameter (Stout et al., 1985) to important properties of engineering surfaces. The curves shown in Figure 4-18 were obtained on finely ground rocker arms within one dressing cycle of the diamond wheel. The rocker arms are collected in pallets, pallet Nr. 0 corresponding to the freshly dressed wheel. The dispersion of the stylus-measured data is much larger than for the S_N data, the reason being that the stylus probes only a limited, linear area as compared to the surface averaging light-scattering technique. A similar configuration was recently used by the Canadian company Diffracto to inspect the surface finish of machined shafts in the automotive industry (German, 1985).

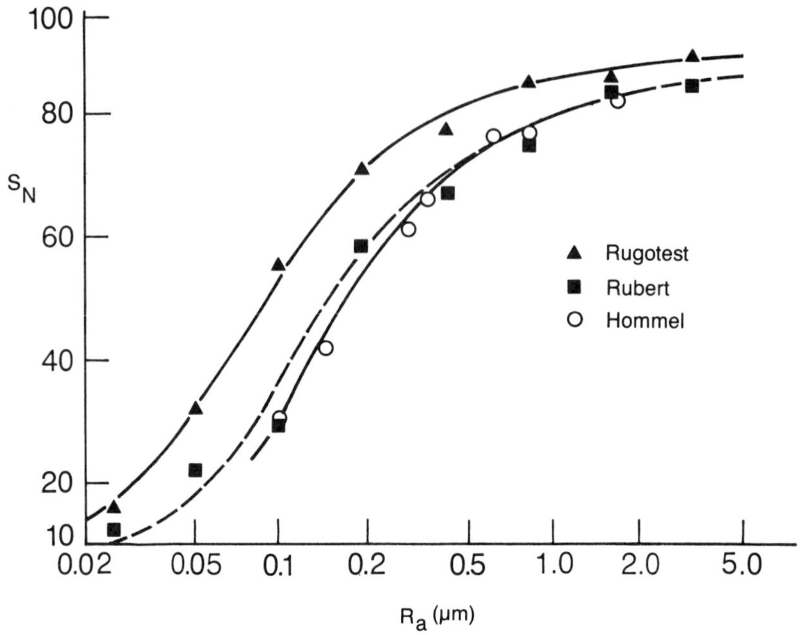

Figure 4-17 Comparison of the S_N roughness readings with the R_a parameter values for different roughness-standard samples.

It must, however, be kept in mind that, as mentioned in Section 4.1.1 in relation to the definition of S_q, the S_N value heavily depends on the high spatial frequency cutoff of the instrument, which is essentially determined by the wavelength of the projected beam. If only a large-scale roughness measurement is of interest without interference from light scattered by micro-roughness in the submicron amplitude range, a medium-IR light wavelength should be chosen for such a measurement.

Object Recognition by Light-Scattering Methods

Light-scattering analysis may be used not only to characterize the surface finish of a workpiece, as described above, but also to recognize specific surface features of a given object. This approach has been followed for the detection of gross defects or of part misalignment in the case of mass produced components (Dewhurst and Swift, 1983).

A typical application is the recognition of part integrity or orientation in the stream of identical components which are fed to an automatic assembly

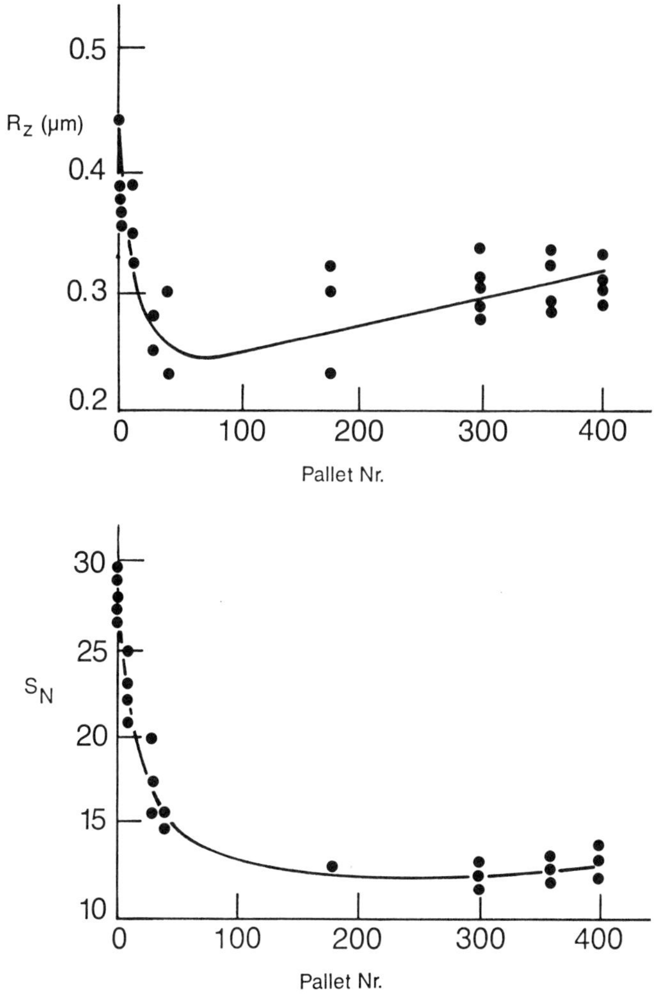

Figure 4-18 Variation of surface roughness within one dressing cycle of the diamond wheel, as measured by a stylus instrument (left) and by the Rodenstock light-scattering instrument (right).

machine. Imaging devices equipped with suitable pattern recognition algorithms can be used to inspect the components, as described in the next chapter. Such algorithms tend to be relatively expensive in terms of computation time. If the ports have only a small number of possible configurations, a faster and less expensive light-scattering technique may alternatively be used. In this case, the parts pass through an inspection station comprising a collimated light source and a single optical detector which monitors the light scattered by the object moving under the parallel beam. The time-resolved signal is compared with a relatively small number of possible signatures, each of which is typical of a given part shape or orientation. Defective or misaligned parts can thus be identified and recycled. Examples of the described applications include the inspection of a stream of adjusting screws supplied by a bowl feeder at rates of 100 parts per minute (Swift and Dewhurst, 1983) as well as a wide variety of other components (Dewhurst and Swift, 1985).

4.2.2 *Polarization and Coherence Properties of the Scattered Light*

In the last section the angular light-scattering properties of rough surfaces were described irrespective of the polarization or the spatial and temporal coherence properties of the incident light beam. This section shows that by taking into account such properties, additional information on the surface topographical features can be obtained.

Polarization Properties

The state of polarization of light may be modified after reflection at an interface (see, e.g., Born and Wolf, 1970, Chapters 1.5 and 13.2 or Meyer–Arendt, 1972, Chapter 3.2). An analysis of the polarization properties of the reflected light can be performed by instruments called ellipsometers (Azzam and Bashara, 1977) and provides information on the complex refractive index of the reflecting material.

The basic apparatus required for ellipsometry is shown in Figure 4-19. The incident beam may be linearly polarized with the electric field component vibrating in a direction parallel to the plane of incidence (p-polarization in Figure 4-19) or perpendicular to such a plane (s-polarization) or at any intermediate direction. The state of polarization of the reflected beam, usually in the specular direction, can be determined by rotating the $\lambda/4$ plate and the analyzer. The analyzer is a second linear polarizer which, similar to the other two components, can be rotated around the light beam axis to select different polarization components. If the incident beam is linearly polarized in either the p or the s direction, the reflected beam will conserve the same direction of linear polarization. The reflectivity of s-polarized light is usually larger than the reflectivity of the p component. Consequently, if the incident beam is initially linearly polarized at 45° with respect to both the s and the

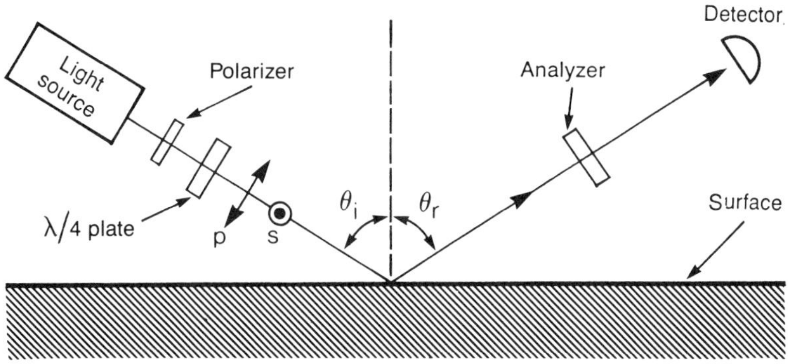

Figure 4-19 Apparatus required to analyze the polarization properties of scattered light.

p directions, the polarization direction of the reflected beam will generally be displaced nearer to the s-direction, as can be determined by rotating the analyzer. Moreover, the s and p electric-field components are phase-shifted with respect to each other by an amount gradually varying from π to 0, depending on the angle of incidence, after reflection by a metallic surface. Such a phase shift results in a change of the reflected light polarization from linear to elliptical, or from elliptical to linear if the opposite phase shift is introduced before reflection by conveniently rotating the $\lambda/4$ plate. Both the ratio of the s and p reflectivities and the phase shift, measured at different angles of incidence and for different wavelengths, provide information on the complex refractive index of the reflecting material.

Because surface roughness affects the local value of the angle of incidence and may produce multiple surface scattering if the surface slope distribution is sufficiently wide, ellipsometric analyses are strongly perturbed by surface roughness (Church and Zavada, 1976; Aspnes et al., 1979). Depolarization may also be introduced by internal scattering in partially translucent materials (Kortüm, 1969, p. 66–70). Conversely, consideration has been given to the possibility of using polarized light techniques to characterize engineering surfaces. An example of such an application is shown in Figure 4-20 (Cielo, Cole and Vaudreuil, 1985). A hot-rolled steel plate with rolled-in scale patches, typically 3 cm × 1 cm in size and 0.5 mm thick was analyzed under crossed polarizers, i.e., with the analyzer oriented perpendicularly to the polarizer. The s-polarized He-Ne laser beam was focused on the moving surface at an angle of incidence of 10°. The right portion of Figure 4-20 shows the signal obtained over the smooth, uniformly oxidized metallic surface without scale. Reflection

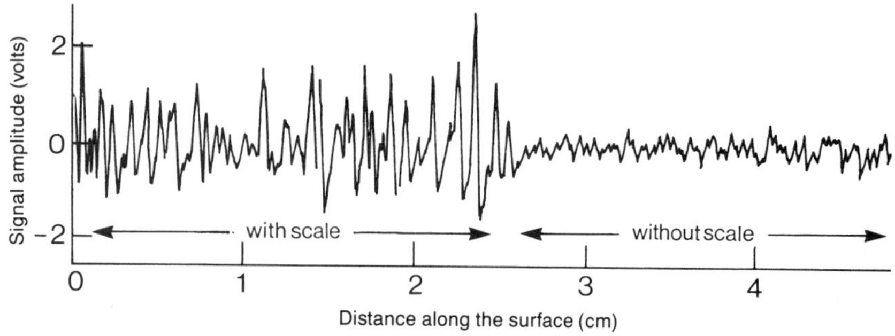

Figure 4-20 Scanning trace over a hot-rolled steel sheet including a rolled-in scale defect, obtained by an apparatus similar to the one shown in Figure 4-19 under crossed polarizers. The depolarization produced by the scale-covered surface is clearly evident.

from this part of the surface does not appreciably modify the light beam polarization state so that the signal transmitted through crossed polarizers is very small. When a scale patch is scanned (left portion of Figure 4-20), depolarization by the rough scale surface produces a substantial increase of the detected signal. A similar experiment was performed under parallel polarizers, and no significant difference was observed between the scale-free and the scale-covered surface. This shows that the observed signal variation is not produced by variations in the specular reflectivity of the scale-covered metal. Such an approach is being considered for the on-line detection of surface scale defects in rolling mills.

Attempts to quantitatively correlate the surface-roughness statistical parameters with the variations in the polarization properties of the reflected field have found limited success. Good correlation has been found between the R_a parameter and changes in the polarization state after reflection by ground and polished steel surfaces in the R_a range of 0.01 to 1 μm (Lonardo, 1978). However, other measurements performed on aluminum surfaces whose roughness was produced by acid etching (Smith, 1976) show that the induced polarization changes may vary considerably depending on the angle of incidence and on the presence of oxide films. Similar difficulties were found by other workers (Vorburger and Ludema, 1980) who analyzed the phase shift between the polarization components versus the R_a roughness of the reflecting surface for samples of different materials at different wavelengths and angles of incidence. Although substantial variations of the phase shift were observed, the results did not vary monotonically with either R_a or S_a, and no simple model could be found to evaluate the roughness parameters from a polarization analysis. It can be concluded that, at the present time, polarization techniques appear to be a powerful tool for defect detection or surface evaluation by empirical correlations, but the exact relationship between the surface roughness parameters and the polarization properties of the reflected beam is not yet completely clear.

Coherence Properties

When a coherent source such as a laser beam is focused on the inspected rough surface, a speckle pattern can be seen in the angular distribution of the reflected light, as described in Section 3.10. Such a pattern generates a high contrast noise when a detector of limited aperture is scanned over the angular spectrum or, equivalently, when the surface is translated under a stationary light source and detector. An example of the signal fluctuations produced by speckle noise is shown in Figure 4-20.

Although speckle is usually a source of noise which must be minimized by increasing the spot size or the aperture of the detector, it is not noise in the proper sense because it is deterministically related to the surface topography. Speckle can thus be considered as unwanted information rather than noise. In many cases, techniques have been found to extract part of the information carried by speckle patterns and to use such information to solve practical engineering problems. One such application is the evaluation of the surface roughness from an analysis of the speckle pattern.

If an apparatus of the kind shown in Figure 4-21 is used to observe the speckle-generated time fluctuations of the detected signal, the speckle contrast can be determined as the ratio of the standard deviation and the average intensity of the recorded signal. The speckle contrast should vary with the surface roughness, as shown in Figure 4-22 (Vorburger and Teague, 1981). With a perfectly coherent source, the speckle contrast should be equal to unity (see Section 3.10) except when $R_q \ll \lambda$, in which case the surface acts as a mirror and the reflected light beam is perfectly uniform with no speckle contrast. This is shown by curve a in Figure 4-22. Conversely, with a spatially coherent source of coherence length $\Delta L \approx \lambda$, such as a white light source filtered by a small pinhole, the speckle contrast evolution should be of the kind shown by curve b. After reaching a maximum, the speckle contrast decreases if the average roughness is larger than the coherence length. In these conditions, light waves scattered by the peaks and by the valleys of the surface profile are mutually incoherent and sum up in intensity, resulting in a decrease of the speckle contrast as discussed in Section 3.10.

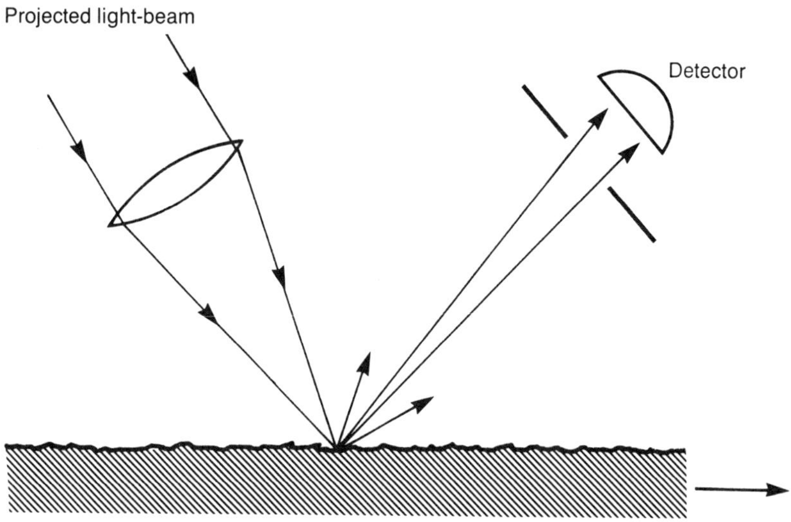

Figure 4-21 Typical apparatus for the observation of speckle statistics in the light beam reflected by a rough surface. The detector aperture is smaller than the speckle grain size.

The increase in speckle contrast under coherent illumination for $R_a \ll \lambda$ is in agreement with experimental investigations (Asakura, 1978). Using an apparatus of the kind shown in Figure 4-21 with a coherent He-Ne laser source, a roughly linear relationship was found between the speckle contrast and the roughness amplitude for $R_a < 0.15\ \mu m$. This technique, as compared to the glossmeter approach described in Section 4.2.1, has the advantage of not requiring a previous knowledge of the total surface reflectance or a simultaneous measurement of the light intensity scattered at an angle other than the specular direction. Abrupt reflectivity variations along the surface may, however, interfere with the speckle-related signal. Moreover, the intensity level of the detected signal is relatively low because of the requirement for the detector aperture to be smaller than the speckle grains. Sharper focusing of the projected beam in order to enlarge the speckle grain leads to a reduced depth of field.

Using partially coherent illumination, a decrease of the speckle contrast in a surface roughness range of the order of or larger than the coherence length has been observed in some experimental investigations (Sprague, 1972; Parry, 1975). With a pinhole-filtered tungsten-zirconium arc source of coherence length $\Delta L \simeq 1.5\ \mu m$, a monotonic decrease in the speckle contrast was observed in an imaging configuration with surface roughness ranging from 0.2 μm to 3.2 μm. This is a particularly interesting range for engineering materials. However, subsequent experiments (Nakagawa and Asakura, 1978) disagree with these results, the speckle contrast being found to increase rather than decrease for $R_q > 1$ μm. Such an increase is attributed to a reduction in the number of random components (see Figure 3-74) within the point spread function of the optical system for large values of the roughness amplitude and spatial period. The influence of the surface autocorrelation length on the speckle contrast has been demonstrated in other investigations (Fujii and Asakura, 1974).

Similar results have recently been reported with an interferometric white light speckle approach (Leonhardt et al., 1984; Ahlers, 1985). The experimental apparatus is shown in Figure 4-23. A spatially coherent white light beam illuminates the inspected sample which is in lateral motion. The scattered light field is superposed to a reference beam and probed by a small-aperture detector situated a short distance from the image plane of the inspected surface. Curves of the speckle contrast versus measured surface roughness are shown in Figure 4-24. Curve a was obtained with the reference mirror completely masked out. This corresponds to the noninterferometric configuration described above and shows that the signal correlation with the surface roughness level is very weak in this case. Conversely curve b shows a good

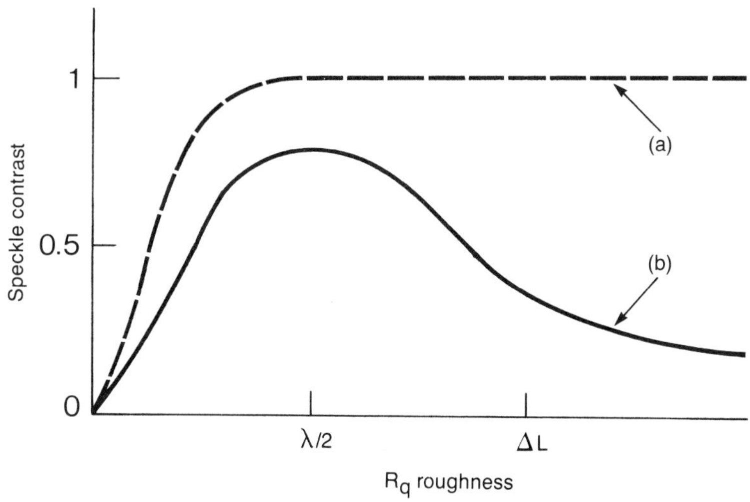

Figure 4-22 Trends of the speckle contrast in the light beam reflected by a surface of variable roughness under conditions of (a) coherent illumination or (b) partially coherent light with coherence length equal to ΔL.

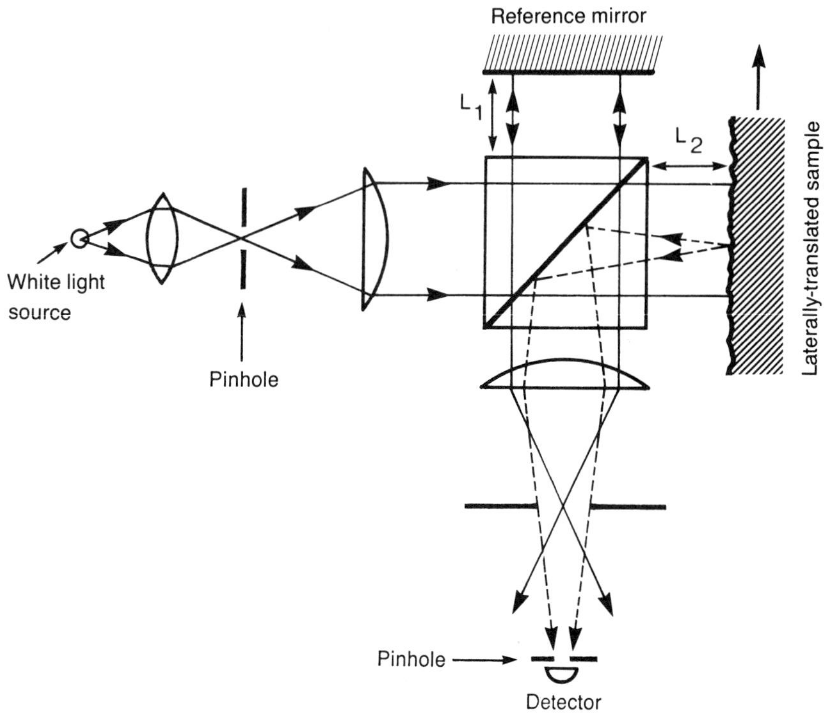

Figure 4-23 White-light interferometric arrangement for the observation of the speckle contrast under partially-coherent illumination.

correlation between the measured speckle contrast and the surface roughness when the reference beam is allowed to reach the detector.

Another approach showing promise for large amplitude roughness analysis is shown in Figure 4-25 (Fujii and Lit, 1978). The configuration is similar to that shown in Figure 4-21, with the difference that the light source is now a double-line Argon laser which emits simultaneously at two wavelengths λ and $\lambda + \Delta\lambda$ in the 0.5 μm range. Each wavelength produces a speckle pattern, which is separately analyzed using a prism spectrometer. It can be shown (Parry, 1975) that if $R_q/\lambda \ll \lambda/\Delta\lambda$, the two speckle patterns are substantially similar. If such an inequality is not satisfied, the two patterns become different so that the two signals sampled with the same pinhole position are no longer an exact replica of each other. By recording the two detected time signals and calculating the rms difference of the two normalized intensities, the authors obtained excellent correlation of such a parameter with the R_q roughness in the range from 0.13 to 6 μm. Similar results were obtained by

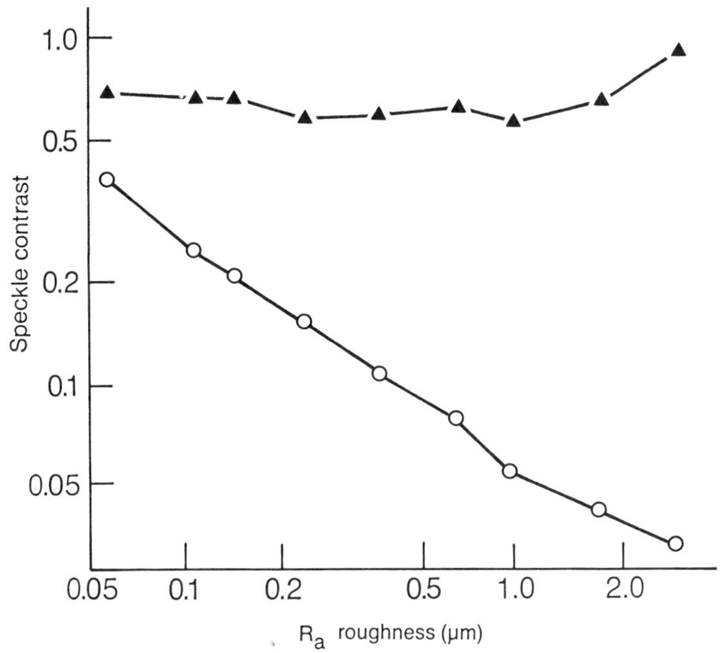

Figure 4-24 Speckle contrast versus surface roughness obtained with the apparatus shown in Figure 4-23 without the superposed reference beam (top curve) and with the superposed reference beam (bottom curve).

other investigators (Giglio et al., 1979) using a commercial cross-correlator to calculate the degree of speckle decorrelation. As compared to the glossmeter approach of Section 4.2.1, this technique has the advantage of covering a larger range of roughness and does not require a knowledge of the light intensity scattered away from the specular direction. However, the speckle decorrelation technique requires sharper focusing, results in a lower SNR because of the low intensity level transmitted through the pinhole and is sensitive to spectral reflectivity variations at the two wavelengths of operation.

4.3 On-Line Surface Inspection

Industrial needs for topographic surface inspection are of two types: measurement of the average surface roughness characteristics of the product and detection of abnormal surface characteristics in localized areas which are referred to as surface defects. This section reviews some on-line applications

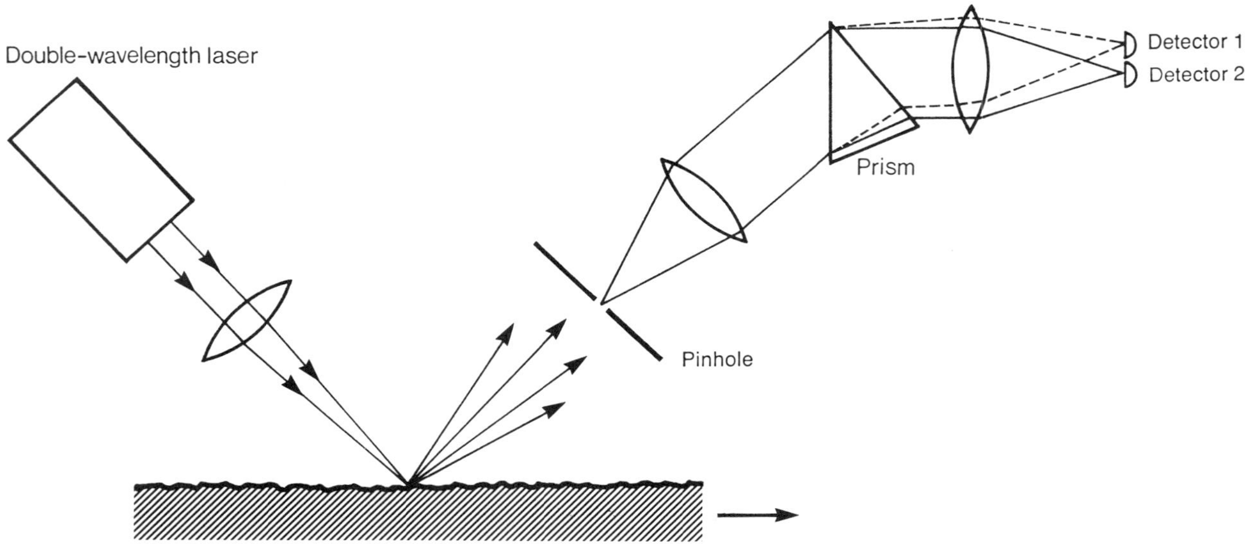

Figure 4.25 Apparatus used to perform double-wavelength speckle decorrelation roughness analysis.

of optical inspection methods for both of these requirements. Applications of surface-average monitoring methods are treated in Section 4.3.1, while Section 4.3.2 describes a few examples of on-line defect detection.

4.3.1 *Roughness Analysis in Rolled Products*

An analysis of the surface roughness properties of continuously rolled sheets may be important both for product quality monitoring and for process control. Product properties such as printability of rolled paper or coating adherence and corrosiveness of cold-rolled sheets are related to surface roughness. Gloss and brightness is an important quality parameter for a number of products, such as aluminum (Fairlie et al., 1986) and stainless steel sheets (Shibata et al., 1986). On-line detection of abnormal surface condition can originate timely interventions of the supervision personnel with a substantial reduction of the amount of product to be scrapped or reworked. Excessive surface roughness may also be related to a faulty process parameter such as roll wear or inadequate applied load, which can be manually or automatically corrected. Similar techniques are being applied to surface finish analysis of other manufactured products, such as the ring grooves which are to receive mating cast iron seals in the automotive industry (Pryor and North, 1985, p. 25).

Laser Roughness Sensing of Cold-Rolled Steel Sheets

A continuous surface roughness sensor using the light-scattering technique shown schematically in Figure 4-26 has been installed in a cold-rolling mill of the Kawasaki Steel Corporation (Asano, 1985; Shibata et al., 1986). Two He–Ne laser beams are projected on the moving steel sheet surface at angles of incidence θ_i of 10° and 75°, respectively. The two angular patterns of the scattered light are recorded by two scanning optical detectors over a measuring cycle of two seconds. As mentioned in Table 4-2, the light-scattering properties for a normally distributed rough surface profile are determined by the value of the average roughness R_q when $R_q \cos \theta_i \ll \lambda$, while they are determined by both the average roughness and the spatial frequency distribution of the surface profile when $R_q \cos \theta_i$ becomes comparable to the light wavelength λ. By monitoring the angular spectrum of the scattered light at two angles of incidence, each corresponding to a different light-scattering regime, both the average roughness and the average spatial frequency of the steel surface are evaluated simultaneously. By comparison

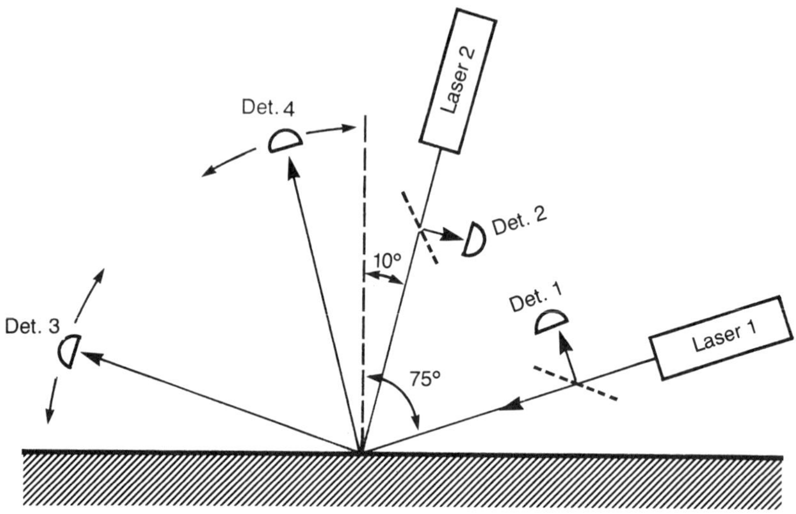

Figure 4-26 Diagram of the on-line surface roughness monitoring system mounted in a Kawasaki Steel cold-rolling mill.

with stylus-profilometry measurements, accuracies of $\pm 10\%$ for the R_a roughness values and of $\pm 20\%$ for the spatial periods were reported.

The roughness monitoring instrument is used for quality and process control at the temper processing line of the Chiba Works plant of Kawasaki Steel. Real-time surface roughness data are obtained and compared to roll-wear characteristics. Some process parameters such as the rolling load appear to be strongly correlated with surface roughness, as Figure 4-27 shows. This figure shows the output of the optical roughness monitoring instrument while the roll-applied load is varied by steps during the process. The surface roughness of the steel sheet shows a marked increase when the roll pressure is released. The possibility to perform such real-time controls in a hostile environment such as a rolling mill, with sheet velocities of the order of 20 m/s, random vertical excursion of several centimeters and strong vibrations, is a remarkable demonstration of the possibilities offered by noncontact optical inspection tools.

Surface Texture Monitoring in Paper-Rolling Mills

Good paper printability requires an ink transfer which is uniform over the paper surface. Areas of noncontact between the paper and the printing

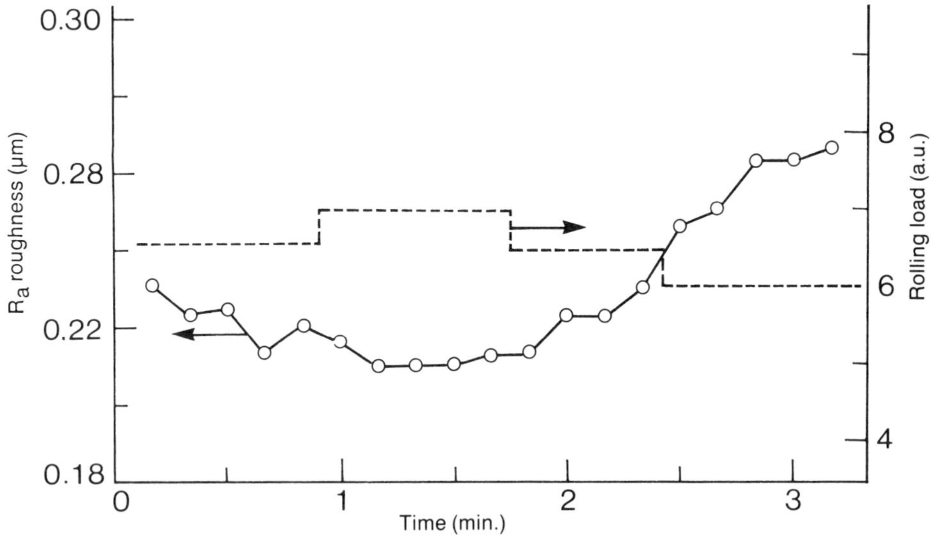

Figure 4-27 Optically monitored roughness variations of the steel sheet surface as the rolling load is varied during the rolling process. (Courtesy of Kawasaki Steel Corp.)

medium produced by random fiber distribution on the paper surface should be smaller than the visual limit of resolution, nearly 100 μm. Consequently, paper surface texture is strongly correlated to its printability performance. Surface roughness is also important for coated paper, where the base stock roughness is one of the factors which determine coat weight requirements.

Glossmeters are widely used to obtain some information on paper texture from a measurement of the specular reflectivity (Lavigne, 1977). Such instruments are also being used for on-line applications (Cook, 1980). More quantitative methods based on internal reflection at a contacting medium have been used in laboratory investigations (Lyne, 1976; Cielo, 1983). A technique which is much more sensitive to paper texture than the glossmeter approach and which is compatible with on-line applications will now be described.

The basic principle is shown in Figure 4-28 (Lucas et al., 1978; Alguard and Lucas, 1981). Light from an incoherent source is projected on a small spot at the surface of the paper sheet. A transparent window in contact with the paper assures a constant operating distance. Two fixed detectors are situated on both sides of the source along a line perpendicular to the paper machine direction, to sample the scattered light at equal angles from the normal. If no asymmetric roughness is present, the two detector outputs will

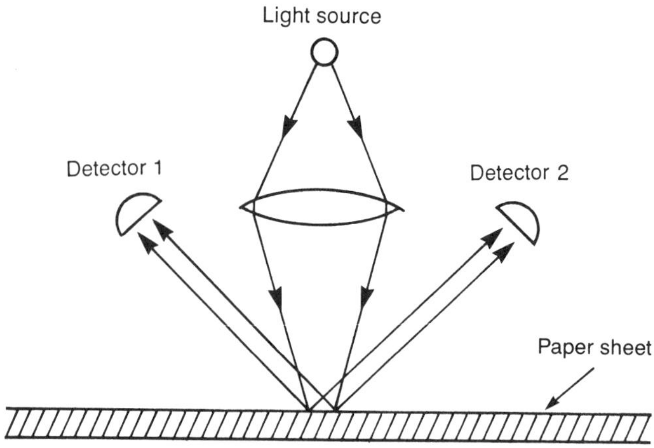

Figure 4-28 Differential light-scattering configuration used for on-line paper-texture inspection in a paper mill.

be equal. The differential output from the two detectors is thus a measure of abnormal texture conditions, and the frequency of this signal (above the critical frequency $f_c = v/w$, where v is the speed of the sheet and w is the diameter of the focused beam) is related to the length of the abnormal surface region. Excessive graininess in the paper surface corresponds to the presence of ridges along the paper-rolling direction typically 1 to 10 mm-long. As Figure 4-29 shows, graininess varies for different positions across the paper sheet.

Constant calibration conditions for the differential paper-texture monitoring device are assured by two mechanisms. First, the DC sum of the two detectors is kept at the same level by controlling the light source power. This

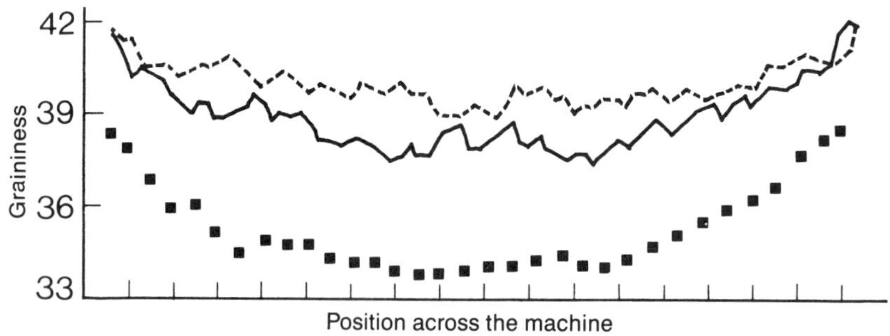

Figure 4-29 Results obtained from the differential light-scattering sensor with different paper qualities and at different positions across the paper sheet.

compensates for paper brightness and color variations, source aging or the building up of opaque deposits on the window. Additional tuning is provided by comparing different portions of the power spectrum of the recorded time signal. The amount of abnormal longitudinal features is weighed against the general paper roughness level (high frequency signal) to discriminate against signals produced by uniform roughness (Lucas, 1980). The paper texture sensor has been developed in the Montreal plant of the paper manufacturer Domtar and is now marketed by Measurex.

4.3.2 *On-line Surface Defect Detection*

A large number of optical systems have been reported for the automated detection of defects in a wide range of products as described in Table 4-3. Inspection needs for fast moving strips in rolling or coating mills have generated most of the driving force for the development of fast optical inspection devices because of the strong economic motivation produced by the large production rate and of the inadequacy of human inspection at high rolling speeds. As a result, several companies are now offering on-line defect detection optical devices, and a number of such devices are in current use in the production line.

While the measurement of the average surface characteristics of the product is usually performed at one position, or at a small number of positions across the web, the detection of surface defects requires a full inspection of the product. Defect detection thus requires surface scanning, often at very high rates when webs of some meters of width moving at some tens of meters per second must be fully inspected. The available scanning devices, either mechanical or electronic, have been reviewed in Sections 3.4.1 and 3.8. This section describes some typical applications of such devices to the on-line 100% inspection of a number of products.

Scanning Configurations

A number of configurations have been developed over the past two decades for the optical scanning of fast moving webs (Baker, 1976; Carlisle, 1980). One of the earliest devices for the inspection of a moving web surface is shown in Figure 4-30. A fluorescent tube illuminates a narrow area across the width of the moving web. Light reflected by the web surface is sensed by an array of detectors, typically at a distance of 1 cm from the surface. Using a narrow slot in front of the array, the field of view of each detector is limited to a small portion of the web, so that some degree of defect localization over

Table 4-3 On-line optical surface defect detection

Inspected industrial product	Surface defect	References
Paper webs and plastics sheets	Pinholes, tears, dark or transluscent spots	Carlisle, 1980 Giers, 1981 Campus et al, 1985 Baker, 1986
Textile fabrics	Thread defects, tears, stains	Cowhig, 1969 Wolf, 1980 Ward et al., 1985 Hyman and Burns, 1985
Glass components	Inclusions, surface defects, dark spots, geometry	Molesworth, 1982 Jones, 1985
Ball and roller bearing components	Surface integrity, geometry	Sevick, 1982 Barton & Kusenberger, 1982
Internal surface of cylinders	Scratches, pits	Batchelor & Williams, 1979
Semiconductor devices	Geometry, connections	Knox et al., 1977 Steigmeier & Auderset, 1983
Hot slabs	Indentations, pits, scratches, scale patches	Shiraiwa et al., 1979 Suresh et al., 1983 Nakakura, 1983 Ferriere & Jaillard, 1983 Schneider et al., 1984 Matsubara et al., 1986
Cold-rolled metal strips	Cracks, tears, pits, blisters, scabs, scratches, lines	Akutsu et al., 1976 Barker and Brook, 1978 Saridis & Brandin, 1979 Dumont-Fillon et al., 1980 West, 1982 Ferrière & Savarit 1985

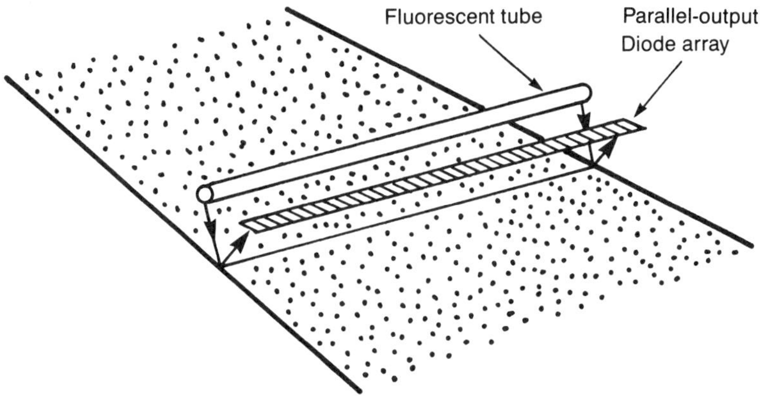

Figure 4-30 Nonimaging configuration for optical defect detection across a moving web.

the web surface is obtained. The spatial resolution across the web is smaller than the resolution along the moving direction with such a configuration. This approach is thus inadequate for the detection of narrow defects oriented along the web-moving direction, which are often present in such materials. Moreover, the requirement for a close proximity to the inspected strip is hardly compatible with practical industrial conditions which may include strip wobbling, high temperature and the presence of fumes.

A natural successor to the nonimaging configuration of Figure 4-30 is the line-imaging approach shown in cross section in Figure 4-31. Again, a strip lamp illuminates a narrow area across the web, over which a linear diode array sensor (see Section 3.4.1) is imaged by a single lens. Spatial resolution better than 1 mm across the web can be obtained by this approach (Purll, 1978). A variation of this configuration consists in replacing the line array by a line-scanned single detector. See Sections 3.4.1 and 3.8 for a comparison of mechanical scanning with electronic scanning by a detector array.

An alternative to the use of a strip lamp and a line-scan camera is to choose a two-dimensional camera with a stroboscopic shutter or a continuously staring camera with a stroboscopic lamp (Dumont-Fillon et al., 1980). For a maximum web speed of 50 m/s, a resolution of 1 mm along the moving direction requires a stroboscope pulse duration of 20 μs, which is within the reach of commercial stroboscopes. This approach has the advantage of making it possible to use the highly developed frame-grabbing and image-processing commercial equipment for two-dimensional cameras. However, the lateral resolution is limited to nearly 500 pixels/line in this case, to be compared to more than 1000 resolved spots in the case of a line array.

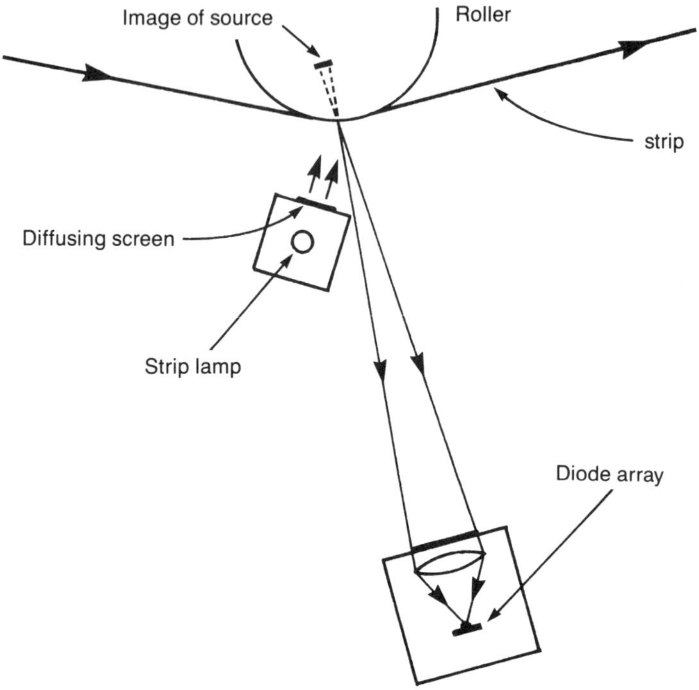

Figure 4-31 Imaging line-scanning system across a moving strip.

If a remotely focused light source such as a laser beam is used for the illumination, both the light source and the detecting device may be situated at a convenient distance from the inspected strip. The signal level is correspondingly increased because each surface spot is strongly illuminated only during the spot observation time. A typical laser-scanning device for the inspection of a moving web is shown in Figure 4-32 (Clarke and Bedford, 1978). The laser beam is post-objective scanned by a rotating polygon, and the light reflected back from the strip is detected by a diffusing cavity photomultiplier mounted on the scanner head. Focused spots as small as 25 μm have been reported with this device, although limited depth of field and speckle noise problems become important for such spot sizes. More recently, a similar laser-scanning apparatus using an Argon laser and four fixed-location detectors was reported for the inspection of 1.2 m-wide slabs at temperatures up to 1000°C, at a line-scanning rate of 720 Hz and a resolution of 0.2 mm (Matsubara, 1986). As discussed in Section 3.8, laser scanning has the advantages of higher brightness and possibility of spectral filtering as compared with incoherent illumination. Disadvantages are the requirement for a larger signal bandwidth, the presence of speckle noise and, most importantly, the need of a mechanical high-speed scanning device of limited lifetime.

As can be seen from Figure 4-32, the angle of incidence of the scanning beam, as well as the observation angle, vary across the strip. Such a configuration is mainly useful for tilted-angle surface inspection, looking for defects which produce abnormal off-specular light scattering. In order to increase the light gathering capability with shiny surfaces and to improve defect detectivity over wide scanning angles, some modified designs have been developed. One such modification, adopted by Intec, is shown in Figure 4-33. The laser scanner is similar to the one shown in Figure 4-32, but the receiver is now situated out of the scanner unit, over the strip for a surface-scattering analysis or under the strip for the inspection of transparent materials. An example of the application of such a system to the inspection of a plastics sheet in transmission with a resolution better than 100 μm is

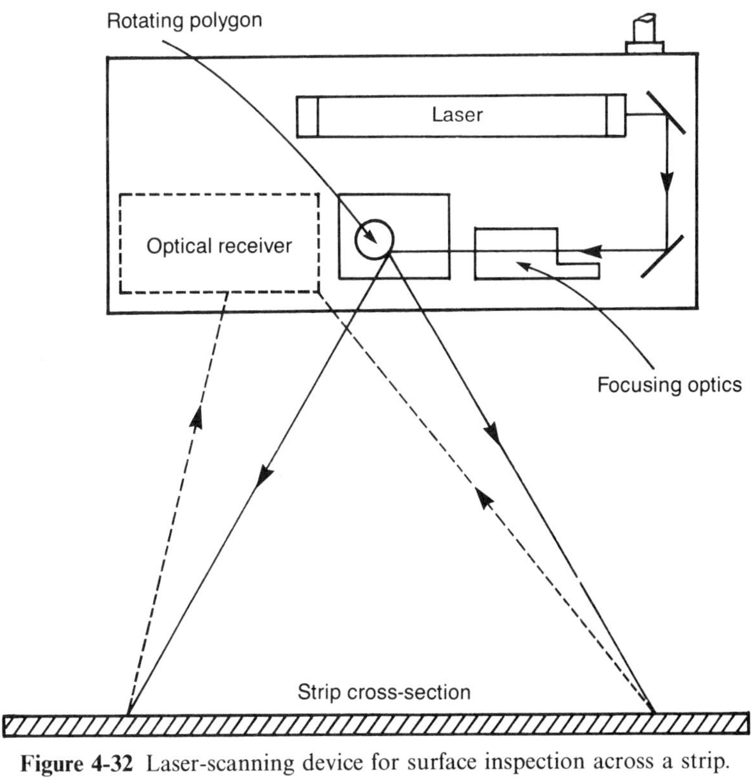

Figure 4-32 Laser-scanning device for surface inspection across a strip.

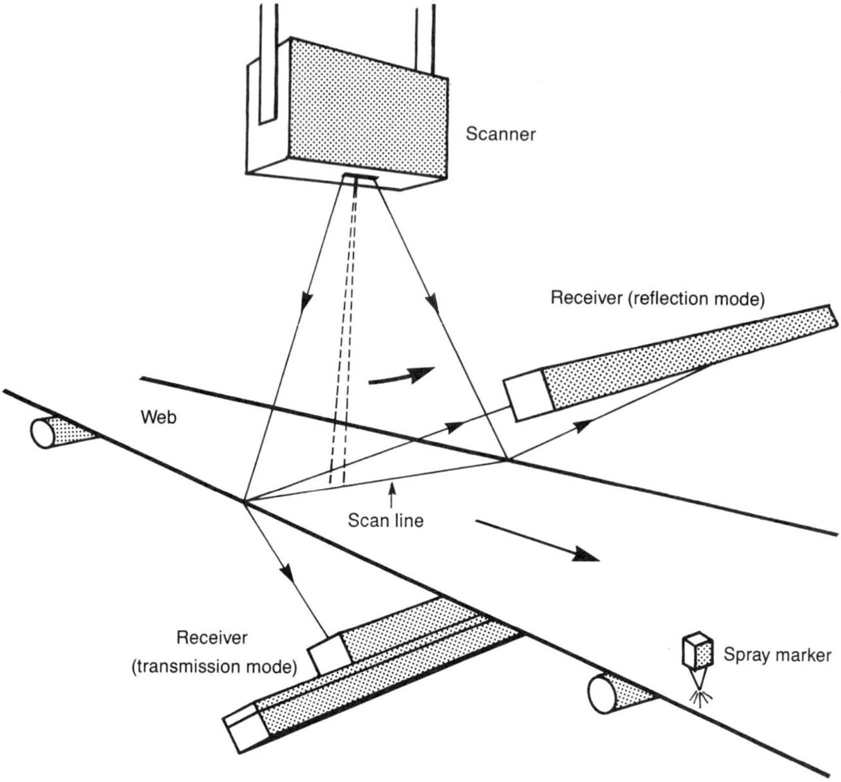

Figure 4-33 Industrial laser web-scanning system with extended linear receiver. The spray marker pinpoints the position where a flaw has been detected. (by Courtesy of Intec.)

described in Campus et al. (1985). The extended linear receiver (see Figure 4-34) is designed so that its sensitivity is approximately constant while the laser beam scans the strip at a variable angle of incidence. Such a receiver comprises two rod lenses which collect the laser light incident on the entrance window and focus it on a light pipe lying on the bottom of the receiver. The light pipe is a glass rod with an unpolished strip on its top or bottom which scatters the incident light in all directions. A substantial amount of the scattered light is guided within the rod by multiple total internal reflections and finally reaches two photomultipliers situated at the ends of the rod. The detected signal is thus relatively independent of the incident light beam position or orientation.

A similar approach for web-scanning makes use of a retroreflective mirror, as shown in Figure 4-35. Transmissive configurations are also in use to inspect

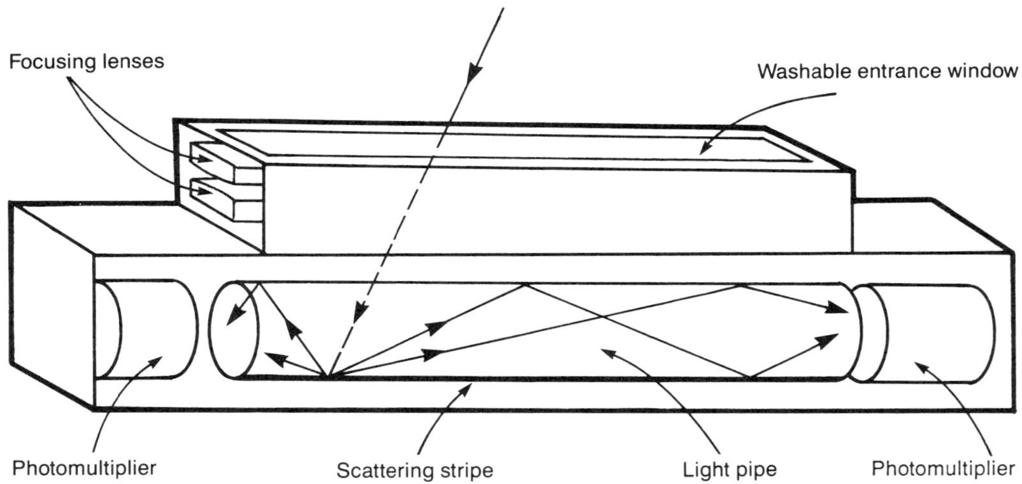

Figure 4-34 Detail of the Intec light-pipe receiver.

transparent materials (Baker, 1986). Such a mirror may be a large strip of retroreflective tape of the kind used for road signs or an array of small corner cube elements. Such an array reflects back the scattered beam to the same point on the inspected surface, so that a good portion of the light scattered in the near specular direction is returned toward the original direction. Both the light pipe and the retroreflective scanning systems are most sensitive to deep indentations or to surface scratches oriented in a direction normal to the web movement.

As discussed in Section 4.2, the intensity of light scattered by a reflecting rough surface, such as a cold-rolled metal strip, is strongly dependent on the angle between the specular direction and the direction of observation. In the previously described scanning configurations, such an angle varies across the scanned line. To avoid this problem, constant angle scanning optics has been developed for the inspection of metallic surfaces (Baker, 1976). The optics of such a configuration is shown schematically in Figure 4-36. The scanned laser beam is successively reflected by an auxiliary plane mirror strip, by a cylindrical mirror and by the inspected surface before reaching the detector. It can be seen by unfolding the successive reflections that if both the scanning polygon and the detector are equally spaced along the axis of the cylindrical surface, the detector will always sample the specularly reflected beam as the laser beam scans the surface. By displacing the detector along a circular path

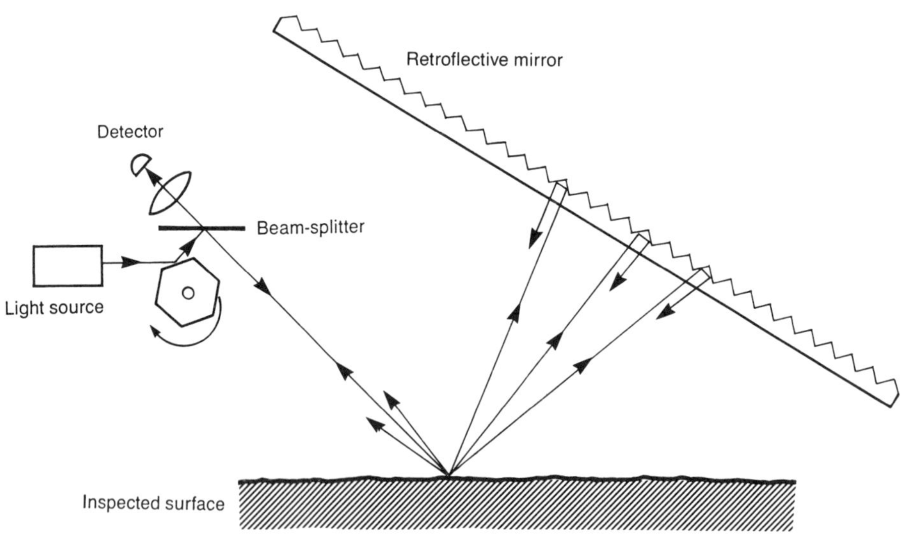

Figure 4-35 Web scanner using a retroreflective mirror.

centered on the scanned line on the inspected surface, nonspecular light-scattering directions can be probed while the polar angle of observation is kept constant across the laser-scanned strip. The detector aperture is chosen to be small enough for an adequate angular resolution but large enough to minimize speckle noise (Barker and Brooke, 1978; see also Section 3.10).

Signal Processing

The extraction of defect-related information from the detected signal is a difficult task in web inspection, given the high scanning rate. Signal fluctuations produced by defects such as scratches, dents, bruises or stains must be differentiated from noise of intrinsic or environmental origin (see Section 3.5). Noise reduction techniques such as synchronous detection are

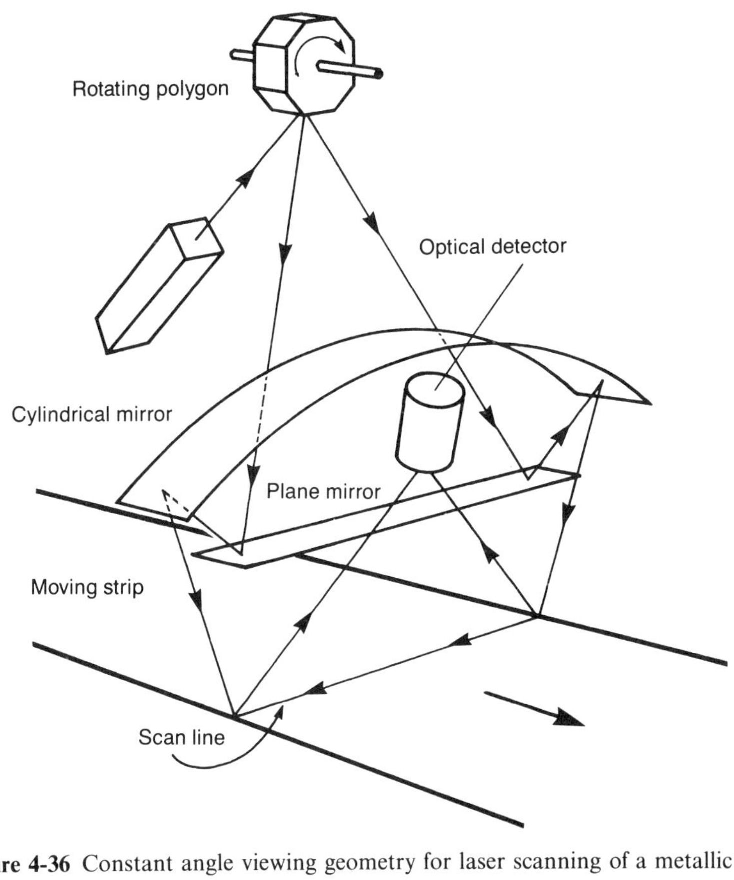

Figure 4-36 Constant angle viewing geometry for laser scanning of a metallic strip.

difficult to apply at such high data rates. Additional signal fluctuation sources are introduced by the scanning action, such as speckle noise, lens vignetting effects or spurious reflections at a given observation angle. Nevertheless, the repetitive nature of web scanning offers some signal conditioning possibilities which help signal interpretation.

The classical approach to discriminate a defect-related signal from background noise is to set a threshold for the detected amplitude above which the presence of a defect is assumed. The threshold is chosen, following Bayesian statistics, at a level which maximizes the probability for a defect to be detected while minimizing the probability for a noise spike to be classified as a defect (Cook and Bernfeld, 1967, Section 2.3). As Figure 4-37 shows, there will always be a number of false alarms as well as a number of undetected defects in practical circumstances. Matched filtering may help to improve the signal-to-noise ratio (Norton-Wayne et al., 1977) but requires an *a priori* knowledge of the signal waveform which is usually not available in surface defect detection. Recent developments on defect extraction by stochastic modelling of the surface texture (Suresh et al., 1983; Attali and Cohen, 1986) appear to be better suited to inspect industrial materials such as rough metallic sheets, textile fabrics or lumber but require a relatively high data-processing effort.

Figure 4-38 shows a typical signal pattern obtained over a scanned line. If a fixed threshold is used to select signal features to be classified as defects, only the largest spikes will be detected. The adoption of matched thresholds

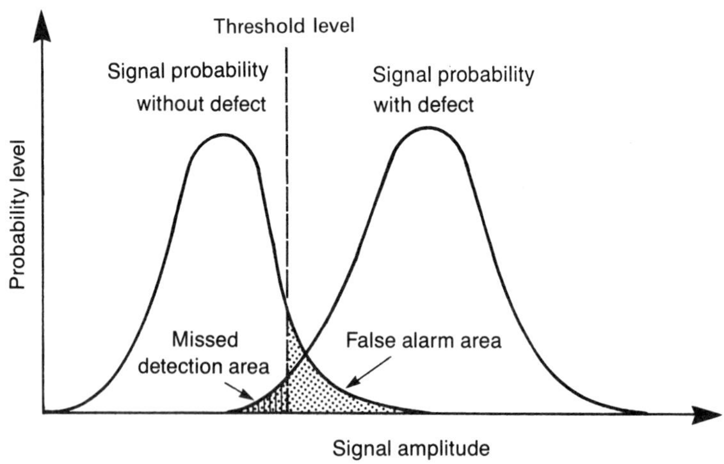

Figure 4-37 Signal statistics in the presence of noise.

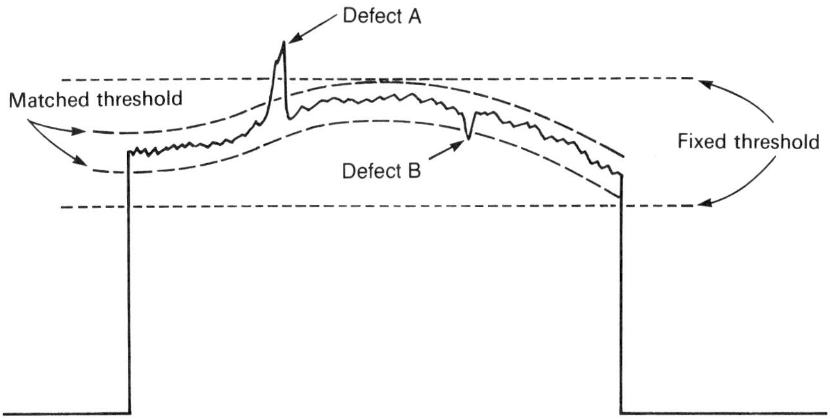

Figure 4-38 Typical line-scan signal output across a web.

(Ejri et al., 1973), taking into account the slow signal variation across the web produced by the variable angle of incidence or by environmental perturbations, strongly improves defect detectablity as Figure 4-38 shows. Such variable threshold lines can be obtained by low-pass filtering the line-scan signal. Equivalently, one can high-pass filter the signal and subsequently apply a fixed threshold. Both schemes are sensitive to defects producing high frequency signal perturbations across the scanned line, such as scratches oriented along the moving direction. For other kinds of defects, line-to-line differentiation may be a more efficient approach. A convenient method to produce finely matched threshold lines is to average the line-scan signal over a large length of the moving web, eventually repeating such an operation periodically to take into account variable strip characteristics or environmental perturbations. Low frequency or transversally-oriented defects can be detected in this way.

Once a first data reduction is operated by recording the coordinates and the signal level of the relatively small number of points exceeding threshold, more sophisticated signal analysis techniques can be applied to interpret such features. Several lines can be held in memory, and two-dimensional image analysis techniques can be used to further select defect-related signals from noise and to classify the defects. Autocorrelation techniques with a data base of possible defect signatures may be used to interpret small isolated features or to reject them as noise. Small feature clustering techniques may be used to recognize long streaks or distributed area bruises. Much progress is possible in this area by capitalizing on the recent availability of fast and inexpensive signal processing equipment.

Inspection of Wire-Shaped Products

The system geometry in the case of a wire-shaped product is quite different from that of a flat web. If surface defects must be detected over the whole wire circumference, the optical design of the scanning device may be quite elaborate.

Electric wires are typically produced by feeding metallic wire to the die of a plastics-coating machine. Several extruders are often used in series along the production line to produce several coating layers of different materials. For some critical applications, surface defects in each of the coatings must be detected for later individual fixing. In other cases, a surface flaw counter is used to monitor the coating average quality in order to timely detect a malfunction of the extruding process. The production of large amounts of imperfect cable which must subsequently be scrapped or reworked is thus minimized. Because of extrusion speeds reaching several tens of meters per second and of the semisoft surface condition of the extruded coating, tactile sensors are usually not applicable to such an inspection problem. High voltage breakdown gauges can detect only the largest flaws, which result in a severe lack of the coating insulation properties. Small surface flaws, which are either visually unacceptable for market release or which are sites for premature cable service failures, are now almost totally detected by optical inspection devices.

One such device is shown in Figure 4-39 (Puffer, 1984). The semiconductor coating in a high-voltage cable is inspected by this apparatus after the extrusion of the conductor screen and before the jacket extruding stage. An annular region around the wire, typically 1 cm-wide, is illuminated by a radial light source at nearly normal incidence. In the absence of defects, light is reflected in the nearly specular direction by the smooth wire surface. Surface defects scatter some of the light, which is collected by a slotted mirror and imaged on the retina of a two-dimensional array camera. The annular image of the illuminated area is processed in real time to obtain the radial and longitudinal position of the flaw.

The need for a real time two-dimensional image analysis makes such an approach relatively slow. If an image scan rate of 30 Hz can be followed, the wire displacement during the 33 ms image scan time must be smaller than half of the 1 cm-long illuminated area around the wire. This limits the maximum wire extrusion speed to 15 cm/s, which is smaller than most extrusion rates. A faster rate is possible by using several detectors and light sources in parallel. One recently reported device (Ortgies, 1985), incorporating

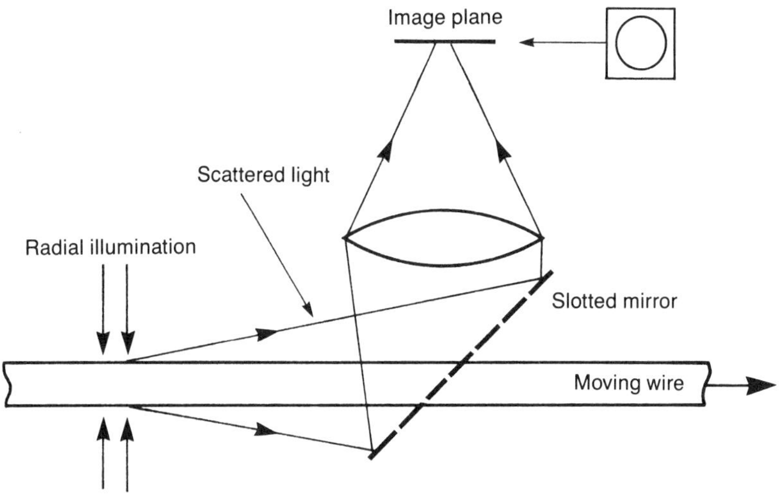

Figure 4-39 Optical configuration for a full circumference inspection of the surface integrity on a coated wire.

a circular arrangement of 64 parallel detectors and 1200 light emitting diodes, can work at wire speeds of nearly 1 m/s, at the expense, however, of rather complex mechanical alignment and electronic switching problems.

If some of the detection requirements can be relaxed, e.g., if a precise knowledge of the radial position of the defect is not required, higher extrusion speeds can be followed. In this case, a configuration of the kind shown in Figure 4-40 can be used (Koizumi and Akiyama, 1985). As in Figure 4-39, defects are detected by normally illuminating a narrow stripe around the wire and monitoring the scattered light level. Here, only two detectors are used to sense the total amount of scattered light, avoiding any requirement for an image scan. Surface defects of longitudinal dimension larger than 1 mm can thus be detected at wire speeds of up to 1.6 m/s.

The optical setup used in the device of Figure 4-40 comprises two ellipsoidal mirrors for maximum collection efficiency. The wire is illuminated uniformly over its full circumference by a single lamp source and a reflecting V-shaped mirror. The ellipsoidal mirrors, each having a small wire feeding hole, are positioned so that one focus of the ellipsoidal surface coincides with the illuminated wire region and the other focus coincides with a detector. All of the scattered light intercepted by each high-aperture mirror is focused on the corresponding detector. As can be seen in Figure 4-40, one detector collects light scattered in the direction of the wire movement and the other collects

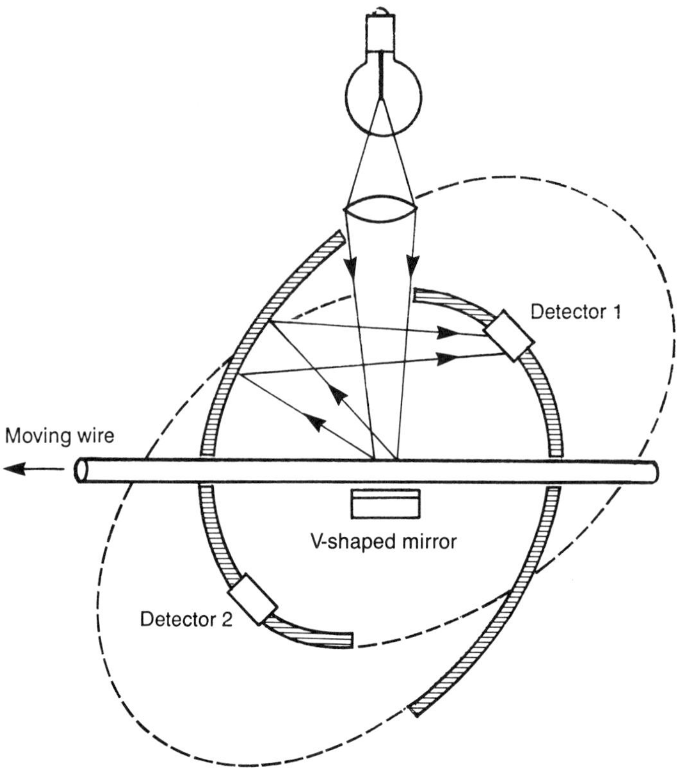

Figure 4-40 Detection of surface defects on an enameled wire by differential detection of scattered light with two ellipsoidal mirrors.

light scattered in the opposite direction. The differential output from the two detectors is mainly sensitive to changes in the average slope of the surface, as produced by "bump" defects whose spatial period is larger than the width of the illuminated region.

Light-scattering sensors are mostly effective for specularly reflecting surfaces. With rough-surface or translucent coatings, this approach is not very reliable because of the large amount of light scattered whether a defect is present or not. In such cases, imaging configurations of the kind shown in Figure 4-39 are required. A similar imaging approach, but much faster because it does not require any image processing, is shown in Figure 4-41 (Koizumi and Akiyama, 1985). As in the case of Figure 4-39, the illuminated section on the wire is imaged by a lens on the image plane. Such an image, rather than being processed by a camera, is in this case filtered by a matched

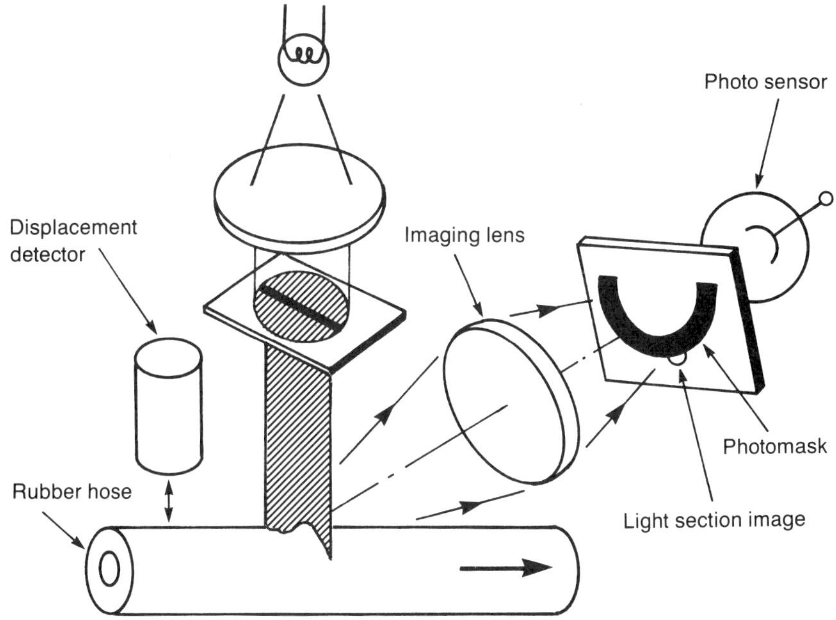

Figure 4-41 Matched-photomask filtering configuration for surface defect detection on a fast moving, rough surface wire.

photomask and the light transmitted by this filter is monitored by a single detector. The filter has a two-dimensional transmissivity which is exactly the negative of the projected image, so that the detector output is strong only when a surface defect produces a deformation of the projected half-circle image. A simple way to produce such a photomask is to expose a photographic film by inserting it in the image plane (Cielo et al., 1977). An obvious problem encountered with the photomask filtering technique is its extreme sensitivity to wire-to-filter alignment. As shown in Figure 4-41, this requires a continuous adjustment of the photomask position to follow transverse wire fluctuations as monitored by an optical wire displacement detector.

Much less sensitive to wire fluctuations are shadow projection techniques (Fuchs-Viniczay and Huber, 1979; Simonson et al., 1983) which are equally effective for either glossy or scattering coating surfaces. The shadow of the wire cross section is projected on a detector array or on a linear fiber optic bundle for an instantaneous reading of the wire diameter. Variations of such a reading correspond to surface defects. Several projection units can be used at different angles to cover the whole circumference. The defect amplitude sensitivity is limited by the number of elements in the detector array, while the wire speed limitations are related to the finite array scanning time. Typical performances are a maximum wire moving speed of 1 m/s for a minimum detectable defect amplitude of 50 μm and spatial period of 1 mm.

A shadow projection configuration which can be used at high extrusion speeds is shown in Figure 4-42 (Cielo and Vaudreuil, 1985). A linear-filament lamp is imaged by two spherical lenses of focal length F_1 on a large-area single detector. Two cylindrical lenses are used to project and recollimate a laminar light beam of uniform intensity, nearly 0.5 mm-wide, across the wire

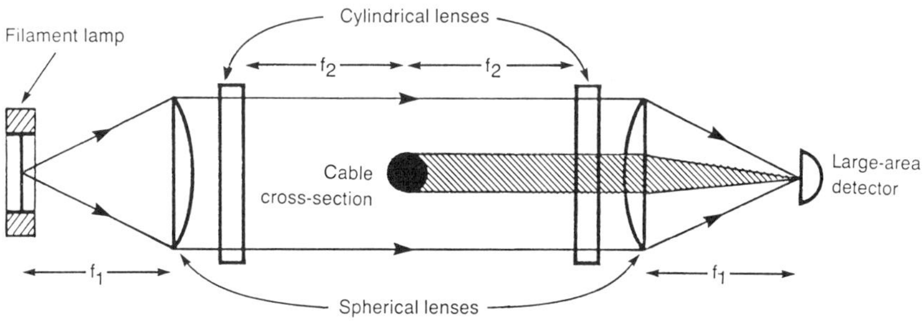

Figure 4-42 Line projection method for monitoring surface roughness on fast moving cables. Top: schematic diagram; bottom: in-plant trials of the IMRI prototype.

situated near their common focal plane. The part of the light beam which is not intercepted by the wire is collected on the detector, which has an AC output that corresponds to the defect-related wire diameter fluctuations. The wire speed is only limited by the detector response time. With a moderate detector bandwidth of 100 KHz, wire extrusion speeds up to 50 m/s can be accepted. Moreover, the uniformity of the nearly collimated projected beam obtained with such a configuration makes the detected signal relatively independent of the random wire excursions in the plane of Figure 4-42. It should be mentioned that the adoption of either a single He-Ne laser or an array of fiber-pigtailed diode lasers proved to be inadequate in this case because of speckle noise, high frequency laser amplitude or mode-to-mode interference fluctuations, and line nonuniformity (Cielo and Vaudreuil, 1984).

An industrial prototype of such a sensor was recently developed at our Industrial Materials Research Institute and tested on the production line at extruding speeds reaching 30 m/s. Figure 4-42 shows the location of the sensor just after the extruder die. Random noise introduced by vapor turbulence could be almost completely suppressed by high-pass filtering. Figure 4-43 shows two examples of signals obtained with a wire of (a) acceptable and (b) unacceptable surface quality. As we can see, a roughness amplitude resolution of a few micrometers can be obtained with such a device. Subcritical surface roughness levels can thus be monitored for real-time control of the extrusion process.

Crack Visualization

As mentioned earlier, cracks are a class of surface defects whose detection is particularly important because of their relation with structure failure mechanisms. A number of nonoptical techniques which are routinely used for crack detection, in particular ultrasonic, eddy current and magnetic particle inspection were described in Chapter 2. Passive optical inspection techniques for crack detection are mentioned in this section.

Visual inspection of the surface is certainly the most widely used approach for the detection of surface-open cracks. Although quantitative angular light pattern analyses have been considered for the detection and characterization of cracks (Sawatari and Mueller, 1977; Pernick and Kennedy, 1980; White and Feldman, 1981), surface inspection for crack detection is normally performed manually by the inspection operator, often in concomitance with other methods.

Narrow-gap cracks are difficult to see with the unaided eye and cannot

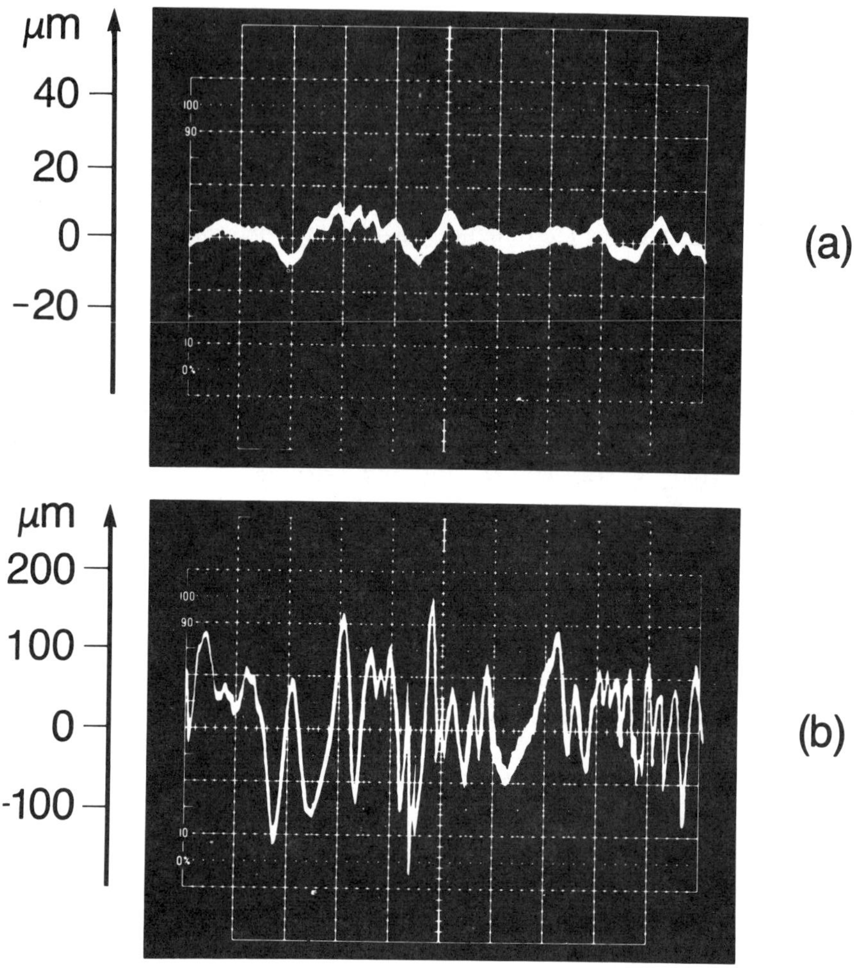

Figure 4-43 Examples of signals obtained with the apparatus shown in Figure 4-42 with a cable of (a) acceptable and (b) unacceptable surface-roughness quality.

be distinguished from scratches or shallow gaps, which are innocuous from the point of view of structural integrity. A technique which is widely used in industrial inspection shops for visualizing deep, narrow cracks is the liquid penetrant technique (Betz, 1979; see also Boyer and Gall, 1985, p. 33.1).

After the surface to be inspected has been cleaned, a suitable liquid penetrant is applied to the surface and allowed to penetrate by capillarity into any surface-open cavities including cracks. The liquid penetrant film is subsequently wiped off or washed away, leaving only the liquid which had penetrated into sufficiently narrow and deep cavities. A developing agent is now applied, which acts as a blotter to assist the natural seepage of the penetrant out of the cavities and to spread it at the edges, so as to greatly magnify the apparent width of the flaw. The developer also provides a uniform background to assist visual inspection. Finally, the surface is visually examined for indications of penetrant bleedback from surface openings. Often, fluorescent penetrants are used, whose visible emission shows up clearly when the workpiece is subject to dark-light illumination from an ultraviolet lamp.

The liquid penetrant inspection technique, as compared to the nonoptical methods described in Chapter 2, is an inexpensive, easy to use and relatively rapid procedure for crack detection. Surface openings as narrow as 0.1 μm can be detected by this method. It can be applied to any material, including nonconducting ceramics, with the exception of very rough or porous workpieces which tend to produce false indications. Continuous, semi-automated inspection of conveyor displaced pieces such as foundry products are performed routinely on liquid penetrant inspection stations. One major limitation of this technique is that it can detect only defects which are open to the surface. Such defects are the most dangerous ones from a failure mechanics point of view because of the stress concentration considerations mentioned earlier.

4.4 Techniques for Automatic Focusing

One of the advantages of optical techniques, as opposed to inspection methods based on nuclear, microwave or ultrasound radiation, is the possibility to tightly focus the probing beam on the inspected surface. This results in high spatial resolution and signal-to-noise improvement because of the possibility of using wide angular aperatures in conjunction with spatial filtering in the image plane. An example of the requirement for sharp focusing was given in Section 4.1.4 when dealing with geometrical optic rough surface profilometers. Several other examples will be encountered in later chapters with reference to gaging and profiling applications or focused sensing techniques.

It was mentioned in Section 3.7.1 that for diffraction limited optics, the diameter of the focused spot can be reduced by increasing the angular aperture of the projecting or of the viewing optics, at the expense, however, of the depth of field. Keeping an optical system sharply focused on the surface to be inspected may appear incompatible with industrial applications where the inspected object, whose surface is not necessarily flat, may be moving along the production line, or where precise workpiece positioning requirements would unacceptably increase the production costs. Indeed, most of the optical devices presently used for industrial inspection tasks are designed with small projection or viewing apertures in order to relax focusing requirements, at the expense of resolution and collection efficiency.

The main purpose of this section is to point out that recent developments in miniaturized optical components and signal processing capabilities are now making automated focusing a realistic and inexpensive alternative for continuous sharpness control of high-aperture inspection devices. Different automatic focusing methods are discussed, and a number of auto-focus mechanisms which have recently been introduced in automatic cameras and optical disk players at costs well below $100 are described in this section.

4.4.1 *Angular Sensing Methods*

As mentioned in Section 4.1.4, several methods for distance evaluation are based on optical triangulation. Because automatic focusing requires a knowledge of the distance of the optical device from the inspected surface, it is not surprising to find that triangulation is also the basis of many focusing techniques.

Techniques Implemented in Auto-Focus Cameras

Active triangulation methods in which a light beam from a small source, typically an infrared diode laser, is projected on the object surface and viewed at an angle by an optical sensor are used in some automatic cameras. Although both the projection and the viewing beam may use part of the camera lens, as shown in Figure 4-44 (Niwa et al., 1981), outside independent windows are usually preferred in order to increase the base length and thus the ranging accuracy, as discussed in Section 4.1.4. A system of this kind, using a pair of windows for the projection and viewing units located on each side of the camera with a base length up to 60 mm, is installed in the Canon AF35M auto-focus camera (Denstman, 1980). If maximum sharpness is to be obtained for a subject in front of a more distant background, it is essential for proper focusing that the subject be centered in the viewfinder target area which coincides with the infrared beam.

Passive triangulation methods are also possible for camera auto-focusing. In this case, no light is projected and the naturally illuminated scene is observed from two angular positions with two sensors such as two line arrays.

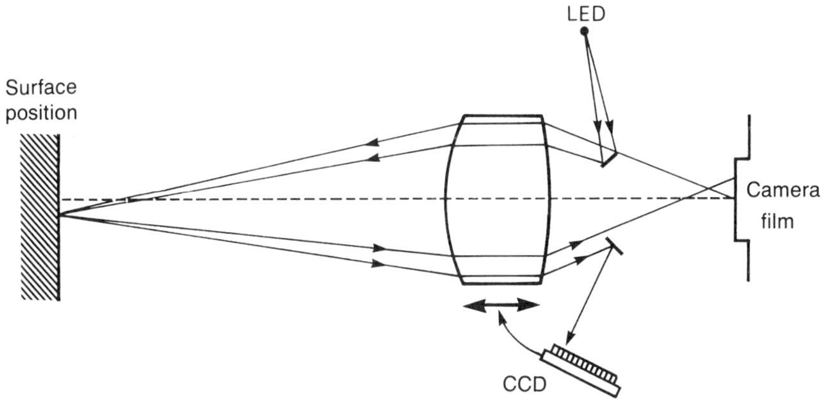

Figure 4-44 Active triangulation scheme for auto-focusing a camera using the peripheral parts of the camera lens.

The signals from the two arrays are correlated to obtain their relative lateral shift, from which the object distance is calculated. In a specific implementation by Siemens (Herbst et al., 1981), the output from two 52-element line arrays is digitized with only one bit per sample to minimize chip area and power consumption. Spatial differentiation of the line-array output improves the signal-to-noise ratio. A distance resolution of 15 ranges between 1 m and infinity is obtained with a 38 mm, *f*/2.8 lens, which is quite satisfactory for photography requirements.

Another angular sensing method for focus locating is the Foucault approach (Dupuy, 1964; Kocher, 1983). The basic principle is shown in Figure 4-45. If a knife edge is introduced near the image plane of a point source, the blur circle on a slightly out-of-focus observation plane appears as a half-moon pattern whose dark portion corresponds to the lower or to the upper half depending on whether the image plane is located before or after the knife edge position. If the image plane is exactly superposed to the knife edge, the blur circle will appear uniformly grey. The image plane can thus be determined by longitudinally displacing the knife edge or the lens.

A passive version of the Foucault focus sensing method has been implemented by Honeywell in its Visitronic HDS-23 sensor which has been adopted by several auto-focus camera manufacturers (Denstman, 1980; see also *Sensor Review*, Oct. '84, p. 172). This sensor is composed of 23 detectors in a linear array. Each detector is divided in two parts and has its own miniature lens, so that when located in place of the knife edge in Figure 4-45, each half senses the amount of light coming from the top or from the bottom

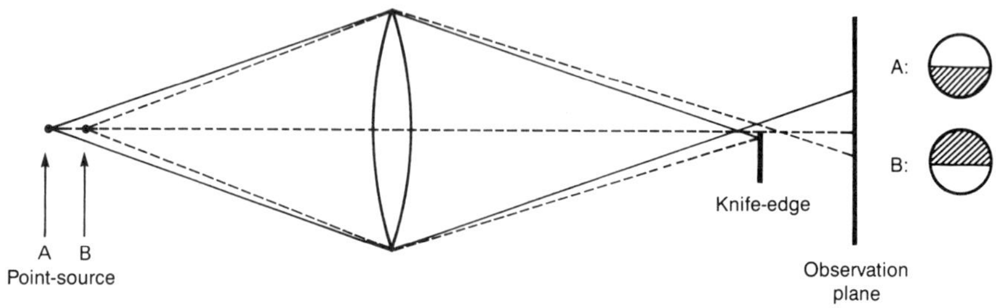

Figure 4-45 The Foucault method for locating the image plane position of a point source.

part of the lens. If the array is situated exactly in the image plane, the signals from the two halves of each detector will be identical, even if the image contains sharp light intensity edges. If, however, the array is located before or after the image plane, the difference between the two half-detectors situated near a sharp edge in the light distribution provides an error signal which is used to correct the objective position. The sensor can give 11 distance ranges between 1.1 m and infinity with a 38 mm, $f/2.8$ camera.

Unlike the Canon active-triangulation method, both the Siemens passive-triangulation and the Visitronic approaches require sufficient ambient light and high contrast image features across the line array for satisfactory operation. The auto-focus function is thus usually not available when flash exposure is required. The performances of the two described passive systems appear to be comparable. The wider popularity of the Visitronic sensor is probably a consequence of its earlier availability at an advanced stage of development. The Canon active triangulation method is quite different: It works properly even in poor lighting situations and on low contrast targets. It may only be fooled when pointing toward tilted shiny surfaces which provide insufficient light backscattering. Both the active and passive methods require that the target, to be kept in sharp focus, be centered in the middle of the picture.

Optical Disk Focus-Tracking Techniques

Although optical disk technology has little connection with industrial inspection, a brief review of automatic focus-tracking techniques in optical disk heads will provide a feeling for the accuracy and speed which can be reached in a closed loop auto-focus mechanism.

The Foucalt method has found several applications in focus-tracking loops for optical disk heads (Lehureau and Bricot, 1978; Earman, 1982; Arai et al., 1983). A typical configuration is shown in Figure 4-46. The polarized laser beam is focused on the disk and the reflected intensity is transmitted to the detecting stage by a polarizing-beam-splitter/retardation-plate combination of the kind described in Figure 3-38. A bi-cell detector collects the blur circle information which is used to continuously reposition the focusing objective to follow minute disk surface displacements due to thickness variations of the rotating disk. A surface-tracking precision of the order of 1 μm over a nearly 250 μm-range with a response time of a few ms can be obtained using electromagnetic actuators (Earman, 1982; Arai et al., 1983; Ohnuki et al., 1982).

A comparison of Figures 4-44 and 4-46 shows the basic similarity between the Foucault and the through-the-lens triangulation configurations: In both cases the lateral position of the peripheral rays is monitored (Johnson, 1979). The high angular aperture of the focusing objective makes double viewing triangulation unnecessary and difficult to implement in this case. Other versions of the basic triangulation method such as those shown in Figures 4-10 and 4-11 of Section 4.1.4 could also be used for auto-focus systems. An integrated system using a glass prism version of the Simon configuration shown in Figure 4-11 has indeed been recently proposed (Gaston, 1980). Still another version of the differential method shown in Figure 4-11, in which the converging and diverging beams are probed by a quadrant detector in the image plane of an astigmatic lens rather than by two separate detectors,

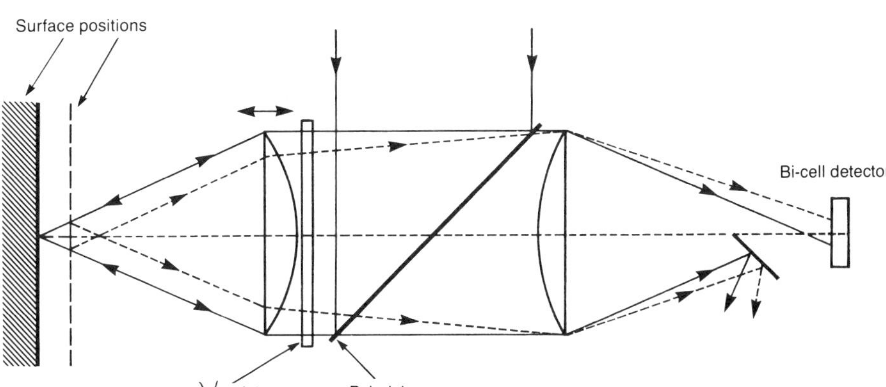

Figure 4-46 An implementation of the Foucault method for focus tracking in an optical disk head.

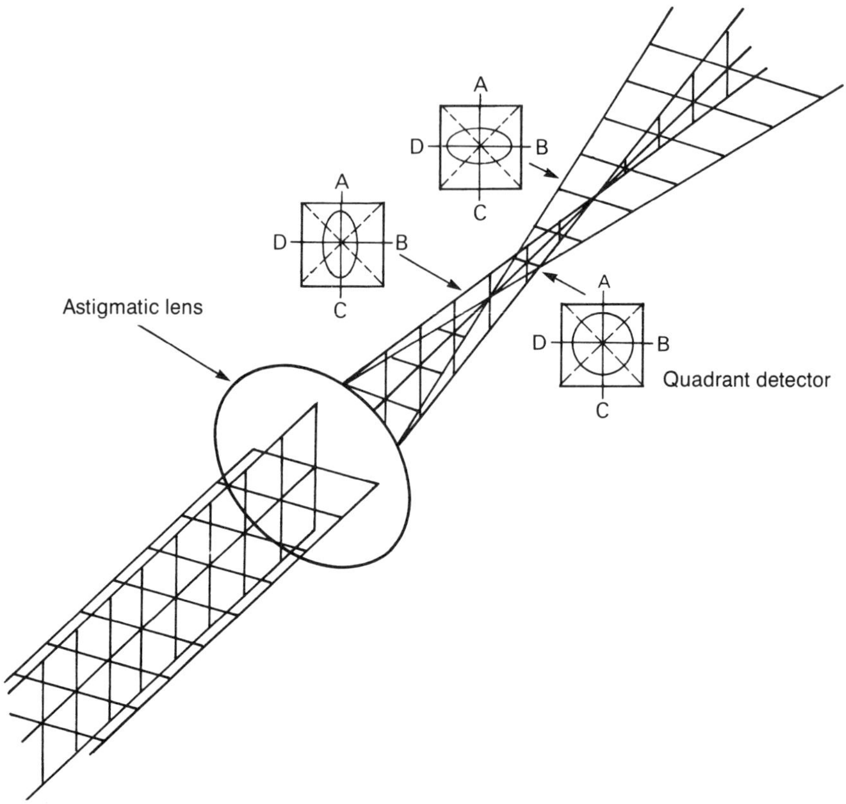

Figure 4-47 The astigmatic lens method for focus sensing.

is shown in Figure 4-47 (Cohen et al., 1984). The difference between quadrants a + b and c + d is related to the distance of the detector position from the image plane.

A different angle-sensing mechanism for automatic focusing using internal reflection near the critical angle in a prism has recently been described (Musha et al., 1982; Ito et al., 1983). The focus-tracking sensor is shown in Figure 4-48. The laser diode beam is focused on the disk surface, and the reflected beam is monitored by a critical-angle prism followed by a quadrant detector. The operation of such a device for a single-reflection prism is shown in Figure 4-48b. The central ray of the reflected beam is aligned so that it is incident at the internal prism surface at the critical angle for total internal reflection. If the disk surface is well in focus (position B), the reflected beam is parallel and produces uniform intensity on the quadrant detector. If the surface-reflected point source is displaced in the position A, the reflected beam is divergent. In this case the upper portion of the beam is incident on the prism surface at an angle slightly smaller than the critical angle so that the light beam is no longer totally reflected. The a and b portions of the detector output will thus be smaller than the c and d outputs, resulting in a focus error signal. The inverse is true if the point source is displaced in position C. Because the angular accuracy of a totally reflected beam is ultimately

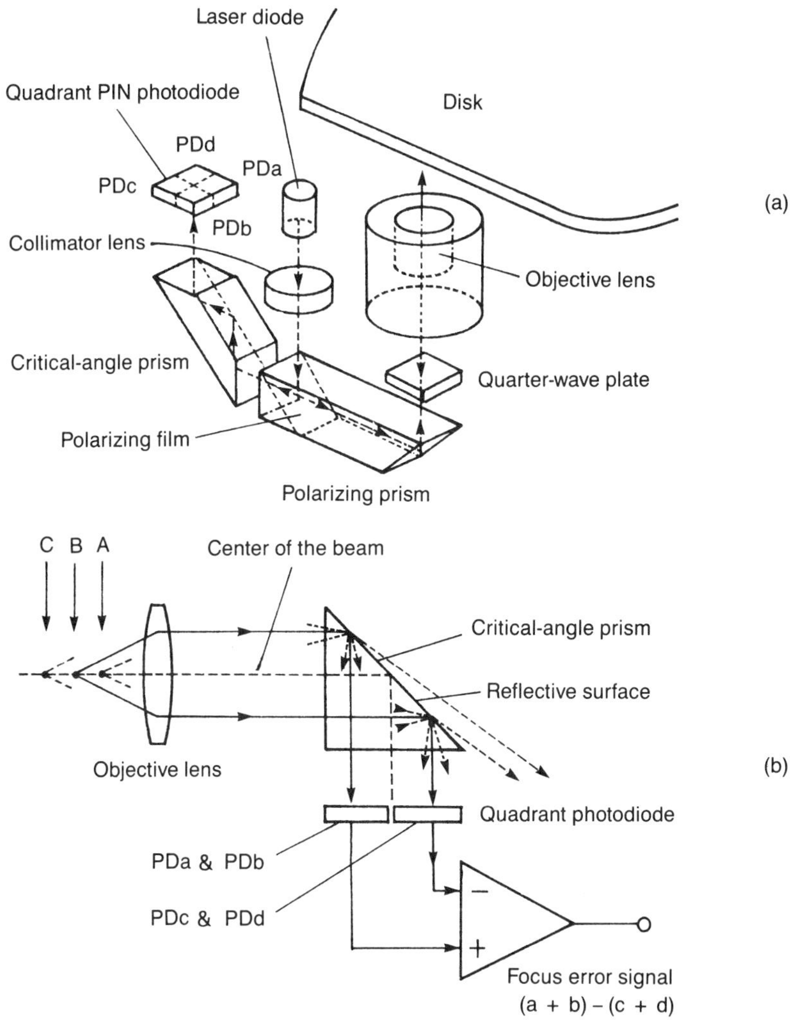

Figure 4-48 Critical-angle method of focus sensing; (a) sensor configuration and (b) basic principle of the quadrant detector output sensitivity to the focused-beam position.

limited by diffraction (Cielo and Delisle, 1975), the potential angle-sensing accuracy of such a device is similar to the classical triangulation method. However, the solidly assembled internal-reflection device is particularly insensitive to vibrations.

4.4.2 *Sonar-Based Approach*

An ultrasonic range finding approach has been chosen by Polaroid for automatic focusing of the SX-70 and Pronto cameras (Denstman, 1980; Gadient, 1983). As shown schematically in Figure 4-49, the object distance is determined by transmitting a short ultrasonic pulse and detecting the time delay corresponding to the arrival time of the first pulse-echo reflected by the object. The object distance can thus be quickly calculated within one of 128 depth-of-field zones in which the focusing range is divided, from 25 cm to infinity.

Like active triangulation methods, the sonar approach performs equally well in bright light or in total darkness. As to the pointing requirements, the angular divergence of the transmitted sonar pulse is relatively broad, as Figure 4-49 shows, making it difficult to miss the subject even if it is not well centered on the viewfinder. However, such a wide aperture may produce spurious echoes from nearby objects which happen to be closer to the camera than the principal subject. As discussed in Section 2.2.1, the directionality of

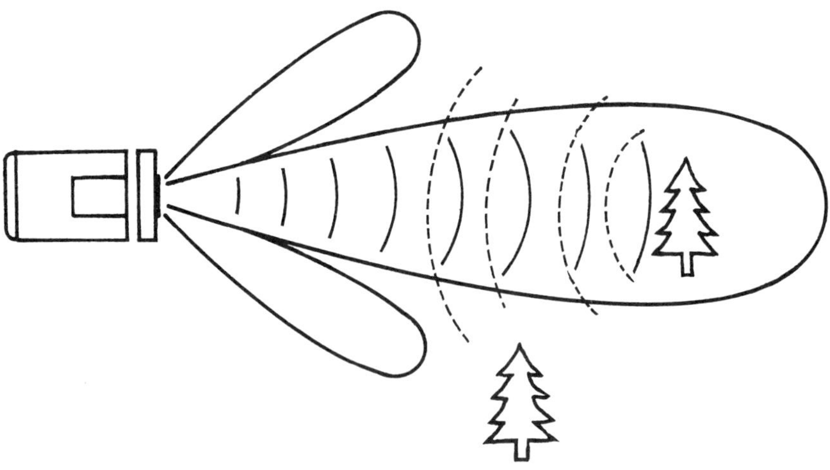

Figure 4-49 The Polaroid sonar based method for automatic focusing by acoustic pulse-echo range finding.

an ultrasonic beam is limited by the acceptable size of the transducer and by the highest frequency which can be transmitted through the air path with an acceptable attenuation.

The adoption of a nonoptical technique to obtain a sharply focused optical image is somewhat incongruous and results in some intrinsic limitations of the sonar-based methods. Objects much smaller than the relatively large acoustic wavelength do not reflect sufficient acoustic power to result in a detectable echo. As an example, obtaining a sharp image of a cobweb would be impossible with the sonar-based autofocus method. Similarly, smooth-surfaced objects specularly reflect away the long wavelength acoustic wave. This may result in spurious echoes or in signal fading when pointing toward a nonnormal, smooth object such as a brightly colored picture on a wall. Finally, the sonar-based auto-focus system is useless when taking pictures through a window.

4.4.3 *Image Sharpness Monitoring Methods*

The techniques for automatic focusing described in the last two sections are indirect techniques, in the sense that their immediate purpose is to localize the object or image plane rather than obtain a sharp, well-focused image. More reliable results under different conditions, such as with interchangeable camera objectives, should conceivably be obtained if the required result, i.e., the image sharpness, can be directly monitored.

The concept of image sharpness is closely linked with that of entropy, or quantity of information, carried by the image. If every picture element may have N distinguishable grey levels, the first order entropy of such an image is defined (in bits per pixel) as

$$H = \sum_{k=1}^{N} p_k \, ld(p_k), \tag{4-12}$$

where p_k is the probability of occurrence of the grey level k, i.e., the ratio of the number of pixels in the image with a grey level k and the total number of pixels, while $ld(p_k)$ is the dual logarithm of p_k. The entropy is maximum if all the grey levels have the same occurrence $p_k = 1/N$, in which case $H = ld(N)$. At the other extreme, if the image is uniformly grey with only one grey level whose probability of occurrence is equal to one, $H = ld(1) = 0$. According to Shannon's theorem, the entropy H is the quantity of information carried by the image, i.e., the lower limit of the average coding length in bits per pixel which can be reached by an optimum coding scheme without any loss of information.

An out-of-focus image generally contains less information and, consequently, has a lower entropy value than a well-focused image. This is easily understood with reference to Figure 4-50, showing the light intensity fluctuations across the image of a sinusoidal grid. If the image is out of focus, the light intensity range is compressed resulting in a lower entropy level. The entropy level can thus be used to find the best focusing position of an optical instrument (Dahne and Lanzl, 1980).

The first-order entropy is not necessarily the best parameter to monitor the degree of sharpness of an image. In some cases, the entropy level may decrease very slowly or even increase, as in the case of a square grid, when the image is slightly defocused. Better results are generally obtained with the second-order entropy, which is related to the probability of finding a grey level k_2 in the pixel adjacent to one having a grey level k_1 (Brillouin, 1962). This reflects the fact that the information in adjacent pixels of an image is not independent, particularly for blurred images.

Second- or higher-order entropy calculations require very long computation times. A more efficient approach for quantifying the degree of independence between adjacent pixels is to calculate the first-order entropy of the differentiated image, i.e., of the difference between adjacent pixels along each image line (Thum, 1984). Parameters other than the entropy may also be used to obtain a more sensitive and univocal correlation with the degree of image sharpness. Examples which have been considered are the notions of energy, local homogeneity, inertia, and contrast of the absolute or differentiated image (Haralick et al., 1973; Connors and Harlow, 1980; Kenny, 1983).

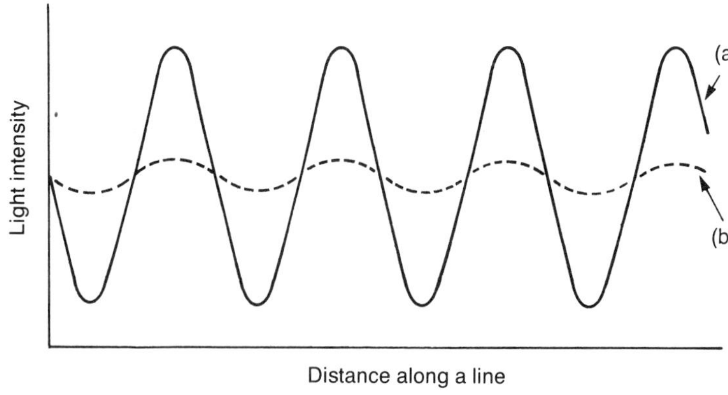

Figure 4-50 Image of a sinusoidal grid under (a) well focused conditions and (b) out-of-focus conditions.

Matched line correlation filters may give excellent results when some *a priori* information is available on the texture of the imaged objects and of the background noise, such as when analyzing the microscope image of fibers on a filter (Kenny, 1983).

The choice of the image sharpness monitoring algorithm must be made taking into account not only the theoretical performance of the different possible methods but also their speed and implementation cost. A particularly fast and simple approach for monitoring image sharpness as seen by a video camera is to high-frequency filter the line-by-line output. The high frequency portion of the power spectrum of such an output signal corresponds to the mean square light intensity difference between adjacent pixels in a line, a parameter which is strongly dependent on the image sharpness. The variation of the high-frequency cutoff in the output signal from a 384 × 485 pixels, 1/60 seconds frame period Hitachi camera as the image plane moves away from the focal plane is shown in Figure 4-51 (Hanma et al., 1983; Hanma and Saito, 1983).

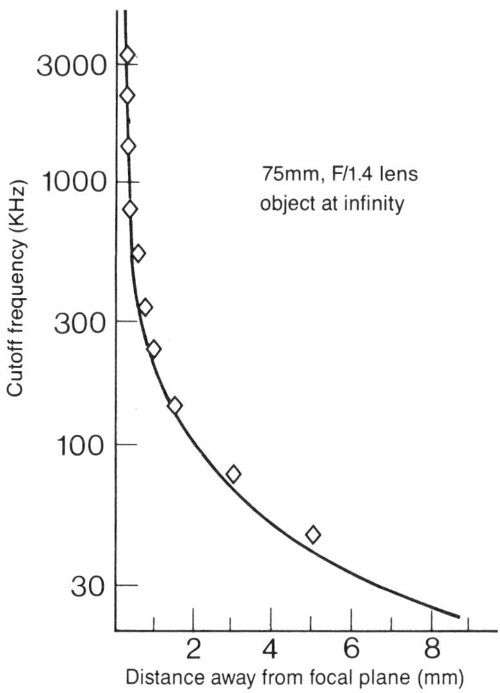

Figure 4-51 Cutoff frequency in the power spectrum of the output signal from a video camera as the lens is moved away from the best focusing position.

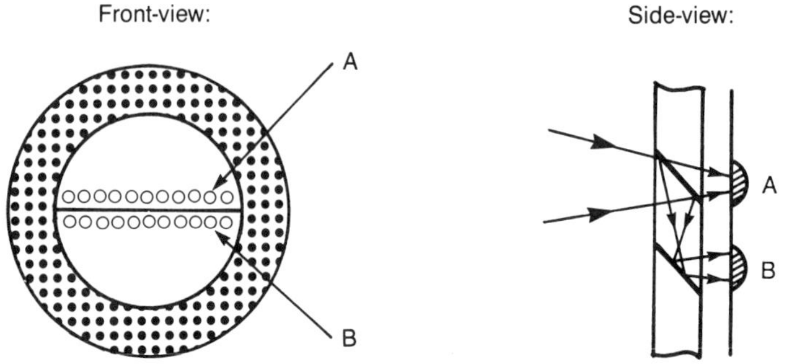

Figure 4-52 Close-up of the sharpness sensing sensor in the Pentax ME-F camera.

The absolute value of the sharpness dependent parameter such as the high frequency cutoff in the output of the Hitachi camera is obviously dependent on the contrast of the imaged scene. On the other hand, the focus adjusting time is normally limited by the mechanical inertia of the lens moving mechanism. For these reasons, the position of optimum focus is normally determined through a relative approach, moving the focusing lens until the sharpness sensor output has reached the maximum and starts decreasing, at which point the lens is moved back to the maximum position. It should also be noted that the bulkiness and mechanical inertia of the lens moving mechanism is an important cause of practical concern for consumer oriented auto-focus cameras (Mannheim, 1981). Agile image-plane displacement methods, such as moving only the rear focusing zoom lens (Hanma et al., 1983) are actively searched.

In a photographic camera there is no electronic signal output to directly monitor for image sharpness. Nevertheless, image sharpness sensing methods are being considered for photographic camera automatic focusing as an alternative to the angle-sensing or distance-ranging methods described in the previous sections. An image sharpness sensor has indeed been mounted in the Pentax ME-F camera and is illustrated in Figure 4-52 (Mannheim, 1981). The sensor consists of a double array of detectors which sample the center of the image, just above and below the image-split in the viewfinder circular area. During the pointing and focusing period nearly 35% of the incident light goes to the sharpness sensor while the rest goes to the viewfinder screen, as Figure 4-53 shows. The beam-splitting system is subsequently lifted during the film exposure period.

The right side of Figure 4-52 shows a cross section of the sensor arrays. A differential method similar in principle to the one illustrated in Figure 4-11

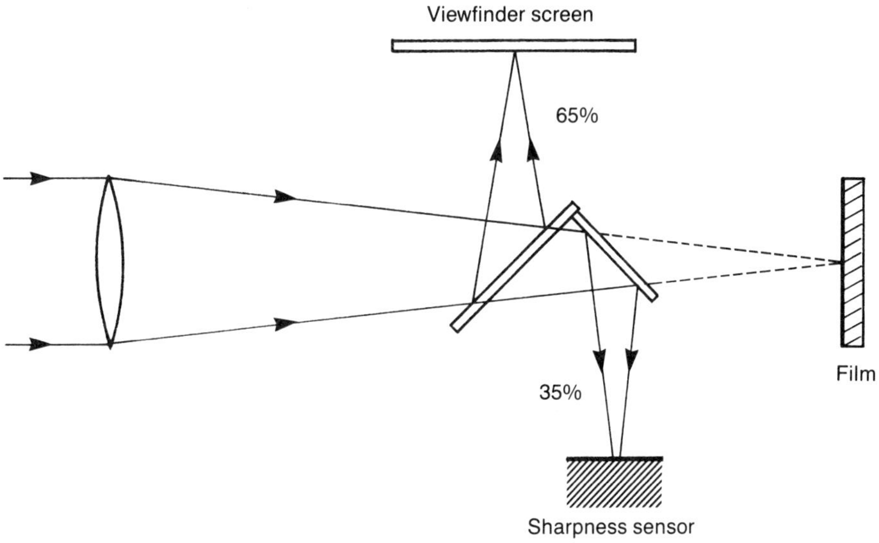

Figure 4-53 Movable beam-splitting system used in the Pentax camera for the pointing and focusing functions.

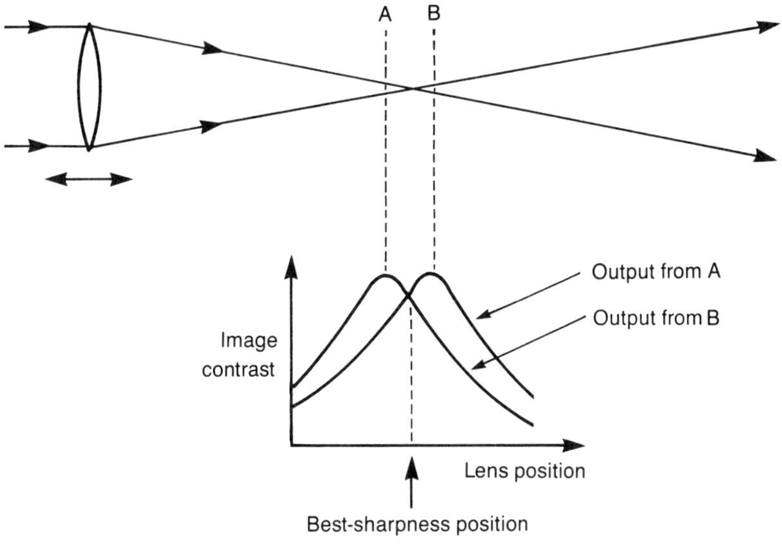

Figure 4-54 The differential approach used for the auto-focus mechanism in the Pentax ME-F camera.

is used to find the position of maximum sharpness. As shown in the unfolded diagram of Figure 4-54, the image contrast along the horizontal direction is sensed by the two line arrays which are situated at two different longitudinal positions through the beam-splitting prism shown in Figure 4-52. The output from each sensor is maximum when the image is well focused on the sensor plane, A and B respectively, in Figure 4-54. By setting the best focus point at the position where the outputs from the two sensors are equal, the maximum-locating sensitivity is greatly improved, as discussed in Section 4.1.4.

Similarly to all the passive auto-focus methods, image sharpness sensing techniques require sufficient ambient illumination. Moreover, this approach gives best results when pointing to images having a good light intensity contrast in the horizontal direction. This method is nevertheless interesting because the sensed image is exactly the same as the one reaching the film, so that parallax errors which may arise with double viewing triangulation methods are avoided. The image sharpness sensing method is particularly convenient with TV cameras because here the auto-focus function may be performed by processing the output signal without the need for an additional auto-focus sensor.

Chapter 5

Optical Gaging and Profiling

A number of optical techniques for the inspection of surface topography, including surface roughness analysis, profilometry of random surface fluctuations and on-line detection of surface defects were reviewed in the last chapter. This chapter is devoted to optical sizing techniques when the product dimensions have to be continuously monitored. This includes a variety of techniques, ranging from simple on-off sensors to determine the longitudinal dimension of the product on a conveyor belt, up to sophisticated three-dimensional imaging techniques for part sorting or robot guiding.

Optical metrology is the subfield of optical inspection which is experiencing the largest growth. Although general-purpose 2-D or 3-D artificial vision systems are attracting a lot of attention (see, e.g., *Report on Electro-optical Inspection Systems* by Frost and Sullivan, London, 1986; see also *Machine Vision Markets*, MIRC, Palo Alto, CA, 1986), there is a growing feeling that simpler sensors, requiring a smaller amount of data processing, are more effective for specific industrial automation and control tasks, at least in the short run. Examples are one-dimensional dedicated diode-array sensors for inspecting products moving along the production line, or straightforward but cheap and reliable object-presence sensors (see, e.g., Control Engineering, February 1987, pp. 79–82 and p. 168). Sensors of this kind are described with particular emphasis in this chapter.

5.1 On-line Optical Metrology

Photoelectric sensors and controls are among the most versatile tools found in industrial sensing instrumentation. Their applications to on-line product monitoring for quality and process control cover quite a wide range (Pfoutz, 1975; Stapleton, 1983; for a list of commercially available optical profile sensors, see the review *Measurement and Inspection Technology*, July 1982, p. 15–23). This section reviews single spot and one-dimensional scanning techniques for industrial sensing. Two-dimensional and three-dimensional imaging techniques will be covered in the next two sections.

5.1.1 *Part Sizing*

Inexpensive on-off light gauges are widely used in industrial control instrumentation. Examples are the detection of breakouts in rolled sheets, bar-code reading to identify and sort a given product, or the verification of the filling level in transparent containers (Considine, 1985, Chapter 9). Figure 5-1 shows some typical applications of simple transmission sensors to process control.

Figure 5-1a shows the operation of a slab cutting machine. A sheet of metal or other material is fed through a conveyor belt to a guillotine cutter. To precisely cut the slab pieces at a given length 1, a photoelectric device is inserted down the line at a distance 1 from the cutter. When the sheet edge intercepts the light beam, a signal is sent to the cutter and the sheet is sectioned. Such an approach ensures that the cut-slab length will not depend on the conveyor speed, which is subject to vary with the load weight.

Figure 5-1b shows another application of a photocell device. The internal surface of a series of box covers on a conveyor is sprayed with glue during package assembly. A transmissive source-detector device detects the presence of the box to turn the glue spray nozzle on and off in order to avoid glue dispersion when no box is present. Similar mechanisms are used to control automated painting systems.

Width Gaging by Reflection

A slightly more complex function of optical gaging systems is related to part sizing by a line array camera (see Section 3.4.1). A typical configuration is shown in Figure 5-2. The width of a strip, such as a paper web or a steel sheet, is imaged by the camera lens on a linear detector array. Typical photodiode or CCD line-arrays have 512 or 1024 elements, so that a resolution of the order of 1 mm over a 1 m-wide strip can normally be achieved. Higher

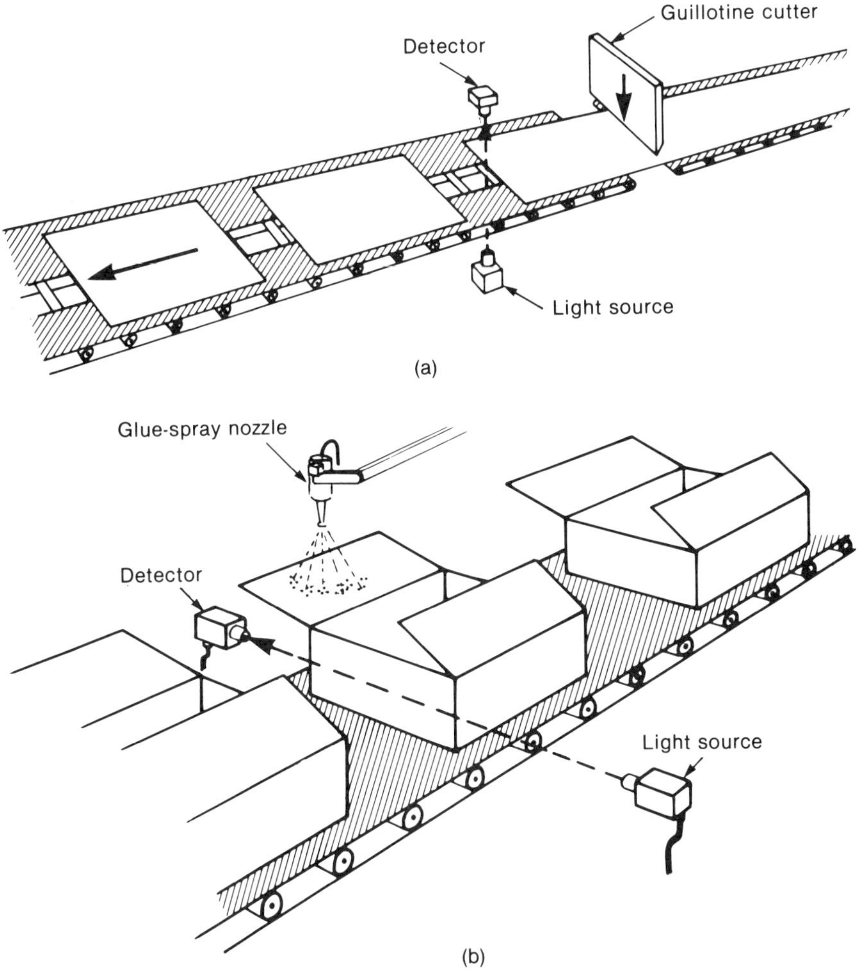

Figure 5-1 Some examples of optical monitoring devices for process control:(a) a slab-edge detection system for precisely guiding a guillotine to cut the slabs at a given length; (b) an optical box detection system to turn a glue spray nozzle on and off.

resolutions can in theory be obtained by interpolating between adjacent elements and subsequently applying a convenient threshold to the continuous curve thus obtained. In practice, signal fluctuations resulting from variations in the illumination distribution, lens aberrations, sensitivity nonuniformity across the detector array, as well as variations or inaccuracy in the camera-to-strip distance leading to parallax errors limit the precision to nearly

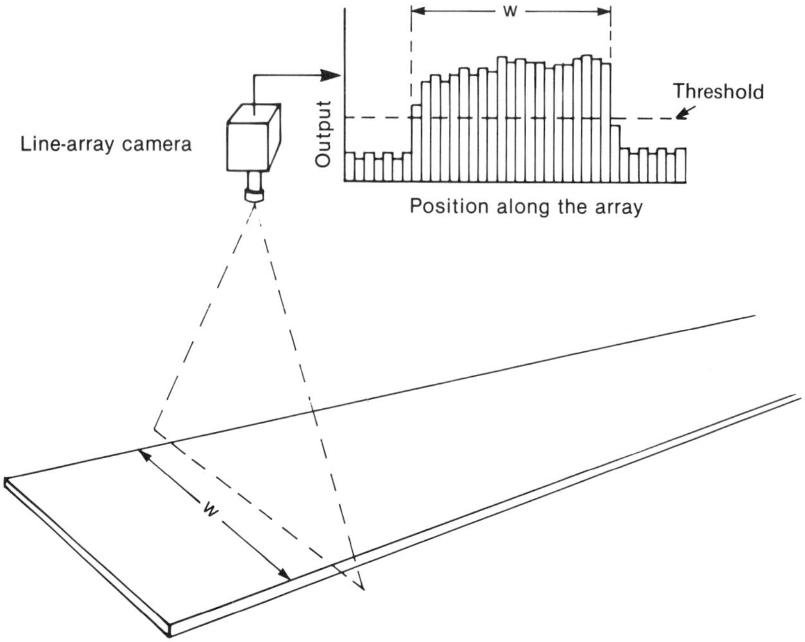

Figure 5-2 Strip sizing by a line-array camera.

one part per thousand in this configuration. In a typical application, hot H-beams of cross-sectional dimensions up to 0.5 m × 1.1 m are optically measured on-line with a precision of 1 mm (Pirlet et al, 1986).

The illumination of the sheet must be carefully chosen in an application of the kind shown in Figure 5.2. If the strip has a glossy surface, specular reflections should be avoided to eliminate the risk of local bright spots which could be interpreted as strip edges by the scanning camera. Several lamps at nonspecular incidence should ideally be distributed along the strip width to produce a well-balanced illumination across the sheet resulting in a uniform signal across the illuminated line array elements. Care must also be taken to ensure a dark background on each side of the strip edges by conveniently orienting the illuminating lamps.

Illumination problems are avoided when the inspected part is self-radiating. This is the case when inspecting steel sheets at incandescent temperatures (Burk and Simcic, 1978). By choosing suitable long wavelength detectors, sheets at temperatures of a few hundreds of degrees centrigrade over a cool background can be inspected by this method.

Transmission Measurements

Strip sizing as shown in Figure 5-2 may be implemented as well in transmission by back-illuminating the strip or by backing it with an elongated light source (Burk and Simcic, 1978). A configuration of this kind for the dimensional control of a wire is shown in Figure 5-3a. The wire is illuminated on the back by a uniform light source obtained by interposing a wide diffusing screen between the incoherent light source and the wire.

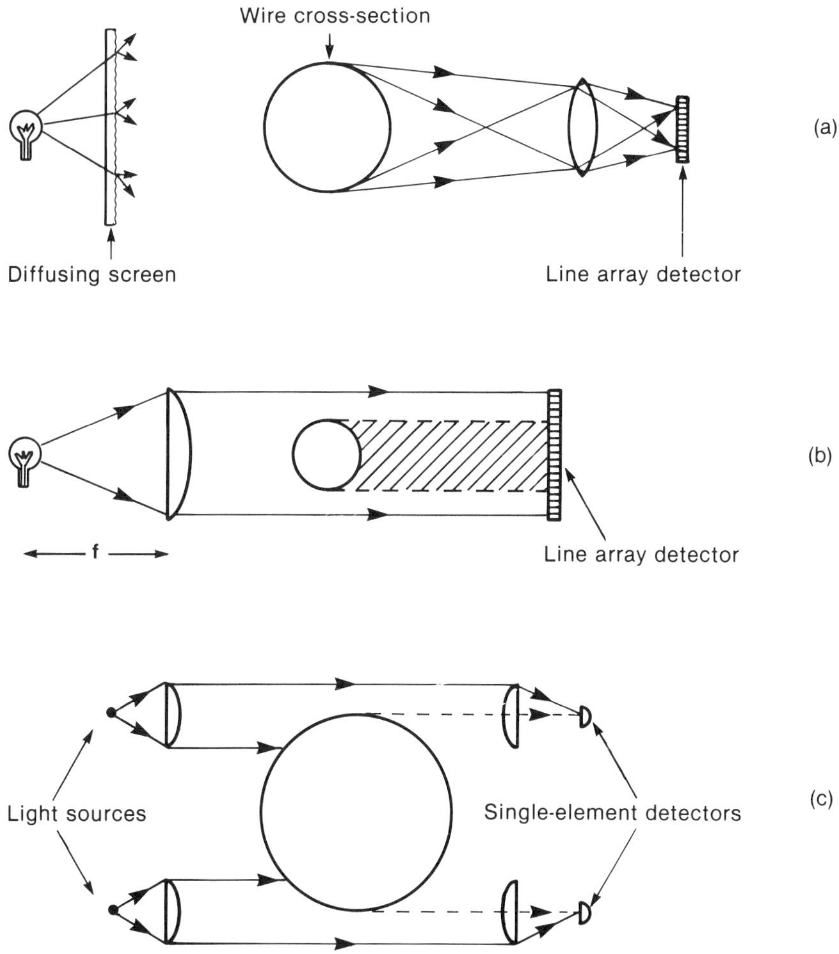

Figure 5-3 Different approaches for wire-diameter monitoring: (a) wire imaging in transmission; (b) shadow-projection technique; (c) double-projection method.

Similar results are obtained by placing a fluorescent tube close to the wire and perpendicular to its axis. The dark region corresponding to the wire is imaged by the lens on the detector array.

It should be noted (Pfoutz, 1975) that with a configuration of the kind shown in Figure 5-3a the apparent wire diameter recorded by the line array detector is strongly dependent on the longitudinal position of the wire (i.e., from left to right in Figure 5-3), a parameter which is often difficult to control in practical situations. Moreover, side reflections of the wide angular distribution light beam from the wire edges are also imaged by the lens on the linear array, causing a reduction of the apparent wire diameter which is more pronounced for large values of the wire surface reflectivity.

The configuration shown in Figure 5-3b avoids most of these problems. By placing the source in the focal plane of a lens, a quasi-collimated beam is projected on the wire. The size of the shadow area projected on the line-array detector is thus, to a large extent, independent of the longitudinal wire position, while side reflections are reduced because of the limited angular aperture of the beam projected by the lens. Best results are obtained with a very small source size, at the expense of signal level. It should also be noted that the size of the wire which can be inspected by this method, in contrast to the configuration in Figure 5-3a, cannot exceed the lens diameter.

For the inspection of larger wires, the approach shown in Figure 5-3c can be applied (Swing, 1975). In this case, two sources are pointed toward the edges of the cable, which must have a relatively constant position. Two single element detectors with added outputs can be used to evaluate the cable diameter. If the two projected light beams have uniform intensity across their cross section, the sum of the two signals is independent of wire oscillations but decreases as the wire diameter increases. Such a configuration is, however, affected by differences in the detector sensitivities as well as by light source aging. A similar configuration using a reference detector to compensate for common-mode noise was recently reported to evaluate wire temperature by measuring its diameter before and after induction heating (Dyos and Smith, 1986).

Precise wire diameter measurement can be obtained by using telecentric configurations (see Section 3.8) of the kind shown in Figure 5-4. Figure 5-4a shows an incoherently illuminated array detection system which is made telecentric on the image side by inserting a small diaphragm at the focal point of the collecting lens. Such a diaphragm restricts the view of the detector to the rays which are parallel to the axis. The result is equivalent to having a point source at the focal point of the transmitting lens.

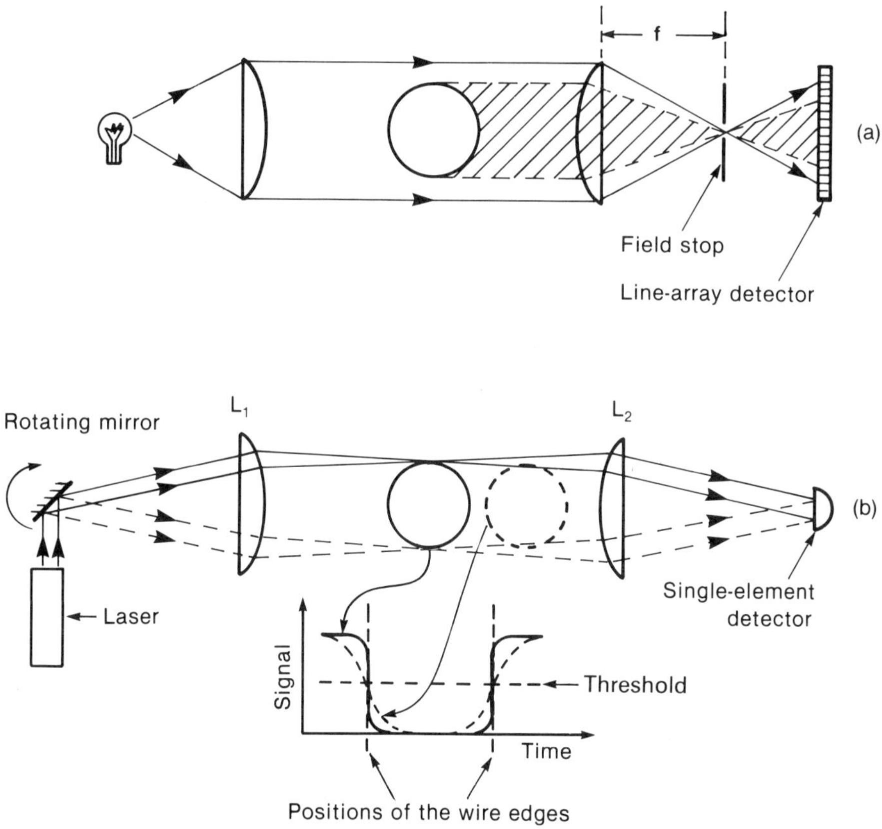

Figure 5-4 Telecentric configurations for wire diameter monitoring independently of its position: (a) incoherent illumination, optical system telecentric on the image side; (b) scanning laser illumination, single element detection.

Figure 5-4b shows a laser scanning wire diameter gaging system (Spalding, 1971; Swing, 1975; Zanoni, 1977). A narrow, collimated laser beam is angularly scanned by a rotating mirror or polygon situated in the focal plane of lens L_1. Such a lens should be an $F\theta$ lens (see Section 3-8) for good measurement accuracy. A standard collecting lens L_2 concentrates the emerging laser beam to a large area, single element detector. The shape of the detected signal versus time is shown in the insert for a wire position well centered at the waist of the focused laser beam as well as for a position displaced along the optical axis. As we can see, the transition from light to darkness is much sharper when the wire is well positioned. Nevertheless, an accurate wire diameter reading is also obtained in the case of a misplaced wire position if the threshold level can be set at exactly half the signal level outside the shadowed region. Typical accuracies of commercial laser-scanning wire inspection devices (see e.g., the publication *Inspection, Gauging and Testing* by SME, Dearborn, MI) are of the order of 2 μm for a 2 cm-diameter wire. Stable surfaces can be located with a repetitivity of 0.2 μm in certain cases (Pryor and North, 1985, p. 3, 11 and 60). Models comprising two scanning heads at right angle to each other for the evaluation of wire ellipticity are also available.

Fiber Sizing by Coherent Techniques

As seen above, shadow projection or laser-scanning techniques are appropriate for optically gaging relatively large objects. When smaller objects, such as fibers in the 10 to 100 μm range, are to be inspected, such geometrical optic methods become inaccurate because of the diffraction effects as well as the difficulty of finely focusing the beam with a convenient depth of field. Light-scattering methods may be used effectively to monitor the average properties of inhomogeneous materials containing randomly oriented small particles, as will be described in Section 8.4.3. Some typical coherent optical methods for the inspection of small diameter fibers are reviewed here.

Several fiber diameter gaging techniques have been developed for on-line control of optical fiber drawing machines (Smithgall et al., 1977; Ashkin et al., 1981; Podmaniczky, 1983). Figure 5-5 shows two typical configurations. The left diagram shows the basic operation of the forward scatter technique for diameter monitoring. The interference between the light beam reflected by the outer fiber surface interferes with the beam refracted through the fiber, as well as with multiple reflections at the core-cladding interface for step-index fibers, to produce a characteristic angular scatter pattern which is monitored in real time to control the fiber drawing parameters. A beam aligning cube allows quick displacement of the beam without angular deviation to follow fiber displacements within a 1 cm window. Accuracies of the order of 0.25 μm for a 100 μm-diameter fiber have been obtained (Smithgall et al., 1977).

Figure 5-5b shows the backscattered beam paths for a different optical approach to optical fiber diameter monitoring. Interference takes place between the front-reflected and back-reflected beams incident at the center of the fiber, as well as between the surface waves counterpropagating around the fiber circumference. By suitably positioning a near-field slit to stop either the R or the S beams, either of the two interference fringe patterns can be

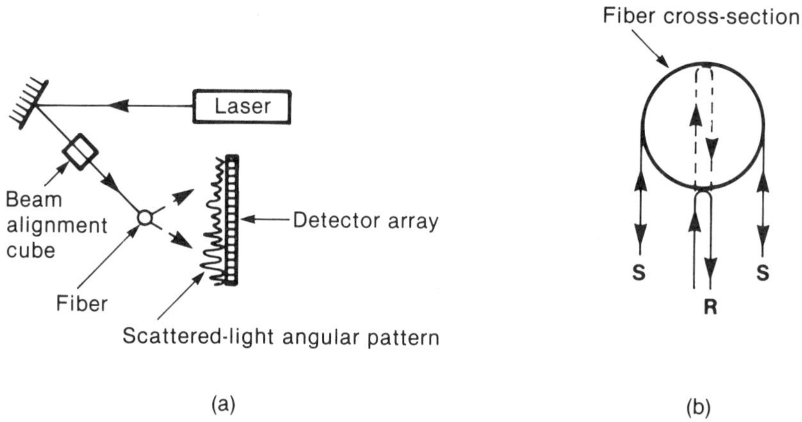

Figure 5-5 Typical configurations for optical fiber diameter monitoring by coherent optical methods: (a) experimental setup for scanning the angular pattern of the scattered light; (b) schematic diagram of the front and back reflected (R) as well as of the surface wave (S) interfering beams.

monitored. Fringe observation is accomplished on the backscattered field within a narrow angular aperture by illuminating the fiber with a variable wavelength dye laser beam and by monitoring the interference fringes across the optical spectrum. Accuracies as high as $10^{-3}\,\mu$m have been reported using such a technique (Askin et al., 1981).

For monitoring opaque, specularly reflecting rods of larger diameters, a different method has recently been reported (Langlois et al., 1985). The approach is based on the use of diffraction edge waves generated by illuminating the edge of a metallic cylinder by a grazing-incidence beam. The ratio between the light intensity diffracted at positive and at negative angles is shown to be strongly dependent on the cylinder radius of curvature at the illuminated point. Preliminary experimental results indicate that an accuracy of the order of one part per thousand is possible for radii of curvature in the 5 mm range.

5.1.2 *Alignment Techniques*

The axis of a narrow collimated beam provides a convenient straight line reference along the chosen direction. Such a property has long been used for the alignment of parts, particularly since the advent of high brightness collimated laser sources.

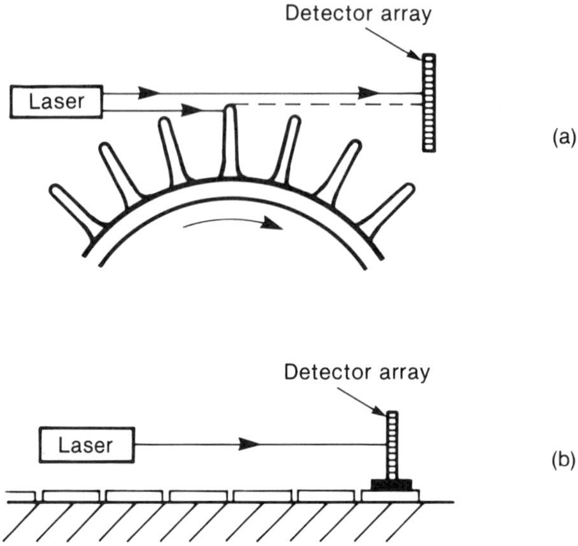

Figure 5-6 Two optical-alignment applications with a laser beam: (a) blade tip clearance evaluation in a turbine engine; (b) checking the height uniformity of a pavement by reference with a narrow laser beam.

Figure 5-6 shows two examples of the application of a coherent light beam to alignment needs (Baumbick and Alexander, 1980; Davinson, 1984). Figure 5-6a illustrates the use of a collimated beam to check the uniformity of blade clearance in a turbine engine. The blade tip clearance must be checked very accurately to optimize the efficiency and avoid risk of damage in such high performance engines. By measuring the center of gravity of the shadow projection on an array detector during rotation of the turbine rotor, the blade-tip uniformity can be verified to within a few micrometers. Figure 5-6b shows another simple application of the alignment capabilities offered by a laser source. In this case, a narrow collimated beam is used as a fixed reference against which a number of components can be aligned. The height of each component can be adjusted by measuring its position through a detector holding stand. Several applications of such a technique can be found in the construction and manufacturing industries.

The Autocollimator

The autocollimator is a classical instrument which has been in use for a long time for aligning tasks (Young, 1967). A schematic diagram of such

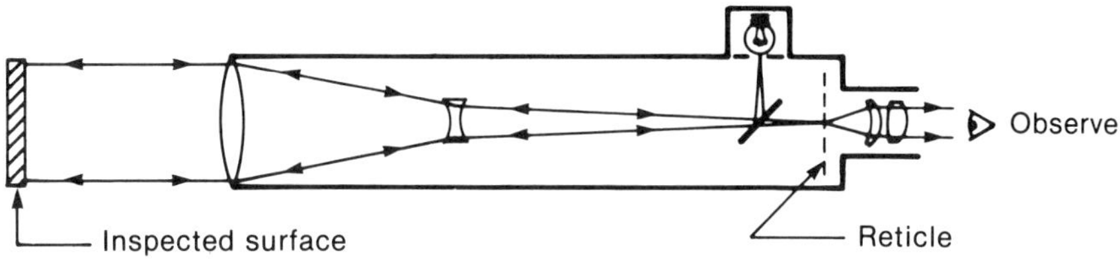

Figure 5-7 Schematic diagram of an autocollimator.

a device is shown in Figure 5-7. The autocollimator is basically a telescope focused at infinity with an internal pinhole point source. The parallel beam is projected on the specular surface to be inspected and is reflected back to form a point image on the reticle. The position of the point image is observed visually through an eyepiece or it is electro-optically monitored by a position sensing or line-array detector. The position of the focused spot on the reticle depends only on the relative angle between the reflecting surface and the autocollimating telescope. Once the instrument has been pointed in a given direction using a reference surface, the orientation of other surfaces can be precisely determined by comparison with the axis of the autocollimator with a precision which may reach 1/50 arc sec (Wyant, 1975).

Some applications of the autocollimator are illustrated in Figures 5-8 and 5-9. Figure 5-8a shows how this device can be used to precisely verify the flatness of a plane surface. The autocollimator is first well aligned normal to a mirror stand placed on the center of the surface. Variations from flatness of the surface can subsequently be monitored by sliding the stand along the axis of the autocollimator. A two-dimensional surface scan may as well be performed without readjusting the collimator by tilting the mirror out of the

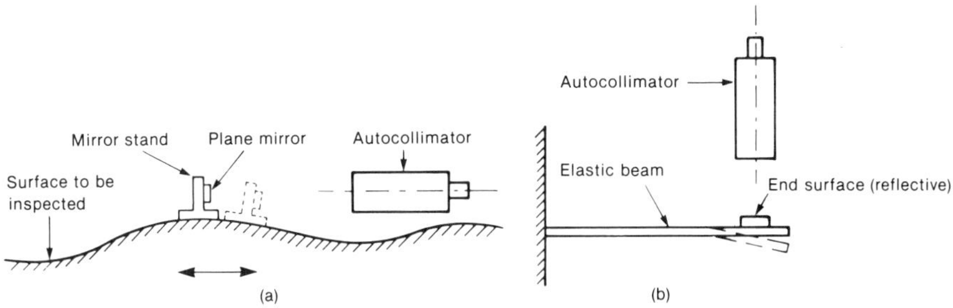

Figure 5-8 Some applications of an autocollimator device to (a) checking the flatness of a surface by a sliding mirror stand; (b) measuring the elastic constant of a horizontal beam by monitoring its deflection under load.

plane normal to the autocollimator axis and using a second mirror stand to scan along the newly defined axis.

Figure 5-8b shows an application to the measurement of the elastic deflection constant of a horizontal beam. A mirror is placed at the end of the beam and the autocollimator is aligned normal to the mirror plane. When a load of given weight is laid on the beam tip, the corresponding slight angular deflection of the beam can be precisely evaluated from the autocollimator reading. The elastic constant of the beam can thus be measured and then used to evaluate useful parameters such as the normal strain or the resonance frequency of the structure.

The autocollimator is widely used to inspect the accuracy of high precision manufactured parts. The base angle of a prismatic piece can be routinely inspected by setting the autocollimator axis normal to a standard angle block and subsequently measuring angular deviations of the manufactured workpieces placed on the same standard surface, as Figure 5-9a shows. Figure 5-9b shows how the presence of pyramidal errors (i.e., angles between the facet and the rotational axis) can be detected on a rotating polygon. As mentioned in Section 3.8, pyramidal errors are a major cause of signal jitter when such polygons are used for beam scanning applications and must be limited to values smaller than 10^{-4} radians. By slowly rotating the polygonal mirror around its axis and observing the displacements of the focused spot on the autocollimator reticle, such deviations can be evaluated with high precision.

5.1.3 *Thickness Measurement*

The measurement of sheet or web thickness is a problem of considerable economic importance in the material processing industry. If a minimum sheet

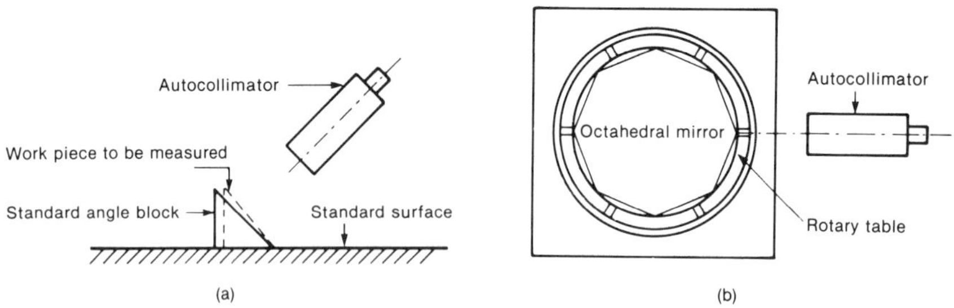

Figure 5-9 Some applications of the autocollimator to angle measurement of manufactured pieces: (a) comparing the angle of a workpiece to a standard angle block; (b) evaluating pyramidal errors in the facet orientation of a rotating polygon.

thickness of, e.g., 1 mm is to be assured, the manufacturer may have to adopt a 1.1 ± 0.1 mm thickness in order to be sure to meet the minimum requirements. If, however, the sheet thickness can be controlled to within ± 0.01 μm by proper on-line gaging and feedback control of the extruding or rolling machine, the average product thickness can be reduced at 1.01 mm, with a raw material savings of nearly 10%. At continuous production speeds of meters per second, this represents an important reduction in costs. It is not surprising that control instrumentation companies may expend considerable effort to develop precise and reliable thickness monitoring devices.

A number of nonoptical thickness gaging devices have been described in Chapter 2. The best known approaches are the radiographic and the ultrasonic techniques. Both of these approaches have their limitations for on-line applications. Radiographic methods, including X-ray, gamma-ray and beta-ray devices, require safety precautions during their installation and operation. Moreover, because they rely on the measurement of an absolute attenuation signal, radiographic devices require frequent recalibration to avoid the effect of source aging or dirt build-up on the windows. In the case of backscattering devices, the probe-to-workpiece distance has to be maintained to within a fraction of a millimeter to avoid r^2 signal fluctuations (see Section 2.3.2), a requirement which is hardly compatible with strip-wobbling conditions in high-speed sheet production mills. Single-side ultrasonic pulse-echo devices are mainly limited by the requirement of good liquid coupling between the transducer and the workpiece, which limits their application to relatively low-temperature materials. Moreover, both the nuclear radiation attenuation and the ultrasonic velocity depend not only on the sheet thickness but also on the physical characteristics of the material such as density, phase composition or microstructure. This is a major source of inaccuracy and unreliability in practical working conditions.

This section describes some typical thickness monitoring optical techniques for both opaque and partially transparent materials. The accuracy and reliability of such methods in an industrial environment is discussed.

Inspection of Opaque Materials

Slab inspection is an important application of noncontact optical position monitoring devices in the steel industry (Williams, 1975; Pirlet et al., 1986). Before entering a hot-rolling mill, steel is reduced from ingot dimensions to slabs, which have a typical thickness of 10 to 30 cm. The noncontact, remote sensing capabilities of optical sensors make them suitable for operation under difficult industrial conditions, namely high slab temperature often in the

1000°C range and hostile environmental conditions often found in the rolling mill. Figure 5-10 shows a schematic diagram of an optical slab thickness sensor (Williams, 1975; Morander, 1980; see also *Optocator Application Report* No US-1, Selective Electronics, Valdese, N.C.). It is basically composed of two optical triangulation systems (see Section 4.1.4) whose outputs are subtracted to obtain the slab thickness. Similar techniques have recently been applied to wall thickness measurement for quality control of can ends for the beverage industry (Asher and Jackson, 1986). It is important to note that each sensor gives the relative position of the illuminated spot, rather than an absolute signal as in X-ray attenuation thickness evaluation devices. Changes in signal amplitude due to source aging or moderate dirt accumulation on the viewing windows do not affect the beam position signal. Typical accuracies are in the range of 0.1 to 1 mm depending on the operating distance.

A similar application for on-line steel inspection is in the field of hot strip flatness monitoring (Wiese et al., 1980; Pirlet et al., 1986). The departure from flatness of the steel sheet after hot rolling, including center buckle and edge

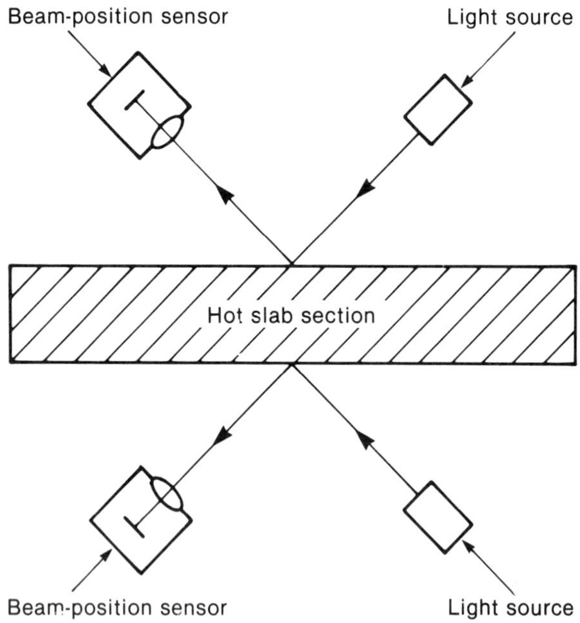

Figure 5-10 Thickness monitoring of a hot steel slab by two optical triangulation devices.

wave undulations, is an important perameter in hot rolling for product quality as well as for rolling process control. Figure 5-11 shows a typical optical sensing station mounted above the hot-rolled sheet for open-loop manual control of the process by the supervising operator. Three triangulation based sensors are mounted above the center and the two sides of the strip to detect the amplitude of longitudinal surface deformations as well as the presence of transverse buckling across the sheet width. As Figure 5-11 shows, a number of precautions must be taken when installing such devices on a hot-rolling mill. The assembly package must be water cooled to avoid overheating of the sensor and associated electronics by the hot strip radiation. Fiber optic line-array probes may eventually be employed to transfer the electro-optical devices to a less hostile environment at a certain distance from the mill. A robust metallic shield is provided to protect the probes from physical damage by the wobbling sheet as well as to increase thermal insulation from the radiating hot metal. An air-jet purging system assures proper long-term protection of the optical windows from lubrication oil residuals or other aerosol components in the mill atmosphere. A system of this kind is installed at the Domnarvet plant of Swedish Steel AB to supervise the production of 0.65 to 1.65 m-wide strip at rates of 750,000 tons/year.

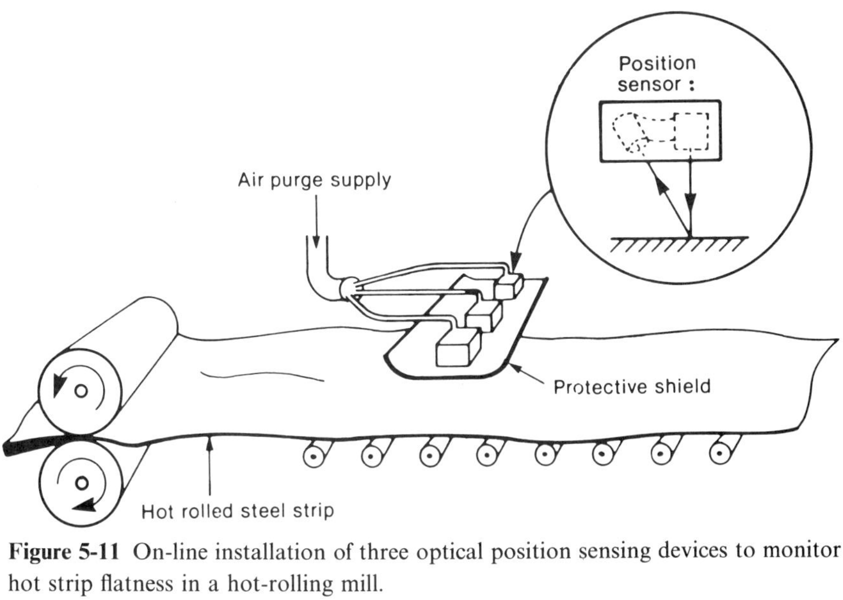

Figure 5-11 On-line installation of three optical position sensing devices to monitor hot strip flatness in a hot-rolling mill.

Monitoring the Thickness of Transparent Films

The thickness inspection of thin organic sheets, such as extruded polymeric sheets or blown film polyethlene materials, is a problem of great concern for the plastics and rubber industry. As mentioned in the previous section, a precise machine-direction thickness control is an important requirement for material savings considerations. Moreover, a departure from a uniform cross-machine-direction thickness profile indicates the presence of a process malfunction such as an eccentricity of the concentric extruding dies in a cylindrical film blowing machine.

Although a wide variety of physical approaches are possible for thickness measurement including electric capacity, microwave attenuation and ultrasonic pulse-echo (Lenk, 1980, Section 7-4), thermal response time monitoring (see Section 5-1) and optical triangulation (Corey, 1986), the most widely used methods for thickness measurement of thin plastic films are nuclear and infrared absorption techniques. Nuclear sensors of the gamma-backscattering type can inspect plastic sheets from 5 μm to 100 μm in thickness, while higher power X-ray transmission devices can be used for thicker sheets. However, such instruments are relatively expensive and have a number of limitations: They require frequent recalibration, must be installed in close proximity to the film, have a relatively slow response time, which is normally inadequate for machine-direction thickness fluctuation monitoring, and must be handled with care because of radiation hazards.

Infrared techniques (Edgar and Hindle, 1980; Burk, 1984; Hassel, 1985; Hindle and Edgar, 1986) are increasingly used for monitoring the thickness of transparent films. Two typical configurations are shown in Figure 5-12. In both diagrams, a light beam is transmitted through the film. A similar configuration is used in reflection for monitoring the thickness of coatings on opaque substrates. The spectral distribution of the light transmitted through the film is given by the Lambert absorption law

$$T(\lambda)/T_0 = \exp[-k(\lambda) \cdot d], \tag{5-1}$$

where $k(\lambda)$ is the absorption coefficient in cm^{-1} and d is the film thickness in cm, $T(\lambda)$ is the transmitted intensity at the wavelength λ and T_0 is the intensity transmitted in a spectral region where there is no appreciable absorption (although Fresnel reflection losses may occur). The absorbance of the film is defined as

$$A(\lambda) = -\log_{10}[T(\lambda)/T_0] = 0.4343 \cdot k(\lambda) \cdot d. \tag{5-2}$$

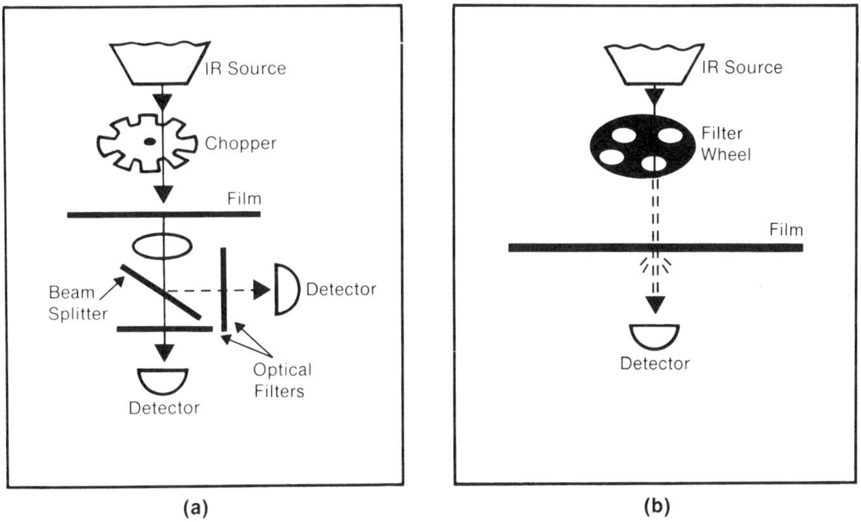

Figure 5-12 Typical infrared-absorption configurations for monitoring film thickness by (a) simultaneous comparison of transmissivity of two wavelengths and (b) sequential analysis of spectral transmissivity.

From a measurement of the transmittance $T(\lambda)/T_0$ at the chosen wavelength, one can then evaluate the absorbance and thus the thickness from equation (5-2) after the factor $k(\lambda)$ has been determined from a calibration with samples of known thickness. It should be emphasized that such a ratio measurement is unaffected by common-mode noise effects such as source aging or spectrally neutral dirt accumulation.

Figure 5-13 shows the transmittance spectrum of a moderately dyed, 15 μm-thick polyethylene film (Cielo et al., 1986c). Absorption regions are clearly seen around 4200, 2900, 1420 and 730 cm^{-1} which are characteristic of the polyethylene molecular vibration bands. By choosing as T_0 and $T(\lambda)$ the transmittance values at about 2.1 μm (4750 cm^{-1}) and 2.35 μm (4250 cm^{-1}), the absorbance and thus the thickness can be evaluated with rugged low wavelength detectors and without being affected by the 1.9 μm and 2.7 μm moist air absorption bands (see Figure 7-11). The presence of strong film pigmentation usually has only a moderate effect on the IR transmission spectrum.

One of the problems encountered with films containing a large amount of light-scattering pigments or having a relatively rough surface finish is the wavelength dependence of the scattering related attenuation in the transmitted or reflected light beam. As shown in Figure 5-14, scattering losses are

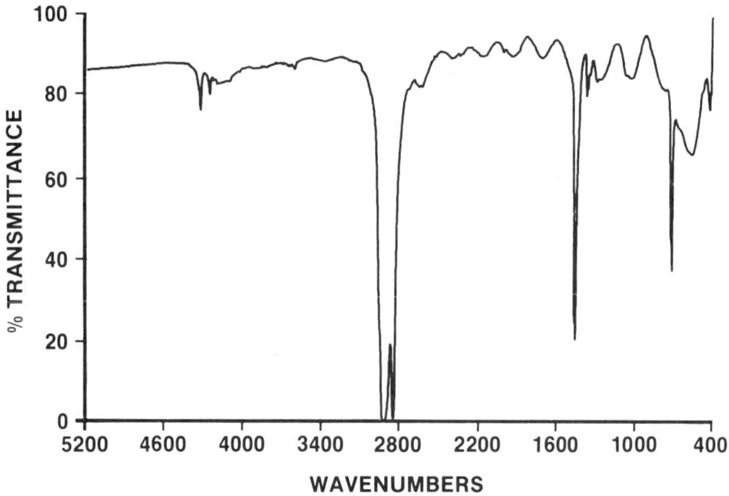

Figure 5-13 FTIR spectral scan of light transmitted through a 15 μm-thick polyethylene film.

higher at low wavelengths when a large amount of pigment is present in the film. An expression of the attenuation suffered by the unscattered light beam as a function of the ratio between the root-mean-square roughness of the reflecting surface and the light wavelength is obtained from equation (4-10) in the last chapter.

If the thickness is evaluated by simply taking the ratio between the transmittance at, e.g., 2.1. μm and 2.35 μm, Figure 5-14 shows that the estimated value will be smaller than the real value when the scattering related attenuation is of the same order of magnitude as the absorption related attenuation. Better thickness evaluation can be obtained independently of the scattering level by measuring the transmittance at three wavelengths, e.g., 2.05 μm, 2.2 μm and 2.35 μm, and using the two signals detected at the lower wavelengths to extrapolate the transmittance before absorption at 2.35 μm, as shown by the dotted lines in Figure 5-14. Figure 5-15 shows the improvement in absorbance versus film thickness correlation linearity obtained with a number of extruded polyethylene samples by using the three-wavelength correction (Cielo et al., 1968d).

Figure 5-13 shows the presence in the 2800 to 1000 cm^{-1} spectral region of some interference fringes due to multiple reflections within the film. Such fringes are particularly evident in the longer wavelength spectral region where the surface roughness amplitude becomes small with respect to the optical

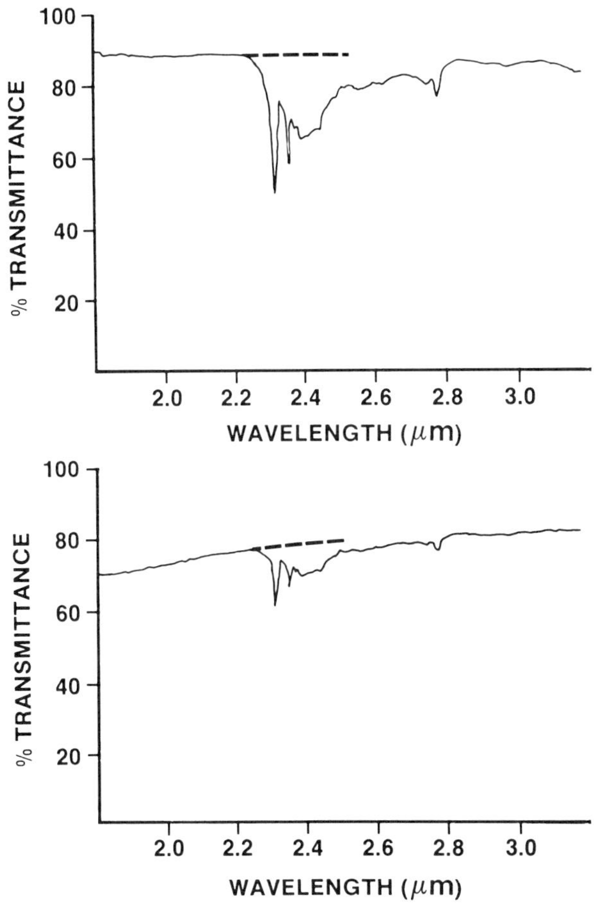

Figure 5-14 Short wavelength spectral distribution of the light transmitted through a 65 μm-thick transparent film (top) and a diffuse-white pigmented, 20 μm-thick film (bottom).

wavelength. The contrast and position of the fringes is also dependent on the angle of incidence and on the angular aperture of the inspecting device. For smoother films, the presence of fringes in the low wavelength region is a major cause of noise in the evaluation of the absorption peak depth. Several methods can be used to minimize the interference fringes (Kiedron, 1986), such as by using larger angular apertures or oblique angles of incidence or by superposing both the transmitted and reflected light beams which have reversed fringe patterns. Another approach is to use the fringe information

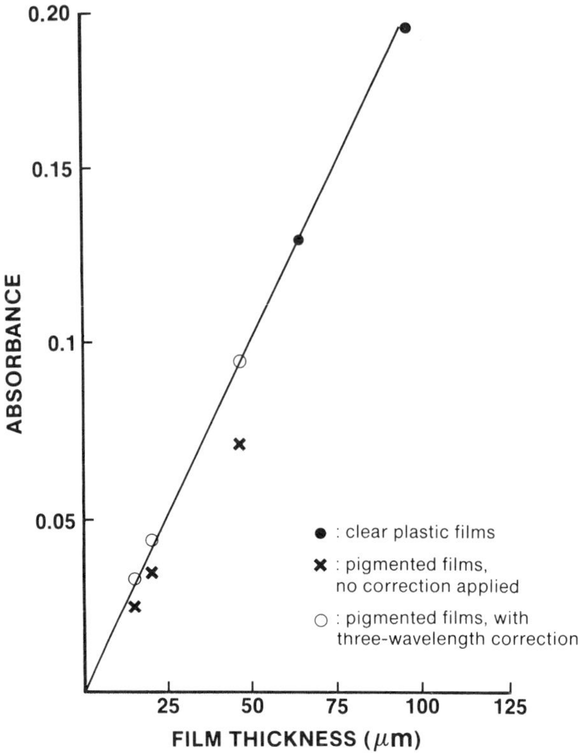

Figure 5-15 Correlation between the experimentally measured absorbance and the film thickness for several polyethylene films, with or without scattering pigments, before and after three-wavelength correction.

to better estimate the film thickness from the expression (Harrick, 1971)

$$d = \frac{1}{2\Delta\nu(n^2 - \sin^2\theta)^{1/2}}, \tag{5-3}$$

where $\Delta\nu$ is the interference fringe separation in cm^{-1}, n is the refractive index of the film material and θ is the angle of incidence.

The fringe visibility can be improved by properly processing spectrograms obtained separately with light polarized in the two normal planes. An example of such an approach is illustrated in Figures 5-16 and 5-17 (Cielo et al., 1986d). Figure 5-16 shows a typical reflectance spectrum from a plastic-coated steel sheet. This spectrum is difficult to interpret because of the presence of

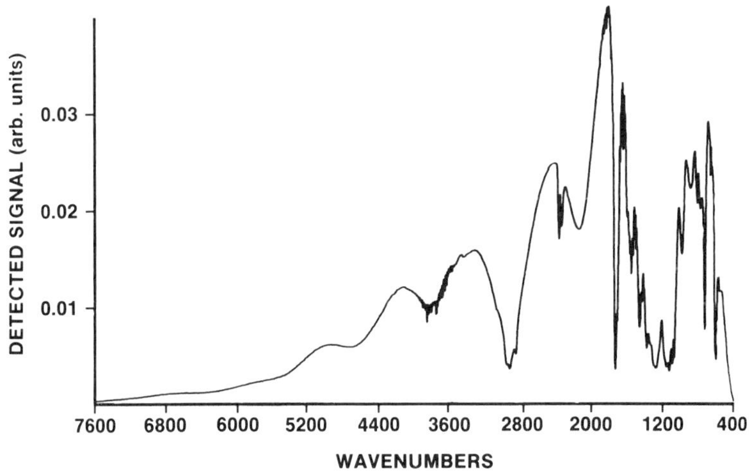

Figure 5-16 Spectral distribution of the light reflected from a steel sheet with a 4.75 μm-thick polymeric coating at 57° incidence.

multiple-reflection interference fringes which are superposed to the absorption peaks, as well as of a number of distortions related to instrumental parameters such as the spectral radiance curve of the source, the spectral sensitivity of the detector and the presence of absorption bands from water vapor and carbon dioxide in the atmosphere. A substantial improvement can be obtained by operating in polarized light conditions near the Brewster angle of incidence (57° in this case). As Figure 5-17 shows, the fringe visibility is much improved in the perpendicular polarization spectrum, while in the parallel polarization spectrum the fringes are removed because of the absence of any reflection at the air-to-plastics interface. It is therefore possible to evaluate the film thickness with better precision by measuring the depth of the absorption peak at 2900 cm^{-1} from Figure 5-17b and using equation (5-2) or else by measuring the fringe period from Figure 5-17a and using equation (5-3).

One of the most interesting aspects of infrared thickness gaging is the possibility of evaluating the thickness of each component in a coextruded double-coating (Burk, 1984; Hassel, 1985; Sneller, 1986). Each of the two materials must have some distinctive spectral features which allow separate recognition of the amount of each component. The increasing availability of powerful and inexpensive industrial spectrometers coupled to the signal processing capabilities of modern microprocessors will certainly make such an approach more and more performable and affordable in the near future.

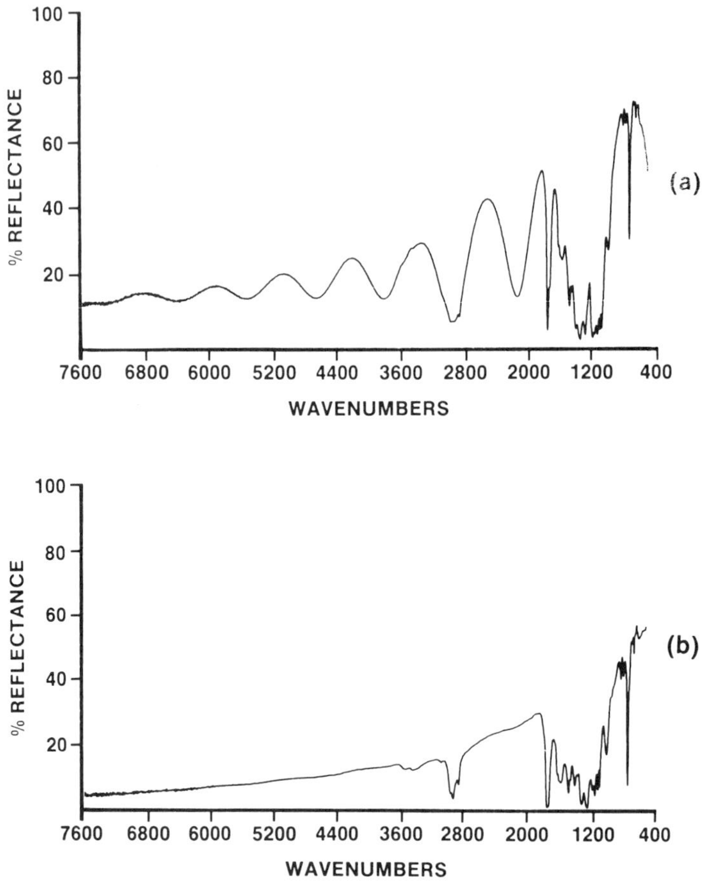

Figure 5-17 Spectra obtained with the same sample as for Figure 5-16 but using light polarized in a direction (a) perpendicular and (b) parallel to the plane of incidence. The spectra were normalized through division by the respective neutral-mirror reference curves.

5.2 Two-Dimensional Imaging

One of the most interesting aspects of optical inspection techniques as compared to other techniques is the convenience with which optical images can be obtained. While ultrasonic imaging requires a mechanical point-by-point transducer scanning and X-ray imaging can only be obtained by transmission shading projection on a large film, optical imaging can simply be performed through a lens under artificial or ambient illumination. Once an image has been recorded, however, its evaluation and the extraction of the useful information carried by the image is a formidable task for an automated recognition system. This section gives an introduction to the basic principles of two-dimensional image processing and its relevance to industrial inspection tasks. More extensive treatments of image-processing and image-interpretation methods can be found in the numerous textbooks which have been written in this field (Rosenfeld and Kak, 1976; Pratt, 1978; Castleman, 1979; Ballard and Brown, 1982; Miller, 1983; Batchelor et al., 1985).

5.2.1 *Principles of Image-Processing Techniques*

When an image as seen by a camera is recorded and stored into memory, it is represented as a first step by an array of m by n integer values, each of which may assume 2^h grey-level values, where m is the number of lines, n the number of columns and h is the number of bits of the analog-to-digital converter.

Information Contents of an Image

A recorded picture may contain a much lower amount of information than its ultimate information carrying capability. The information contents of a digitial image can be evaluated using the concept of image entropy H (in bits per pixel) as defined in Section 4.4.3

$$H = -\sum_{k=1}^{N} p_k \operatorname{ld}(p_k), \tag{5-4}$$

where N is the number of distinguishable grey levels, e.g., 256 for an image obtained from an 8-bit digitizer, and $\operatorname{ld}(p_k)$ is the logarithm in the dual base of the probability p_k of occurrence of the grey level k. If all the grey levels have an equal probability of occurrence $p_k = 1/256$, equation (5-4) gives a value of $-256(1/256)\operatorname{ld}(1/256) = 8$ bits/pixel for the entropy value of such an image. For a typical image of $m = 380$ lines and $n = 488$ columns, we obtain a total information mnH of more than 1 Mbit at every frame period of typically 1/30th of a second from the camera. This is a huge amount of information to be handled by a compact industrial inspection device, and it is normally not required for product evaluation tasks.

The actual entropy level of a digital image depends on the intensity distribution among the possible grey levels. Figure 5-18 shows three examples of images having low entropy contents because they contain only pixels with intensity levels narrowly distributed around I_0 (the "black" areas) and I_1

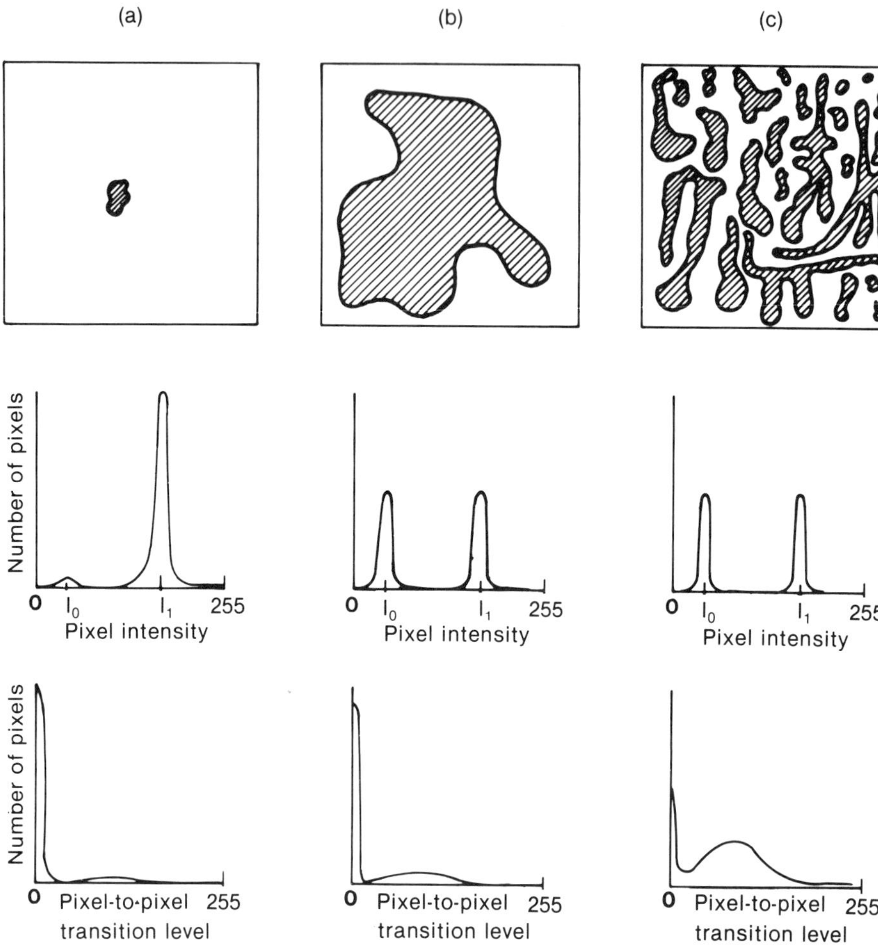

Figure 5-18 Three low entropy images with the histograms of their pixel intensity and of their intensity-gradient distributions.

(the "white" areas). Their pixel-intensity histograms thus present only two peaks at I_0 and I_1. Image (a) in Figure 5-18, almost completely white, contains very little information, as it can be verified by applying equation (5-4) to its histogram. Indeed, in the limit of a completely uniform image containing only one grey level whose probability of occurrence is thus equal to $p_k = 1$, equation (5-4) gives $H = -\mathrm{ld}(1) = 0$ for its entropy.

Images (b) and (c) of Figure 5-18 have a slightly higher entropy, their surface being nearly equally distributed between pixels with intensities I_0 and

I_1. If the grey level of such images assumes only two discrete values, e.g., 50 and 150 on the scale of 0 to 255, then equation (5-4) would give an entropy value $H = -\frac{1}{2}\mathrm{ld}(\frac{1}{2}) - \frac{1}{2}\mathrm{ld}(\frac{1}{2}) = 1$ bit/pixel for such pictures: For every pixel we need only to know whether it is "white" or "black," i.e., a 1-bit information. The intensity histograms of pictures (b) and (c) in Figure 5-18 actually indicate a slightly higher information contents because of the finite spread of the probability plots around the values of I_0 and I_1. Clearly, all of these images could be represented in a two-level grey scale (binary images) rather than in 256 tones with little loss of information, provided that the black-to-white threshold is conveniently chosen between the levels I_0 and I_1.

When observing pictures (b) and (c) in Figure 5-18, one may feel, intuitively, that the latter contains more information, even if their intensity histograms are similar. Indeed, as mentioned in Section 4.4.3, the first-order entropy is only one parameter for quantifying the information contents of a picture. Higher order entropy of the original or of the differentiated image can more effectively assess the amount of information in terms of intensity variations between adjacent pixels. The difference between the information contents of pictures (b) and (c) can be appreciated by considering their intensity-gradient histograms, i.e., the distribution of the intensity differences between adjacent pixels. In the case of image (b), as in the case of image (a), such a histogram has a strong one-level peak near zero, giving an entropy value close to $N\mathrm{ld}(1) = 0$ for the entropy of the differentiated image. Conversely, picture (c) has more frequent black-to-white transitions across the image, leading to a more evenly distributed intensity-gradient histogram and thus to a much higher entropy value for the differentiated image.

Data Reduction in Binary Images

We have seen above that the information contents of many images is relatively low. According to Shannon's theorem, there must exist some coding schemes allowing retention of all the information contained in the image without storing each of the m by n intensity levels of all the pixels in the image. Some of these schemes will now be described.

Figure 5-19 shows some examples of binary images with low entropy values. Picture (a) has only a few black pixels over a uniformly white background. The information contained in such a picture may be stored more efficiently by keeping track of the ij coordinates of the black pixels

$$(i_1, j_1), (i_2, j_2), \ldots, (i_q, j_q),$$

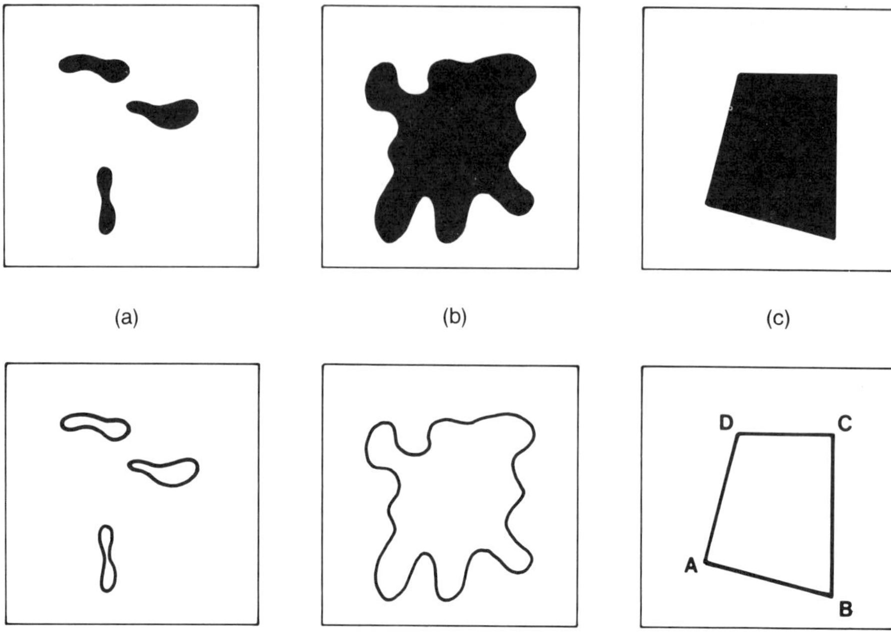

Figure 5-19 Three binary images and their edge-enhanced (differentiated) counterparts.

where i may vary between l and m, j may vary between l and n, and q is the number of black pixels. The storage requirement for such a record is equal to $q\ 1\mathrm{d}(mn)$ bits. For example, for $q = 100$, $m = 380$ and $n = 488$, this amounts to a total of 1750 bits, much less than the $mn = 1.85 \cdot 10^5$ bits required to store the full matrix and not too far from the nearly 1230 bits which represent the absolute minimum mnH where H is given by equation (5-4) with $p_1 = 100/mn$ and $p_2 = (mn - 100)/mn$.

For a picture of the kind shown in (b) in Figure 5-19, a coding scheme of this kind would not be effective, the number of black pixels being of the order of one half of the mn total pixel count. A dramatic reduction of storage requirements are, however, obtained in this case by differentiating first the image (i.e., by recording only the absolute value of the intensity gradient between adjacent pixels) and then storing the positions of the black pixels in the differentiated image. The original image can always be retrieved from the stored differentiated image by integration. For a picture of the kind (b) shown in Figure 5-19, the number of black pixels in the differentiated image is of the order of $m + n$, requiring a storage space of $(m + n)1\mathrm{d}(mn)$ of nearly 15 Kbits, which is an effective data reduction result.

In an industrial inspection environment, the objects to be inspected are usually manmade and are manufactured from well-defined geometric descriptions. This helps reduce image storage requirements if methods are found to take advantage of the object regularities. An example is given by picture (c) in Figure 5-19. The picture represents a quadrilateral whose differentiated image can be perfectly defined by the *i*, *j* coordinates of its four vertices, A, B, C and D. The total storage requirements here are equal to 4 ld(mn) or just 70 bits.

Matrix Operations on a Digital Image

We have seen that extensive data reductions can be obtained by transforming the image through operations such as differentiation or thresholding. We shall now describe how such operations can be performed on the matrix representation of a digital image.

A first class of matrix operations are monadic in nature, where each matrix element (i, j) of the input image is transformed into the corresponding (i, j) element of the output matrix by a given mathematical operation. Examples are the reduction of all the pixel intensities by a fixed amount to correct for overexposure, a logarithmic transformation to expand the low intensity range in underexposed images or thresholding to transform a multiple grey-level image to a binary image. An example of a monadic transformation for range compression is shown in Figure 5-20. In this case, all the pixels with intensity smaller than 110 in the input matrix are given the value 0 in the output matrix, while the pixels with intensity higher than 155 are given the fixed

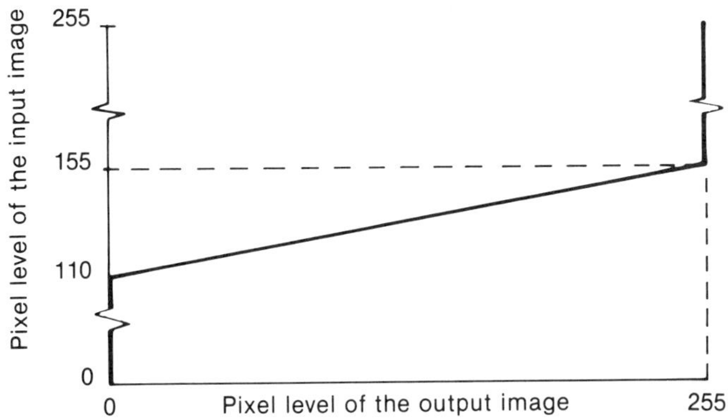

Figure 5-20 Typical contrast-enhancement function for range compression of the output image relative to the input image.

value 255. The range between 110 and 155 in the input matrix is linearly transformed to the range 0 to 255 in the output matrix. The result is a contrast enhancement on an input image whose intensity level is mainly between 110 and 155.

Another class of transformations have a dyadic form, i.e., the output (i, j) matrix element is obtained from a mathematical expression involving the intensity levels of the (i, j) elements of two input matrices. Typical examples are the addition and subtraction of two images. Image subtraction is a particularly useful operation for the elimination of common-mode noise in repetitive images. An example is the detection of moving intruders in a still background for intrusion alert systems. Another example is the enhancement of an object in a highly contrasted but repetitive background by subtracting the picture with object from the same picture taken without the object (background suppression). An application of such a technique to subsurface defect detection by thermal imaging will be described in Section 6.4.2.

A third class of matrix transformations involves local operators, where the element (i, j) of the input matrix is transformed into the element (i, j) of the output matrix by a mathematical expression involving a number of elements of the input matrix as well as a third matrix. An important example is represented by filtering operations, which are based on a spatial convolution with the operator matrix.

The convolution $h(x)$ of two one-dimensional functions $f(x)$ and $g(x)$ is defined as (see, e.g., Davenport and Root, 1958 or Castleman, 1979)

$$h(x) = \int_{-\infty}^{\infty} f(\tau)g(x-\tau)\, d\tau. \tag{5-5}$$

An example of an input function $f(x)$, operator function $g(x)$ and output $h(x)$ are shown in Figure 5-21. A substantial amount of smoothing is obtained in this case. Such an operation can be applied, e.g., to reduce high frequency electronic noise from a camera output: If $f(x)$ corresponds to the intensity level across a line scanned by the camera and the object width is known to be larger than the width of the gaussian window $g(x)$, most of the high frequency noise can be eliminated by such a convolution.

The relation of the convolution operation with the frequency contents is apparent by considering the frequency spectra $F(s)$, $G(s)$ and $H(s)$ which are the Fourier transforms of the respective functions $f(x)$, $g(x)$ and $h(x)$ in Figure 5-21. The input function contains a low frequency signal (the narrow peak centered at $S = 0$ in the graph showing $F(s)$ in Figure 5-21) flanked by two broad wings corresponding to the nearly white noise. The transform of the

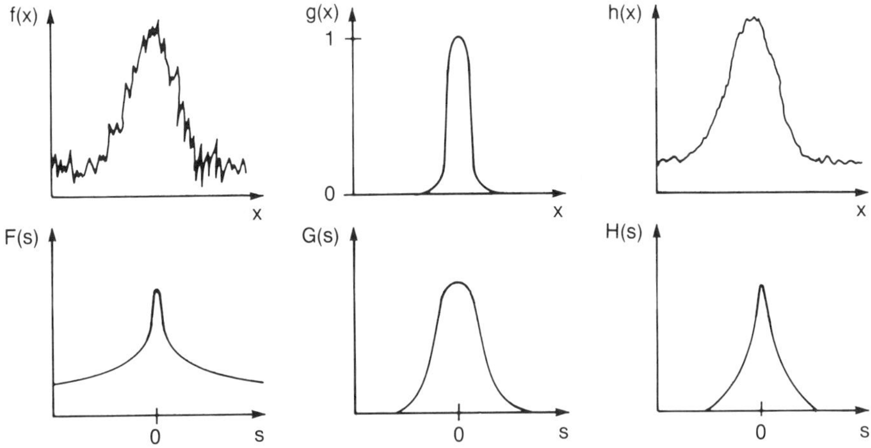

Figure 5-21 Example of a smoothing operation performed by the one-dimensional convolution of an input function $f(x)$ with a gaussian window $g(x)$ to obtain the smooth output $h(x)$. Also shown are the corresponding frequency spectra.

gaussian window $g(x)$ is a gaussian function $G(s)$ whose width is inversely proportional to the width of $g(x)$. The power spectrum of $h(x)$ is given by

$$H(s) = F(s)G(s) \tag{5-6}$$

because of the convolution theorem (see, e.g., Castleman, 1979, Chapters 9 and 10). The output spectrum $H(s)$ thus corresponds to the input spectrum $F(s)$ from which the high frequency components have been suppressed by multiplication with the narrow-banded power spectrum $G(s)$.

This basic approach has been widely applied to the restoration of images having suffered degradation from a number of factors such as out-of-focus conditions, blurring due to movement or fixed aberrations of the imaging optics (Castleman, 1979, Chapter 14). If the signal spectrum $P_s(s)$ (such as the modulated transfer function of a lens system, see Section 3.7.1) are roughly known, the optimum transfer function of a matched filter is that of the Wiener filter (Castleman, 1979, Chapter 11)

$$G(s) = \frac{P_s(s)}{P_s(s) + P_n(s)}. \tag{5-7}$$

Such a filter can be directly multiplied by the spectrum of the input signal, or else its inverse Fourier transform can be convoluted with the input signal.

Such concepts can be applied to image processing by considering the two-dimensional intensity distribution over the image to be processed as the input function, the processed image as the output function and a conveniently chosen compact matrix as the convolution operator. Two examples of such operators corresponding respectively to a smoothing and to an edge enhancing operation are shown in Figure 5-22. The operator on the left, when convoluted with the input matrix, gives an output matrix whose element $h_{i,j}$ is obtained from the elements of the input matrix following the expression

$$h_{i,j} = \frac{1}{9}(f_{i-1,j-1} + f_{i-1,j} + f_{i-1,j+1} + f_{i,j-1} + f_{i,j} + f_{i,j+1} + f_{i+1,j-1} + f_{i+1,j} + f_{i+1,j+1}). \tag{5-8}$$

The result is a two-dimensional average across the nine adjacent elements of the matrix, with a strong reduction of the high frequency noise similarly to the one-dimensional convolution shown in Figure 5-21.

The operator shown on the right of Figure 5-22, after convolution with the input-image elements $f_{i,j}$, produces output elements following the expression

$$h_{i,j} = -f_{i-1,j-1} - 2f_{i-1,j} - f_{i-1,j+1} + f_{i+1,j-1} + 2f_{i+1,j} + f_{i+1,j+1}. \tag{5-9}$$

The result is an enhancement of the horizontal edges in the input image, as can be verified on a specific input-image matrix. An example of the application of a two-dimensional edge enhancing operator to an input image with a square edge is shown in Figure 5-23. After application of the operator, the output image differs from zero only around the edges. The result is a differentiation or high-pass filtering operation of the input digital image.

Image Segmentation, Recognition and Interpretation

Let us consider the inspection function of an operator watching the parts being fed on a conveyor to a machine, verifying whether each part is being

1/9	1/9	1/9
1/9	1/9	1/9
1/9	1/9	1/9

−1	−2	−1
0	0	0
1	2	1

Figure 5-22 Two simple spatial-convolution operators for filtering a digital image to obtain a smoothing (left operator) or a horizontal edge enhancing (right) function.

5	5	5	5	5	5	5	5	5
5	5	5	5	5	5	5	5	5
5	5	5	5	5	5	5	5	5
5	5	5	5	5	5	5	5	5
5	5	5	5	10	10	10	10	10
5	5	5	5	10	10	10	10	10
5	5	5	5	10	10	10	10	10
5	5	5	5	10	10	10	10	10
5	5	5	5	10	10	10	10	10

(a)

-1	-1	-1
-1	8	-1
-1	-1	-1

(b)

0	0	0	0	0	0	0
0	0	0	0	0	0	0
0	0	-5	-10	-15	-15	-15
0	0	-10	25	15	15	15
0	0	-15	15	0	0	0
0	0	-15	15	0	0	0
0	0	-15	15	0	0	0

(c)

Figure 5-23 Image differentiation by convolution with a two-dimensional gradient operator: (a) original image; (b) convolution operator; (c) differentiated output image.

submitted with the proper orientation for the machine handling requirements. The decision to accept the well-oriented parts and remove the misoriented ones is based only on the orientation angle of the part axis, an amount of information extremely small as compared to the total amount of visual information in the scene which he is seeing. If a general purpose inspection machine is to execute a similar task, it must be able to reduce the input pictorial information to such a simple evaluation. This is a relatively difficult inference problem in artificial vision, requiring successive steps of image segmentation, recognition and interpretation.

Image segmentation is the process by which different parts of an image corresponding to different objects or parts of an object are singled out and given some specific features. This is usually implemented first by locating the part edges by techniques of the kind described above, obtaining a number of edge segments which are analyzed and associated to obtain a subdivision of the image into relatively homogeneous regions. This step is closely related to the recognition step, where the different regions are assigned to given objects or object components. Interpretation follows to determine the value of the parameter required for the decision making task, such as the evaluation of a part position, orientation or size.

The choice of the characteristic segment features to be attributed to the different regions of an image must be made with a view to facilitating the recognition task. Several shape descriptors have been considered in an attempt to capture invariant geometrical properties of the object. As an example, let us consider the quadrilateral object shown in image (c) of Figure

5-19. If the features chosen for characterizing the quadrilateral are the coordinates of the four vertices, the object representation will be sensitive to camera positioning. If instead we assume as characteristic features the values of the four consecutive angles together with the relative proportions of the segments joining each angle, this representation will be invariant with respect to the object position, orientation in a plane normal to the line of sight and distance with respect to the camera. Examples of invariant representations which have been proposed include two-and three-dimensional moment invariants, texture, perimeter or area, and graph representations (Gonzalez and Safabkhsh, 1983).

Recognition and interpretation is finally achieved by model matching or by higher level inference procedures (see, e.g., Ballard and Brown, Part 4). For the inspection of industrial parts, comparison with regular geometric shapes can be an efficient procedure for part recognition and interpretation (Perkins, 1978). In this approach, connected curves (called concurves) are used to form models of industrial parts. Such concurves may be straight segments, areas, circles, parts of polygons, etc. which are easy to describe and to match with image-extracted region boundaries. Part interpretation is conveniently performed by a global analysis of the features of the component concurves and comparison with the concurve-specified model.

5.2.2 *Industrial Applications*

The application of two-dimensional pattern recognition algorithms to quality control and part sorting on the production line is a formidable task at the present state of artificial vision technology. Nevertheless, restricted applications to relatively simple or highly repetitive objects are becoming feasible (Pfoutz, 1975; Agin and Dudea 1977; Holland et al., 1979; Bible and Bible, 1979; Kruger and Thompson, 1981; Wagner, 1982; Koch, 1983; Chin and Harlow, 1982; Gonzalez and Safabakhsh, 1983; Miller, 1983, vol. 2; Considine, 1985, p. 9–95).

A schematic diagram of a possible industrial inspection station for quality control by automated two-dimensional imaging is shown in Figure 5-24. Manufactured parts are being fed by a conveyor belt to the inspection station where a camera with appropriate software is used to analyze the parts and determine whether some unacceptable defects are present such as chipped sides, lack of thread or unconformity with size specification. Rejected parts are sorted out for reworking. A high-speed system of this kind, using a stroboscopic light source to inspect parts on the fly, has recently been installed in an automative assembly line. By eliminating defective units which represent

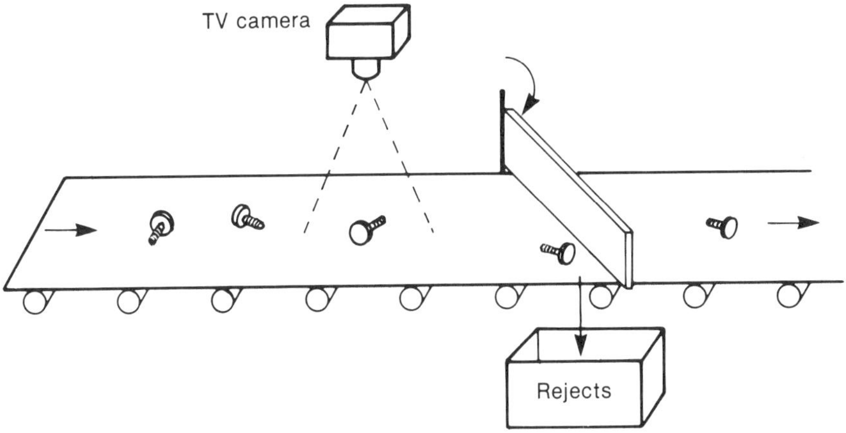

Figure 5-24 A possible quality control station sorting defective manufactured components by two-dimensional imaging automated inspection.

1% to 3% of the purchased fasteners, this $40,000 investment is expected to yield an annual savings of $250,000 (Pryor and North, 1985, p. 24).

Preliminary investigations of machine vision capabilities for the accomplishment of such tasks have been reported (Kruger and Thompson, 1981; Wagner, 1983). Figures 5-25 and 5-26 show some laboratory results obtained by image processing with industrial workpieces (Birk et al., 1979; Nakagawa and Rosenfeld, 1979). Figure 5-25 shows a typical application of edge localization algorithms to the acquisition of the silhouette of a manufactured

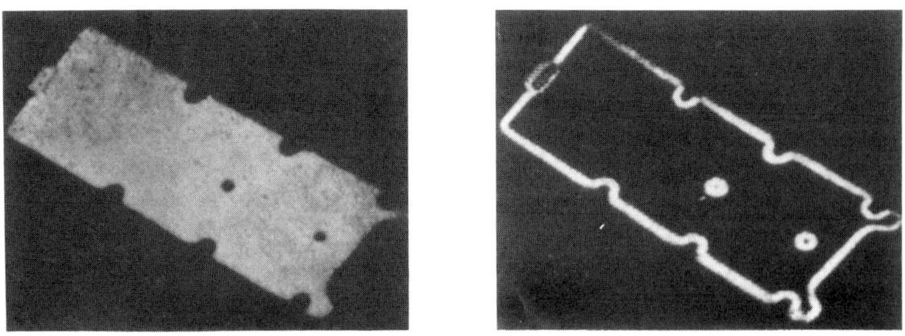

(a) (b)

Figure 5-25 (a) A two-dimensonal image of an industrial workpiece; (b) image obtained after processing through a gradient operation.

part. After application of a gradient operator and thresholding, the part borders and the holes are clearly in evidence. Figure 5-26 shows the results of segmentation for the image acquisition of different industrial components. The image shown in Figure 5-26 is processed using differentiation and thresholding algorithms to extract the edges. One of the difficulties encountered during such an operation is related to variations in the illumination level across the image due to nonuniformity of the light source, inappropriate angle of illumination and viewing, or variations in the surface reflectance of the imaged objects which are particularly strong in the case of shiny surfaces. To reduce such problems, dynamic thresholding techniques of the kind shown in Figure 4-38 in the last chapter can be used to adapt the threshold to the local illumination level across the image. Figure 5-26b was obtained by applying both dynamic thresholding and a differentiation operation on the original image. Figure 5-26c shows the results of a part description technique using polygonal approximations. The polygonal fitting approach is based on the examination of the curvature along the perimeter of an object and on the approximation by a straight-line segment of any sequence of contiguous points whose curvature is below a given threshold. This is a typical approach for simplifying the descriptive features of the image in order to facilitate object recognition.

Much more work is still required before general purpose image-interpretation algorithms can be implemented in industrial inspection systems with the required speed, reliability and straightforward utilization by nonspecialist operators. Specific tasks can, however, be handled by special purpose software, such as counting the number of plastic cups packaged in stacks of 50 to 100 elements which are fed by conveyor at rates of 120 stacks per minute (Hudson, 1983) or inspection of valve rocker cover assemblies for automotive engines (Koch, 1983).

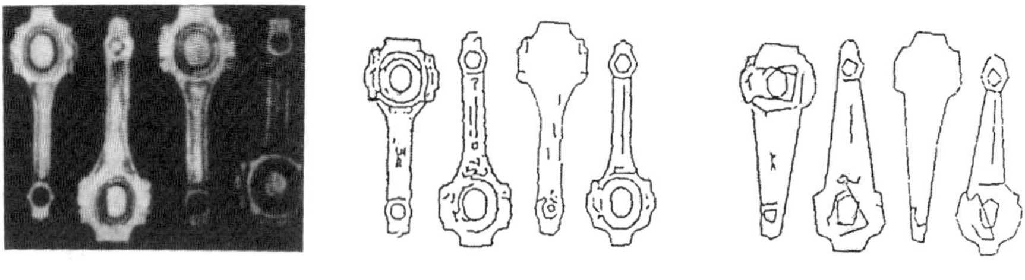

(a) (b) (c)

Figure 5-26 (a) Images of industrial parts; (b) processed images obtained after differentiation and thresholding; (c) polygonal approximation of edge points in the differentiated image.

Automated Inspection in the Electronics Industry

The need for automated inspection of electronics components is well illustrated by considering the dramatic increase in semiconductor circuitry complexity in recent decades as shown in Table 5-1 (Harding, 1981).

Automated inspection of printed circuits and wiring boards (Sterling, 1979; Nakagawa, 1982; Corman, 1983; McIntosh, 1983; Zeller and Doemens, 1983, Macfarlane and Woodruff, 1984; Besl et al., 1985; Abramovich, 1985) and of integrated circuit wafers (Hsieh and Fu, 1980; Yu et al., 1982; Hirata and Tanaka, 1983; Harris, 1983; Kayaalp and Jain, 1983; Kochan, 1986) is a very active field of two-dimensional imaging applications. Insertion of integrated circuits in PC boards also requires intelligent vision control (Gahinet et al., 1986). Several of the research efforts reported in this area have led to production systems.

Most of the inspection algorithms developed for electronic circuit inspection are based on a prior knowledge. This knowledge concerns the stratification of the limited scene models which are possible because manmade objects are under inspection. Such a stratification produces regular geometric structures, such as straight lines and right angles, and greatly reduces the possible universe of "normal" structures.

Table 5-1 Complexity and Productivity Comparisons of Two IBM Semiconductor Logic Technologies

	1964	1980
Components per circuit	9	5
Minimum feature size (μm)	12.5	3
Gates per chip	0.25	3234
Wireable circuits per chip	0.25	1500
Chip size (mm^2)	0.66	22
Circuit density (circuits/mm^2)	0.38	68
Major process steps	16	38
Lithography steps	5	14
Interconnection levels	1	3
Terminals per chip	3	122
Wafer size (mm^2)	800	5280
Circuits per wafer (unyielded)	300	273,00

A number of conventional machine vision techniques, such as pattern matching and binary imaging, cannot be used in the analysis of PC boards and IC wafers because they are inherently slow and insensitive to small features or to low contrast. On the other hand, the volume of data to be analyzed makes global inspection techniques unrealistic. The number of possible alternative configurations is reduced by examining predetermined areas as defined by windows. Such windows are of arbitrary size in the field of view, which usually define the location of a component in an IC or of a lead on a PC board.

A typical application of the window technique to the inspection of PCBs was implemented using local pattern matching (Jarvis, 1980). Under appropriate lighting conditions, a binary image can be obtained where the 1s represent the conductor and the 0s represent the substrate. Briefly, the inspection method consists of the preselection of a list of local 5-by-5 pixel template regions that describe all possible bit patterns observable on an errorless PCB. It was found that 99% of the possible 5-by-5 template patterns could be enumerated by storing only 170 such binary template patterns. When the PCB is inspected, any 5-by-5 region that does not exactly match one of the prestored templates is marked as suspicious, and supplemental tests are performed on those regions. The supplemental tests include, among others, the computation of conductor area, the length of the conductor-substrate boundary, and the ratio of area to length within each suspicious 5-by-5 region.

Different approaches are also under investigation for the automated inspection of electronic components. An optothermal technique for solder joint inspection has been implemented in a commercially available inspection system (Vanzetti et al., 1981; see also Section 6.4.1). The solder joint is thermally stimulated by a short laser pulse, while its temperature is probed by a focused infrared sensor. Defects such as porosity or inadequate bonding surface give rise to abnormal thermal signatures. Such an approach is more powerful than optical-imaging techniques (Besl et al., 1985) because it is sensitive to subsurface features. It is, however, inherently slow because of the requirement for sequential point-by-point scanning. Similar considerations are valid for wafer inspection using photoacoustic (Rosencwaig, 1980) or photothermal (Rosencwaig et al., 1985) techniques.

Passive thermal-imaging techniques where the electrically powered circuit is imaged by an infrared camera are inherently faster because they do not require external energy injection by a scanned laser beam, while a region of a given size can be inspected in parallel. Such an approach has been

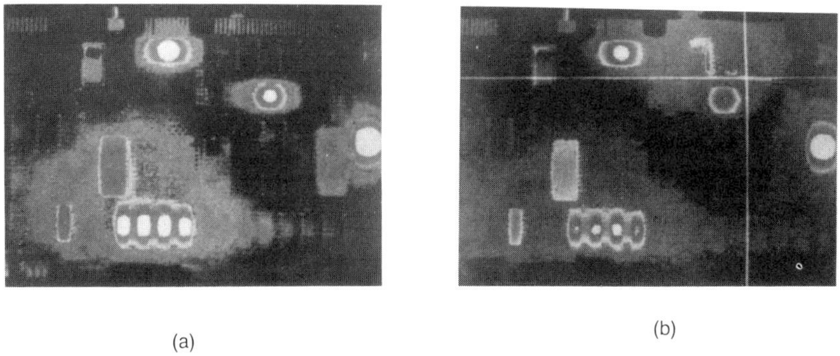

Figure 5-27 Thermal images of a powered-up printed circuit board obtained by an infrared camera, showing heat-dissipation spots in the same board under (a) normal conditions and (b) with a short in the area above and on the left of the intersecting crosshairs. (Courtesy of UTI Instruments.)

implemented in a number of commercial devices. Figure 5-27 shows two thermal images of the same PCB under (a) normal conditions and (b) presence of a short in the area just to the left and above the intersection of the crosshairs. Defects producing abnormal heat dissipation patterns can thus be sorted out by simple image differentiation techniques. The application of advanced image processing algorithms to thermal-imaging is a relatively unexploited area of research and development.

Finally, an electro-optical approach to circuit probing using built-in photo-conductive bridges should be mentioned (Poulsen, 1979). Figure 5-28

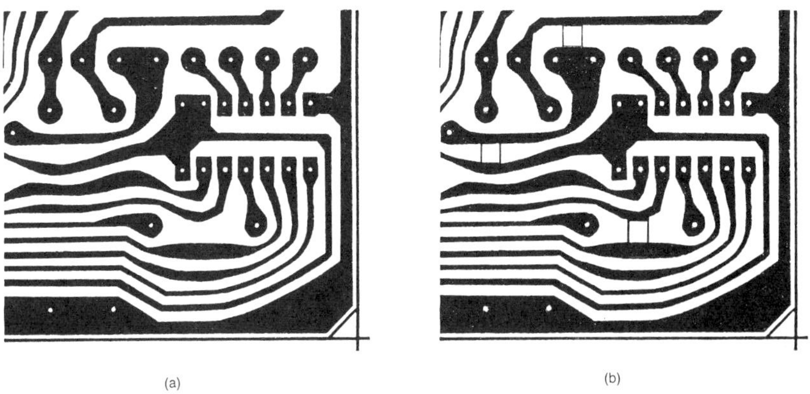

Figure 5-28 Portion of circuit board layout being tested using photoconductive probing: (a) circuit without photoconductive bridge; (b) with photoconductive bridges.

shows a typical circuit board being inspected by such a technique. After insertion of photoconductive patches on suitable positions, artificial shorts can be temporarily created by addressing each of these bridges using a scanning laser source or a fiber optic illuminating matrix. The electrical response of the circuit to such actions is recorded and compared with standard response to detect any malfunction. Such a noncontact testing approach is faster to execute than the classical method using an array of pneumatically driven contact pins, and it can be implemented without the need to remove protective conformal coatings from the circuit boards.

5.3 Three-Dimensional Imaging by Incoherent Techniques

Man and a number of animals use binocular vision to gain range perception. A certain degree of three-dimensional viewing is thus implicit in human vision using triangulation based stereo vision. Other depth-sensing mechanisms which are found in nature include the sonar based sensory organs which are used by bats for in-flight range finding. Computer vision is following similar approaches to gain depth resolution for three-dimensional modelling of the inspected scene.

This section reviews the most important techniques for three-dimensional imaging using incoherent optical methods (Parthasarathy et al., 1982; McFarland and McLaren, 1983; Jarvis, 1983; Strand, 1985; McCollum et al., 1985; Pugh, 1986). Nonoptical approaches such as ultrasonic ranging or electromagnetic proximity sensing are described in Chapter 2, while coherent optical techniques using interferometry or holographic displacement sensing are reviewed in Sections 4.1.3 and 6.1, respectively. Applications of 3D vision to industrial inspection include the automated measurement of engine castings, naval propellers, airplane wings, turbine blades, electronic components, automobile parts and robot guidance (Miller, 1985, Chapter 2).

5.3.1 *Flying Spot Scanning Methods*

The simple triangulation technique (see Section 4.1.4) can be used to obtain a three-dimensional map of surface morphology by scanning the focused light beam over the inspected surface (Keller, 1967; Himmel, 1978; Kanade and Asada, 1981; Brunk, 1982; Parthasarathy et al., 1982; Zhao et al., 1983; Pipitone et al., 1983; Bowyer et al., 1985; Maldague et al., 1986). The basic principle is illustrated in Figure 5-29. A focused laser beam is scanned over

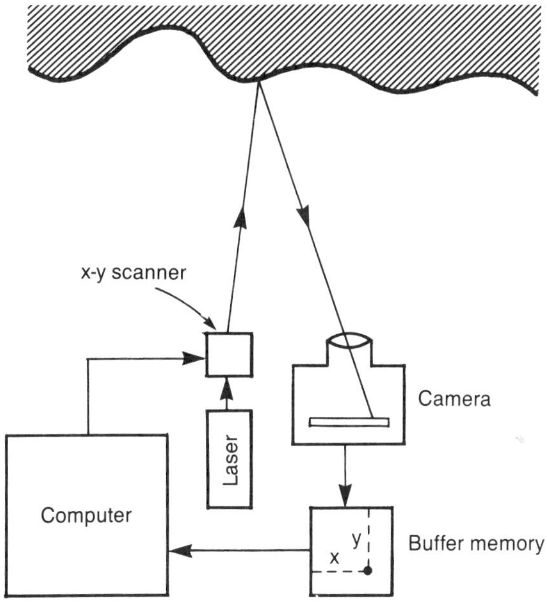

Figure 5-29 Schematic diagram of flying-spot three-dimensional imaging by triangulation.

the surface to be inspected while its *x-y* image position is recorded by a camera. The *x-y* position of the spot image for each deflection angle of the scanner is known for a given distance of a plane surface from the inspection device. Small differences of the recorded *x-y* position with respect to the position of the plane reference surface thus allow evaluation of the distance of the surface from the reference plan at such a position. Proper signal-processing techniques are used to build the three-dimensional computer model from the range information (Laurendeau and Poussart, 1986; Chi, 1986). Variations of this approach include the use of two synchronously-rotating mirrors to obtain sequential time pulses from which the triangulation angle can be evaluated (Nimrod et al., 1982) or the adoption of a holographic scanner (see Section 3.8) and of a binary coded multiple-strip mask for accurate spot-image positioning (Haugen et al., 1984).

An application of this technique in a particularly hostile industrial environment is illustrated in Figure 5-30 (Neiheisel, 1983). A difficult problem in the steelmaking industry is to determine the degree of wear of the refractory lining in a steelmaking vessel of the kind shown in Figure 5-30a. Such a refractory layer is formed of approximately 1 m-thick ceramic bricks which

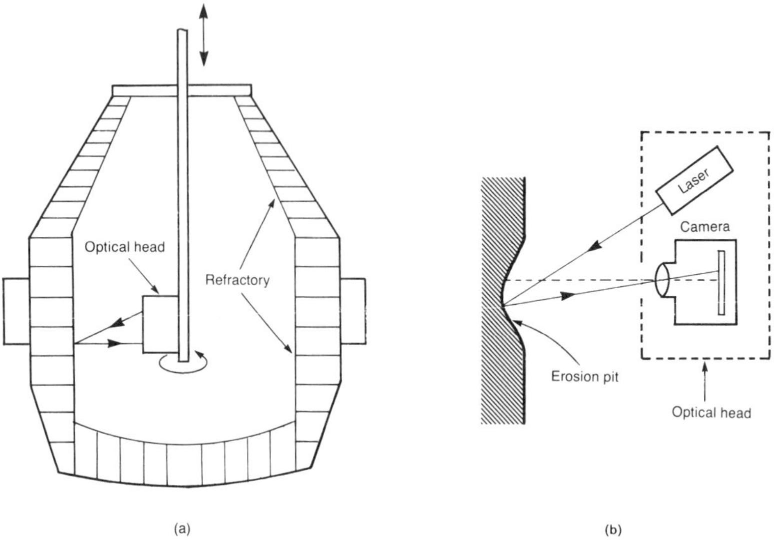

Figure 5-30 Three-dimensional imaging within a steelmaking vessel at 1300°C to determine refractory-lining wear between heats: (a) cross section of the steelmaking vessel; (b) schematic diagram of the optical head.

thermally insulate the exterior vessel body from the molten steel at nearly 1600°C. The refractory lining eventually wears away because of scrap charging and the corrosive nature of the molten steel/slag combination. The brick lining around the vessel must be periodically inspected to determine average residual thickness as well as the presence of localized wear pits which are normally repaired by spraying a refractory slurry onto the damaged areas of the vessel in the process shutdown period between heats. Such an operation must be performed within less than one hour while the refractory material is still at temperatures of the order of 1300°C.

A triangulation approach was selected for such an industrial inspection problem as Figure 5-30b shows. A water-cooled optical head including a He–Ne laser and linear phodiode array camera is lowered within the vessel. This system is rotated 360° for each vertical position of the lance to provide radial-range measurements at a rate of approximately 15 readings per second. The full inspection process takes nearly five minutes for 15 vertical levels and 48 circumferential points per level. Figure 5-31 shows a typical radial-range scan at a given vertical position in the case of a new lining (no wear) and of a severely worn-away lining, respectively. The measurement accuracy is of the order of a few centimeters. Both water and cold-air cooling were used

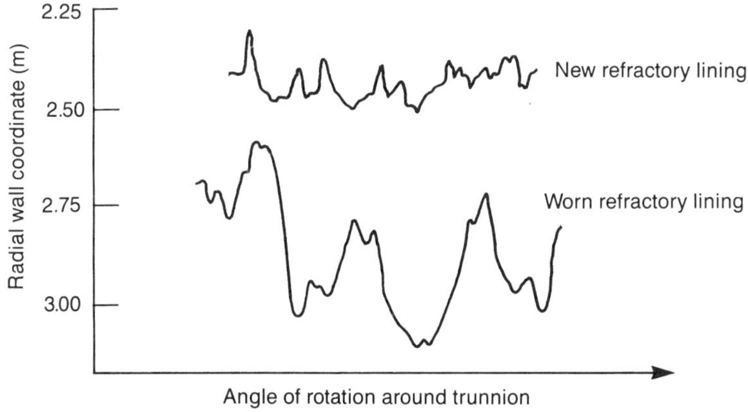

Figure 5-31 Radial-range display of the sensor-to-lining distance with the system shown in Figure 5-30 at a given vertical position. The difference between a new lining (no wear) and a severely worn lining is evident.

to keep the temperature of the electronics within the box under 43°C. Fiber optic extension bundles could eventually be used to reduce temperature criticality during operation.

A scanning configuration of the kind shown in Figure 5-30 where the complete optical head is rotated and vertically displaced has the advantage over the scanned-beam system in Figure 5-29 of keeping the image of the focused spot always close to the optical axis of the camera. The angular excursion of the image point on the camera retina is limited to the differential value between the image positions of laser spots projected on a surface of different depth levels. The accuracy which can be obtained in the measurement of such a differential angular range is much better than the accuracy which can be obtained with the configuration of Figure 5-29, where the full x-y angular-scanning range must be covered by the imaging camera (Lin and Chi, 1983). The obvious limitation of the rigid-body scanning approach of Figure 5-30 is, however, the lower rate of image acquistion of such a relatively cumbersome scanning system.

Fast three-dimensional imaging with reduced angular-range excursion of the detected beam is possible using two synchronized scanning mirrors (Rioux, 1984; Bickel et al., 1985; Häusler and Maul, 1985). Similar concepts have also been applied with one-axis scanning over a moving surface for welding robot guiding (Oomen and Verbeek, 1983) and for hot-strip flatness monitoring in a rolling mill (Pirlet et al., 1983). The basic principle is illustrated in

Figure 5-32. Figure 5-32a shows a simple one-axis implementation of the synchronized-scanning approach (Oomen and Verbeek, 1983; De Keijzer and De Groot, 1984). In contrast with the staring-camera approach shown in Figure 5-29, both the projected and the reflected beams are now deflected by two mirrors which are synchronously rotated by mounting them on the same shaft. As a consequence, the axis of the reflected beam is always situated close to the axis of the viewing lens, the small angular excursion corresponding to the depth fluctuations of the probed object. A small CCD line-array is thus sufficient to sense such a depth-related angular excursion with high accuracy. The angular sensitivity is higher than with the staring-camera approach of Figure 5-29 because only the angular excursion corresponding to the depth range is sensed by the CCD aray, while in Figure 5-29 the camera must follow both the scanning displacement of the flying spot and the superposed small displacement related to depth variations.

Figure 5-32b shows a fast double-axis scanning configuration using the synchronized scanning approach (Rioux, 1984). To achieve a higher scanning speed than would be possible with the relatively cumbersome long-shaft

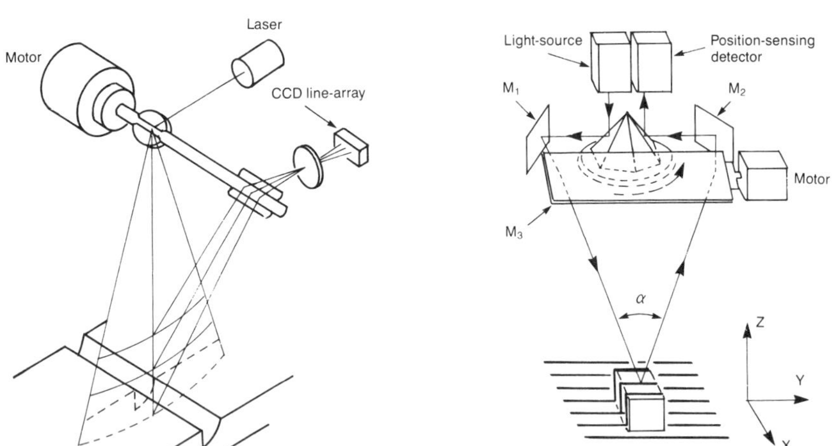

(a) (b)

Figure 5-32 Three-dimensional imaging by synchronized scanning. (a) One-axis scan of the projecting and viewing mirrors mounted on the same rotating shaft; (b) double-axis scanning using a pyramidal prism rotating around its axis to scan along the y-direction and a long mirror M_3 rotating around its longitudinal axis for the x-direction. M_1 and M_2 are fixed mirrors.

configuration of Figure 5-32b, a rotating pyramidal prism with two fixed mirrors, M_1 and M_2, is now used for scanning in the y-direction. As mentioned in Section 3-8, several thousands of line-scans per second can be achieved with compact rotating prism scanners, while the viewing angular aperture α which determines the depth sensitivity (see equation 4-9) can be conveniently varied by displacing the distance between mirrors M_1 and M_2. A slower rotation of mirror M_3 along its longitudinal axis provides the x-direction scan. Using analog position sensing detectors which may have response times smaller than 1 μs, depth-range images of 256 pixels/line with 128 lines could be recorded in less than one second, the depth at each pixel being calculated in real time. This basic approach was recently used for 3-D image acquisition of large-diameter propeller blades (Livingstone and Rioux, 1986) as well as for robot-arm guiding in adaptive arc welding (Beranek et al., 1986).

Variations of this basic approach include telecentric scanning and adoption of suitable optical components for an expanded range of depth (Häuser and Maul, 1985; Bickel et al., 1985). An interesting optical design is shown in Figure 5-33. Two axicons (lenses of conical cross-section) are used to obtain a donut-shaped converging beam which is projected on the surface to be inspected. A high intensity focal line surrounded by a low intensity halo is obtained over the full depth range Δz along which beam superposition occurs. This results in an extended range of depth with high spatial resolution as compared with the depth of field obtained by the conventional approach using a spherical focusing lens (see Section 3.7.1).

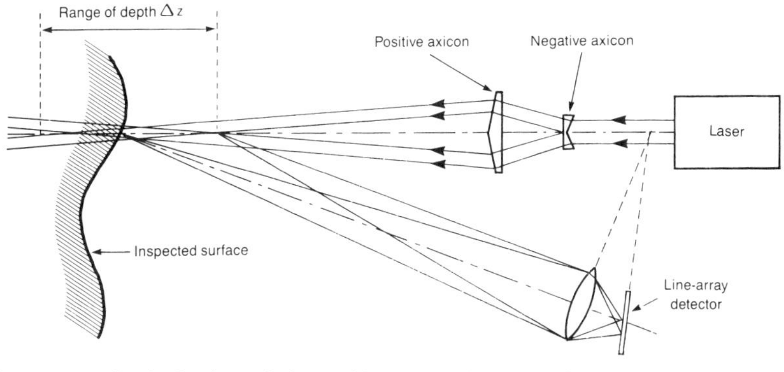

Figure 5-33 Optical triangulation with extended range of depth. Two axicons are used to obtain an extended focal line along the depth range Δz, while the line-array detector is suitably oriented to keep the spot image well focused over the full range of depth.

Another useful detail shown in Figure 5-33 is the choice of the line-array detector alignment. If the longitudinal axis of such an array, as well as the imaging lens plane, are aligned on the same point along the projected beam axis, good focusing properties are obtained for a point source whose position varies along the axis of the projected beam (see also Oomen and Verbeek, 1983). Consequently, full advantage is taken of the projected beam by keeping the image spot well focused on the line-array detector for any surface position within the depth range.

5.3.2 *Pattern Projection Techniques*

As discussed in Sections 3.4.1 and 3.8, laser-scanning techniques have the advantage over the video-imaging approach of providing higher instantaneous intensity and hence better immunity to environmental noise, while reaching an increased spatial resolution and dynamic range as well as a smoother pixel-to-pixel response. However, video imaging methods have other advantages such as the convenience of parallel, rather than serial, picture acquisition and, most importantly in industrial applications, the absence of any high-speed mechanical scanning devices which are subject to wear and long-term balancing problems. Moreover, increasingly fast and inexpensive frame grabbing hardware and standard image processing software are making the video equipment more and more competitive with mechanical scanning devices. Video-based triangulation techniques where a light pattern is projected over a surface which is observed at a slanted angle with a two-dimensional camera are thus a field of intense development (Agin and Binford, 1973; Popplestone et al., 1975; Oshima and Shirai, 1979; Binford, 1979; Hall et al., 1982; Porter and Mundy, 1982; Levine, 1983; Dufour and Cielo, 1984; McCollum et al., 1985; Halioua and Liu, 1986).

A typical configuration is shown in Figure 5-34. A laser beam is expanded by a cylindrical lens to obtain a light strip and the illuminated surface is observed from a different angle through a two-dimensional camera. The shape of the recorded light strip provides depth information instantaneously over the full length of the strip without any moving parts. Appropriate spectral filtering or modulation techniques of the kind described in Section 3.5 can be used to provide high immunity against ambient light noise. Fiber-pigtailed laser diode arrays with suitable strip equalization optics can be employed as an alternative to the gaussian-distributed expanded laser beam to obtain a nearly rectangular intensity distribution along the strip for optimum contrast (Dufour and Cielo, 1984). Obviously, only the upper part of the objects which

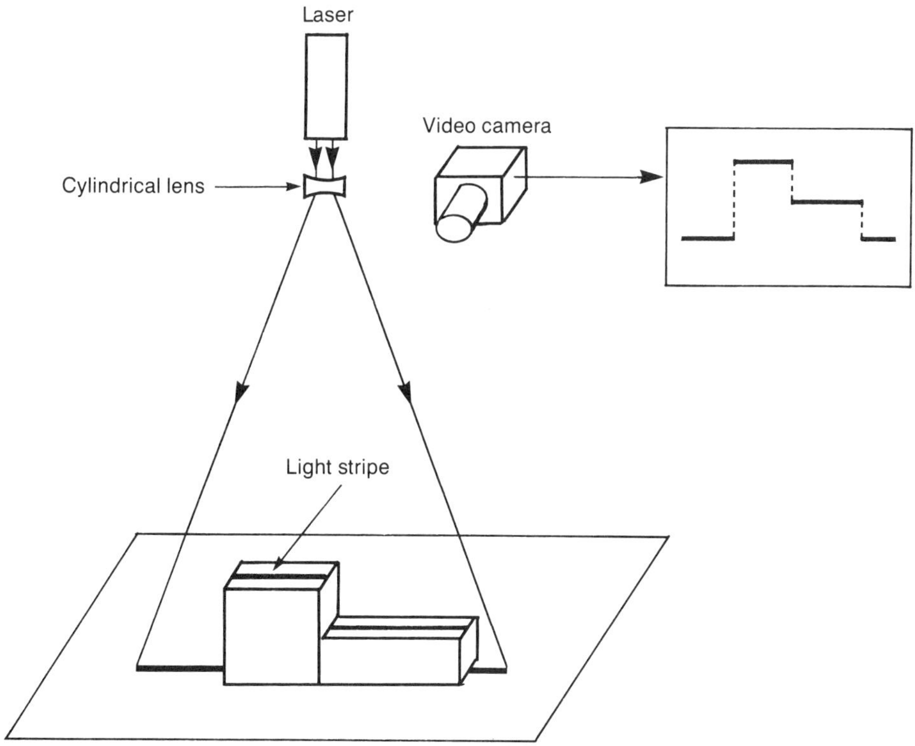

Figure 5-34 Three-dimensional imaging by light stripe projection and tilted angle two-dimensional viewing.

are exposed to both the projected beam and the line-of-sight of the viewing camera can be probed by this method or by any other light projection technique.

The single stripe projection method shown in Figure 5-34 is mainly convenient for the inspection of surfaces moving in a direction perpendicular to the stripe, thus providing scanning in the longitudinal direction. Typical applications are surface topography inspection of moving sheets or robot-head guiding in automated welding (see Sections 4.3.1 and 7.5, respectively). For the inspection of a still surface, projection of a two-dimensional light pattern such as an array of parallel stripes (Will and Pennington, 1971; Jalkio et al., 1985) or a dot matrix (Altschuler et al., 1981) is more convenient. Depth range mapping of the full x-y area imaged by the camera can be performed in this way, taking full advantage of the bidimensional imaging capabilities of the video camera. A periodic multiple-stripe pattern can be conveniently obtained with an extended depth of field by superposing two slightly tilted coherent beams to produce an interference fringe pattern (see, e.g., Thalmann and Dändliker, 1985 or Figure 7-20).

A slightly different approach is shown in Figure 5-35 (Porter and Mundy, 1982). A double-wavelength multiple stripe pattern is here projected on the workpiece by splitting a wide-spectral-bandwidth light beam into two spectral regions, one in the visible and the other in the infrared, and recombining the two beams by a stripe-patterned mirror. The two stripe patterns projected on the surface are thus shifted by half a period, so that the total light intensity is constant across the illuminated area. The double-wavelength pattern is subsequently divided into its two spectral components by a dichroic beam splitter mounted within the viewing device, each component being monitored by a separate line array detector. By comparing the periodic patterns detected by the two sensors, the effects of surface reflectivity variations across the

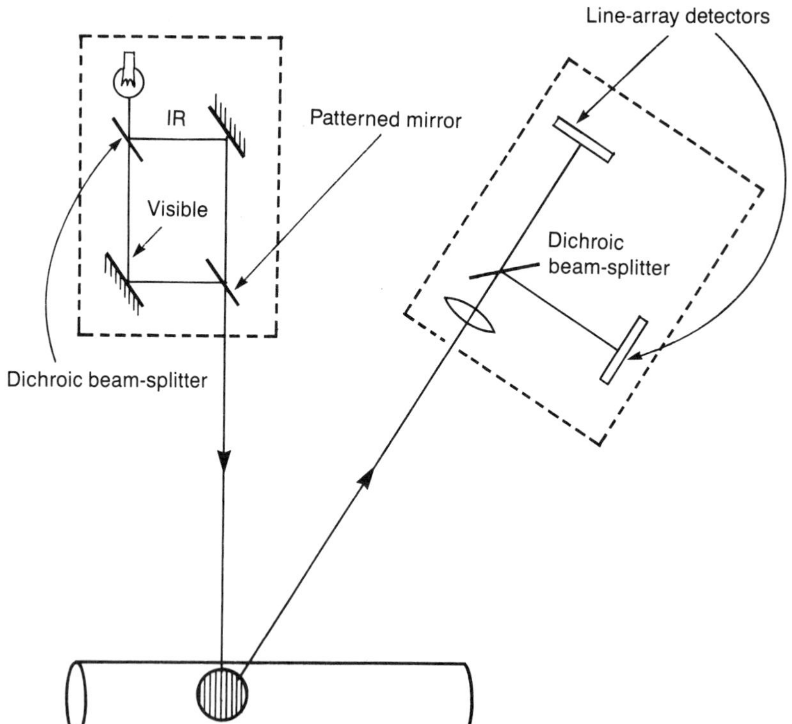

Figure 5-35 Three-dimensional imaging by projection of a double-wavelength matched multiple-stripe pattern and balanced detection.

scanned surface are minimized. Metal turbine parts with a reflectance range of 10:1 could be depth-range imaged by this method with a resolution of 25 μm.

Moiré Contourography

The multistripe pattern projection and slanted camera viewing technique for depth ranging can be considered as a comparison of the projected multistripe pattern image on the camera retina with a reference linear pattern which may be recorded by pattern projection on a perfectly plane surface or which may be assumed in advance and inserted as a digital pattern in the camera electronic buffer. The optical equivalent of such an electronic comparison, i.e., the superposition of a multistripe pattern to an actual grid to obtain depth-related optical effects, has long been used for surface contouring before modern image-processing techniques were available (Weller and Shepherd, 1948; Takasaki, 1970; Meadows et al., 1970; Varner, 1974; Pirodda, 1982; Reid, 1984; Reid and Rixon, 1986).

Figure 5-36 shows a typical apparatus to perform such a superposition. As in the multistripe projection method, a grid pattern or an interference fringe pattern is projected on the surface to be depth-ranged and is viewed under a different angle by a camera. In this case, however, a grid is physically inserted in front of the camera retina. The grid size and position are chosen to perfectly match the image of the projected pattern if a perfectly plane surface was inspected. In such a case, the observed camera image would be uniformly dark or uniformly clear, apart from the high-frequency grid modulation, depending on whether the dark stripes of the reference grid are exactly superposed to the clear or to the dark stripes, respectively, of the imaged pattern. If the inspected surface is not plane, the two patterns will locally match over the image producing depth-related clear or dark areas which are called moiré fringes. Apart from an ambiguity on the direction of height variation, moiré fringes represent equal height contours of the surface which can be extracted automatically if a number of geometric conditions are fulfilled (Idesawa, et al., 1977; Yatagai et al., 1982).

A slightly different approach is shown in Figure 5-37. In this case, the grid is located in front of the inspected surface to serve both as the spatial filter for casting a patterned shadow on the surface and as reference for the viewing camera. The grid and its shadow produce moiré fringes on the camera image which represent equal height contours in a manner similar to the pattern projection technique. Geometrical alignment problems are minimized in the

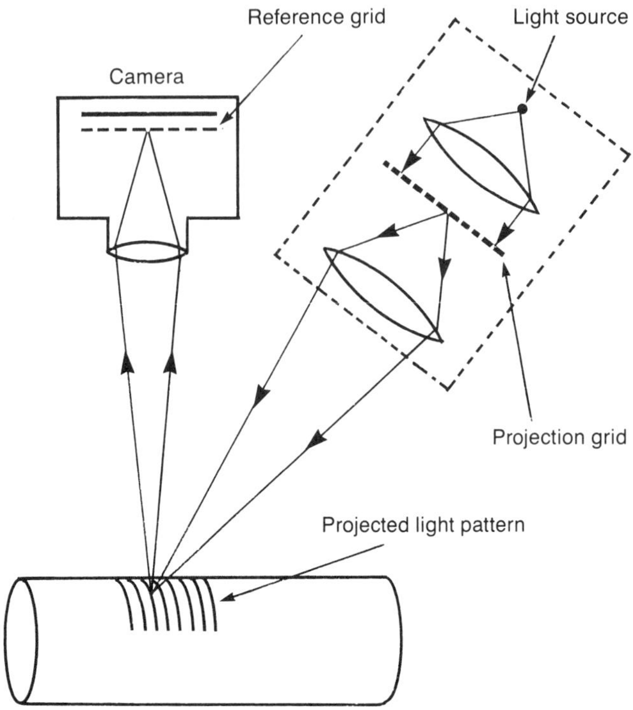

Figure 5-36 Moiré contouring by the pattern-projection method.

shadow technique, but the pattern projection approach is preferred when it is inconvenient to place a grating close to the object or when relatively large objects are to be covered. In both cases, height resolutions of 0.1 mm up to several centimeters are typical.

Variations of this basic approach are the use of pseudo-random gratings to reduce obscuration by dark fringes (Kathyl, 1972), different methods to remove the sign ambiguity in height variations (Chiang, 1975; Livnat et al., 1980; Ibrahim, 1983), optical transformations to increase the field of view (Duncan et al., 1980) or more sophisticated heterodyne methods (Reid, 1982) of a kind similar to the heterodyne holography approach which will be described in Section 6.1.4.

The objective of the moiré technique is to visualize the equal height contours by a set of fringes which are convenient for visual interpretation. For machine vision, though, such fringes must be further scanned and analyzed to extract the depth range map. Under such circumstances, one-step

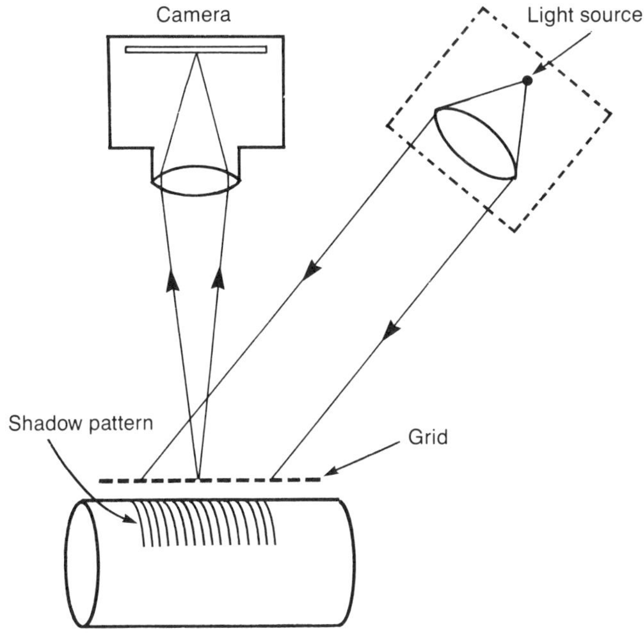

Figure 5-37 Moiré contouring by the shadow technique.

approaches to obtain the depth map such as multistripe pattern projection with no physical reference grid may perform equally well and be less cumbersome to implement than the moiré approach.

5.3.3 *Passive Techniques*

The requirement for an active projection of a light pattern over the inspected object is not an absolute necessity for 3-D imaging. As mentioned in Section 4.1.4, passive triangulation methods have long been used in land surveillance and navigation. Figure 5-38 shows an application of such a stereo-photogrammetry approach to aircraft model observation in a wind tunnel (Danielson, 1980). The model aircraft is observed from two angles of view, and the two stereo images are analyzed to determine three-dimensional object position and deformation. To help in localizing positions along the model surface, a number of target marks are distributed along the aircraft wings and body prior to the test. Computer-processed photogrammetry is a powerful and versatile metrology tool in part design, manufacture and quality control.

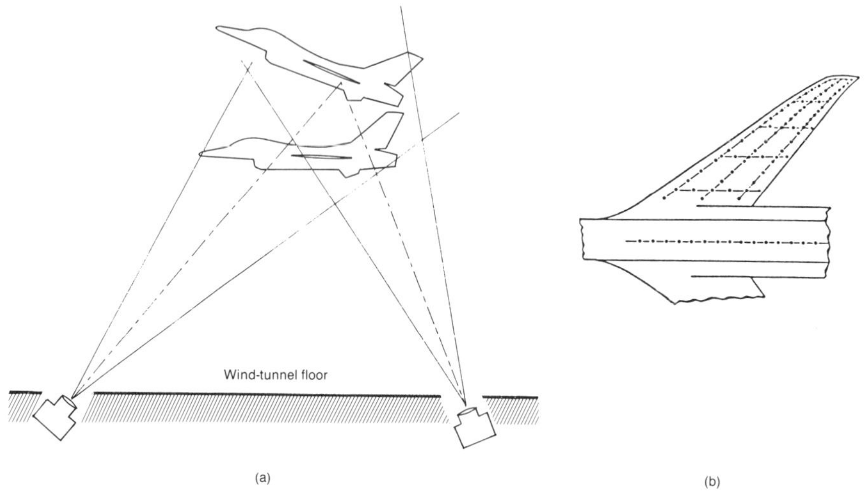

Figure 5-38 (a) Stereo photogrammetry for aircraft model observation during wind-tunnel testing; (b) detail of target point distribution over the aircraft model body.

Passive triangulation by stereo viewing under ambient illumination may also be used for three-dimensional imaging of objects having no artificially inserted targets along their surface by matching the two stereo images and recognizing disparities between the two images (Levine et al., 1973; Yakimovsky and Cunningham, 1978; Jarvis, 1983). The correspondence problem of locating matching points is, however, particularly difficult (Duda and Hart, 1973; McVey and Lee, 1982; McFarland and McLaren, 1983) because of problems generated by surface specularity, surface discontinuities at sharp edges and different occlusion patterns under different viewing directions. One way of simplifying this task is to precede the matching algorithm by an edge detection algorithm to reduce the number of points involved. This is particularly useful when simple geometrical objects are inspected. Another means of reducing computation requirements is to use a hierarchical approach that first matches images at a very coarse resolution and then proceeds to feature matching at progressively finer resolution scales. As a general rule, however, the post-detection data processing requirements constitute a large burden in passive stereo imaging as compared to active projection techniques, which is generally not justified by the slight simplification of the optical apparatus.

Even more complex in terms of computational and pattern interpretation requirements are three-dimensional imaging techniques with monocular view using a number of clues such as texture, shading or sharpness of focus (Jarvis, 1983). If a surface is uniformly textured, surface orientation and relative range can be related to changes in coarseness of texture (Bajcsy and Lieberman, 1976; Rosenfeld, 1983). This method, which originated from psychophysical studies of known human depth cues, is more appropriate for three-dimensional analysis of natural outdoor scenery than for precise measurements. Absolute range measurement is only possible if texture elements are of constant known size.

Surface orientation may be evaluated from shading perception if certain characteristics of object reflectivity, together with information on the light source, are available (Ikeuchi and Horn, 1979; Pentland, 1984). The required computations are often complicated and usually a single two-dimensional image cannot uniquely define orientation. To gain more information, several light source positions may be used sequentially to give different images with different shading. This photometric-stereo approach (Ballard and Brown, 1982, Section 3.5.3) is similar to active pattern projection methods.

Finally, surface depth may be calculated from the focal setting of the lens required to sharply focus the image (Jarvis, 1976). A statistical estimate of image sharpness can be applied throughout the image for a number of different lens positions. Absolute range measurement is available over the imaged surface by this method only if the inspected object contains sufficient texture upon which to focus. As discussed in Sections 4.1.4 and 4.4, such an approach for range finding is normally less sensitive than the optical triangulation approach.

5.3.4 *Time-of-Flight Ranging*

Sonar-based air propagation ultrasonic ranging systems were described in Section 2.2.1. A commercial device of this kind is mounted on Polaroid auto-focus cameras as mentioned in Section 4.4.2. Typical characteristics of such a transducer are a depth resolution of the order of 1 cm over a range extending from 25 cm to nearly 10 m, which is sufficient for several applications, but an angular resolution of the order of 10°, which is unacceptable for most 3-D imaging applications. The reason for such a limited transverse resolution is related to the relatively long ultrasonic wavelength of operation for such a device (larger than 5 mm in air at the operating frequency of nearly 50 KHz) as compared to the 3 cm-diameter of the transducer, leading to such a wide diffraction lobe. Decreasing the angular divergence of the ultrasonic beam is difficult to do with a reasonable transducer size unless the operating frequency is substantially increased, in which case the higher ultrasonic attenuation in air leads to a further reduction of the operating range. Another problem faced by sonar-ranging systems as well as by microwave radars is the specular reflectivity of the projected beam from smooth surfaces. As long as the surface rms roughness is smaller than the wavelength, the surface behaves like a mirror, specularly reflecting the incident beam at any angle equal to the angle of incidence. The beam is thus reflected away from the incident direction giving no pulse-echo return when pointing toward non-normal plane surfaces. Such problems are avoided in laser rangefinders, whose wavelength is sufficiently small to allow sharp focusing with little attenuation through the air and to avoid specular reflections from any surface which has not been optically polished to a mirror finish.

Pulse-echo laser rangefinders have long been known as a possible approach for distance sensing and 3-D imaging (Stitch, 1972; Lewis and Johnston, 1977; Jarvis, 1983). A typical configuration of a time-of-flight ranging system is shown in Figure 5-39. The basic principle is very simple: A short laser pulse is directed toward the surface to be inspected. The backscattered light is collected by a synchronously scanned rotating prism and directed to a photomultiplier with associated signal processing. The time delay between the transmitted and received pulses is measured to evaluate the range distance to the target surface at each angular direction. Using a system of this kind, 64×64 depth range images could be obtained within 40 seconds with a depth resolution of 0.5 cm over a one to four meter range (Jarvis, 1983a). Similar performances have recently been reported with 5 ns-long pulses from a diode laser as the transmitter source (Heikkinen et al., 1986). The recent advances in ultrafast diode-lasers (pulses in the subpicosecond range are now possible), and fiber optic communication detectors suggest that high resolution and inexpensive laser rangefinders using such an approach will be available in the near future. Indeed, a recent paper reports sub-millimeter range resolution (Lettieri, 1986).

As with radar and sonar devices (see Section 2.2.1), a frequency modulation technique may also be applied for distance evaluation with a laser rangefinder (Nitzan et al., 1977, Duda et al., 1979; Page and Hassan, 1981). A typical apparatus is shown in Figure 5-40. A CW laser is modulated at 9 MHz and directed toward the target by an angularly-scanned prism of the kind shown in Figure 5-39. The detected signal is compared to the modulation signal to determine its phase shift from which the object range is determined at each angular direction. The low laser energy used in this case makes it necessary to integrate over several readings for each position to reduce the uncertainty

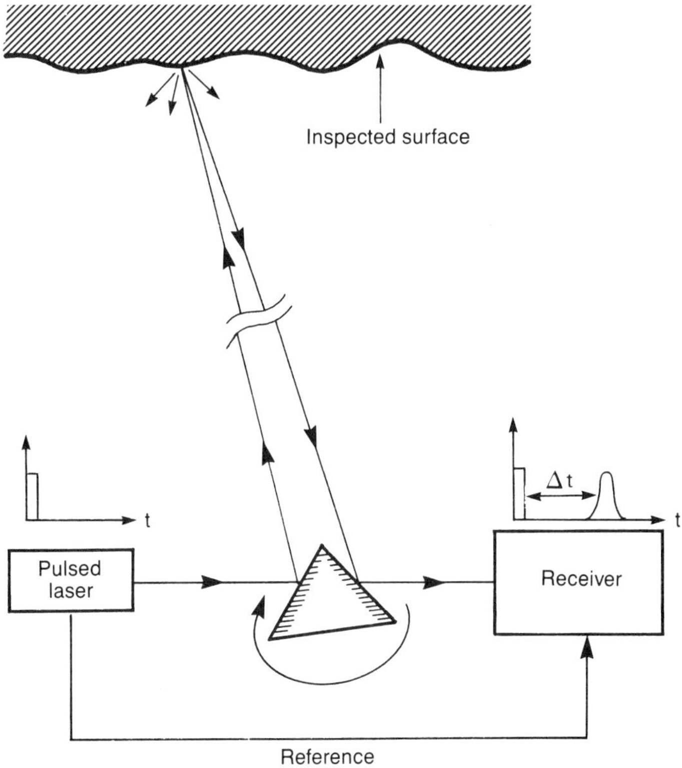

Figure 5-39 Basic configuration of a time-of-flight laser rangefinder by reflected pulse-echo monitoring.

of measurement to an acceptable degree. This results in a relatively long scanning time with the CW modulation approach as compared to the pulsed approach (Jarvis, 1983a). Nevertheless, recent results obtained with a frequency-modulated diode laser demonstrating subcentimeter resolutions over a 1.5 m range show that such an approach is quite promising (Beheim and Fritsch, 1986).

Advantages of the time-of-flight method as compared to other ranging techniques include: (a) high accuracy even at large values of the range, as compared to triangulation; (b) the optical scanning head is particularly compact, in contrast with the triangulation method where the lateral dimensions must be large at long ranges to maintain a sufficiently wide angular viewing aperture; and (c) the receiver and transmitter are nearly

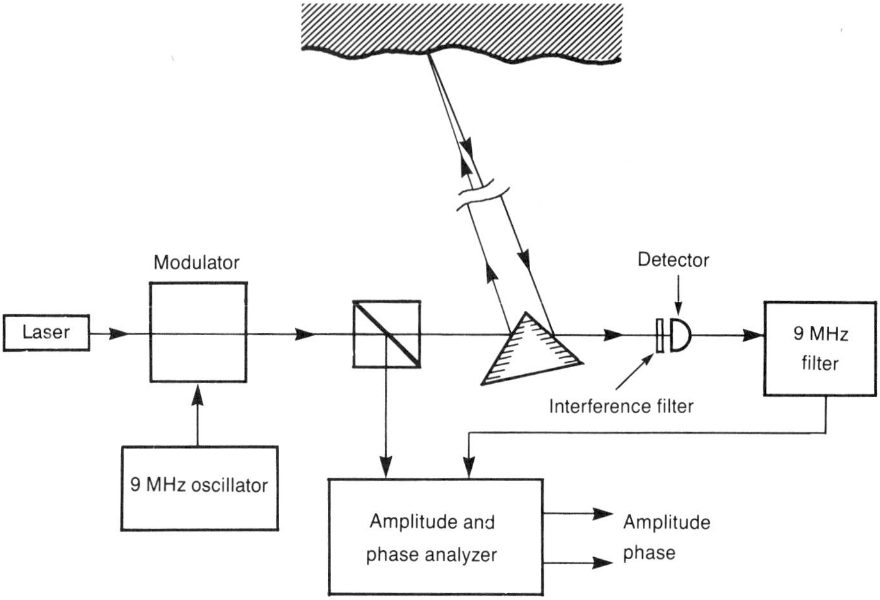

Figure 5-40 Laser rangefinder using CW modulation and phase detection of the backscattered laser beam.

axial. Consequently, no missing-beam problems are encountered because of occlusions of the field of view by sharp edges on the object surface.

On the other hand, current time-of-flight systems tend to be less accurate, more expensive and slower compared to triangulation techniques. Moreover, movable parts are required for mechanical scanning, reducing the sturdiness of the system compared to pattern projection methods. Nevertheless, the recent availability of faster and inexpensive optoelectronic components will probably improve the performance of time-of-flight systems in the near future.

Chapter 6

Optical Techniques for Nondestructive Testing

As mentioned in Chapter 1, manufactured products must often be submitted to specific tests to verify their integrity and suitability for the intended use. Such controls may take place during the fabrication process or during the service life of the item.

Proof-testing is performed on finished products to detect any gross anomaly before release to the user. Examples include the intensive high-regime test procedures to which aeronautical engines are submitted after final assembly, or the high-pressure leak testing of vessels or other containers. Failures during final proof-testing are, however, quite expensive in terms of rejected components and should be minimized by preventive controls. Proof tests are performed where physically and economically possible on single components before assembly or after partial assembly. But destructive tests cannot, by definition, be used for 100% product testing: They can only provide statistical information on the fitness of a given production batch or new product.

The successful completion of a proof-testing procedure is not sufficient to ensure product fitness for a normal service life. The product lifetime may be reduced not only by externally induced damage during service, but more fundamentally by the gradual growth of fatigue related microdefects. A structure may survive during severe initial tests such as static high-pressure or shear-strain loading but may fail after a limited number of relatively low amplitude mechanical solicitations. This may be caused by the gradual propagation of a small surface crack within a metallic component or by the expansion of a delaminated area in a layered composite material.

Accelerated aging tests are thus performed on statistically selected samples to evaluate the expected service lifetime of the product. In the case of safety sensitive products such as aircrafts or nuclear vessels, the part is systematically retired from service at a suitable period of time before the projected failure. However, such an approach is expensive in terms both of the amount of products which must be tested to failure in order to obtain statistically significant data and of the service life reduction of possibly sound products.

Nondestructive testing (NDT) is a different approach by which the part is inspected and the eventual defects localized before they result in catastrophic failure. Such an approach can be applied to 100% inspection of the whole production. Additionally, in-service NDT makes it possible to periodically verify the integrity of the structure and prevent failures produced by externally induced or fatigue initiated damage. If such procedures are sufficiently reliable, they can reduce the need for extensive destructive tests while helping to extend the service life by identifying the parts needing repair. Ideally, if the period between successive controls is sufficiently small, defects such as small cracks may be detected before they grow to an extent where they affect the structure integrity. In this case an evaluation of the average lifetime is no longer necessary because the product will be withdrawn only when it becomes unfit for use. This is known as the "retirement for cause" approach.

Several nonoptical techniques for nondestructive testing were reviewed in Chapter 2. In this chapter a number of optical methods are reviewed and compared to competitive counterparts. Optical NDT techniques have their limitations and, apart from some notable exceptions such as fluorescent-penetrant crack detection which was described in Section 4.2.3, are not yet recognized as standard NDT procedures for routine inspection in the industrial environment. They have, however, received increasing consideration during the last decade, mainly because of their noncontact nature and suitability to rapid scanning of large surfaces.

6.1 Holographic Nondestructive Testing

Most mechanical testing procedures involve submitting the tested material to static or recurrent stress. A defective part will assume an abnormal strain configuration when loaded which amplifies the stress level in certain critical regions, such as near the tip of a crack, eventually leading to failure. If the strain distribution of the part can be recorded with sufficient precision, regions of abnormal deformation can be related to the presence and location of structural defects. Such a visualization of the spatial strain distribution over a loaded structure is possible by holograhic NDT (Erf, 1974; Vest, 1982).

6.1.1 *Optical Holography*

Section 3.7.3 described how holographic optical elements can be obtained by exposing a holographic plate with two mutually coherent beams. Similar considerations can be used to explain how the three-dimensional reconstruction of a laser illuminated surface can be obtained by holography.

A typical setup for the recording and observation of a holographic image is shown in Figure 6-1. The holographic plate is simultaneously exposed to the reference light beam and to the light beam scattered by the object to be inspected. If these two beams are mutually coherent, they generate a closed-space interference fringe pattern which is recorded on the plate emulsion and fixed by development. In the reconstruction phase, Figure 6-1b, such a fringe pattern acts as a complex diffraction grating redirecting the reconstruction beam toward the observer, who can see a three-dimensional image of the inspected object at the same position where the object was originally located during the exposure phase.

To understand the reconstruction process, let us consider a single scattering point A on the illuminated object. The spherical wavefront scattered by such a point interferes with the exposure beam producing a sinusoidal fringe pattern of variable period on the holographic plate. Considerations similar to those illustrated in Figure 3-50 of Section 3.7.3 show that the fringe spatial period increases from top to bottom of the hologram in Figure 6-1. As the diffraction angle is larger for closer fringe periods, the diffracted beam appears to the observer to originate from a virtual point source A′ located in the same position as point A during exposure. Similar considerations repeated for all the other scattering points on the object surface show that the reconstructed virtual object is an exact three-dimensional image of the original part.

Some straightforward precautions must be observed to obtain good quality holograms. For best interference fringe contrast, the reference and object-scattered beam must have similar intensities when superposed on the holographic plate. Suitable beam attenuators must eventually be inserted along the reference beam path to assure such a condition (Erf, 1974, p. 50).

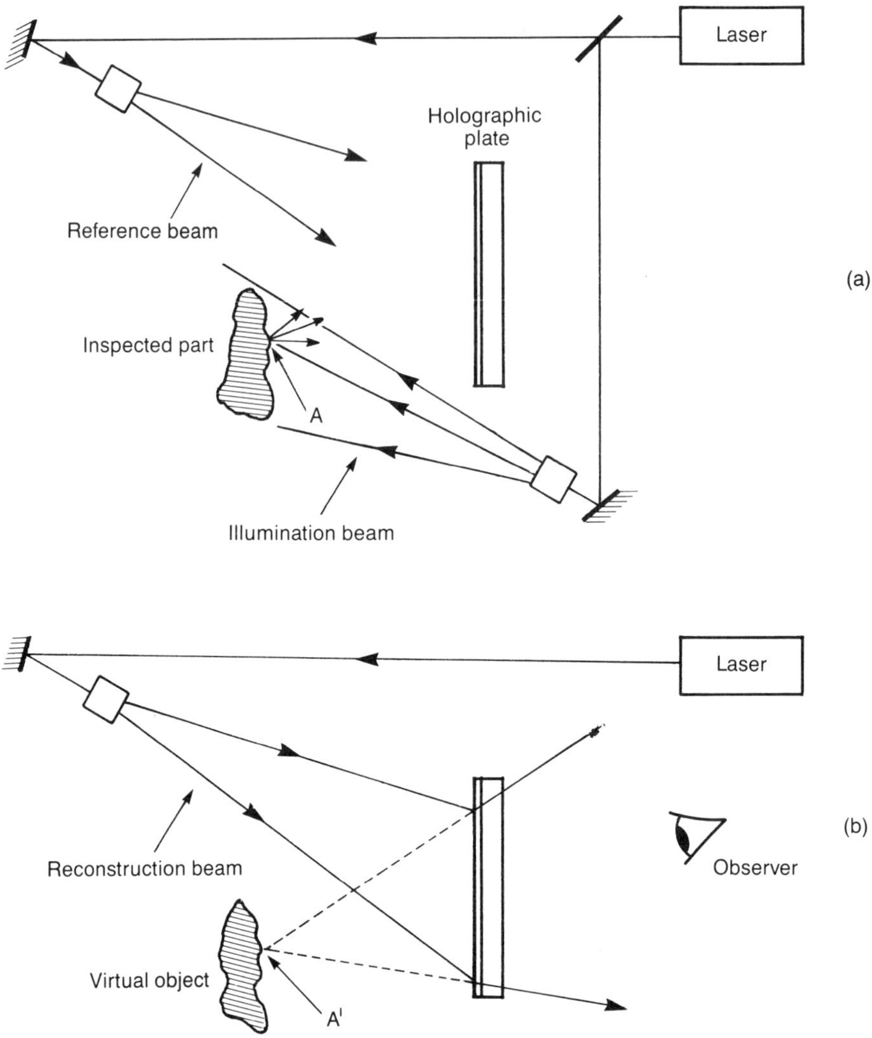

Figure 6-1 A typical configuration for the exposure (a) and observation (b) of a hologram.

The angle between the two beams must be sufficiently small to avoid producing fringe-spacings exceeding the resolution capabilities of the plate. Typical holographic plates have maximum resolutions of 2000 lines/mm corresponding to a maximum angle of 30° between the interfering beams for $\lambda = 500$ nm (see equation (3-16) in Section 3.7.3). Finally, the path difference between the two beams must be kept smaller than the coherence length of the laser beam used for the exposure.

Other technical requirements make holographic techniques difficult to be implemented in an industrial plant environment. The main problem is related to ambient vibrations (Erf, 1974, p. 43–45). Relative displacements of the order of 1 μm between the illuminated object and the optical components during the relatively long (typically one second) exposure time produce fringe-blurring eventually leading to a complete loss of the information. To avoid such a problem, all the optical apparatus as well as the inspected object are normally mounted on a vibration-isolated table, while care is taken to avoid excessive air currents during exposure. Such requirements constitute a serious obstacle to the implementation of routine holographic inspection in the industrial floor. A more convenient approach is to use a pulsed laser for the holographic exposure (Gagosz, 1974). If a sufficiently powerful, typically 20 ns-long laser pulse is used to expose the hologram, the fringe movement due to the low frequency (mainly below 100 Hz) ambient vibrations is effectively frozen during such a short period. However, the operation of such high power lasers on the industrial floor requires appropriate shielding to avoid risk of eye damage, as discussed in Section 3.2 (see also Gagosz, 1974, p. 77). A different approach to limit object-to-plate movement during exposure or between the two exposures in holographic interferometry (see below) is to clamp the plate holder directly onto the inspected structure (Van Renesse and Burgmeijer, 1985). Such an approach is, however, rather cumbersome for routine industrial inspection applications.

If relatively large industrial pieces, such as an engine or a propeller blade, are to be inspected, the coherence length of the laser must be at least as large as the depth of field required to inspect the part. A single-longitudinal-mode exposure laser is thus necessary for such applications. Another practical problem is created by ambient light illumination which may overexpose the photographic plate. Large angular aperture interference filters in conjunction with large area shutters must be used in front of the holographic plate to avoid such problems (Gagosz, 1974, p. 74; Erf et al. 1976, p. 26).

The requirement to develop the holographic plate after exposure has long discouraged many industrialists from considering holography for routine NDT tasks. The recent introduction of photoconductor-thermoplastic recording media is modifying this attitude. Such materials can be automatically dry-processed *in situ* in less than ten seconds and are erasable and reusable several times, making hologram recording simple and inexpensive.

6.1.2 *Strain Analysis by Holographic Interferometry*

Holography is a unique tool for the reconstruction, both in amplitude and phase, of the distribution of light scattered by an object. This capability can be coupled to the sensitivity of interferometric inspection (see Section 4.1.3) to obtain a map of the deformation of an object under a mechanical load.

The basic principle of double-exposure holographic interferometry is illustrated in Figure 6-2. A hologram is produced by exposing twice the plate, before and after subjecting the object to a mechanical deformation. At reconstruction, the virtual three-dimensional images of the object in the undeformed and deformed configurations are both seen by the observer through the plate, superposed on each other. Because of the coherence of the source, light coming from the two virtual surfaces of the object interfere producing a pattern of fringes from which the object displacement between the two exposures can be determined. Best fringe visibility is obtained when the object has a uniform reflectivity across its surface. Dark or specular surfaces should be covered with a diffuse white paint for best results.

An example of the interference fringe pattern obtained by double-exposure from a hydraulic valve before and after pressurization is shown in Figure 6-3 (Cielo et al., 1983a). A 1 m-tall hydraulic valve, of a kind used in the nuclear industry and for which accurate testing is required because of safety

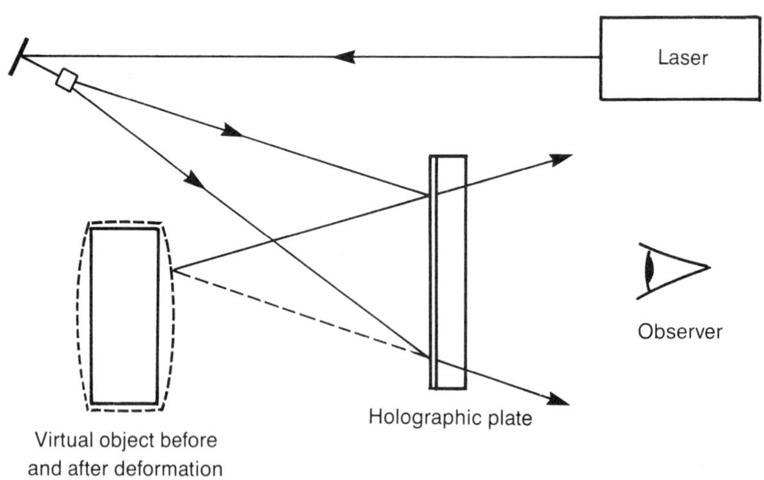

Figure 6-2 Holographic interferometry: Reconstruction of a double-exposure hologram produced with the same object before and after mechanical deformation.

Figure 6-3 Double-exposure hologram obtained from a 60 cm × 80 cm hydraulic valve before and after pressurization at 100 psi (0.7 MPa).

considerations, was mounted on a vibration-isolated table and pressurized at 100 psi (0.7 MPa). Although such a pressure load is much smaller than the typical operating pressure for such valves, a detailed fringe pattern can be observed across the reconstructed image, each fringe corresponding to a surface displacement of approximately $\lambda/2$. To obtain the required depth of field and illumination power level for such a relatively large object, a single-longitudinal-mode Argon-ion laser ($\lambda = 514.5$ nm, 1.6 W, more than 1 m of coherence length) was used. The exposure time for the Kodak 649 F plate was two seconds per exposure in this case.

The concentric ring pattern on the neck of the valve indicates, as expected, that this part of the valve has suffered the largest strain. By counting the fringes from the bottom pedestal which did not move during pressurization, one can evaluate the maximum displacement of the neck surface at nearly 20 half-wavelengths, or 5 μm. A qualitative observation of the hologram may also enable identification of the bolts which are tight (such as the second from the bottom in the right side of the image) and those which are loose. Well-tightened bolts show a larger fringe density which is proportional to their tractional displacement. From patterns of this kind a valve prototype can be evaluated and the parts needing reinforcement can be identified without testing the valve to failure. The presence of hidden flaws such as inclusion defects within the metal may also be localized nondestructively by comparing the observed strain distribution with that of a sound valve. Weak regions show abnormal deformation patterns under pressure.

Interferogram Production Methods

The *double-exposure holography* approach described above is one of several methods by which a holographic interferogram can be produced. Such an approach is particularly convenient for pulsed hologram recording in the presence of strong ambient vibration noise, because the two plate-exposing laser pulses can be fired at a time interval short enough to avoid significant random movement of the object during the pulse period. Typically, the two laser pulses may be separated by a time-period of 50 μs if a sufficiently fast loading technique, such as the vibration loading approach, described later, can be found to produce a sufficiently large controlled strain during such a short period. Double-exposure interferograms are, however, more difficult to analyze quantitively, as discussed below.

Another approach for the production of an interferogram is *real-time holography*. Here the holographic plate is exposed only once, and after development it is replaced in exactly the same position as during the exposure. Interference can thus be observed between the holographically reconstructed virtual object and the laser-illuminated real object which is left in the same position as during exposure. If the plate can be restored to the same position within a fraction of a wavelength and if the object has not moved, the superposition of the wavefronts from the real and the virtual objects will yield an image with no interference fringes. If the real object is now perturbed with a convenient loading force, a set of interference fringes can be seen in real time as the strain field is produced. Real-time interferograms are easier to produce if the hologram can be developed *in situ* as in the case of photoconductor-thermoplastic plates, avoiding the need for a time consuming plate repositioning process after development.

A third approach for interferogram production is *time-averaged holography*, where the hologram is exposed to light scattered by a moving object.

The best known configuration is that of a sinusoidally vibrating object. If the vibration amplitude is equal to a small number of wavelengths, interference fringes can be observed in the hologram corresponding to the two extreme positions of the vibrating surface, where the dwell time of the object is longest (Abramson, 1981, Section 3.4). However, the fringe visibility is poor compared to that obtained with double-exposure or stroboscopic illumination (Vest, 1979, Section 4.3.4).

Quantitative Fringe Interpretation

Although the motion of a simple-shaped object may usually be roughly evaluated from a qualitative examination of the interference fringe pattern, the exact reduction of an interference fringe pattern to the corresponding strain distribution is not a straightforward task in the case of a complex structure.

The relationship between surface displacement and the number of observed interference fringes is illustrated in Figure 6-4 (Vest, 1979, Section 2.2; Abramson, 1981, Section 4.2). If after a small deformation the point A on the inspected surface is displaced to A′, the number N of clear fringes which can be seen in a real-time holographic interferometry experiment as A gradually moves to A′ is related to the path difference $\delta = \mathrm{IAO} - \mathrm{IA'O}$ between the two beam paths before and after displacement. Simple trigonometric considerations show that

$$N = \frac{\delta}{\lambda} = (\mathbf{K}_2 - \mathbf{K}_1) \cdot \mathbf{D}, \tag{6-1}$$

where λ is the light wavelength and the vectors $\mathbf{D}$, $\mathbf{K}_1$ and $\mathbf{K}_2$ are shown in Figure 6-4, with $K_1 = K_2 = 1/\lambda$. The number of scanned fringes for a given displacement amplitude D will thus be maximum when the displacement takes place along the bisector between the illumination direction IA and the observation direction AO. The same number of fringes is, however, produced by displacements of different amplitude and direction if the displacement vector projection along the bisector is the same. This leads to a basic ambiguity when the strain field is to be determined from fringe counting techniques.

Similar considerations are valid when the object deformation field is reconstructed from a double-exposure hologram. In this case, the displacement of a surface point A is determined by counting the fringes visible in the interferogram between A and a zero-motion point Z such as the valve pedestal

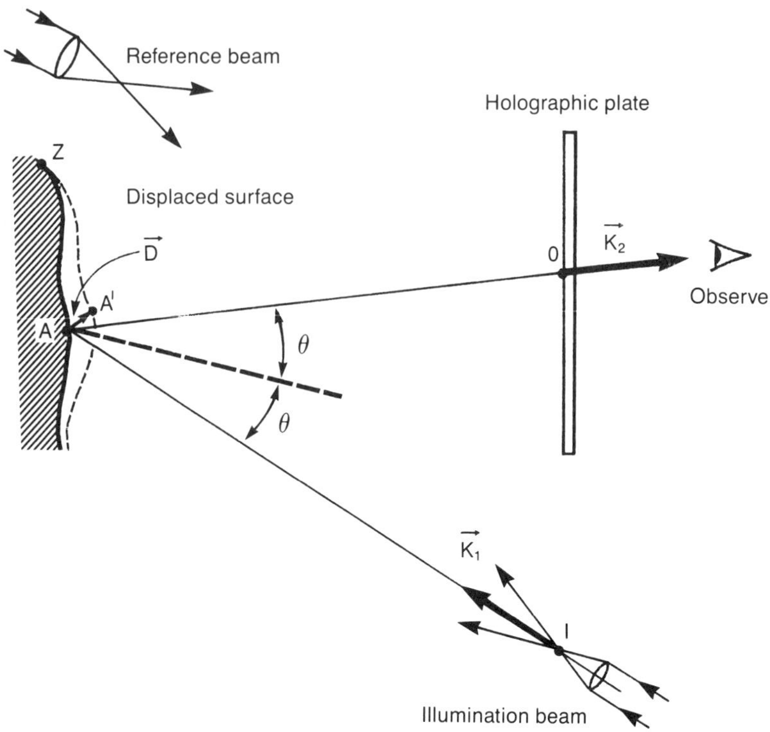

Figure 6-4 Geometrical parameters showing the relation between surface displacement and fringe count in holographic interferometry.

in Figure 6-3. Apart from the above-mentioned error source related to the unknown orientation of the displacement vector, further ambiguity arises in this case if the surface displacement cannot be assumed to vary monotonically between points Z and A.

Such ambiguities may be removed if the observation point O is moved across the viewing plate. This results in a variation of the three-dimensional orientation of the bisector in Figure 6-4 and thus in a movement of the fringe pattern as the observer position moves behind the plate. By counting the number of fringes passing through A while the observation point 0 is moved to three independent positions across the plate, the displacement vector at A may be completely determined (Vest, 1979, Section 2.2; Jones and Wykes, 1983, Section 2.4). If the fringe density is not sufficiently large to produce a convenient fringe count during such an operation, three different holograms at different angular positions may be simultaneously exposed to obtain a sufficiently large angular aperture. However, such strain-field reconstruction operations are rather time consuming. Efforts to automate such analyses are in progress (Vest, 1979, Section 2.5; Müller et al., 1983; Trolinger, 1983).

6.1.3 *Part-Loading Methods and NDT Applications*

Mechanical loading to produce convenient microdisplacements of the inspected part is probably the most critical aspect of holographic NDT in terms of its industrial applicability. Surface displacements of several wavelengths are required for good fringe production efficiency, while whole-body motion producing spurious interference fringes should be minimized between the two exposures (Van Renesse and Burgmeijer, 1985). Moreover, stressing techniques which can be implemented rapidly and which require minimum sample preparation are preferred for industrial applications. The most widely used loading methods are briefly reviewed and discussed in this section with reference to some typical NDT applications. More details may be found in the copious literature on this subject, including more than 150 papers during the last two decades (Vest, 1982).

Mechanical Static Loading

The most direct approach to produce a mechanical deformation of the workpiece is to apply a static mechanical stress between the two hologram exposures or between the first exposure and the time of observation in real-time interferometry. A static load may be produced by pressurization, as in the case of pressure valve inspection referred to in Figure 6-3, or by direct mechanical stressing. The latter technique has been applied, for example, to crack visualization (Vest, 1974; Campbell and McLachlan, 1979; Ennos and Virdee, 1982). A double-exposure hologram obtained on the surface around a hole-initiated, 1.1 cm-long crack is shown in Figure 6-5 (Vest, 1974). To obtain such a hologram, a tapered-shank bolt was inserted in the hole and the holographic plate was exposed before and after a torque was carefully applied on the bolt to widen the crack. The resultant strain field has a discontinuity along the crack because of stress release at the crack's edges, producing a discontinuity in the fringe pattern under appropriate viewing angle conditions. Similar techniques can be used to study crack propagation conditions by monitoring the strain field around the crack tip (Hsu et al., 1978; Ennos and Virdee, 1982).

Another method for mechanically stressing the inspected object, which is particularly effective for finding delaminations in layered materials (Kersch, 1974; Klappert, 1976; Vest and Sweeney, 1977), is the vacuum-stressing

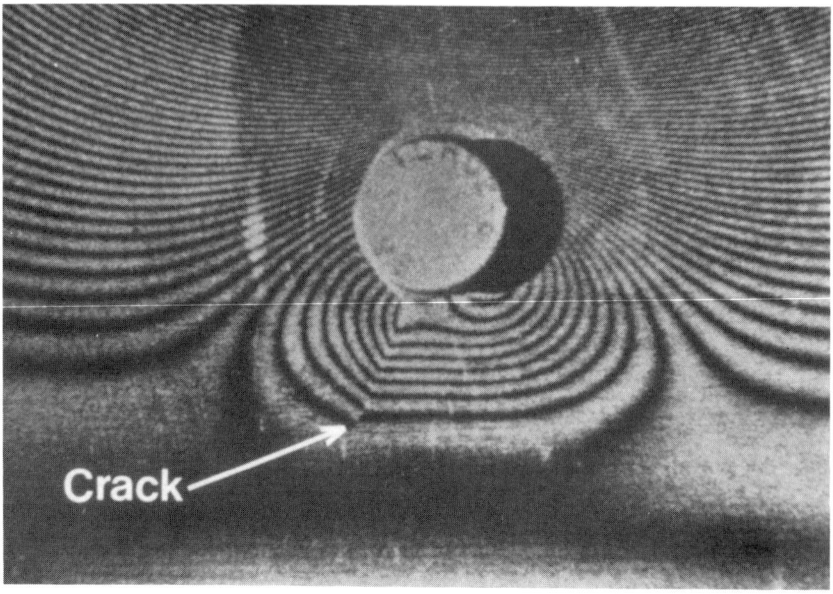

Figure 6-5 Fringe pattern obtained from a double-exposure hologram of a metallic plate with a 1.1 cm-long crack.

technique. The basic principle is shown in Figure 6-6a. A vacuum box is clamped to a layered sheet such as an adhesively bonded aluminum-to-aluminum or graphite-epoxy laminate of the kind used in the aeronautical industry. Such materials are subject to strict controls to make sure that no layer-to-layer unbonded regions are present which could propagate under stress and result in catastrophic failure. A slight vacuum suction is applied before the first holographic exposure to seal the box against the plate, and a larger suction is applied before the second exposure or the real-time observation to obtain a set of fringes of the kind shown in Figure 6-7a. In this case a 0.75 mm-thick aluminum sheet was adhesively bonded to a honeycomb structure with two lack-of-bond artificial defects. The unbonded areas are subject to a larger deformation under vacuum suction resulting in the two localized fringe patterns which are clearly visible in the interferogram of Figure 6-7a, superposed to a regular pattern of background circular fringes. The circular fringes are produced by the uniform lifting of the sheet under suction (see Figure 6-6a) and may mask the visibility of smaller disbonds. Techniques are available to eliminate such background fringes by optically producing a reverse interference fringe pattern (see, e.g., Kersch, 1974,

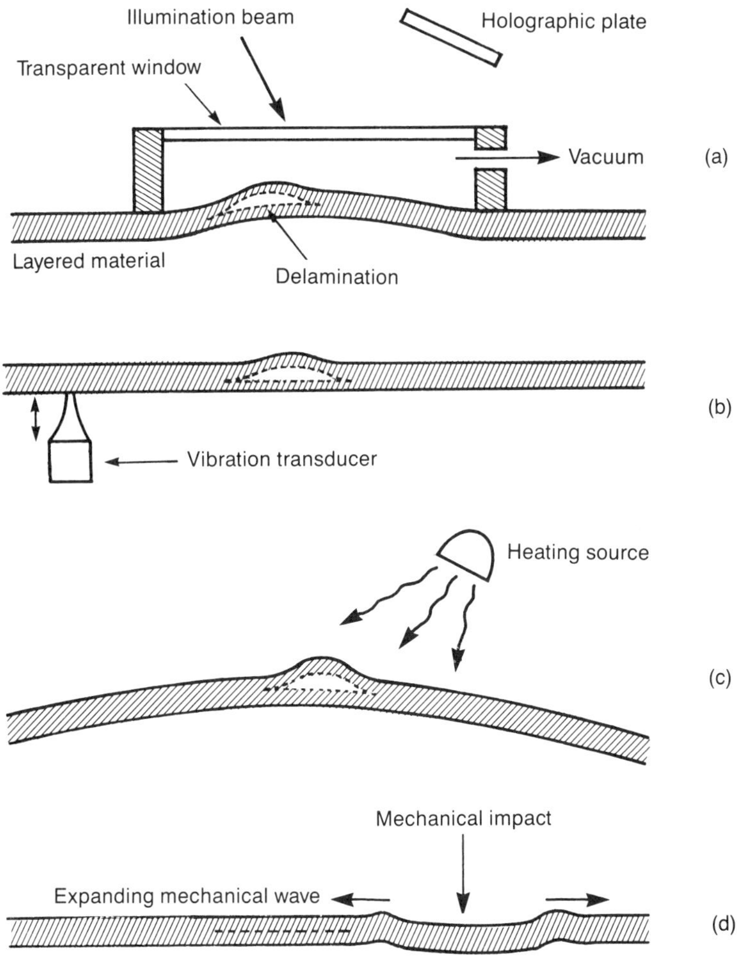

Figure 6-6 Different part-loading techniques for holographic NDT of a layered sheet: (a) vacuum stressing, (b) vibration excitation, (c) thermal stressing and (d) impact loading.

p. 309–310) or by sandwich holography methods (Abramson, 1981, Section 3-5), but such techniques are somewhat artificial and time-consuming. Balanced stressing techniques by two vacuum boxes clamped on each side of the sheet have also been considered to avoid background fringes (Winkler et al. 1981).

Clamping a vacuum box to each portion of the structure to be inspected is a cumbersome procedure which is hard to apply to nonflat surfaces.

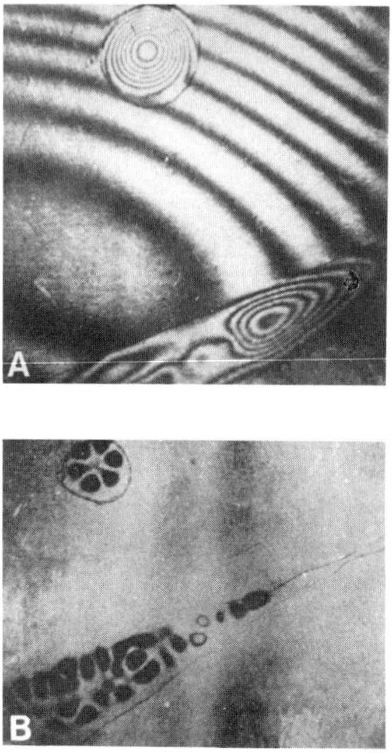

Figure 6-7 Double-exposure hologram of an adhesively-bonded aluminum-to-honeycomb laminate with (a) vacuum stressing and (b) vibration excitation at 25.6 kHz.

Moreover, vacuum stressing is limited to a maximum applied pressure differential of 1 atm. Consequently, adhesive-unbond defects (i.e., when the two adherents are in physical contact but no bond strength exists at the bond line) cannot be detected by this method unless an air-filled gap at least a few micrometers thick is present at the interface (Kersch, 1974, p. 308).

In spite of these limitations, holographic NDT by vacuum loading has found an important industrial application to the routine inspection of tires (Brown, 1974; Schoerner and Rottenkolber, 1983; Fisher, 1985). Retreaded tires for aircraft use are routinely inspected for ply-to-ply delaminations by installing the tire in a vibration-isolated vacuum chamber of the kind shown in Figure 6-8. The tire is mounted on a rotating fixture inside a vacuum bell, together with the laser and holographic material, and is sequentially inspected at different angular positions. Problems such as air ventilation or ambient light are at the same time eliminated by the tight enclosure. Advanced automated systems are capable of inspecting up to 20 tires per hour (Vest, 1982).

Figure 6-8 An industrial system for routine inspection of retreaded aircraft tires using vacuum-stressing holographic interferometry.

Acoustic Excitation

As illustrated in Figure 6-6b, an alternative approach to produce a controlled microdisplacement of the holographically inspected structure makes use of acoustic vibration. In this case, an acoustical wave is coupled to the structure by a vibrating transducer. As a result, the unbonded area will vibrate in a set of normal modes as the driving transducer is scanned through a range of frequencies, typically in the 10 to 100 KHz region. The normal vibration modes of the whole structure, such as a compressor blade (Erf, 1974, Section 8.8), may be observed by double-pulsed, stroboscopic or time-averaged holography, and hidden defects can be located by comparison with the resonant frequencies and mode patterns of undamaged structures. Disbonds in laminates can be detected from the abnormal fringe patterns obtained on the unbonded regions. Such fringe patterns are particularly pronounced when the driving frequency corresponds to a normal flexural resonance of the unbonded layer, as Figure 6-7b shows (Kersch, 1974). Advantages of this approach compared to the standard pulse-echo ultrasonic inspection methods described in Section 2.2.3 include the possibility to operate at frequencies which are two orders of magnitude lower than in ultrasonic testing. This results in a much lower ultrasonic attenuation, while dry-coupling of the transducer to the surface is conveniently performed at such low frequencies (Erf, 1974, p. 331).

A variation of the acoustic loading technique is represented by impact loading (see Figure 6-6d), where a point impulse is produced on the structure surface by impact with a steel ball or a pendulum, and the diverging flexural wavefront is recorded by pulsed holography after the impact. Subsurface delaminations produce anomalous fringe patterns when the acoustic wavefront reaches the unbonded area (Chu et al., 1972). Transient excitation makes it possible to observe the response of the layered structure to the expanding wavefront before the structure strain field is strongly affected by normal-mode resonances of the whole structure. Increasing attention is now being devoted to this excitation scheme (Vest, 1982, p. 193).

Acoustic excitation is a particularly effective approach to produce relatively large strain fields in a short time period thus reducing spurious displacements of the whole structure during such a period. Consequently, acoustic loading is usually preferred for double-pulse holographic inspection of large structures in a factory environment. Spectacular examples of applications of this technique are the inspection of full automotive or aircraft bodies and of engines on the industrial floor (Erf et al., 1976; Felske and Happe, 1976; Murata and Kuroda, 1983; Brown and Wales, 1983). Acoustic stressing can be achieved by rivet gun excitation or by exploiting the natural vibration of an operating piston engine. Fringe interpretation is, however, a difficult problem in this case, because of the presence of strong whole-body resonance, the presence of widely different fringe patterns at different excitation frequencies and the unavailability of real-time interferometric holography techniques to remove ambiguities in the interpretation of the fringe pattern.

Thermal Stressing

Thermal stressing, as illustrated in Figure 6-6c, consists of the application of heat to the inspected surface with an infrared lamp or a heat gun to produce a longitudinal thermal stress field within the first layer of the laminate. The thermoelastically-induced bending moment generated during heating by the thermal gradient across the heated sheet produces a whole-body curvature of the laminate if it is free to move, as Figure 6-6c shows. If the surface layer is locally unbonded to the substrate, the unbonded region may experience additional thermal buckling under the longitudinal stress. This shows up as a local perturbation of the interference fringe pattern as Figure

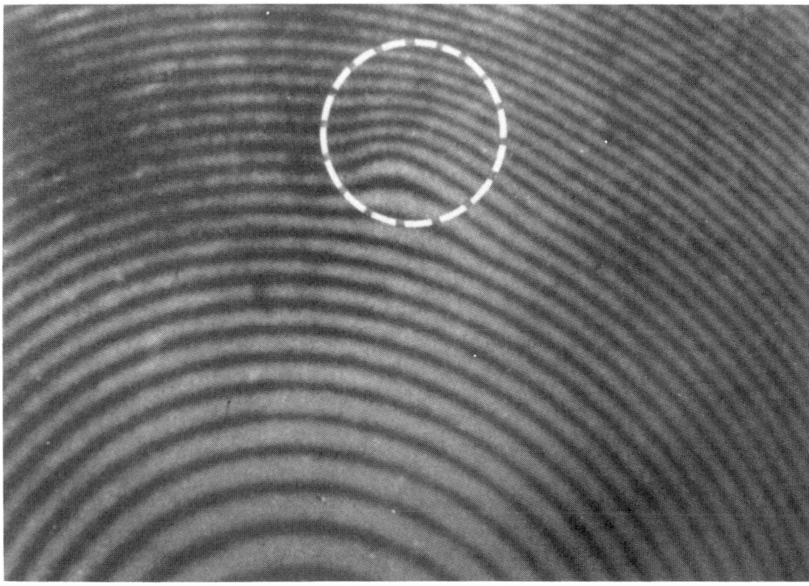

Figure 6-9 Interference fringe pattern produced by thermal stressing on a laminate structure, showing the circular fringe pattern produced by whole-body plate bending as well as the local perturbation of the fringe pattern above a disbond defect.

6-9 shows (Kersch, 1974). Several examples of laminate inspection by thermal-stress holographic interferometry have been reported in the past 15 years (Harris and Claus, 1971, Harris and Wood, 1974; Maddux and Sendeckyi, 1979; Schliekelmann, 1982).

Heat stressing has some very attractive characteristics for industrial applications: It is a noncontact technique and thus rapid to apply and easy to automate; it produces little whole-body-deformation spurious fringes if applied quickly on a solidly fixtured or thick structure; and it is appropriate to both double-exposure or real-time holographic interferometry. However, such an approach has not always been found as effective as, for example, vacuum stressing to reliably produce the required strain field (Harris and Wood, 1974; Winkler et al. 1981). This can be understood by considering a simple model of the thermal expansion of the unbonded layer (Kersch, 1974, Section 8.7.2; Birnbaum and Vest, 1983; Rousset et al., 1985).

When a surface layer is uniformly heated at a temperature higher than the substrate to which it is bonded, longitudinal thermoelastic stress is generated within the layer in a direction parallel to the surface. Such a stress field is insufficient to produce thermal buckling of the unbonded portion of the layer unless the layer is initially in a curved configuration with the convex side up, as Figure 6-6c shows. If the sheet is perfectly flat and uniformly heated, no thermal buckling can be produced (Cielo et al., 1984; Rousset et al., 1985). Conversely, if the sheet is initially curved with the convex side down, the unbonded portion of the layer will tend to increase its curvature under thermal stress, thus compressing further against the substrate.

Such considerations explain why thermal stress is sometimes unreliable for holographic NDT of layered materials. An initially convex configuration may be found only for specific structures or can be produced by a thermal gradient across the sheet or by a greater thermal expansion coefficient of the layer as compared to the substrate only if the sheet is freely supported. A more reliable mechanism for the bending of an unbonded portion of the layer is related to the larger thermal resistivity of the unbonded interface as compared to the surrounding well-bonded interface. This results in higher temperature of the layer in the delaminated region compared to the surrounding area. Under such circumstances, a three-dimensional shear stress field develops within the layer near the boundaries of the unbonded area. This produces an uplifting bending moment even below the buckling threshold (see Section 6.3.2). However, such a mechanism requires a substantial temperature differential above the delaminated area, which can more efficiently be observed by thermography (see Section 6.4).

6.1.4 *Heterodyne Holographic Interferometry*

One of the limits of holographic interferometry for NDT applications is the need for surface displacements of several wavelengths in order to produce a conveniently dense interference fringe pattern which can be unambiguously interpreted to obtain the object strain distribution. Fringe interpolation for the measurement of displacements within a fraction of a wavelength is possible but difficult to implement because of the previously mentioned displacement direction ambiguity as well as the superposition of variable surface reflectivity features on the fringe-related intensity variations across the holographic image. Heterodyne holography (Crane, 1969; Dändliker, 1980) was developed to allow unambiguous fringe interpolation within a small fraction of a fringe.

An earlier experimental configuration for heterodyne holography is shown in Figure 6-10 (Dändliker et al., 1973). More recent systems use acousto-optic modulators to produce the frequency shifted beams (Dändliker et al., 1980).

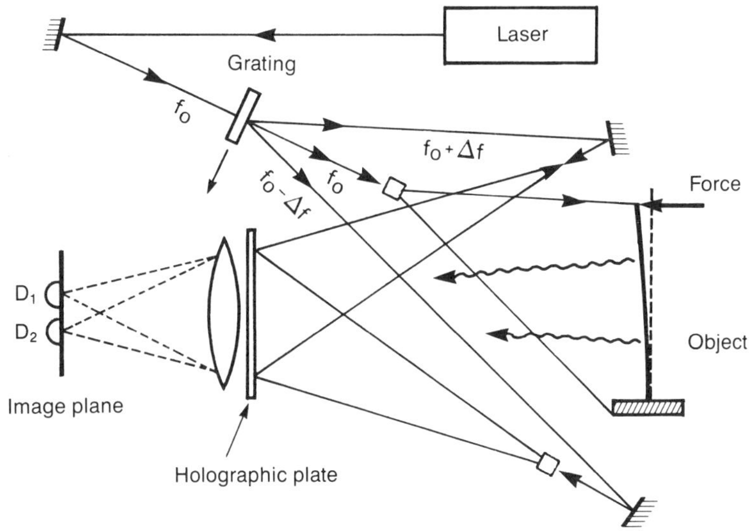

Figure 6-10 Experimental apparatus for heterodyne holographic interferometry of a bent cantilever. The grating is static during exposure, while moving at constant speed during reconstruction so as to produce two diffracted beams of offset frequencies $f_0 + \Delta f$ and $f_0 - \Delta f$.

The light beam from a 0.5 watt single-mode Argon-ion laser is split into three beams, corresponding to the 0, +1 and −1 diffraction orders, by a diffraction grating. During the holographic exposure phase, the grating is kept stationary so that all three beams have the same optical frequency, f_0, of the original laser beam. The zero-order beam illuminates the object, while the two diffracted beams are alternatively used as reference beams during the two holographic exposures. The holographic plate is first exposed with the cantilever at rest and with one of the diffracted beams blocked, after which the cantilever is slightly bent and the plate is exposed again with the other diffracted beam masked out. The plate is subsequently developed and set back in its initial position. The use of a different reference beam during the two exposures makes it possible to reconstruct the holographic image of the object in either its initial or its displaced position by illuminating the developed plate with one or the other of the two diffracted beams.

During the reconstruction phase, the grating is moved at constant speed in a direction perpendicular to the grating ruling lines, and the developed plate is simultaneously illuminated by the two diffracted beams. Because of the grating movement, the frequency of the diffracted beams is Doppler-shifted by an amount $+\Delta f$ and $-\Delta f$, respectively, which is proportional to the grating speed (see, e.g., Drain, 1980, p. 48; the physical mechanism is similar to the frequency shift experienced in a Bragg cell, see Section 3.8). In the specific experiment mentioned here, such a frequency shift was of $\Delta f = 40$ KHz. To process the double-exposure hologram, a lens is placed close to the hologram to form an image of the two reconstructed object surfaces on the same plane. This image plane is scanned by two optical detectors. Each detector receives light from a given area of the two superposed holographic images of the cantilever in its undisplaced and displaced positions. As the object is reconstructed in the two positions with light of different frequencies, the superposition signal recorded by each detector is strongly modulated at the beat frequency $2\Delta f$. The phase of the beat signal is related to the relative position of the object in its initial and displaced configurations, similarly to the dependence of the beat-signal phase upon the surface height in the heterodyne profilometers described in Section 4.1.3. By monitoring the phase of such a signal with a zero-crossing phasemeter and by scanning one of the detectors across the image while the other detector is kept stationary as a reference, a map of the surface displacement with respect to the reference point can be plotted with high accuracy.

Figure 6-11 shows schematically the fringe pattern which would be obtained at a given instant of time across the holographic reconstruction of the cantilever inspected with the apparatus of Figure 6-10. If the two detectors, D_1 and D_2, each having an aperture smaller than half a fringe spacing, are positioned along the same fringe, the two detected beat signals appear to have the same phase, as shown in the bottom insert. The other two inserts show how the phase of the D_1 signal changes with respect to the stationary reference D_2 as the D_1 detector is moved up along the fringe pattern. The three examples show a phase displacement of 0°, 90° and 180°, sequentially corresponding to movement of D_1 along a quarter and one-half of a fringe spacing. If the relative phase of the two detectors can be measured with an accuracy of 0.3°, the cantilever displacement can be reconstructed with an accuracy better than 1/1000 of a fringe. However, such precision, which is much greater than with classical holographic interferometry, requires point-by-point mechanical scanning of the holographic image by the phase-sensing detectors. This operation is performed by a stepper motor apparatus and requires typically one second per point (Dändliker and Thalmann, 1985).

The overall accuracy and reproducibility of heterodyne holographic interferometry depends on the strain configuration, electronic noise and mechanical stability of the optical setup (Dändliker, 1980). For a precise

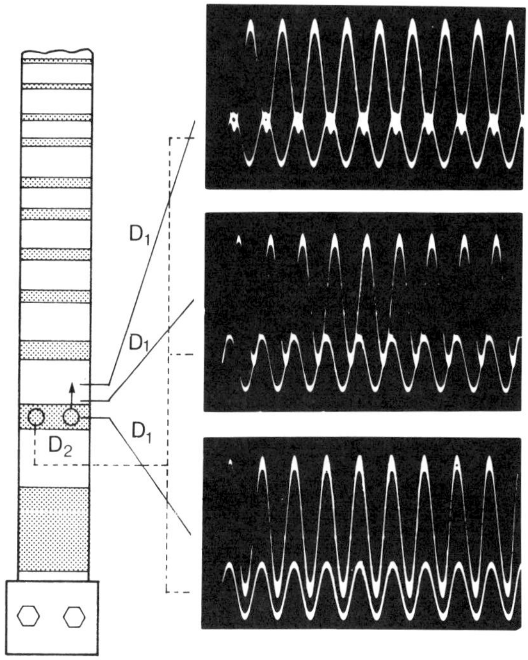

Figure 6-11 Diagram of the reconstructed holographic image of the cantilever and of the output from the two detectors D_1 and D_2 shown in Figure 6-10. The inserts show the beat-frequency signals detected as D_1 is moved to the three positions indicated on the holographic image. The constant-phase beat signal obtained from the stationary D_2 is shown in each case as a reference.

phase readout, the aperture of the detectors must be smaller than the smallest fringe half-period across the hologram. This is in conflict with the other requirement that the number N of speckle grains within the detector aperture must be as large as possible to minimize statistical noise. On the other hand, if the size of the speckle grain is made smaller than the transverse displacement of the strained surface, the phase readout error will increase because of partial speckle decorrelation. Best results are thus obtained with relatively small strain gradients giving wide fringe patterns. Electronic and mechanical noise depend on the specifications of the phasemeter, the SNR of the detector signals, air turbulence and mechanical oscillations during readout. Such noise sources can be reduced using electronic phase control modules (Martin et al., 1983), working with well-stabilized mounts and in a shielded atmosphere, while, in general, the SNR is improved by increasing the measurement time.

It should be noted that a readout time of the order of 1 s/point, i.e., a scanning rate of $0.1\ cm^2/s$ for a spatial resolution of 3 mm, is relatively slow for industrial NDT requirements.

To avoid the electronic-noise and long-scanning-time problems of heterodyne holography, quasi-heterodyne techniques have been developed (Dändliker and Thalmann, 1985) using a PZT-stepped interferometric scheme of the kind described in Section 4.1.3 dealing with the Mirau-type profilometer. A typical apparatus is shown in Figure 6-12. Similarly to the setup shown in Figure 6-10, two beams are used as the reference, but in this case the two beams are angularly very close to each other in order to produce a pattern of parallel microscopic interference fringes of typically 1 to 2 mm spatial period in the plane of the holographic plate. The relative phase of one of the reference beams with respect to the other, and thus with respect to the object if the whole system is adequately stabilized mechanically, can be piezoelectrically shifted by an amount which is controlled by a twin-photodiode sensor tracking the fringe maxima. Rather than reconstructing the holographic interferogram with two optical frequencies as in the heterodyne-holography approach, a single wavelength is used for reconstruction in the quasi-heterodyne technique and the fringe pattern is

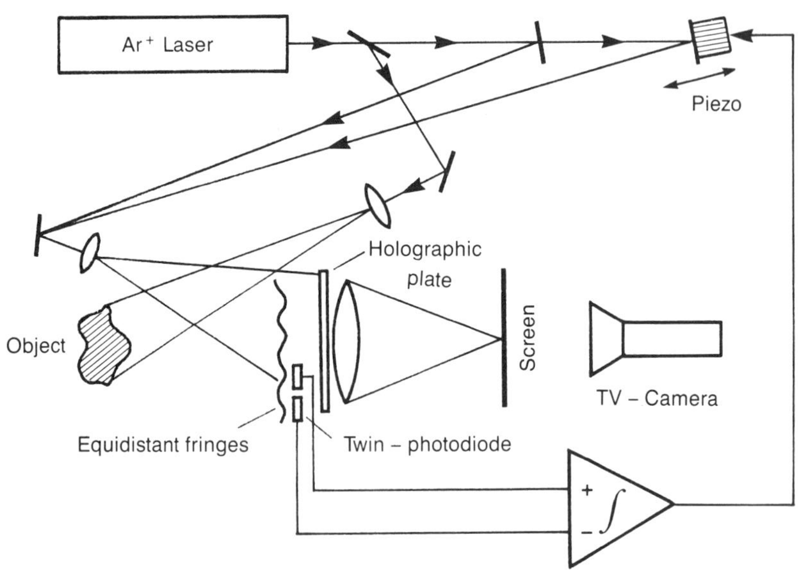

Figure 6-12 Experimental arrangement for quasi-heterodyne holographic interferometry with two reference beams at a small angle to each other and fringe stabilization.

interpreted using video techniques. To interpolate between the interference fringes, multiple fringe patterns are video recorded and computer processed with different PZT-produced and twin-sensor controlled phase shifts, typically three patterns with $\frac{1}{3}$ fringe separation shift between each other. This approach has actually little resemblance to heterodyne holography, being equivalent to a classical holographic interferometry approach with multiple recording during real-time stressing, or else with multiple observations from different observation directions through a double-exposed hologram as discussed in relation to the quantitative fringe interpretation problem in Section 6.1.2.

The phase-stepped holographic approach avoids the long mechanical scanning requirements of heterodyne interferometry by making it possible to process by video means the phase-shifted fringe patterns. If the phase distribution varies smoothly across the image, sign ambiguities and intensity fluctuations induced by variations of the surface reflectivity across the fringe pattern can be overcome by this method. Resolutions of $\frac{1}{100}$ of a fringe have been reported (Dändliker and Thalmann, 1985). This technique faces the same practical problems as classical holographic NDT in terms of the required optical resolution of the holographic plate, mechanical stability needs, sensitivity to displacement variations rather than to the absolute displacement, and difficulty in rapidly applying uniform stress fields with minimum whole body motion.

6.2 Speckle Metrology

Coherent speckle patterns are an important source of noise in optical imagery under laser illumination. Nevertheless, speckle properties may be used to advantage in certain cases. A useful application of speckle phenomena to surface roughness characterization was described in Section 4.2.2. This section describes some applications of speckle techniques to microdisplacement analysis.

The main advantage of speckle metrology techniques as compared to holographic interferometry is the lower spatial resolution which can be accepted for the recording medium. Moreover, the directional and magnitude sensitivity of the speckle approach can be varied over a substantially larger range than can be done with holographic interferometry. However, speckle correlation fringes usually have a lower contrast as compared to the holographically produced interference fringe patterns.

6.2.1 *Speckle Photography*

The most direct method to produce an image-plane speckle pattern is shown in Figure 6-13. In this speckle-pattern photography approach, a viewing lens collects light scattered from an object in an undisplaced and in a displaced position, and a recording is made on a single photographic plate. An alternative method implies superposing a negative of the undisplaced speckle pattern upon a positive of the speckle pattern after object displacement (Groh, 1970). This approach requires a more cumbersome double-step process and time-consuming repositioning accuracy, and consequently it is less appropriate to industrial applications.

The speckle pattern obtained after each exposure in the image plane of a uniformly illuminated surface is of the kind shown in Figure 6-14a (Fourney, 1978). A high-contrast speckle distribution can be observed with an average grain size depending on the recording geometry and on the diameter of the aperture stop, as discussed in Section 3.10. If the object is uniformly displaced by a small distance in the upper direction (along the y-axis in Figure 6-13) and a second exposure is taken, a double-pattern of the kind shown in Figure 6-14c will be obtained. The second exposure results in a second random-speckle pattern, strongly correlated to the first one but slightly shifted in the

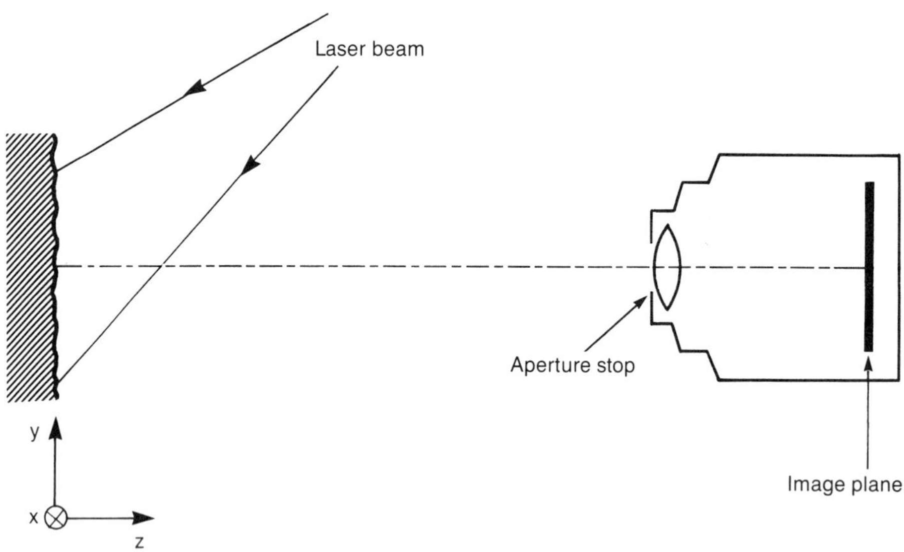

Figure 6-13 The basic arrangement for the recording of an image-plane speckle pattern in a speckle photography setup.

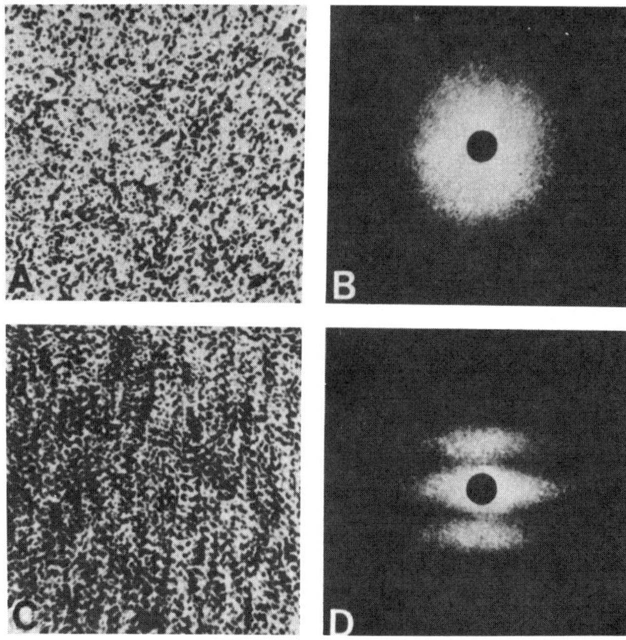

Figure 6-14 Speckle patterns (A, C) and their corresponding diffraction patterns (B, D) for a single-exposure photo (top) and a double-exposure photo after a uniform linear displacement (bottom). The dark center spots in the diffraction patterns correspond to the zero-order block.

direction of the surface displacement vector, which is superposed to the first speckle pattern in the photographic plate. Such a displacement of the image-plane speckle pattern is similar to the image displacement in the case of an ordinary photograph of the displaced object. Speckle photography has, however, the advantage over ordinary photogrammetry of a much higher contrast even in the case of an object surface of uniform reflectivity.

In-plane Strain-Gradient Visualization by Spatial Filtering Techniques

The object displacement between the two exposures can be quickly and precisely measured by placing the developed photographic film negative in a Fourier-transform setup of the kind shown in Figure 6-15. The film is illuminated with a converging beam and its Fraunhofer diffraction pattern is obtained in the Fourier plane of the focusing lens (see, e.g., Goodman,

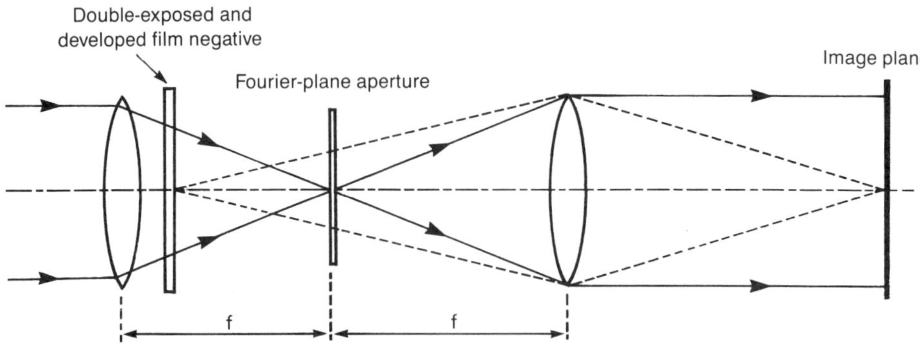

Figure 6-15 An optical arrangement for the observation of the displacement distribution over a coherently illuminated object by optically processing the double-exposure speckle pattern impressed on the image-plane photographic film.

1968). Such a pattern looks like the circular halo shown in Figure 6-14b for a single-exposed film. This circular speckle pattern has a diameter which depends on the maximum spatial frequency of the film-recorded speckle and thus on the viewing angular aperture of the recording lens (see Section 3.10).

When the film is double-exposed to obtain a speckle pattern such as shown in Figure 6-14c, its Fourier-plane diffraction pattern is of the kind shown in Figure 6-14d. A series of parallel fringes appear across the diffraction pattern, with a fringe wavenumber proportional to the displacement vector on the object surface. The formation of such fringes can be simply understood with reference to the interference pattern produced by two close slits illuminated by a coherent source (Young's fringes). A similar pattern would be obtained by masking out all but a single speckle grain and its displaced copy on the film negative. As in the double-slit experiment, the angular width of the fringes (ratio of the fringe spacing and the distance from the negative to the Fourier plane) is equal to λ/d for a small speckle grain displacement d. If we now open up the whole negative, all the other speckle-grain pairs will produce fringes with the same spacing (because they are displaced by the same distance d). A fringe pattern of the kind shown in Figure 6-14d will thus result, with a random fine-speckle structure which is produced by cross-interference between uncorrelated speckle grains.

If the strain distribution varies across the object, rather than being a uniform lateral translation as in the case of Figure 6-14a, the speckle grain displacement d will change across the film. The magnitude and direction of the local speckle grain displacement, and thus the component of the displacement vector in the xy-plane of Figure 6-13, can be obtained by scanning the film negative in Figure 6-15 with a small-aperture diaphragm. The diameter of the diaphragm aperture must be large enough to contain a large number of speckle pairs, but small enough to have only speckle pairs with essentially the same displacement within the open aperture. If such conditions are satisfied, fringe patterns of the kind shown in Figure 6-16 can be observed (Erf, 1978, Section 9.1.2). Such fringes were obtained by locally illuminating different portions of a double-exposure speckle photograph which was obtained on the image-plane of a cracked concrete beam. The fringe spacing and orientation differs if the film is sampled at different positions around the cracks, reflecting the stress release action of the cracks at the surface of the strained structure. Using the above relation between the fringe spacing and the displacement vector, in-plane displacements perpendicular to the finge orientation of 16, 17 and 40.8 μm were calculated for the sampled positions corresponding to the three patterns shown left to right in Figure 6-16.

A full image of the dispacement field over the imaged surface can be obtained by spatial filtering in the Fourier plane using a setup of the kind shown in Figure 6-15. Typically, an out-of-center slit aperture is inserted in the Fourier plane, and the filtered image of the film negative is observed in the image plane. Assume first that the speckle displacement vector is constant over the whole speckle pattern as in the case of Figures 6-14c and 6-14d. If the slit coincides with a dark fringe of the diffraction pattern, the full field observed in the image plane will be uniformly dark. If we now replace the speckle pattern with one corresponding to a variable-displacement distribution, only the regions having a displacement vector producing a dark fringe

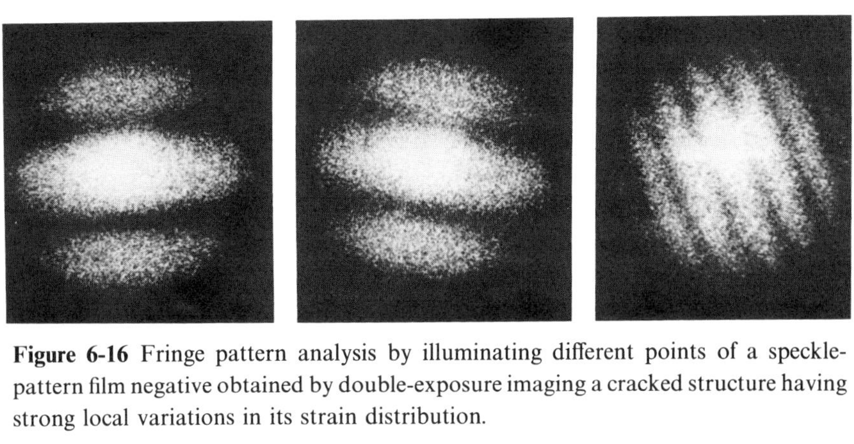

Figure 6-16 Fringe pattern analysis by illuminating different points of a speckle-pattern film negative obtained by double-exposure imaging a cracked structure having strong local variations in its strain distribution.

superposed to the slit aperture will be imaged as dark regions in the image plane. The filtered image will thus be crossed by dark fringes which indicate regions having correlated speckle patterns which are displaced by multiples of a given dispacement vector, i.e., regions of equal displacement in a given direction. Several applications of such methods to strain analysis and flaw detection in industrial structures have been reported (Gregory, 1976; Erf, 1978; Jones and Wykes, 1983).

Interference fringes can be observed in the Fourier plane or in the image of a Fourier filtered speckle negative only if the two speckle patterns recorded during the double exposure are well correlated. If this is not the case, because, for example, the surface microstructure of the imaged object has changed between the exposures, no fringes will be obtained. This behavior can be used to advantage by imaging the speckle pattern from a given surface on different negatives before and after a local surface modification. If the two films are well aligned one against the other and inserted on a spatial-filtering setup of the kind shown in Figure 6-15 with a slit along a dark fringe in the Fourier plane, the filtered image will be uniformly dark except for the modified regions where the two speckle patterns are uncorrelated resulting in brighter areas. This technique has been used to detect surface texture modifications resulting from corrosion, wear or plastic strain (Köpf, 1973; Cielo et al., 1977; Sigler and Haworth, 1981).

Out-of-Plane Displacement-Gradient Monitoring by Defocused Speckle

We have seen above how the in-plane component (i.e., along the xy-plane parallel to the image plane in Figure 6-13) of the surface-displacement vector can be measured by taking a double exposure negative of the speckle pattern in the image plane of an imaging lens. The out-of-plane component (along the z-axis) of the displacement vector cannot be evaluated with such an optical configuration. Out-of-plane displacement variations can, however, be monitored by analyzing the speckle pattern out of the image plane.

A typical apparatus for defocused speckle monitoring is shown in Figure 6-17. In the example shown, the illuminating beam is parallel and the speckle is recorded in the focal plane of the lens. Other configurations are also possible, such as a diverging illuminating beam with speckle recording in any other plane, provided it is not the image plane of the lens. If the inspected surface is subject to a uniform small displacement in the z-direction, no change will be observed in the speckle pattern. If, however, the surface is

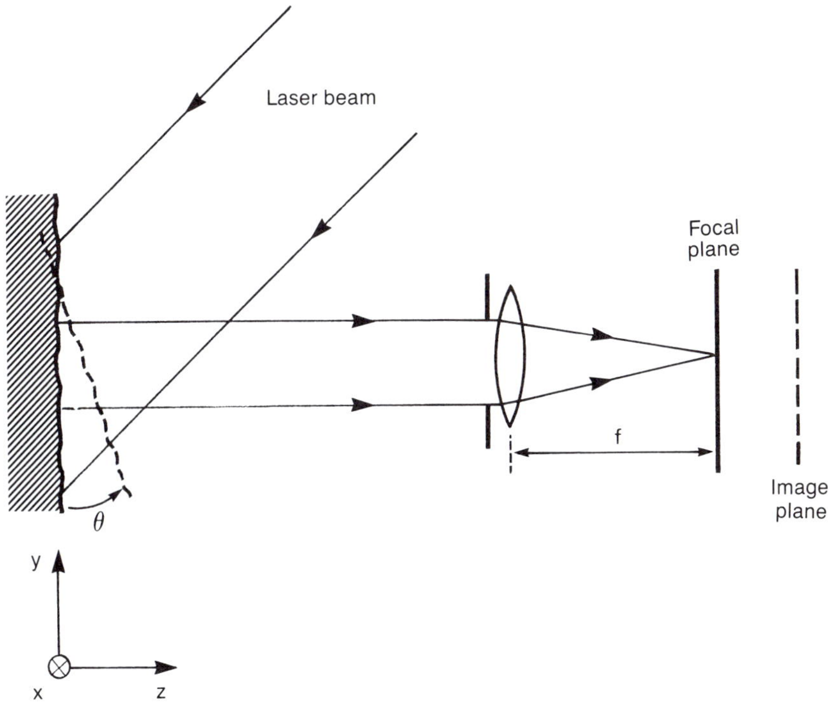

Figure 6-17 Defocused speckle pattern recording for surface rotation monitoring.

rotated around an axis laying in the xy-plane (see Figure 6-17), it can be shown (Jones and Wykes, 1983, Section 3.3.1) that the speckle pattern will correspondingly rotate by twice the rotation angle of the surface, similarly to the light specularly reflected by a mirror. The speckle displacement is thus a very sensitive probe of surface tilt, or, in other terms, of the out-of-plane displacement distribution. In-plane and out-of-plane displacements can be resolved by monitoring the speckle pattern both in and out of the image plane.

The high sensitivity of the speckle pattern position to surface tilting limits the applicability of speckle photography to small whole-body rotations. This can be understood with reference to Figure 6-17. If a surface tilt by an angle θ causes the speckle pattern to be displaced out of the lens aperture (i.e., if 2θ is larger than D/l, the ratio of the lens aperture and the lens-to-surface distance), the displaced speckle pattern will show no resemblance with the undisplaced pattern and no correlation will be possible.

Speckle photography has some advantages and some limitations as compared to holographic interferometry. The stability requirements for laser photography are not so severe as for the interferometric methods. The illumination direction is relatively unimportant, so that the laser does not have to be mounted on the same stabilized table. The coherence requirements for the laser are not so strict, since good coherence is only required within the volume which contributes to a speckle element (Ennos, 1976). Multimode lasers are thus usually adequate for speckle recording. The quality requirements for the camera lens are quite relaxed. The spatial resolution required in speckle photography is much coarser than in the case of holographic interferometry, making it possible to use fast camera films or electronic imaging techniques (see next section).

As to limitations, the first one concerns the minimum displacement which can be resolved. While holographic interferometry has a resolution of one fringe, or a displacement of the order of $\lambda/2$, and even smaller using interpolation techniques, speckle photography is limited to minimum displacements of the order of one speckle grain, or $\lambda l/D$ from equation (3-23), where l is the lens-to-surface distance and D is the lens aperture. As l is usually much larger than D, the sensitivity of speckle photography is much poorer than the sensitivity of holographic interferometry, unless a very large lens aperture is used. Note that the speckle photography resolution is the same as the resolution of an ordinary imaging camera with the same aperture. Speckle photography has, however, the advantages over standard photogrammetry of a higher contrast and a larger depth of field.

The maximum displacement which can be analyzed, apart from the above mentioned vignetting problems induced by rigid-body rotation, is limited by the visibility of the fringes in the Fourier plane. If the spatial period of the fringes of the kind shown in Figure 6-16 becomes comparable to the average speckle grain size, the statistical noise becomes very important or, in other terms, the displacement becomes comparable to the correlation distance of the speckle pattern. Speckle decorrelation problems are also generated in high-aperture arrangements by object motion other than the type of motion to which the system is most sensitive (see Jones and Wykes, 1983, Section 3.5.2).

6.2.2 *Speckle Interferometry*

The limits to the sensitivity of speckle photography methods for a given lens aperture can be overcome using speckle interferometry techniques. However, stability requirements are correspondingly increased. Speckle interferometers are basically similar to the interferometric profilometers described in

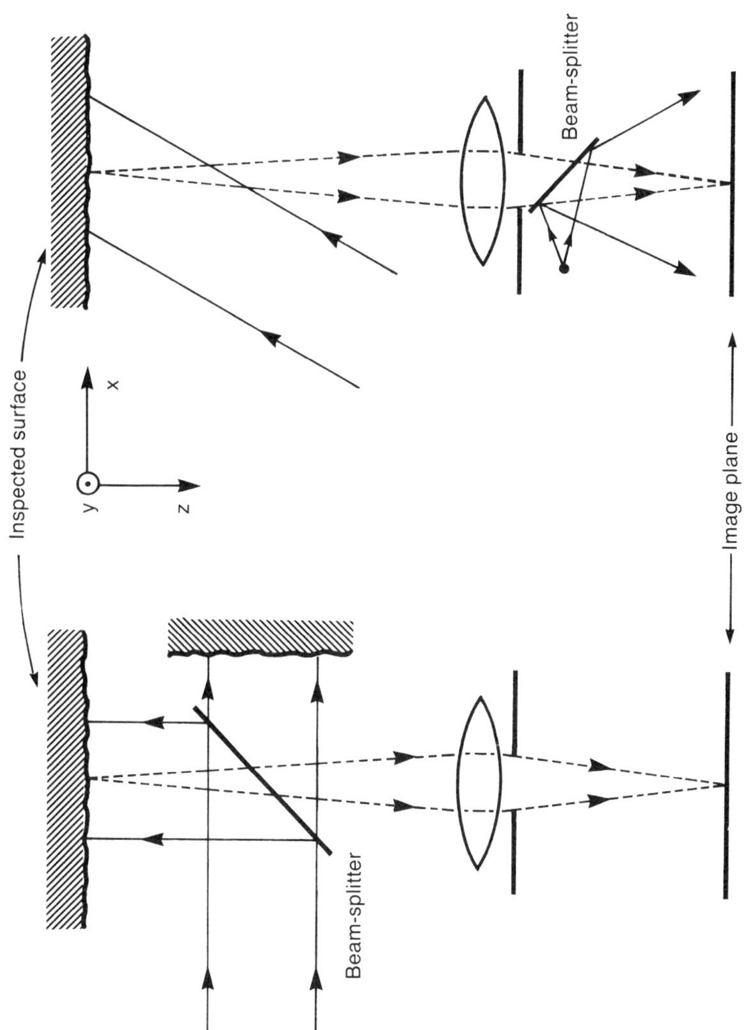

Figure 6-18 Arrangements for speckle pattern correlation interferometry using the superposition of light reflected from two scattering surfaces (left) or the superposition of the surface scattered light to a mutually coherent smooth-wavefront reference beam.

Section 4.1.3, but they are used to monitor the phase distribution of the surface-scattered coherent light with a resolution element much larger than the single scattering centers rather than to obtain a deterministic surface profile. Such a random phase distribution contains enough information if only the displacement of the surface with respect to its undisplaced configuration is to be determined.

Some typical speckle interferometry arrangements for monitoring out-of-plane (along the z-direction) displacements are shown in Figure 6-18 (Stetson, 1975; Jones and Wykes, 1983, Chapter 3). The configuration shown on the left corresponds to a Michelson interferometer arrangement where the two mirrors have been replaced by two rough surfaces. Both the wavefronts scattered by the two surfaces and imaged by the lens in the image plane have random intensity and phase distributions. They give speckle patterns whose general appearance is the same whether one, the other or both surfaces are present or whether either surface is translated or rotated by a small amount. The precise speckle pattern changes, however, in an unpredictable manner when either surface is moved. Now, if one surface is translated in the z-direction by exactly $\lambda/2$, not only will the general statistical properties of the speckle pattern be the same, but the pattern itself will be identical to that before the translation. This comes from the fact that the pattern is the result of the superposition of two mutually coherent, but randomly modulated, wavefronts. When the relative phase between the two lightwave components at each image point changes by a mulitple of 2π, the resultant superposition will be exactly the same (provided that the surface displacement is small with respect to the depth-of-field of the imaging lens).

The above arguments imply that if one of the two rough surfaces is slightly rotated around an axis situated on the xy-plane, the speckle pattern taken after the rotation will periodically match and differ from the pattern taken before rotation, the matching regions being situated along the axis of rotation, as well as along the axes parallel to this axis, and corresponding to an out-of-plane displacement of $\lambda/2$, λ, $3\lambda/2$, etc. By comparing the two patterns, one can tell in which direction and by what angle the rotation has taken place, or, in general, what is the component of the surface displacement in the z-direction.

Techniques similar to the ones seen in the last section can be used to visualize the fringes of equal displacement. A negative of the undisplaced speckle pattern could thus be superposed to a positive of the displaced pattern (or to the image of the displaced surface for real-time observation) to obtain dark fringes in the regions where the two patterns match. Alternatively, both patterns could be recorded on the same plate with a slight lateral displacement between the two exposures, and the developed film would be inserted in a spatial filtering setup of the kind shown in Figure 6-15. Only the regions where the two speckle patterns are correlated show fringes in the Fourier-plane diffraction pattern. Regions where the two patterns are uncorrelated give a uniformly distributed diffraction pattern. By superposing the aperture slit to a dark fringe in the Fourier-plane, dark fringes will be seen in the filtered image over the regions of good speckle correlation.

The alternative configuration shown in the right part of Figure 6-18 is substantially similar to the Michelson interferometer arrangement except that the reference beam is a smooth wavefront spherical wave and the illumination beam is not parallel to the z-direction. Using a smooth reference beam (which must be obtained from the same source as the illumination beam for coherence requirements) is essentially equivalent to using a speckled reference, except that it generally gives better fringe visibility (see Jones and Wykes, 1983, Section 4.4). Oblique illumination results in the interferometric setup being sensitive to both the in-plane and the out-of-plane displacement components. Other configurations would obviously be possible, such as a Michelson interferometer arrangement with a smooth reference mirror.

A configuration which is sensitive only to in-plane displacements is shown in the left part of Figure 6-19. In this case, two mutually coherent parallel beams are projected on the surface to produce a close pattern of optically unresolvable interference fringes. As Figure 3-49 shows, the interference fringe planes are parallel to the bisector of the two beams (i.e., to the yz-plane), and their spacing is equal to $d = \lambda/(2 \sin \alpha)$ from equation (3-16). If the surface is displaced in the x-direction by an amount exactly equal to the fringe-spacing d, or to a multiple of it, the regions of the surface which are illuminated by the bright or by the dark fringes will be exactly the same as before the displacement. The resultant speckle pattern of the displaced surface will thus be identical to that of the undisplaced surface, apart from a corresponding translation of the pattern in the negative x-direction in the image plane. Consequently, a uniform surface displacement by a multiple of d does not produce any loss of correlation of the speckle pattern, while a displacement of $d/2$, $3d/2$, etc. produces a complete change of the illuminated surface areas and thus a completely uncorrelated speckle pattern. If a variable stress field is applied to the inspected surface, the above mentioned techniques for spatial-filtering a double-exposure speckle pattern negative may be used to visualize the fringes of equal displacement along the x-direction.

The alternative configuration shown in the right part of Figure 6-19 is

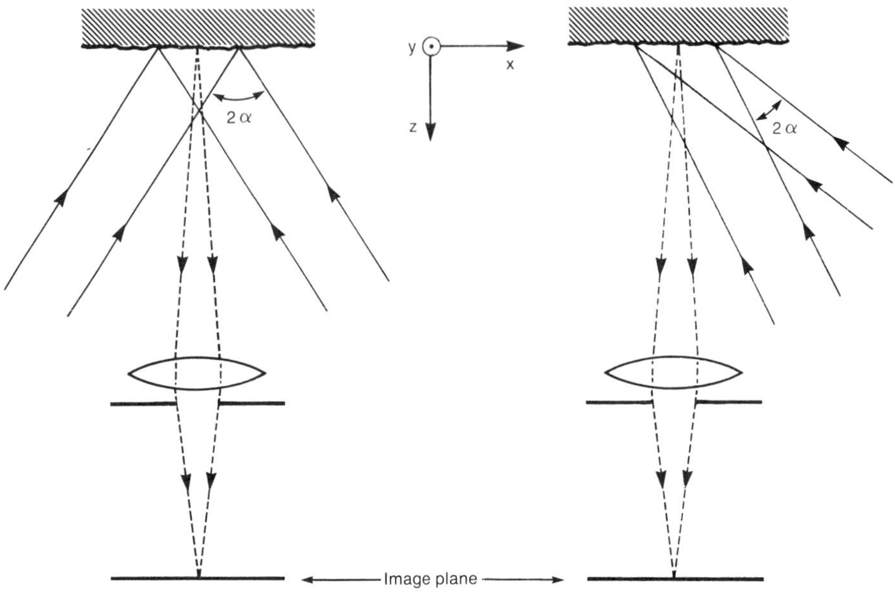

Figure 6-19 Speckle pattern correlation interferometry by projection of an interference beam pattern normal to the surface (left) or oblique with respect to the surface (right).

similar to the left arrangement, except that the fringe planes are now at an oblique angle to the surface, and the angle 2α between the interfering beams can be made as small as desired. This arrangement may be made more sensitive to in-plane or out-of-plane displacements depending on whether the projected beams are closer to the z-axis or to the x-axis, respectively. Furthermore, the displacement sensitivity can be varied by varying the angle 2α between the interfering beams. According to equation (3-16), the fringe spacing can be varied from nearly $\lambda/2$ to values as large as wanted by choosing an appropriate value for α. This may be useful when relatively large displacements, in the range of several tens of micrometers, are to be monitored. An example is shown in Figure 6-20 (Hung, 1978). A centrally loaded rectangular plate is illuminated by a double-beam system of the kind shown in the right part of Figure 6-19. The double-exposed speckle pattern negative is processed by the usual method to obtain the speckle correlation fringes shown in Figure 6-20. The direction of the two projected beams is close to the xy-plane so as to be mainly sensitive to the out-of-plane displacement of the surface, while the angle between the beams is sufficiently small to obtain a spacing between the interference fringe planes, and thus an incremental

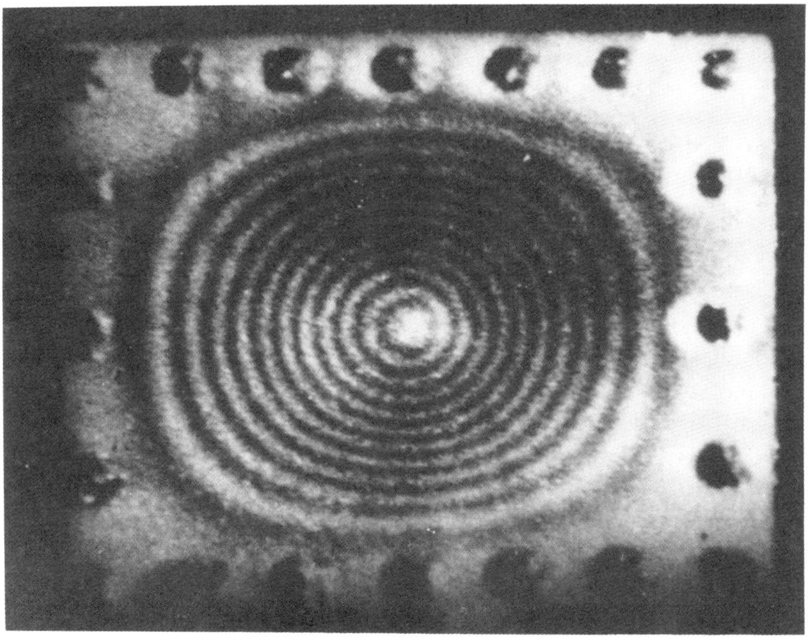

Figure 6-20 Speckle-correlation fringes of equal out-of-plane displacement of a centrally loaded rectangular plate. The incremental displacement between two adjacent fringes is 10 μm.

displacement between two adjacent dark fringes in the filtered speckle pattern, as large as 10 μm. This approach can thus provide a sensitivity which is intermediate between the sensitivities offered by holographic interferometry and Moiré techniques. However, the maximum range of displacement is limited to values of the order of 100 μm by speckle-pattern decorrelation problems (Jones and Wykes, Section 3.7.1).

A configuration equivalent to the one shown in the right part of Figure 6-19 may be obtained by using a single illuminating beam and two viewing directions by placing a double-aperture in front of a wide-open camera lens (Archbold and Ennos, 1972). Different sensitivities can be obtained by varying the separation distance between the apertures. Quadruple apertures may also be used to obtain the in-plane displacement components along the two orthogonal axes. Alternatively, a speckle pattern can be recorded with a wide aperture, and the different components of the displacement vector can be visualized by varying the position of the filtering aperture in the Fourier plane of the optical spatial-filtering processor (Hung, 1978).

Another variation of the speckle pattern correlation approach, called speckle pattern shearing interferometry, is shown in Figure 6-21. Let us first consider the configuration shown on the left. A prism is introduced over half of the lens aperture to deflect the light beams passing through this surface by a small angle. As a result, a given point in the image plane simultaneously receives light scattered from two nearby points separated by a distance Δx on the inspected surface. The light field received on the image plane is the superposition of two speckle patterns, similar to the configuration shown in the left part of Figure 6-18. Similar results are obtained if one of the speckle patterns is modified as a result of surface deformation, and the double-exposed speckle recording negative is optically processed by the usual techniques.

In the case of Figure 6-21, however, the surfaces producing the two

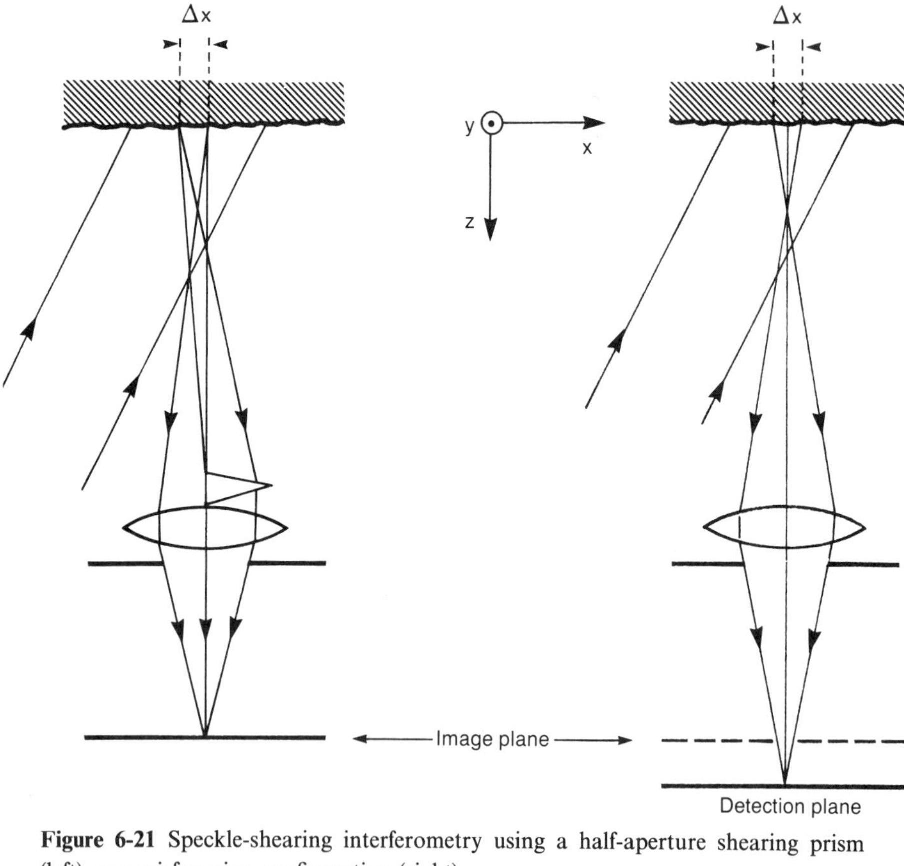

Figure 6-21 Speckle-shearing interferometry using a half-aperture shearing prism (left) or a misfocusing configuration (right).

superposed speckle patterns are actually the same surface sampled at slightly different positions. In contrast to the case of Figure 6-18, here the speckle patterns cannot be modified by loading the surface without simultaneously modifying the other pattern. The loss of correlation which can be observed after surface loading in this case must be related to a variation in the displacement amplitude over the shear distance Δx. In other terms, the fringes which can be observed after double-exposure and optical processing of the speckle recording film are fringes of equal displacement gradient in the x-direction, the displacement being mainly in-plane or out-of-plane depending on the angle of the illuminating beam with the z-axis. Such a configuration, which is analogous to shear interferometers such as the Nomarski interferometer mentioned in Section 4.1.3, is particularly convenient to evaluate the strain tensor, whose in-plane components are directly related to the displacement gradient.

The other configuration shown in the right portion of Figure 6-21 is equivalent to the one shown on the left, except that in this case the superposition is obtained by deliberately misfocusing the imaging lens. The components of the displacement derivatives can be evaluated by spatial filtering with different aperture-filter positions in the Fourier plane. This makes it possible to change the derivative direction as well as the sensitivity after recording. However, the misfocused arrangement results in poorer fringe quality and lower tolerance to rigid-body rotation when compared to the prism shearing configuration (Hung, 1978, Section 4.3.3).

Speckle pattern correlation interferometry has been shown to be applicable to a number of industrial metrology problems, including vibration monitoring, pressure vessel inspection and nondestructive testing (Hung, 1978; Erf, 1978, Section 9.1; Hung, 1982). Figure 6-22 shows the fringe pattern obtained by speckle-shearing interferometry with a fiber-reinforced composite cylindrical structure, 30 cm-long and 7.5 cm-large, utilized in solid propellant rocket motors and launch tubes (Erf, 1978, Section 9.1.6). The 1 mm-thick wall contained an artificial 15 cm × 1.25 cm lack-of-bond defect producing a discontinuity in the pressurization. It can be appreciated by comparison with holographic interferograms of the kind shown in Figure 6-7 that speckle-produced fringes have a lower visibility when compared to holographic interferometry.

Electronic Speckle Pattern Interferometry

Speckle techniques require a recording medium with lower spatial resolution as compared with holographic interferometry requirements. Speckle

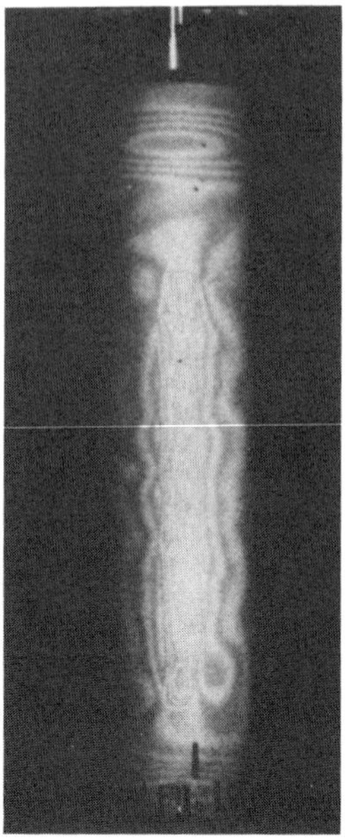

Figure 6-22 Disbond detection in a composite structure using speckle-shearing interferometry with pressure loading.

grain sizes are typically in the 5 to 100 μm range, which is compatible with the resolution offered by commercial television cameras. Video processing may thus be used to generate correlation fringes equivalent to those obtained photographically. This method is known as Electronic Speckle Pattern Interferometry, or ESPI (Butters and Leendertz, 1971; Jones and Wykes, 1983, Chapter 4), and is particularly useful for rapid quality control.

ESPI can be used in any of the speckle correlation interferometry configurations of the kind shown in Figures 6-18, 6-19 or 6-21 where the photographic recording medium is replaced by the light-sensitive surface of a video camera. However, the configurations shown in the right part of Figure 6-18 and in the left part of Figure 6-19 are most often preferred because of their good sensitivity to, respectively, out-of-plane and in-plane displacements. Double exposures cannot be made with television cameras unless a relatively complex procedure is followed by which the camera scan is first switched off, the camera retina is exposed to the two speckle patterns and then the camera field is scanned, as in the recording of pulsed speckle patterns (Jones and Wykes, 1983, Section 4.7.4). In the normal operation, however, the two speckle patterns are compared by storing them in a computer memory and subsequently processing them electronically or else by storing the first and comparing it with the real-time signal obtained from the modified surface.

Two electronic processing methods are most frequently used to obtain correlation fringes by comparing two speckle patterns: One involves image subtraction and the other the addition of the two patterns.

In the video signal subtraction approach, the two locally correlated speckle patterns are subtracted pixel-by-pixel. Those areas of the two images where the speckle pattern remains the same will give a resultant signal equal to zero, while uncorrelated areas will give nonzero signals. Speckle pattern correlation fringes are thus obtained whose visibility can be further improved by high-pass digital filtering.

In the signal-addition approach, the two speckle patterns are added pixel-by-pixel. The resultant image has maximum speckle contrast in the areas of maximum correlation, while the contrast is minimum in areas where the two speckle patterns are uncorrelated, as discussed in Section 3.10 regarding the superposition of speckle patterns with different polarizations. The video signal is subsequently high-pass filtered and rectified to produce the speckle correlation fringes, which nevertheless have a slightly lower visibility than those obtained with the subtraction approach.

A Comparative Analysis

ESPI is a much faster and easy-to-use technique than speckle interferometry using photographic film. Hence it is more appropriate for routine industrial inspection requirements. Applications to vibration monitoring and to nondestructive testing have been reported in a number of publications (Butters, 1977; Erf, 1978; Montgomery and Tryer, 1985; Holownia, 1985). ESPI deficiencies are related to the limited dynamic range as well as to the limited spatial resolution of TV cameras. The latter limitation implies that average speckle grain sizes of at least 20 μm are necessary, limiting the maximum number of fringes which can be obtained across the frame. Moreover, such relatively large grain sizes are obtained using small lens apertures thus requiring more powerful illumination sources.

As compared to holographic interferometry, the advantages of ESPI in terms of speed and convenience to use have been somewhat mitigated by the advent of self-developing photoconductor-thermoplastic holographic recording media. Nevertheless, the convenience of using ESPI includes exposure times of a few hundredths of a second for a full frame and to the availability of cameras with good sensitivity in the near-IR spectral region, where powerful and inexpensive diode laser sources can now be found.

In general, speckle techniques are particularly appropriate to the analysis of relatively large displacements, in the 1 to 100 μm range. Higher sensitivities require high-aperture interferometric techniques which are as much affected by ambient vibration and require as good a mechanical stability as holographic interferometry techniques, while producing lower fringe visibility. Other practical problems such as the difficulty to quickly apply uniform stress fields with minimum whole-body motion, the sensitivity to relative rather than to absolute displacements, and the need for subsequent image-processing steps to obtain defect-related information from a sometimes ambiguous fringe pattern, equally affect the holographic interferometry and the speckle-based NDT techniques.

6.3 Focused Beam Testing

Holographic and speckle NDT methods have the advantage of processing all of the pixels of the inspected-part image in parallel so that a complete picture can be obtained quickly and without the need for point-to-point scanning. However, a full-surface inspection of the piece is not always required. Similar techniques may in this case be used to sense strain or vibration amplitude on a single spot of the inspected object (Dändliker and Willemin, 1981; Yamaguchi, 1981; Stetson, 1983). On the other hand, we have seen that the pictorial information contained in an interferogram must often be machine processed to extract the relevant displacement distribution and that such a step may take as much as a second per pixel in the case of heterodyne holographic interferometry. Under such conditions, a point-by-point scanning sequential technique may actually be faster than the parallel approach to obtain all the relevant information across the scanned object while sometimes providing additional information such as the vibration phase distribution or time-resolved information under impulse loading.

This section contains an overview of single-focused-beam laser NDT methods, mainly based on the interferometric approach. As usual, the emphasis is on techniques showing the highest potential for industrial applicability. Section 6.3.1 below reviews point-by-point vibration-sensing techniques, the equivalent of holographic or speckle vibration-loading NDT. Section 6.3.2 deals with thermoelastic testing, the time resolved, single-point counterpart of thermal stressing holography, while Section 6.3.3 is devoted to pulse-echo ultrasonic techniques using a laser beam to produce and to detect the ultrasonic wave. All the techniques reviewed in Section 6.3 are based on the elastic properties of the inspected structure under mechanical and vibrational loading. Nondestructive testing techniques based on the heat transfer properties of the part under localized thermal excitation will be reviewed in Section 6.4.

6.3.1 *Vibration Sensing by Interferometric Devices*

Single-point optical vibration sensing can be performed in several ways, such as by knife-edge, diffraction or interferometric methods (Stegeman, 1976; Birnbaum and White, 1984). Interferometers are the most widely used normal displacement sensors. A popular choice is the Michelson configuration (see e.g., Deferrari et al., 1967 or Palmer and Green, 1979), but other kinds of interferometers such as the Mach–Zehnder (Lavan et al., 1976; Fisher and Warde, 1979), the Fizeau (Khanna et al., 1968; Palmer, 1984) and the confocal Fabry-Pérot configurations (Paul and Jackson, 1971; Bruins et al., 1975; Monchalin, 1985a) have also been used. Miniaturized (Izutsu et al., 1982; Emmony et al., 1983) and fiber-optic based interferometer versions (Ueha et al., 1977; Imai et al., 1980; Bowers, 1982; Garg and Claus, 1983) which promise to take full advantage of recent advances in integrated optics and of the availability of cheap and compact diode laser sources have recently been developed.

A typical Michelson-interferometer surface microdisplacement probe is shown in Figure 6-23. Such a design is a slightly modified version of the classical configuration shown in Figure 4-6. The probe beam is focused on the inspected surface to reduce intensity losses of the reflected beam produced by scattering when the inspected surface is unpolished, as well as to minimize the wavefront deformation of the beam reflected by a rough probed surface. The attenuation of the reflected beam caused by scattering from a Lambertian surface is of the order of the probing-beam numerical aperture, while the wavefront deformation should theoretically be negligible giving a modulation index, $V_{pp}/2V_0$ (see Figure 6-24) equal to unity if the probing beam is well focused on the surface by diffraction limited optics. In practice, SNR reductions by several tens of dB and a decrease of the modulation index by a factor of two are suffered when probing unpolished metallic surfaces (Cielo, 1981; Rudd, 1983).

Different signal processing methods have been developed to assure a high sensitivity and proper stabilization for such an interferometer. Figure 6-24 shows the detector output signal as a function of the path difference between

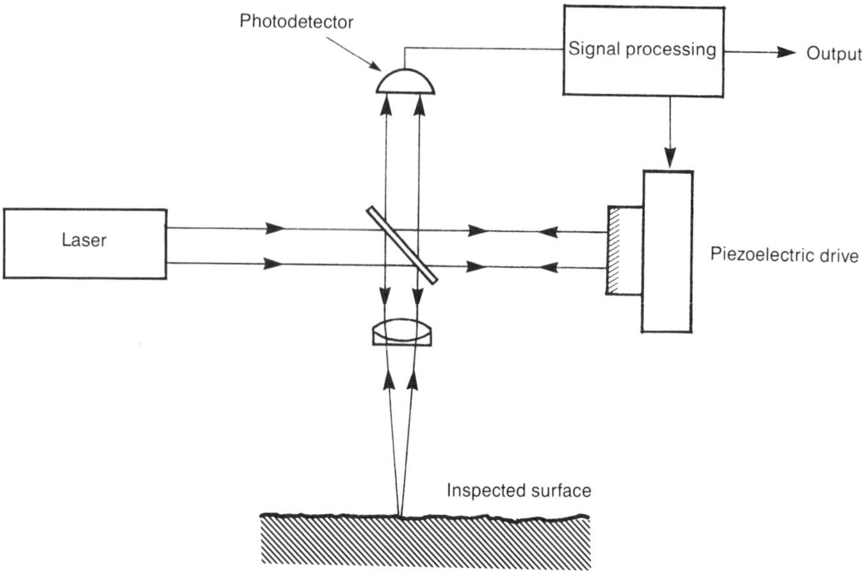

Figure 6-23 A typical laser interferometer configuration for the monitoring of small surface normal displacements by a focused laser beam.

the two arms of the interferometer. The output is maximum when the path difference is equal to $n\lambda$ and minimum when it is equal to $(n + 1/2)\lambda$, where n is an integer and λ is the light wavelength. The modulation index $V_{pp}/2V_0$ is equal to 1 when the two beams superposed on the detector surface are perfectly coherent and of equal intensities. If the surface displacement to be measured spans several wavelengths, its absolute value can be determined by counting the number of fringe maxima in the recorded signal and multiplying by $\lambda/2$. If, however, the surface displacement Δx is much smaller than λ, the path difference must be kept at a value $(n + \frac{1}{4})\lambda$ to ensure maximum output ΔV for a given surface displacement Δx. Such an operating point can be continuously tracked in the presence of low frequency random micro-displacements of the surface relative to the instrument by using a cavity tuning servo-loop of the kind shown in Figure 6-24 (Deferrari et al., 1967; Midgley and Sander, 1971; Palmer and Green, 1979). The output from the photodetector is filtered to extract the high frequency, small amplitude signal output from the near-DC amplitude fluctuations which are produced by random ambient vibrations or refractive index fluctuations along the path of the two interfering beams. The low frequency signal is compared to the reference level V_0, corresponding to the required operating point, and the

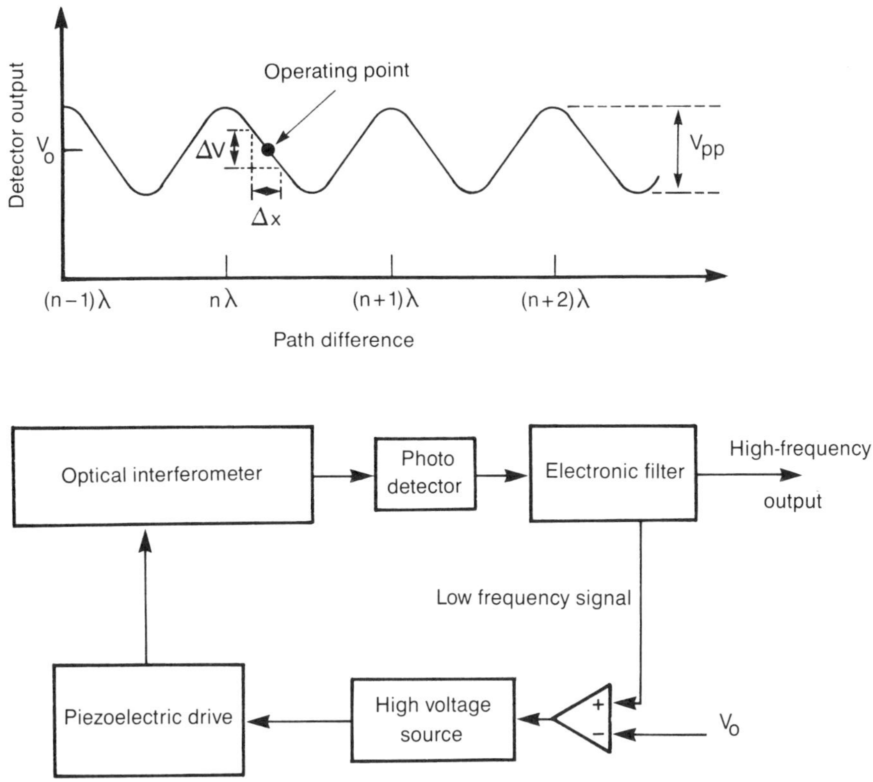

Figure 6-24 Top: output signal from the photodetector as a function of the path difference between the two beams in the two arms of the interferometer shown in Figure 6-23. Bottom: diagram of the cavity tuning servo-loop used to track a suitable operating point for the detection of small amplitude high frequency vibrations.

difference signal is fed to the reference mirror piezoelectric drive to ensure that the path difference is continuously adjusted to a value equal to $(n + \frac{1}{4})\lambda$. The output from the high voltage source may be of interest if low frequency surface displacements are to be measured. An alternative to the use of a piezoelectric drive is to vary the laser wavelength with a frequency shifter (Drain et al., 1977).

An absolute calibration of the interferometer output in terms of nm/volt can be obtained by comparing the output signal magnitude to the maximum peak-to-peak signal amplitude V_{pp} obtained by moving the reference mirror over a distance larger than $\lambda/2$. It can be shown (see, e.g., Cielo et al., 1985a) that the absolute value of the surface displacement Δx is related to the corresponding signal output ΔV by the expression (which is valid for surface displacements much smaller than λ)

$$\Delta x = \frac{\lambda}{2\pi} \frac{\Delta V}{V_{pp}}. \tag{6-2}$$

Such a calibration must, however, be repeated if the V_{pp} value is modified by variations of the surface reflectivity or of the focusing condition of the probe beam. It may also be shown (Cielo et al., 1985a) that the quantum-noise-limited minimum displacement which can be detected by an interferometer of the kind shown in Figure 6-23 with a modulation index equal to unity is

$$\Delta x_{\min} = \frac{(eB)^{1/2}}{(2\eta I_0)^{1/2}} \frac{\lambda}{2\pi}, \tag{6-3}$$

where $e = 1.6 \cdot 10^{-19} C$ is the electronic charge, B is the signal bandwidth, η is the responsivity of the photodetector and I_0 is the average light intensity incident on the detector (producing an output voltage V_0). Assuming some typical values of $\eta = 0.4 A/W$ and $I_0 = 1$ mW for a large-numerical-aperture focused beam, as well as $\lambda = 633$ nm, we obtain a sensitivity of $1.5 \cdot 10^{-15}$ m/Hz$^{1/2}$, which is not too far from experimental sensitivities of 0.02 nm obtained in laboratory setups with a 30 MHz bandwidth (Cielo et al., 1985a).

An alternative approach to detect small surface displacements in the presence of random ambient fluctuations of the path difference is shown in Figure 6-25 (Mezrich et al., 1976; Vilkomerson, 1976; Drain et al. 1977; Reibold and Molkenstruck, 1981). In this case, the reference mirror is fixed and the laser beam is originally linearly polarized at 45° with respect to the figure plane. One of the interferometer beams is transmitted twice through a $\lambda/8$-plate, so that one polarization component of such a beam is phase-shifted by $\pi/2$ with respect to the other component. The two polarization components of the superposition beam are split by a polarizing beam-splitter. Consequently, if the path difference corresponding to one component is equal to $n\lambda$, giving a displacement sensitivity $\Delta V/\Delta x$ equal to zero (see Figure 6-24), the other component will see a path difference of $(n + \frac{1}{4})\lambda$ and thus provide a maximum sensitivity. If the absolute outputs $|V_1|$ and $|V_2|$ of the two detectors are added, the sensitivity will always be conveniently high, at the expense of some signal distortion and a 50% loss of the average detected intensity as compared to the configuration of Figure 6-23. The signal distortion may be avoided by taking the value of $(V_1^2 + V_2^2)^{1/2}$ as the output

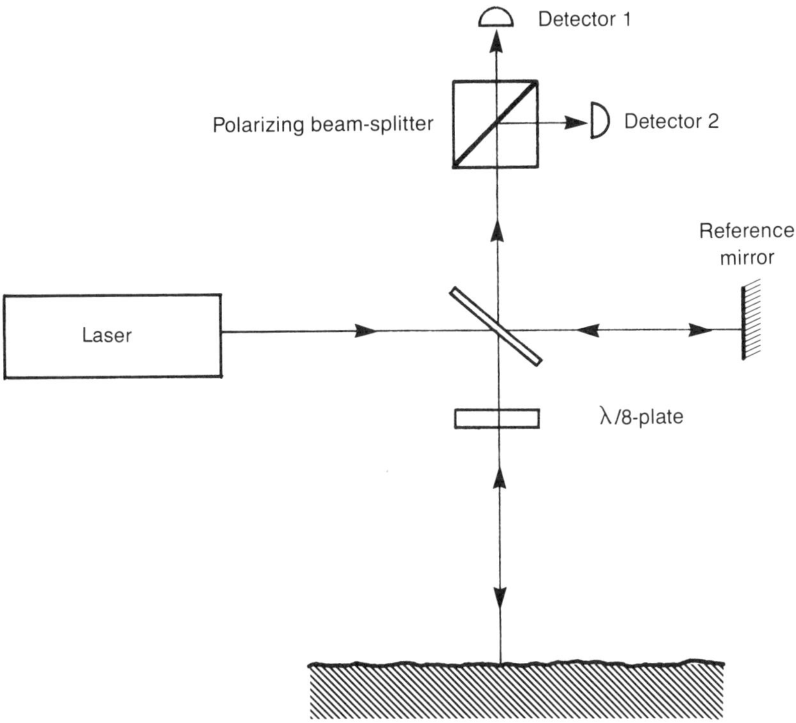

Figure 6-25 Laser interferometer with double-polarization detection.

signal, but such an operation may require too long a computation time if a large bandwidth signal output is requested in real time. Calibration and sensitivity considerations similar to those described above are valid for this interferometric configuration.

Heterodyne interferometry, which was mentioned in Section 4.1.3 in relation to optical profilometry, has also been used to measure micro-displacements (Eberhardt and Andrews, 1970; De La Rue et al., 1972; Buchhave, 1975; Drain et al., 1977; Nokes et al., 1978; Bowers, 1982; Monchalin, 1985). A typical apparatus is shown in Figure 6-26. In this Mach–Zehnder configuration, the laser beam in one of the arms of the interferometer is frequency-shifted by an acousto-optic Bragg cell and superposed to the unshifted beam reflected from the inspected surface to produce a beat signal at the Bragg-cell frequency, as described in Section 3.6.3. The phase of such a beat signal depends on the path difference between the two interfering beams and thus on the position of the inspected surface.

Normal displacements of such a surface can be sensed by monitoring the phase of the beat signal with a phase demodulator of the kind used in radar or broadcasting instrumentation. Equivalently, it can be said that the reflected beam is Doppler-shifted by the moving surface so that the surface velocity information can be recovered by frequency demodulating the beat signal, while the surface displacement can be obtained by time integrating the velocity signal. Apart from the convenience of operating at a higher frequency range in terms of electronic and environmental noise reduction (see Section 3.6.3), heterodyne interferometers have the advantage over homodyne techniques of not requiring any cavity tuning or other mechanisms to continuously assure that the interferometer is operating with optimum sensitivity. Moreover, absolute surface position information can be obtained from the measured phase signal independently of beam positioning or surface reflectivity variations, which only affect the SNR level.

Figure 6-26 also shows some design expedients used to improve the interferometer performance and which can also be applied to other interferometric configurations. The telescopic double-pass focusing device is used to expand the laser beam in order to increase the aperture of the probing beam focused on the surface. Using an off-center double-pass configuration makes it possible to recover the full intensity of the reflected beam, thus avoiding the splitting losses as well as the spurious reflections which would be encountered using an ordinary beam-splitter. This is, however, obtained at the expense of the depth-of-field (which is determined by the angle between the incident and the reflected beams) and of the aberration performance of the lens near its outer edge. Another design improvement to be noted in Figure 6-26 is the use of a differential-detection scheme for common mode rejection, as discussed in Section 3.6.2. As a final point, it should be mentioned that the first beam-splitter should have a reflectivity much higher than the transmissivity to compensate for the large scattering losses of the surface-reflected beam.

A commercial vibrometer using a configuration of the kind schematically represented in Figure 6-26 is shown in Figure 6-27 with its range of operation (Buchhave, 1975a; Chijiima and Morii, 1980; Schurink, 1983). A standard zoom lens is incorporated in the focusing telesope to allow a variation in the instrument-to-surface operating distance from 1.2 to 20 meters. For optimum sensitivity, the zoom lens must be adjusted to obtain a probe beam well-focused on the surface. As discussed in Section 3.10, this implies that the average speckle grain matches the angular aperture of the projected beam. Phase variations over the reflected-beam wavefront superposed to the

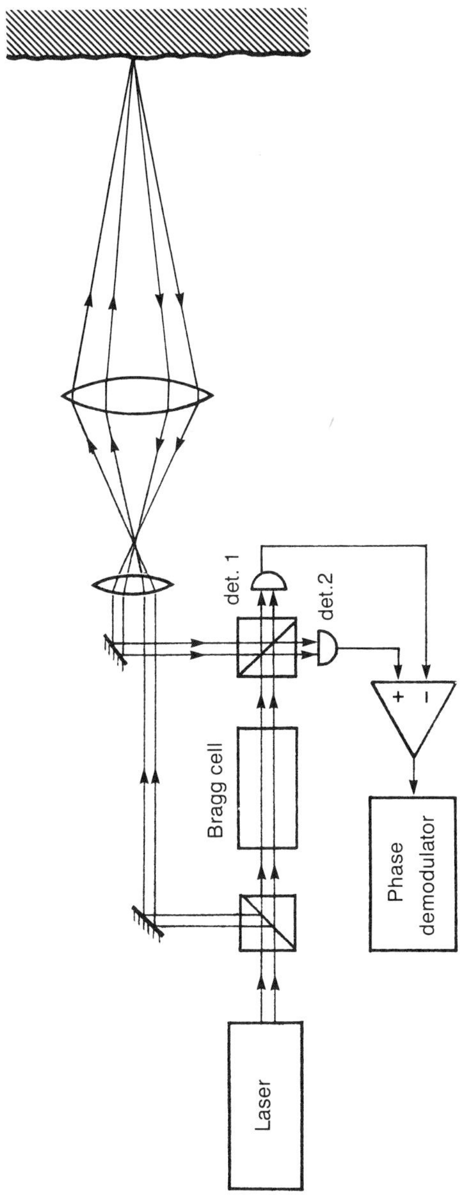

Figure 6-26 A typical optical configuration for heterodyne interferometry.

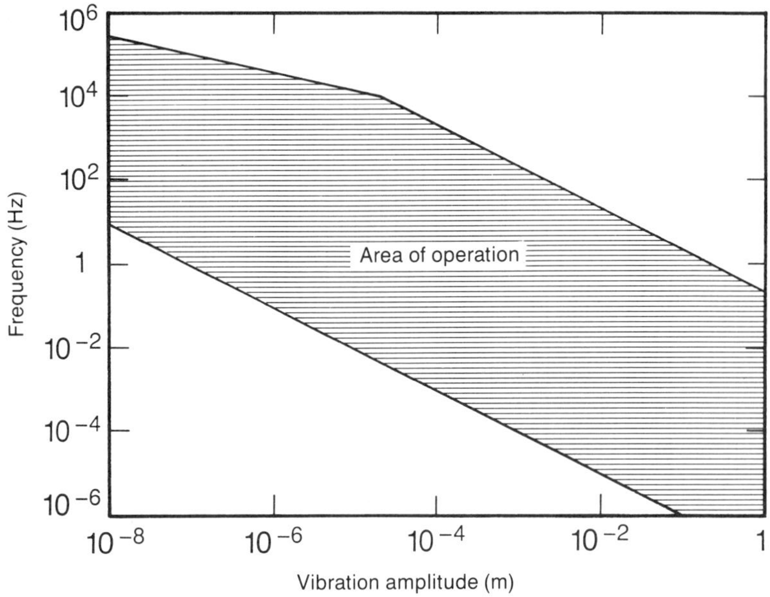

Figure 6-27 Appearance (top) and operating area (bottom) of a commercial heterodyne interferometry vibrometer. (Courtesy of Dantec/Disa, Copenhagen.)

reference beam are thus minimized and a modulation index approaching unity can be obtained if the two interfering beams have equal intensity. The statistical nature of speckle may, however, result in occasional low values of the modulation index, because the speckle grain size obtained on the specific focusing point on the surface may happen to be smaller than the average, or because the average intensity of the obtained speckle grain may happen to be very low, thus mismatching the intensities of the interfering beams. When the vibrometer is manually adjusted, care must be taken to set the zoom lens and eventually one of the pedestal screws to choose a pointing position on the surface providing a maximum amplitude of the beat signal. Such an operation could obviously be performed automatically as required in a fast point-by-point scanning device by including an automatic focusing device of the kind mentioned in Section 4.4, as well as an angular wiggling mirror to choose the best-signal focusing position around the probed spot. Such adjustments could be made in a few ms per sampled position and would have the advantage of always choosing the most favorable speckle grain. The presence of speckle would thus improve the SNR as compared to the signal, which would be obtained with a speckle grain of average intensity and size.

The configuration shown in Figure 6-26 inevitably leads to a large path difference between the interfering beams. The laser must thus have a comparably long coherence length to obtain a modulation index approaching unity. However, the laser used in the vibrometer shown in Figure 6-27 is a standard He–Ne multimode model having a temporal coherence which rapidly deteriorates at a path difference of 5 to 10 cm, going back to a level close to the maximum at path differences which are multiples of the laser cavity path length (see Section 3.1.2: The mutual temporal coherence versus path difference between the two beams is the Fourier transform of the spectral distribution curve which is of the kind shown in Figure 3-12). Optimum operation of the vibrometer thus requires that the operating distance be chosen at one of the positions assuring maximum temporal coherence of the reflected beam with respect to the reference beam. This is an annoying requirement in routine industrial inspection applications. The choice of a single-mode laser such as the model 100 marketed by Tropel would eliminate such a problem while having a lower laser noise, as Fig. 3-29 shows. However, such lasers have a lower output power than a comparable multimode model, while suffering reliability problems and high sensitivity to back-scattered radiation. A well-stabilized and powerful single-mode diode laser may be a better choice for such an application.

Single-point laser interferometric vibrometers have found a number of applications in industrial inspection. One of the most frequent applications is for inspection of loudspeakers and audio or ultrasound components, a classical application of interferometry (Tolanski and Wood, 1958; Massey and Carter, 1968; Speake, 1978; Chijiima and Morii, 1980; Monchalin, 1985). A similar application is remote inspection of vibrating surfaces such as engines or large structures in the transportation or construction industries. Vibration-amplitude scans may provide information on hidden defects, while abnormally large vibrations at specific locations may indicate machinery malfunction at a given rpm regime or after a certain lifetime. Figure 6-28 shows two examples of such applications (Drain and Moss, 1983). The top diagram shows the contour plot of ultrasonic pulse amplitude across a 2 MHz transducer of 15 mm in diameter obtained with an interferometric vibrometer built at AERE Harwell, U.K. The contours are at intervals of 2 nm, the maximum amplitude being 31.6 nm. The bottom plot is a Harwell-vibrometer scan of an engine side block and cylinder head. The vibration velocity is plotted in $\frac{1}{3}$-octave frequency band centered at 3150 Hz. Nearly 1 second/point was required in this case to collect such information over all the frequency bands, so that the total engine scan took nearly 15 minutes. This appears to be a reasonable inspection time by industrial standards.

Other applications include the detection of acoustic emission bursts (Kline et al., 1978; Kline et al., 1981), which may provide important information on the integrity of structures (see Section 2.2.4); the interferometric monitoring of an automatic ultrasonic wire bonder (see *Lasers and Applications*, April 1983, p. 45); *in situ* testing of high voltage insulators along power lines

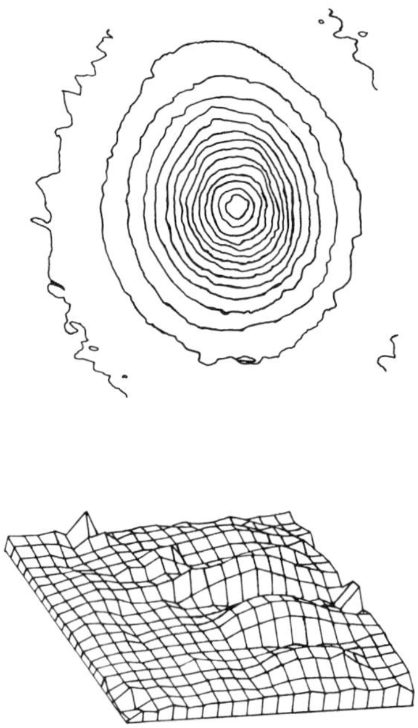

Figure 6-28 Top: countour plot of pulse amplitude across a 2 MHz transducer, bottom: vibration scan of an engine side-block and cylinder head.

(Lachambre, 1983); and probing photothermally generated surface displacements as described in the next sections.

Compared to holographic and speckle vibration testing, point-by-point interferometric techniques require a sequential surface scan to produce a full image of the vibrating structure. Nevertheless, the time required for such a scan may be comparable to the time required to machine process a holographic or speckle interferogram, while point-by-point interferometry has a number of additional advantages. First, single-point probing gives a time resolved signal from which both the frequency and phase of the vibration field can be obtained across the inspected surface, rather than only the vibration amplitude obtainable by double-exposure or time-averaged holograpic or speckle interferometry. Second, the sensitivity of an interferometer is generally much better than that obtainable through the imaging approach. Third, power requirements for the single-point focused beam are much lower than those of holographic or speckle interferometry, so that industrial safety rules are easier to meet. Ambient light shielding is much easier using spatial filtering techniques of the kind shown in Figure 3-33. Finally, focused-beam interferometry can be performed using high performance and inexpensive diode laser sources and silicon detectors, while holographic plates require more cumbersome and fragile visible-light lasers for efficient exposure.

6.3.2 *Thermoelastic Testing*

A review of single-point vibration loading techniques for the testing of materials and structures was presented in the last section. This section describes single-point nondestructive testing methods using thermal stressing techniques.

One of the difficulties encountered in holographic testing with thermal stressing is the requirement to rapidly heat the inspected surface using a powerful and uniform heat source without introducing excessive whole-body motion or air turbulence between the two holographic exposures. Such a task is much simplified if the surface is sequentially tested point-by-point. In this case, a relatively low power source such as a focused laser beam can be used to rapidly heat each inspected point, while problems related to the spatial uniformity of the heat source are avoided.

Thermoeleastic probing is a relatively recent technique based on the localized surface heating of a given structure and sensing the thermoelastically induced surface displacement by a noncontact probe (Ameri et al., 1981; Olmstead et al., 1983; Rosencwaig et al., 1983; Murphy and Aamodt, 1983).

A typical configuration is shown in Figure 6-29. A heating beam, which is typically a focused laser beam, but which could be replaced by other sources such as an electron beam or a microwave, produces a thermal expansion of the heated surface whose displacement is monitored by a probing low-power laser beam. In the configuration shown in Figure 6-29, the surface displacement is monitored by a probing low-power laser beam. In the configuration shown in Figure 6-29, the surface displacement is measured by detecting the angular displacement of the reflected probe beam with a position sensitive detector (see Section 3.4.1). Sharp focusing of the heating beam is required in this case to obtain maximum slope of the laser-probed surface with a given energy input. Alternatively, the vertical displacement of the heated surface may be probed using a focused beam interferometer of the kind described in the last section. In this case, the diameter of the heating beam is not critical, while the heating and probing beams may be collinear. The possibility of using a single laser beam for both heating and probing has also been considered (Ameri et al., 1981).

The thermoelastic displacement of the heated surface may be probed during the absorption of a short laser pulse, a periodic series of pulses as obtained from a chopper modulated laser beam (see Section 3.6.3) or a continuous laser beam (Tankovsky and Ivanov, 1985). If the surface heated material is opaque and homogeneous, the thermal wavefront after a time t from the beginning of heating will have propagated within the material through a

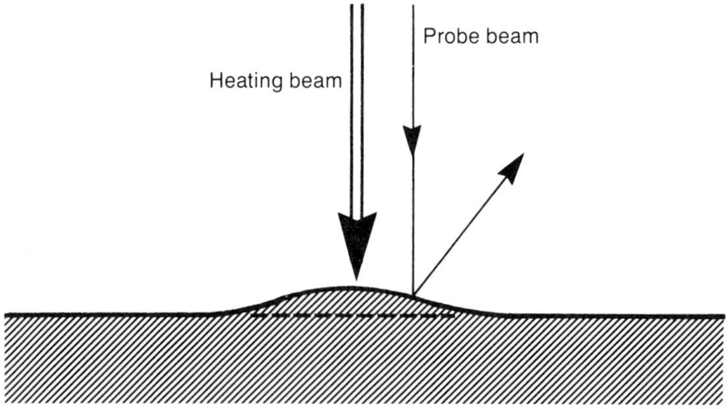

Figure 6-29 Principle of thermoelastic testing by laser probing of the thermally-induced surface-deformation produced by a focused heating beam.

depth z, called thermal propagation depth, which is of the order of (Carslaw and Jaeger, 1959)

$$z = (\alpha t)^{1/2} = \left(\frac{Kt}{\rho C_p}\right)^{1/2} \tag{6-4}$$

where α is the thermal diffusivity, K is the thermal conductivity, ρ is the mass density and C_p is the specific heat of the heated material. If the diameter of the heated surface is much larger than the thermal propagation depth during the observation time, a one-dimensional thermal flow will take place within the material during and after the irradiating period. In such a case, it can be shown that the surface displacement δ produced by the thermal expansion of the surface-heated material depends only on the absorbed energy density, regardless of the time history of the energy injection, following the expression (Rousset et al., 1985)

$$\delta = \frac{a_T E}{\rho C_p}, \tag{6-5}$$

where a_T is the coefficient of linear thermal expansion of the heated material and E is the absorbed energy per unit surface. For typical values of $a_T = 3{\cdot}10^{-5}\,{}^\circ\mathrm{C}^{-1}$, $\rho = 3\ \mathrm{g/cm^3}$, $C_p = 1\ \mathrm{J/g^\circ C}$ (for a material such as an aluminum alloy) and an absorbed laser pulse energy of 10 mJ over a 1 mm^2 area, we obtain $\delta = 0.1\ \mu\mathrm{m}$, easily measurable with an interferometric probe of the kind described in the last section. The surface temperature $T \simeq E/\rho C_p z$ (Carslaw and Jaeger, 1959) will be of the order of 100°C in this case if a pulse duration of 0.1 ms and a value of $\alpha = 1\ \mathrm{cm^2/s}$ are assumed.

The displacement reading from an interferometer may be further complicated by temperature-induced variations of the air density above the heated spot. The ratio of the path-difference ε due to air density variations and the thermal expansion component δ is, under isobaric conditions (Rousset et al., 1985),

$$\frac{\varepsilon}{\delta} = \frac{n_0 - 1}{n_0 T_0 a_T}\left(\frac{\alpha_g}{\alpha}\right)^{1/2}, \tag{6-6}$$

where $n_0 = 1.0003$ is the refractive index of air at STP conditions, T_0 is the ambient temperature in degrees Kelvin and $\alpha_g = 0.2\ \mathrm{cm^2/s}$ is the thermal conductivity of air. The ratio given by equation (6-6) is of the order of 0.01 for aluminum and of 0.1 for steel so that the air density effect is negligible for most materials. Thermally-induced air currents may be a source of noise over long heating periods but are normally not produced if the laser pulse is sufficiently short.

By measuring δ with an apparatus of the kind shown in Figure 6-29 and knowing the absorbed pulse energy E, the ratio $a_T/\rho C_p$ may be calculated from equation (6-5). Such a measurement could be useful as a quick on-line material sorting test. In practical conditions, however, the uncertainty in the value of the surface absorptivity (which may easily change by a factor of five for a polished metal depending on whether a thin oxide film is present on the surface or not) makes the precise evaluation of the absorbed energy E difficult. A more realistic application for this technique is the evaluation of film thickness. If the film is opaque and the heating pulse duration smaller than the thermal propagation time across the film thickness, the thermal displacement will be given by equation (6-5) where the material's properties are those of the coating. After a time period much longer than the thermal propagation time across the film, the thermal expansion will be mainly determined by the substrate thermal parameters. A time record of the surface-displacement signal after the absorption of a short pulse will thus present a gradual discontinuity after a time of the order of the thermal propagation time across the film, from which the film thickness may be evaluated using equation (6-4). An approach of this kind, using the frequency dependence of the AC signal amplitude under chopped-laser heating, was used to evaluate the thickness of an SiO_2 film on a silicon wafer (Rosencwaig et al., 1983). A more extended discussion of thermal-propagation material testing will be found in Section 6.4.

The thermoelastic approach may also be applied to crack detection. Figure 6-30 illustrates the basic principle (Martin and Ash, 1984). A heating laser, in this case a small and inexpensive diode laser, is focused on the surface and an interferometric He–Ne focused beam probes the thermoelastic displacement of the surface at a distance which may be varied from 0 to 50 μm from the heat spot. If no crack is present, the surface probe will measure a constant thermally-induced vertical displacement as shown schematically in the bottom part of Figure 6-30. If a crack is present between the two laser-illuminated points, it will largely release the thermoelastic stress so that the displacement of the probed surface will be substantially reduced. By scanning the specimen under the two focused beams, a transversal crack will thus manifest as a sudden drop in the probed signal. A longitudinal crack (i.e., a crack parallel to the direction joining the two laser-illuminated points at a distance comparable to the laser-spot diameters) will also generate a drop in the probed signal as shown schematically in Figure 6-30. This

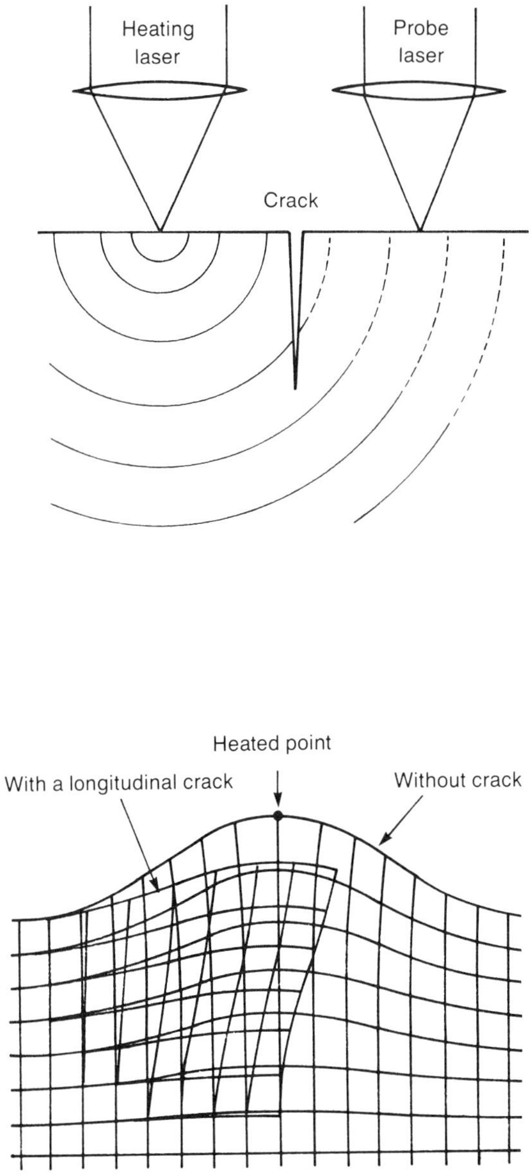

Figure 6-30 Top: schematic diagram of the thermal expansion method for crack detection with an interferometric probe, bottom: qualitative representation of the (highly magnified) surface deformation under spot-heating, respectively, in the absence of a crack and in the presence of a nearby crack parallel to the line joining the heated and the probed points.

technique has been used to detect a thin crack in a glass sample by scanning over a 100 μm × 100 μm area (Martin and Ash, 1984). The thermoelastic approach has the advantage over thermal propagation techniques (see Section 6.4) of not being limited by the relatively slow thermal propagation time across the defect (Murphy et al., 1986). However, compared to the laser generated ultrasound approach (see Section 6.3.3), it requires a closer spatial sampling across the surface.

Thermoelastic Detection of Delaminations in Layered Materials

Disbond detection is probably the most appropriate application of thermoelastic techniques (Martin et al., 1982; Cielo et al., 1984; Aindow et al., 1984; Cielo, 1985; Sontag and Tam, 1985; Cielo et al., 1986f). A typical configuration is shown in Figure 6-31. A pulsed YAG laser is partially focused on the surface of a layered material, such as an aluminum-to-aluminum or aluminum-to-honeycomb adhesively bonded laminate or a metallurgically coated substrate. As discussed in Section 6.1.3 with reference to thermal stressing holograhic interferometry, the surface-heated first layer will be subject in such conditions to thermoelastic stress, eventually leading to thermal buckling if the layer is locally disbonded from the substrate. The surface displacement under pulsed heating is monitored with a sensitive interferometer of the kind shown in Figures 6-23 and 6-24. By point-by-point scanning the surface of the layered structure, delaminations can be found quickly and noninvasively.

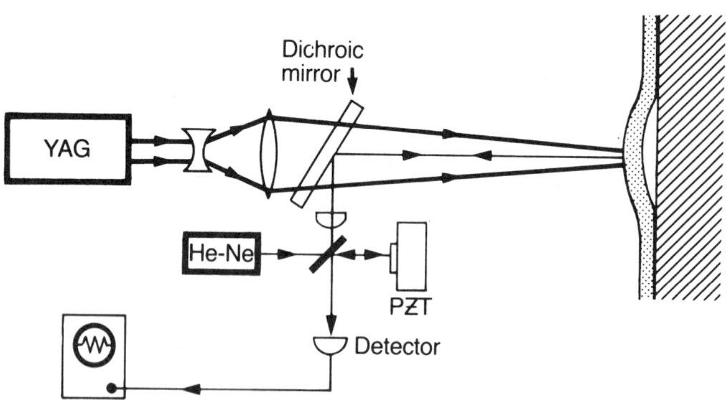

Figure 6-31 Experimental apparatus for the thermoelastic detection of a layer-to-substrate disbond in a layered material.

Compared to thermal-stressing holography, the thermoeleastic focused-beam approach has the practically important difference that each inspected spot is heated by a focused, pulsed heat source. The result, as regards defect visibility, is illustrated in Figure 6-32. If the heating source is either continuous (i.e., with a duration much longer than the thermal propagation time through the layer thickness) or broader than the delaminated area or both, a radial strain field will be established with little buckling effect unless the layer is initially convex (Cielo et al., 1984; Rousset et al., 1985). If, however, the heating laser is both focused and of duration shorter than the thermal propagation time through the layer, the three-dimensional thermoelastic stress field results in a transient bending moment producing efficient lifting of the unbonded layer with limited surface heating. Figure 6-33 shows an experimental verification of such a behaviour: The vertical displacement of a locally unbonded layer decreases as the constant-energy heating beam diameter increases, becoming negligibly small when the heated-spot diameter is larger than the diameter of the unbonded area. It will be recalled that for thermal-stressing holographic NDT, the heated surface area is usually larger than the diameter of the unbond defects so that disbonds are in this case revealed by second-order thermal buckling effects, as discussed in Section 6.1.3.

Figure 6-34 shows a line-scan across a 2 cm-wide delaminated region of an aluminum-honeycomb laminate (Cielo, 1985). Such structures, of which a schematic diagram is shown in the insert, are produced by epoxy-bonding two aluminum sheets (0.3 mm-thick in this case) to an aluminum-foil based

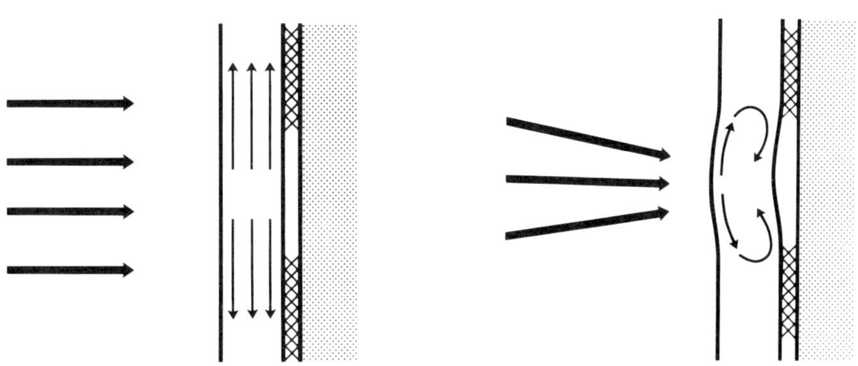

Figure 6-32 Magnified representation of the thermally-induced strain in a locally unbonded layer under conditions of (left) continuous and extended surface heating and (right) pulsed and focused surface heating.

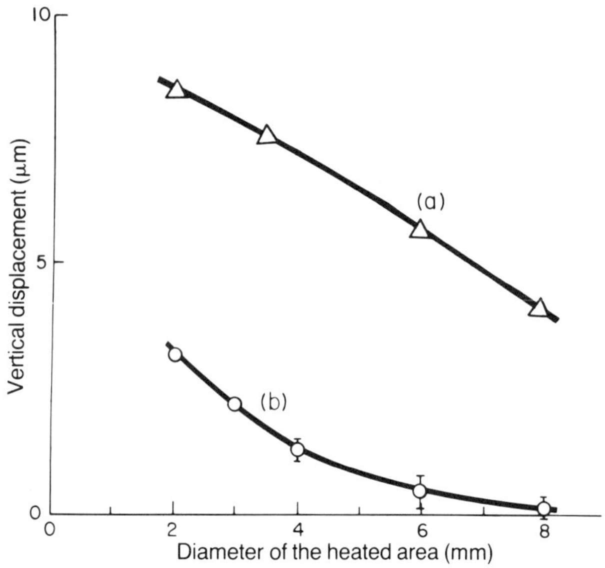

Figure 6-33 Experimental results obtained by focusing to different diameters a pulsed laser beam of constant energy on top of a delaminated area of (a) 20 mm and (b) 7 mm in diameter, using the apparatus shown in Figure 6-31.

honeycomb structure, followed by autoclave curing. Such laminates are widely used, particularly in the aircraft industry, because of their high strength with an extremely low weight. Skin-to-core unbonds may result for a number of reasons including improper temperature or pressure conditions in the autoclave, surface contaminations of the adherents or inadequate storage of the resin film leading to partial pre-cure before assembly. Such defects severely impair the structural resistance of the laminates, which must therefore be 100% inspected at each manufacturing step. The three curves shown in Figure 6-34 were obtained with a heating-beam diameter going from 1 to 6 mm. Only when the heated diameter is smaller than the 3 mm honeycomb-cell size are the single honeycomb cells apparent in the scan signal, in agreement with the discussion above. The 0.75 J, 1 ms YAG pulses produced an estimated maximum surface temperature of the heated aluminum skin going from 2°C to 70°C depending on the heating beam diameter, safely below the temperature level causing thermal damage to the epoxy interface.

Some additional features of thermoelastic single-spot testing which make this approach attractive as compared to holography are that (1) a time resolved signal is obtained for each point, making the absolute displacement

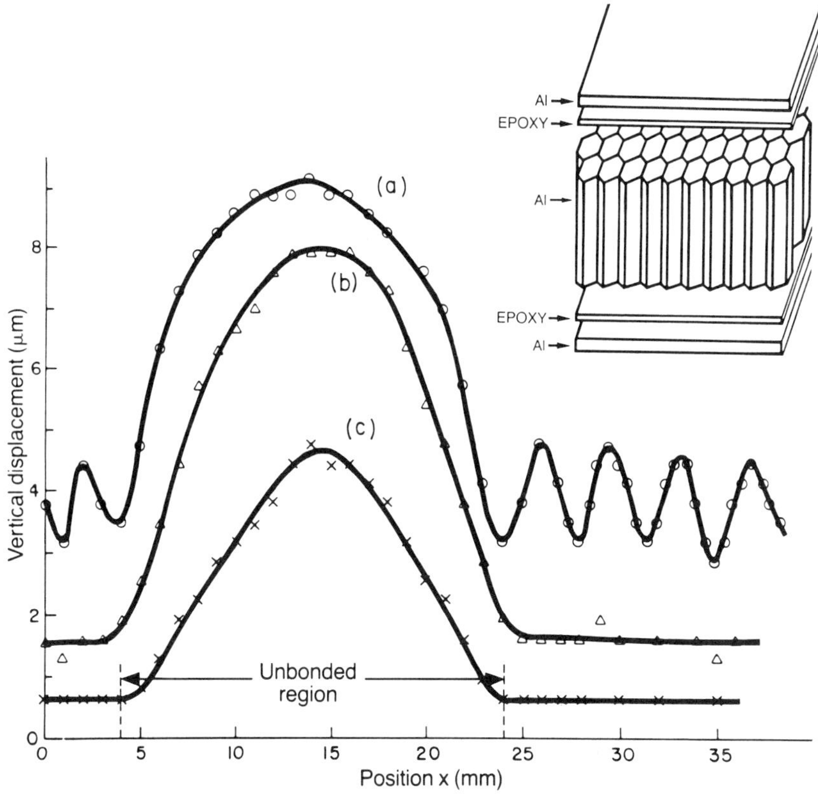

Figure 6-34 Experimental results obtained by spatially scanning the surface of an adhesively-bonded honeycomb laminate (shown in the insert) over a 2 cm-wide delamination. The three curves were obtained with a heating beam diameter of (a) 1 mm, (b) 3 mm and (c) 6 mm. The estimated maximum surface temperature of the heated layer goes from 70°C to 2°C, respectively, for curves (a) to (c).

evaluation directly available without further processing; (2) sensitivities much better than $\lambda/2$ are obtained; (3) good heat input uniformity is achieved across the scanned area; and, quite importantly, (4) the thermal excitation and signal observation take place in a short period of time. The implications of this last aspect, apart from faster scanning rate and the possibility to observe local behavior of the thermally- excited structure before whole-body resonances set up, include good protection against low-frequency ambient vibrations and air fluctuations above the heated area. These considerations are of primary importance if such techniques are to be transferred to the industrial floor, as mentioned in Section 6.1.

Obtaining a time resolved signal following a short heating pulse may provide some additional information on the tested structure. If a layer of thickness l is unbonded over an area of diameter $2a$, it will show membrane-like oscillations after pulsed loading with a frequency which is proportional to l/a^2 (Aindow et al., 1984; Cielo et al., 1985b). A frequency analysis of the transient signal can thus be used to obtain information on the extent of the unbonded area if the thickness of the unbonded layer is known or else on the depth of the delamination if its area is known from spatial scanning. Advantages with respect to mechanical impact testing (Cawley, 1985; Lange, 1985) include noncontact operation, localized analysis and production of extensional, rather than compressional, strain. Such capabilities are demonstrated in Figure 6-35, showing the signal obtained on a clamped aluminum plate 1 mm thick and 5 mm in diameter thermally loaded by a 15 ns laser pulse. The surface position is shown to oscillate, at a frequency related to the plate dimensions, above and below an average position which is determined by the bulk thermal expansion of the heated plate. The transient signal can thus be processed to separate the bulk thermal expansion (which is a noise source in this case, being unrelated to the presence of a delamination but depending on the surface absorptivity) from the plate-bending component (which is the delamination dependent signal). Applications of this techhique to the inspection of Al-epoxy laminates have been reported (Cielo et al., 1985b)

Finally, a question of considerable practical interest for laminate inspection is whether adhesive-unbond defects, when the adherents are in physical contact (0.1 to 1 nm gap) with each other but with zero bond strength, can

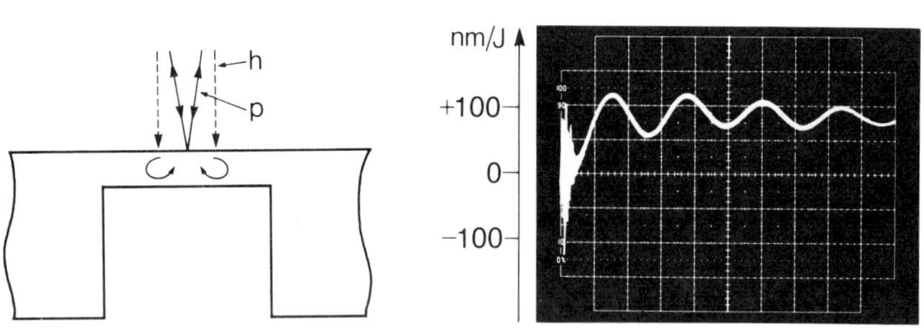

Figure 6-35 Transient thermoelastic excitation of a 5 mm-diameter flat-bottomed hole at 1 mm from the surface. Left: schematic diagram of the hole cross section, where h represents the heating beam and p the interferometric probe, right: recorded signal, with a horizontal division correspond to 2 μs and a vertical division corresponding to 50 nm per joule of incident laser power.

be detected. Classical NDT techniques such as ultrasound cannot detect such defects, because ultrasonic waves are transmitted through such interfaces without any reflection (Chaskelis and Clark, 1980). Thermoelastic techniques may be sensitive to these defects provided that the thermally-induced vertical stress component at the adhesively-unbonded interface is larger than the minimum force required to create a vacuum between the two adherents. This implies that the thermoelastic layer-lifting stress must be larger than 1 atm (10^5 Pa). Finite element models of the kind shown in Figure 6-36 indicate

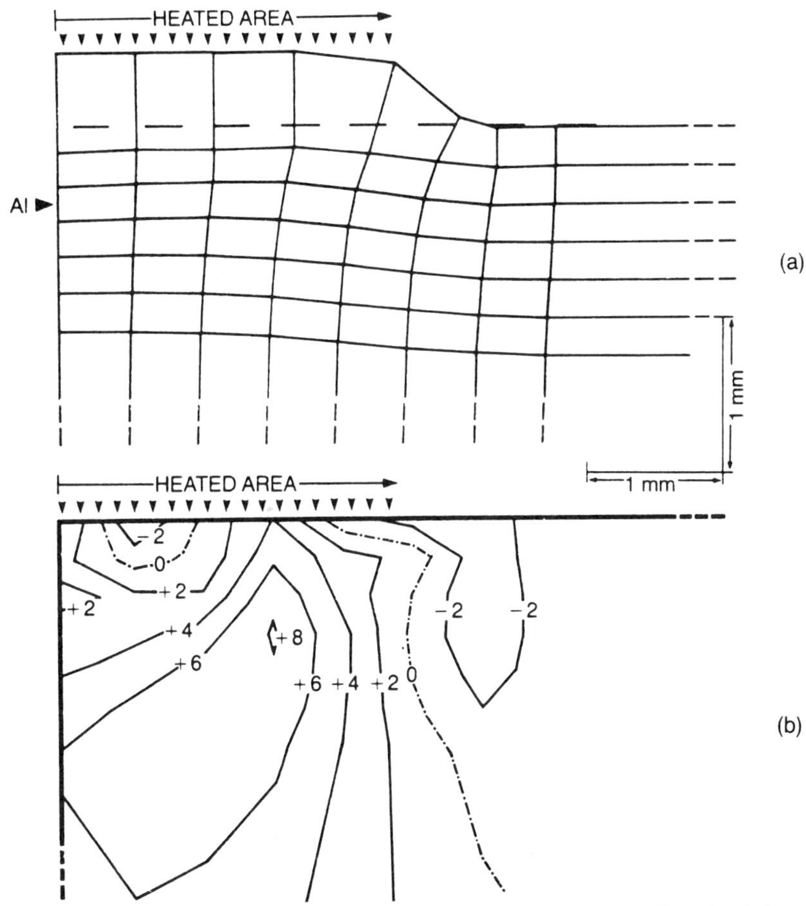

Figure 6-36 Finite-element, cylindrically-symmetric computation of (a) the deformation and (b) the vertical iso-stress contours within a surface-heated bulk-aluminum sample. Positive numbers in the stress display correspond to extensional stress, in 10^5 Pa units.

that extensional thermoelastic stresses much larger than 1 atm can be obtained within a bulk metal sample at depths which are of the order of the heated-spot radius, providing adhesive-unbond detectability. Experimental results obtained with vacuum-clamped plates are in agreement with such an analysis (Cielo, 1985).

6.3.3 *Laser Generation and Detection of Ultrasound*

In the focused-beam thermoelastic technique described in the last section, the deformation of a surface-heated material is probed to obtain information on the structural properties of the workpiece. A surface-heating pulse may also be used to produce an ultrasonic mechanical wave travelling within the material. Pulse-echo reflections from subsurface defects may be optically detected to achieve a completely noncontact ultrasonic inspection. In contrast with the thermoelastic technique, as well as with the thermal techniques described in Section 5, the laser-ultrasound approach can detect subsurface features whose depth is larger than their transversal size.

The generation of acoustic waves from the absorption of a modulated light beam was first observed more than one century ago by Alexander Graham Bell (Bell, 1881). Pulsed laser excitation of elastic waves was investigated shortly after the development of the laser (White, 1963; Parker, 1973). Photoacoustic techniques have recently found several important applications, namely in the nondestructive testing and spectroscopic areas, and a number of reviews in this field have recently been published (Pao, 1977; Rosencwaig, 1980; Scruby et al., 1982; Birnbaum and White, 1984; Hutchins, 1986; Tam, 1986).

Elastic waves are generated in a laser-irradiated opaque material both because of momentum transfer and of thermoelastic stress mechanisms. Momentum is transferred to the irradiated surface from the absorbed or reflected laser photons, but this effect is of negligible magnitude. Much more important is the transfer of momentum from the surface material which is rapidly vaporized by the incident laser pulse. When the surface vaporization threshold is not reached, thermoelastic stress is the main source of the elastic wave. Radial thermoelastic compression (see Figure 6-32) is produced on a bare surface, while surface-normal stress can be generated on a surface covered by an acoustically-thick transparent medium such as a plastics coating or an oil film (Von Gutfeld, 1980). The layer is acoustically thick if its thickness is larger than the acoustic wavelength of the generated elastic wave. The transient stress field produced near the surface by the heating beam generates longitudinal, transverse and surface acoustic waves, each with characteristic directivity patterns (Scruby et al., 1982; Hutchins, 1986), which may propagate through relatively long distances within the material. In the surface-ablation regime, a strong longitudinal wave is generated in a direction normal to the heated surface, while in the thermoelastic regime (without surface vaporization) strong surface waves are generated, as well as transverse waves in out-of-normal directions. The laser-beam spatial distribution may be suitably shaped to obtain a convenient directivity pattern. Cylindrical lens line focusing may be performed to obtain a collimated surface-acoustic wave in the thermoelastic regime (Aindow et al., 1982), while an axicon may be used to obtain a high power point-focused surface wave (Cielo et al., 1985a). A structured-light beam approach may be followed to focus bulk acoustic waves (Von Gutfeld et al., 1983; Cielo and Bussière, 1985).

The laser generated thermal wave can be detected by probing, with a microphone enclosed in an air-filled cell clamped to the inspected surface, the thermally-induced pressure fluctuations of the air above the heated area (Rosencwaig and Gersho, 1976; Wong et al., 1978; Kuo et al., 1982), as shown in Figure 6-37. High-frequency thermal perturbations may also be detected

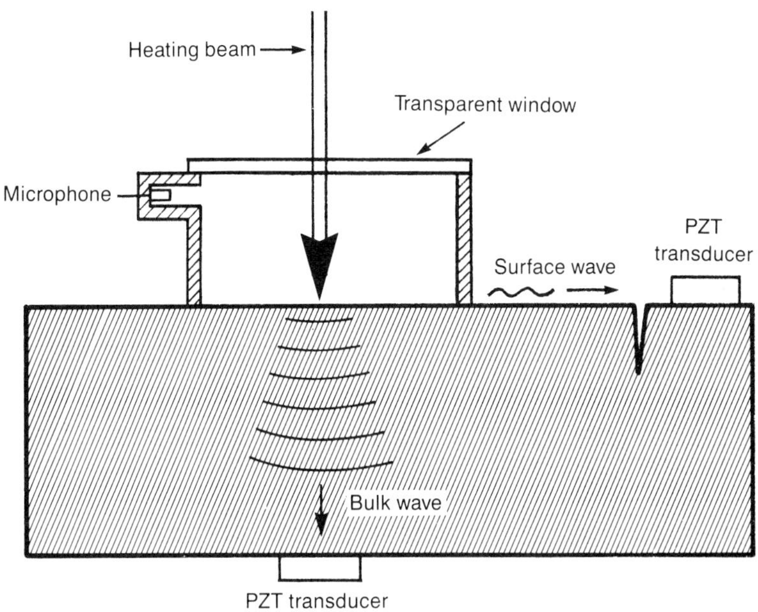

Figure 6-37 Photoacoustic detection by gas-cell or contact ultrasonic-transducer techniques.

by a focused air-coupled ultrasonic transducer (Stearns and Kino, 1985) at the expense of strong air-to-solid coupling losses. In both cases, however, the defect sensing mechanism within the material is related to the thermal-wave propagation. Consequently, such an approach is limited to near-surface defects and its speed is limited by the relatively slow thermal diffusion process. The alternative approach shown in Figure 6-37, detection of a laser generated pulse by a contact PZT transducer, is a more classical ultrasonic shading technique by which the presence of defects such as cracks or subsurface inclusions along the path of the ultrasonic wave can be inferred from the reduction of the transmitted ultrasonic beam intensity. Both techniques shown in Figure 6-37 require previous sample preparation steps and can achieve a high spatial resolution. Consequently, they have found applications mainly in the microscopy field, where they compete with more classical ultrasonic microscopy methods (Lemons and Quate, 1975) where the ultrasonic wave is both generated and detected by ultrasonic transducers, as well as with the scanning-laser acoustic microscope marketed by Sonoscan (Kessler, 1974; Kessler and Yuhas, 1980) where the ultrasonic wave is generated by a transducer and detected by knife-edge optical techniques.

Optical Probing of Ultrasound

Noncontact ultrasonic-sensing methods have more potential than the contact photoacoustic approach for in-plant materials testing applications. A number of optical techniques can be used to optically probe acoustic waves (Stegemann, 1976; Birnbaum and White, 1984; Monchalin, 1986), including the interferometric position-sensing methods described in Section 6.3.1. In all of the latter methods, the inspected surface acts as one of the interferometer mirrors, whose minute displacements generate strong variations in the intensity (in homodyne interferometers) or in the phase (in heterodyne interferometers) of the detected signal. An alternative approach for the detection of surface displacements is offered by velocity monitoring devices (see Figure 6-38), including Michelson (Krautkrämer, 1979), Fabry-Pérot (Johnson and Burgess, 1968; Paul and Jackson, 1971; Bruins et al., 1975; Monchalin, 1985a) or Mach-Zehnder (Clifton, 1970; Barker and Hollenbach, 1972) interferometers. In this case, the surface movement produces a Doppler shift of a light beam reflected by the moving surface, and the interferometer is used as a spectral filter to detect the velocity dependent wavelength shift of the reflected beam.

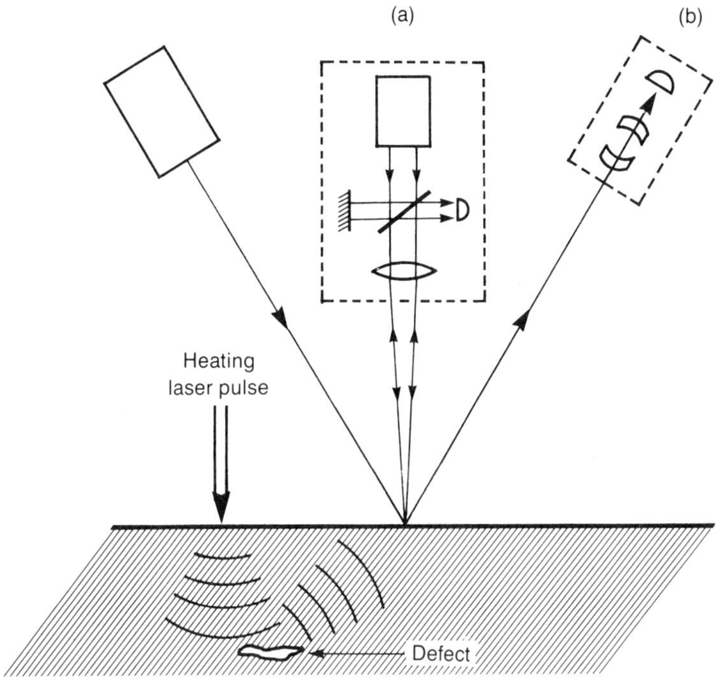

Figure 6-38 Detection of laser-generated ultrasonic pulse echoes using either (a) a position-sensing or (b) a velocity-monitoring interferometer.

Velocity-sensing interferometers are most convenient for monitoring high velocities, such as in supersonic impact-testing experiments or in high-frequency ultrasonic vibration monitoring. To obtain a high sensitivity with lower surface velocities, the interferometer path difference must be increased as much as practical to reduce the free spectral range. This range is the period of the spectral transmissivity curve which is either sinusoidal (in the case of a double-beam interferometer) or comb-shaped (in the case of a multiple-beam interferometer). The instrument frequency pass-band is correspondingly reduced to a fraction of the free spectral range. Multiple-beam interferometers, particularly of the confocal Fabry–Pérot type, are often preferred to obtain high sensitivity with a reasonable interferometer size when the response linearity is not critical (Paul and Jackson, 1971; Monchalin, 1985a). The spectral line width of the probing laser must normally be smaller than the frequency shift to be detected, so that a single-frequency laser is usually required for velocity-sensing interferometry (Paul and Jackson, 1971). Double-cavity filtering approaches have, however, been used for velocity sensing using large-bandwidth incoherent sources (Cielo and Delisle, 1976a). Such a configuration may be attractive with fiber optic interferometers (Cielo, 1979; Bowers et al., 1983) where the path difference may reach several kilometers in a compact device, although strong coupling losses must be suffered in this case because of the small diameter of the required single-mode fiber.

Velocity interferometers for ultrasound detection on industrial-quality surfaces have been increasingly developed in recent years (Kaule and Primbsch, 1983; Monchalin 1985a). The main interest of this approach for industrial applications is related to its relative insensitivity to speckle related noise. This results in a larger optical throughput, defined as the product of the area of a beam at a focus and its solid angle. As discussed in Section 6.3.1, position-sensing homodyne or heterodyne interferometers require precise focusing and diffraction-limited optics to obtain a single speckle grain as well as good mutual coherence of the two superposed beams across the instrument aperture. In velocity interferometers, the presence of a speckle pattern across the reflected probe beam hardly affects the magnitude of the detected signal, which depends only on the amount of the velocity-dependent Doppler frequency shift. Larger out-of-focus projected laser spots can thus be accepted with relatively wide angular apertures, within the limits imposed by the dependence of the interferometer spectral response on the angular position of the input beam. Such limits are particularly wide in the case of a confocal Fabry–Pérot interferometer. PZT-driven cavity tuning mechanisms and calibration techniques of the kind described for homodyne interferometers in Section 6.3.1 can also be used for these kinds of instruments.

An unconventional approach for laser ultrasound detection on rough surfaces has recently been reported using phase-conjugated optics (Paul et al., 1987). The laser beam scattered by the rough surface is collected by a wide-aperture, but not necessarily high quality, lens system and focused on a phase-conjugating mirror. This mirror, a $BaTiO_3$ nonlinear optical crystal, reflects the scattered beam back to the inspected surface with an exactly reversed angular and phase distribution. After a second reflection on the same spot of the rough surface, a coherent and collimated beam is thus reconstructed with little intensity loss. Advantages of this approach include a wide signal bandwidth and the possibility to obtain reflectivities higher than unity by taking advantage of the nonlinearity of the optically pumped phase-conjugating mirror.

NDT Applications

One of the main interests of laser-ultrasound testing is in the inspection of high-temperature materials for which no contact with an ultrasonic transducer is allowed. High-temperature ultrasonic testing to determine the dynamic properties of industrial materials is a typical application (Brammer and Percival, 1970; Calder and Wilcox, 1980). A spectacular demonstration of the laser-ultrasound technique capabilities is the measurement of the ultrasonic properties of a molten metal (Gathers et al., 1977; Calder and Wilcox, 1978; Calder and Wilcox, 1979). In a typical setup, a lead wire 1 mm in diameter and 25 mm long was subjected to rapid heating by a strong electric-current pulse resulting in wire melting. The measurement of the longitudinal-wave propagation time across the wire diameter was taken by exciting one side of the wire with a high-power laser pulse and by monitoring the acoustic pulse arrival on the other side with a Michelson interferometer. Such a measurement was made a few microseconds after the specimen was melted, but before the molten wire had the time to collapse. Figure 6-39 shows the detected singal in the (a) solid and (b) liquid sample. The longer propagation time in case (b) accounts for the smaller acoustic velocity in the molten material.

Other advantages of laser-ultrasound sensing as compared to the ultra-sonic-transducer approach are the possibility to produce ultra-short, single-spike laser pulses (Cachier, 1971) and to detect them without distortion, thanks to the wide frequency bandwidth of laser interferometric sensors. This

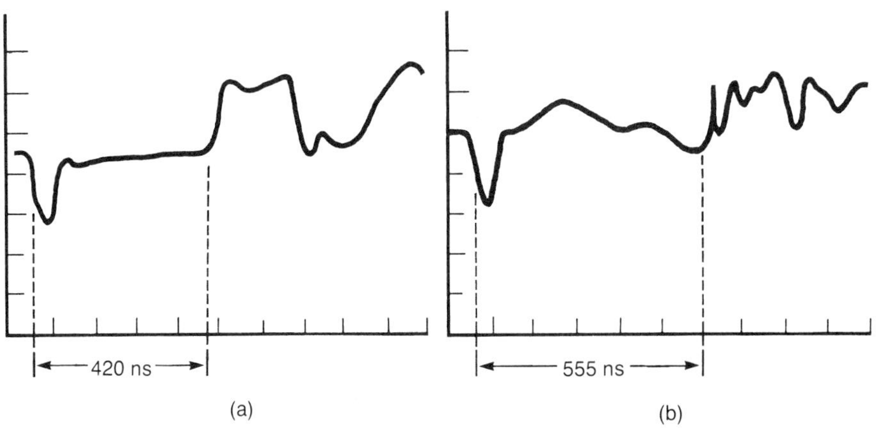

Figure 6-39 Detected signals for laser-ultrasound velocity measurement of a high temperature metallic wire (a) before melting and (b) after wire melting.

makes it possible to probe the thickness of thin films or to detect defects very close to the surface without resorting to high-frequency ultrasonic microscopy techniques (Lemons and Quate, 1975) which require transducer-to-surface distances of a few tens of micrometers to avoid excessive attenuation in the coupling liquid. A demonstration of such capabilities is illustrated in Figure 6-40 (Krautkrämer, 1979; Scruby et al., 1982). Ultrashort ultrasonic pulses were generated on one side of a razor blade by a subnanosecond laser pulse produced with a mode-locked laser. The sharpness of the pulses optically detected on the other side of the blade compares favorably with the signals obtained with high-frequency ultrasonic transducers (Tam, 1984). Such near-surface sensing capabilities are also offered by photothermal techniques (see next section). The choice between these techniques will depend on the required precision (better for the ultrasonic approach), the cost and the eye-safety aspects of the required apparatus (more favorable for the thermal approach), as well as on the physical characteristics of the interface materials (best ultrasonic echoes are obtained when there is a strong discontinuity of the ultrasonic impedance, while the observability of the thermal signal requires a strong discontinuity in the thermal effusivity).

Because of their noncontact nature, laser-ultrasound techniques constitute an attractive candidate for on-line characterization of high temperature materials. Potential applications are the detection of subsurface inclusions

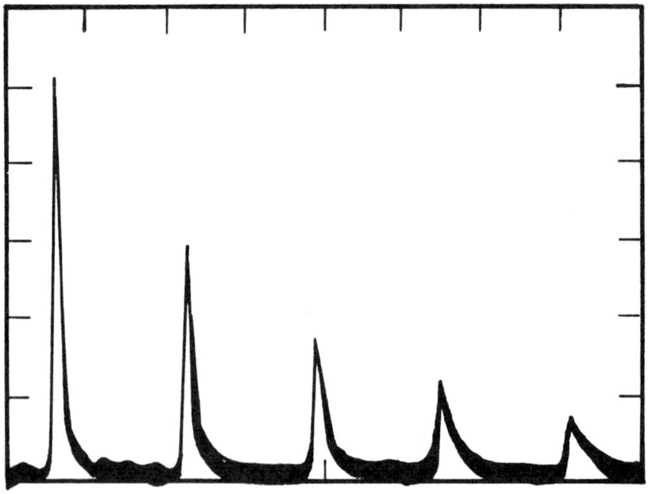

Figure 6-40 Multiple reflections of a short acoustic pulse in a razor blade using a mode-locked laser source.

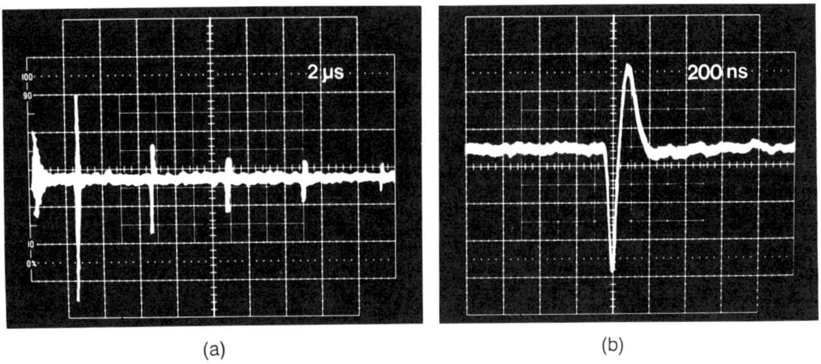

Figure 6-41 (a) Ultrasonic echoes detected with a velocity-sensing interferometer from laser-generated ultrasound in a 12.5 mm-thick steel plate, horizontal scale: 2 μs/div; (b) enlarged first echo, horizontal scale: 0.2 μs/div.

(which are typically detected by flame scarfing, see, e.g., Sakai et al., 1986) or of pipe in hot-metal billets or slabs, as well as the detection of the liquid-solid interface in continuous casting. One important requirement in such applications is operation at a conveniently large distance. Recent results obtained with velocity-sensing interferometers meet such a requirement (Monchalin, 1985a). Figure 6-41 shows a multiple-reflection signal obtained across a 12.5 mm-thick steel sheet with a velocity sensing interferometer at a distance as large as 1.5 meters from the surface. Work is now in progress at the Industrial Materials Research Institute to optimize such an apparatus.

For the inspection of finished products, surface ablation by the laser pulse is usually not allowed. One method for obtaining good acoustic wave generation efficiencies is to spray a thin oil film or an organic paint on the laser-irradiated surface prior to testing. The laser pulse vaporizes such a film producing a strong momentum-transfer longitudinal ultrasonic pulse without damage to the solid surface (Fairand and Clauer, 1976; Bondarenko et al., 1976). Optical sensing must, however, be performed on the other side of the sample in this case to avoid the plasma-produced air instabilities above the laser-irradiated surface (Calder and Wilcox, 1980). One-side reflections from a nonnormal subsurface defect can be detected in the ablation regime only if the propagation time of the plasma-generated acoustic pulse through the air from the irradiated to the probed spot is larger than the propagation time of the defect-reflected pulse echo (Wellman, 1980; Cielo et al., 1985a).

Subsurface defect detection in the thermoelastic regime is more difficult to implement because of the low bulk-acoustic-wave generation efficiency if no ablation takes place. Best sensitivities are obtained by monitoring the shear-wave pulse which is generated more efficiently than the longitudinal-wave pulse in the thermoelastic regime, particularly at angles of the order of 45° to the surface. An example of such an approach to the detection of a disbonded interface between a 4 mm-thick aluminum plate and the substrate is shown in Figure 6-42 (Cielo et al., 1985a). A converging-wave focusing configuration was adopted in this case to compensate for three-dimensional spread losses of the generated acoustic pulse. The small longitudinal (L) pulse and the shear (S) pulse (travelling at a velocity which is typically one-half of the longitudinal-wave velocity) are seen to almost completely disappear when the surface layer is well-bonded to the substrate. The same figure shows that the generation efficiency of Rayleigh (R) surface acoustic waves is much stronger in the thermoelastic regime. Efforts to use the strong Rayleigh pulse to evaluate the thickness of metallic coatings by surface-wave velocimetry (Cielo et al., 1985a) or the presence of subsurface delaminations by an analysis of the resonance properties of the unbonded layer under convergent acoustic-wave loading (Cielo, 1985; Cielo et al., 1985b; Cielo et al., 1986) have been recently reported. Other applications of laser-generated surface acoustic waves include the detection of surface cracks (Nadeau and Hutchins, 1984; Cielo et al., 1985a; Maldague et al., 1986) as well as the characterization of piezoelectric ceramics (Jen et al., 1985).

6.4 Thermal Inspection

Thermally-induced structure deformation or elastic-wave generation is a useful tool for nondestructive testing, as described in the previous sections of this chapter. In this section purely thermal techniques will be reviewed, where the subsurface features of an opaque material are evaluated by observing how they affect thermal propagation within the material.

Quantitative analyses of thermal propagation within most industrial materials have been performed for more than a century to measure thermal conductivity and diffusivity (see the excellent review of measurement techniques in the handbook *Thermophysical Properties of Matter* published by Y. S. Touloukian and coworkers of the Thermophysical Properties Research Center, Purdue University, Volumes 1, 2, 3 and 10). A variety of stationary methods, where the heat flow between the heated end and the cooled end of the sample is measured, have been developed for the measurement of thermal conductivities. Nonstationary methods for the measurement of thermal diffusivities have gradually become more popular than the stationary ones,

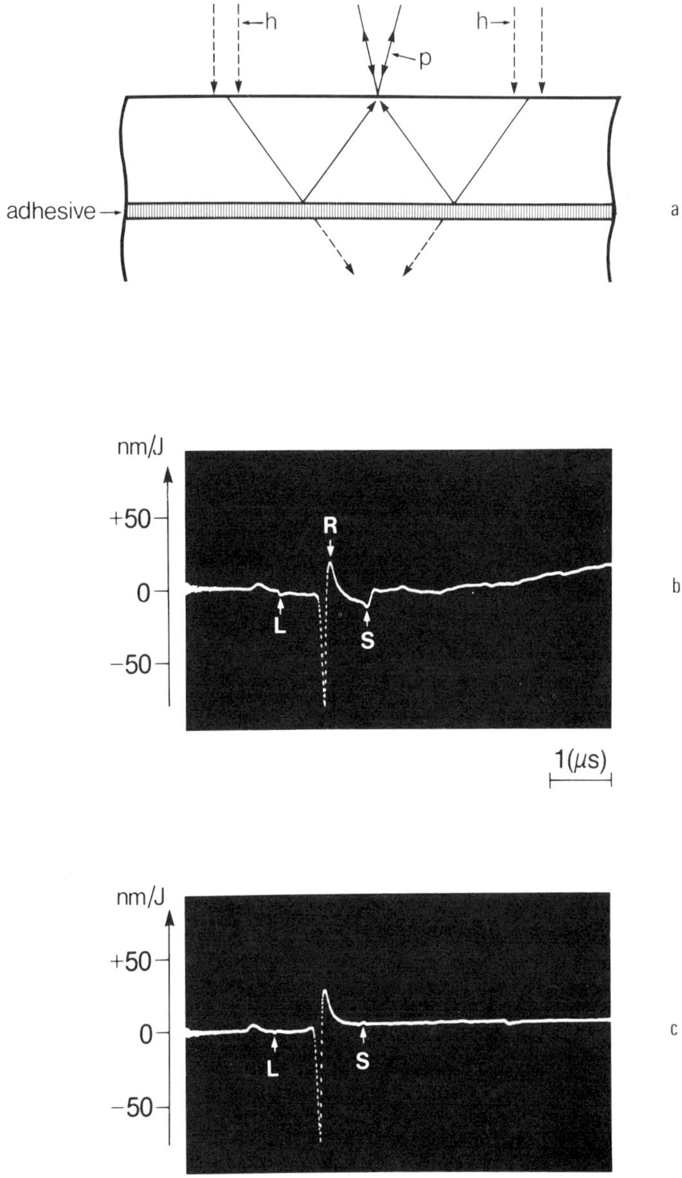

Figure 6-42 Inspection of adhesively bonded structures by laser-generated ultrasonic bulk waves: (a) experimental configuration, where h is the cross section of the annular laser beam and p is the focused probing beam; (b) signal detected in an unbonded area, where L and S indicate the longitudinal and shear reflected echoes, respectively, and R indicates the surface-wave pulse; (c) signal detected in a well-bonded area.

because they are faster to perform and thus less affected by boundary losses. Nonstationary methods include transient techniques, where a given portion of the material is heated for a short time period after which the temperature dissipation rate is monitored to evaluate the thermal diffusivity, as well as periodic techniques, where the material is heated by a modulated heat source and the phase lag of the thermal wave after propagation through a given distance along the sample is measured to evaluate the thermal properties of the material. Heat sources such as heat guns, torches or contact with hot solids or liquids have mainly been used, while temperature monitoring was performed with thermocouples or with optical pyrometers in the case of high temperature materials.

Thermophysicists were quick to capitalize on the availability of pulsed laser sources shortly after the development of the laser (Parker et al., 1961). Periodic sources including chopped light beams or modulated electron beams were also used (Cowan, 1961; Wheeler, 1965), as well as step-heating radiant sources for the one-side measurement of thermal inertia (Schultz, 1968). The flash technique, where a pulsed laser is used to heat one side of an opaque sample while the thermal impulse response is monitored on the other side, is now widely used to thermally characterize materials such as ceramics and composites in terms of their composition, structure or porosity content (Elchinger et al., 1979; Ziegler and Hasselman, 1981; Brandt, 1981; Taylor, 1983; Brittle and Taylor, 1984). Thermoanalytical measurements are also widely used for quality control. Phase transitions can be monitored by tracing the enthalpy versus temperature curve using a variety of time-resolved calorimetry techniques (Fruh and Widmann, 1982; Riesen and Sommerauer, 1983), including photoacoustic methods (Florian et al., 1978; Korpiun and Tilgner, 1980).

Nondestructive testing applications of thermal techniques are attracting increasing attention, thanks to the availability of noncontact heat sources and thermal sensors such as laser beams and infrared devices. The appearance of modern thermographic cameras nearly two decades ago has spurred the development of thermal NDT techniques (Lawson and Sabey, 1970; Vavilov and Taylor, 1982). Applications include the detection of hot spots in printed circuit boards (Wirick and Diede, 1980), the detection of overheated parts such as turbine blades (Beynon, 1982; Smith, 1983), the monitoring of fatigue damage and mapping of the stress field during cyclic mechanical testing (Belgen, 1967; Reifsnider and Williams, 1974; Wilburn, 1977; Pye and Adams, 1981), the detection of surface cracks (Kubiak, 1968; Deiniger and Kupperman, 1980), of surface defects in induction-heated hot-rolled plates

(Randak, 1983; Nakamura, 1983) and of subsurface defects (McLaughlin et al., 1980; Williams et al., 1980; Wilson and Charles, 1981; Hsieh et al., 1982; Reynolds and Wells, 1984).

The thermographic tool has until now been used for mainly qualitative nondestructive testing applications (Griffiths, 1980), observing the temperature distribution in the stationary regime or applying simplified one-dimensional heat-flow models. The application of quantitative thermophysical or thermoanalytical methods to the nondestructive testing field should open new possibilities for industrial noncontact inspection. Such a quantitative approach has recently been followed in the field of focused-beam testing with a chopped laser source (Nordal and Kanstad, 1979; Busse, 1982; Luukkala et al., 1982; Busse, 1985; Almond et al., 1985) and to a smaller extent with a pulsed heat source (Tam and Sullivan, 1983; Imhof et al., 1984; Cielo, 1984). This section briefly reviews such recent developments; a comparison of the different approaches is presented and their applicability to thermographic image processing is discussed.

6.4.1 *Single Spot Techniques*

Thermal-Wave Testing

The basic principle of thermal-wave testing by infrared temperature monitoring is illustrated in Figure 6-43. A laser beam, or an incoherent source, heats the surface of the material to be inspected to a few degrees. The

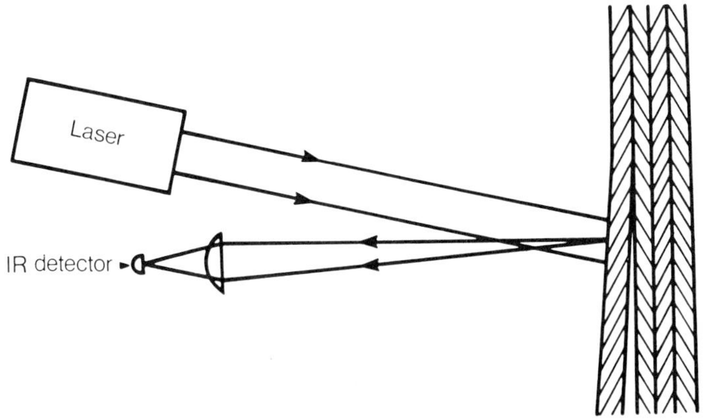

Figure 6-43 Basic principle of single-spot thermal inspection with infrared temperature detection.

temperature-time curve (thermogram) is recorded by an infrared detector pointed to the center of the irradiated area. Other temperature monitoring methods have been used including optical beam deflection or "mirage effect" detection (Boccara et al., 1980; Murphy and Aamodt, 1981), temperature-dependent reflectivity variations (Rosencwaig et al., 1985), as well as the gas-cell photoacoustic microphone detection method represented in Figure 6-37. However, such methods are difficult to implement in an industrial inspection environment because they require careful beam alignment and a substantially flat surface (for the mirage method) or else they perform best when applied to semiconductor materials (for the thermoreflectance approach) or because they require previous sample preparation. We shall focus on infrared detection schemes in this section.

If a short laser pulse heats an area of diameter much larger than the thermal propagation distance during the observation time (see equation 6-4), the surface temperature of an opaque material after a perturbation time of the order of two or three times the pulse duration (Cielo, 1983a, Balageas et al., 1986) follows the expression (Carslaw and Jaeger, 1959, Chapter 10; De Vriendt, 1982)

$$T(z, t) = \frac{Q}{(\pi K \rho c t)^{1/2}} \exp \frac{-z^2}{4\alpha t}, \tag{6-7}$$

where z is the depth below the heated surface, t is the time from heating, Q is the injected energy density, K is the thermal conductivity, ρ is the density, c is the specific heat and $\alpha = K/\rho c$ is the thermal diffusivity, respectively, of the inspected material. The analysis of equation (6-7) shows that (a) the absolute level of the temperature curve is inversely proportional to the thermal effusivity $(K\rho c)^{1/2}$ (other authors prefer the concept of thermal inertia, $K\rho c$, see e.g., Schultz, 1968): Less conductive materials reach higher surface temperatures for the same heating pulse; (b) the temperature of the material at a depth z below the heated surface is unaltered for a time t much smaller than z^2/α: This leads to the concept of the thermal propagation depth $z \approx (\alpha t)^{1/2}$ defined by equation (6-4); (c) the surface temperature ($z = 0$) decays following a $t^{-1/2}$ curve which appears as a straight line in a log-log scale.

The thermograms obtained with an apparatus of the kind shown in Figure 6-43 will thus be of the form shown in Figure 6-44 (Cielo, 1984). If the material has no subsurface defects, the temperature decay follows a nearly straight line with $-\frac{1}{2}$ slope in the log-log scale in agreement with equation (6-7). If a delamination is present at a depth z under the surface as in curves (b) and (c), the large thermal resistivity of the unbonded interface produces a marked

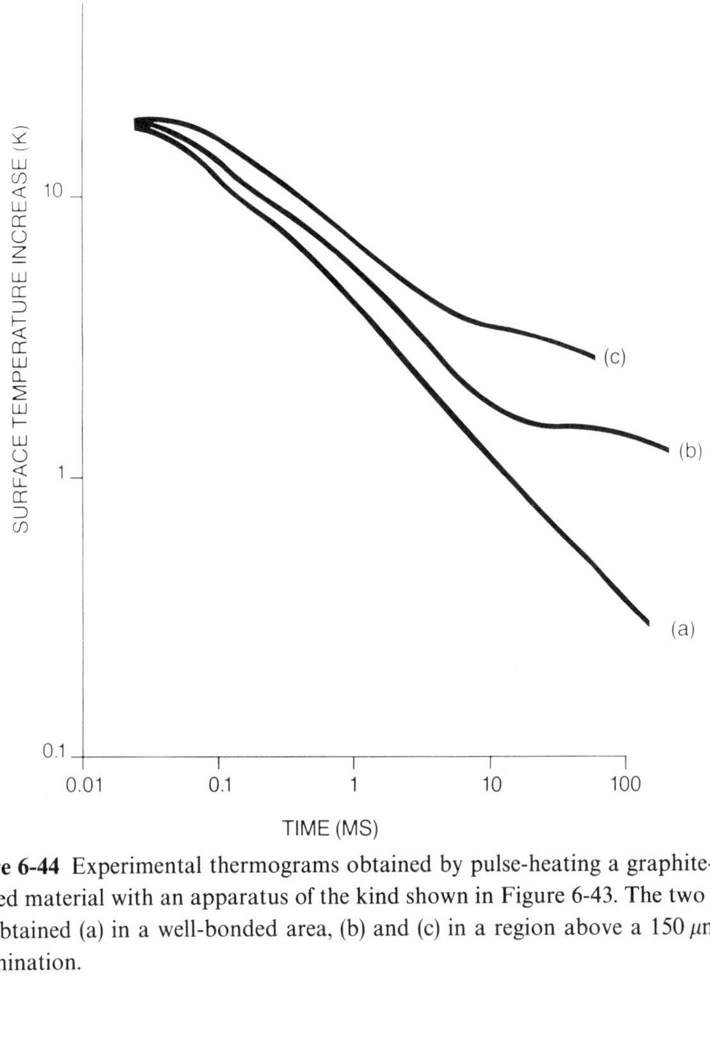

Figure 6-44 Experimental thermograms obtained by pulse-heating a graphite-epoxy layered material with an apparatus of the kind shown in Figure 6-43. The two curves are obtained (a) in a well-bonded area, (b) and (c) in a region above a 150 μm-deep delamination.

reduction of the temperature decay rate after a time of the order of the thermal propagation time z^2/α within the material. This makes it possible to detect a subsurface defect by observing the slope of the thermogram independently of the emissivity-dependent level of the signal, while the depth of the defect can be estimated from a measurement of thermal propagation time.

Thermal NDT is thus similar to ultrasonic pulse-echo in the sense that the delayed signal perturbation produced by the subsurface defect can be considered as a reflection of the thermal wave by the defect at the time-dependent thermal-wave speed

$$v_{\text{th}} \approx \frac{z}{t} \approx \left(\frac{\alpha}{t}\right)^{1/2} \approx \frac{\alpha}{z}. \tag{6-8}$$

The inspection time thus increases with the square of the defect depth: A 100 μm-thick ceramic material ($\alpha \approx 0.01$ cm^2/s) requires an observation time of the order of 10 ms per inspected spot, while a similar material 1 cm-thick requires a relatively long inspection time of 100 seconds per point.

One important difference between ultrasonic and thermal techniques is that thermal propagation is based on a diffusion equation rather than on a wave equation so that the "thermal wavelength" (Carslaw and Jaeger, 1959, Section 2.6; Rosencwaig and Gersho, 1976) is of the order of the thermal propagation depth. A first consequence is that material porosity much smaller than the depth of the defect to be detected will be much smaller than the thermal wavelength when the thermal front reaches the defect and thus will not affect the signal. A second consequence is that the thermal technique will be relatively insensitive to subsurface defects whose diameter is smaller than their depth.

Within this limitation, thermal techniques are attractive because they are noncontact and thus can be deployed on awkward-shaped, high-temperature or difficult-to-access materials. Moreover, both the heating and the IR-pointing beams are easy to scan over large surfaces, so that thermographic testing may often be performed very quickly over a wide surface. An automatic solder joint inspection system for printed circuit boards based on this principle is commercially available (Vanzetti et al., 1980).

Beam Modulation Possibilities

Different choices can be made for the temporal distribution of the heat injection. The three best known approaches are pulse-heating, step-heating and periodic heating, as well as all the intermediate configurations.

Pulsed heating, represented on the left of Figure 6-45, is the approach just described. Step-heating, top right, gives a rising temperature-time curve with a slope equal to $+\frac{1}{2}$ in a log-log scale if the heated material is thermally homogeneous (Schultz, 1968; Cielo, 1983a). As in the case of pulsed heating, the presence of a subsurface defect results in a departure of the curve from the constant-valued slope after a time corresponding to the thermal propagation time for the defect depth.

If subsurface defects such as delaminations are to be detected, pulsed

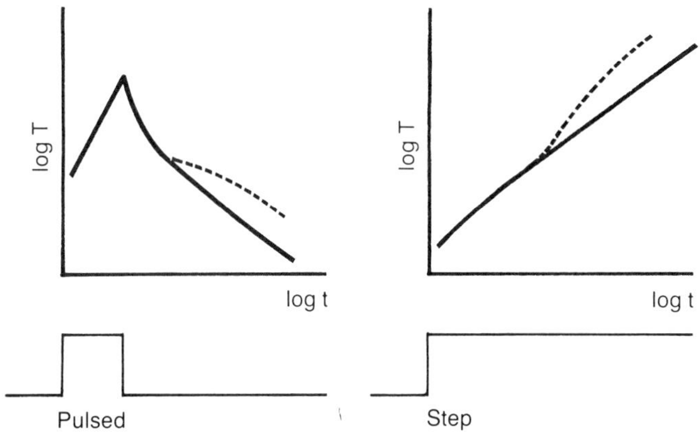

Figure 6-45 Schematic diagrams of the temperature versus time curves obtained with pulsed (left) or step (right) heating for a homogeneous material (continuous line) and for a material with a subsurface delamination (dotted line).

heating has the advantage over step-heating of requiring a smaller dynamic range for the temperature variations, as Figure 6-45 shows. Another argument which may favor the pulsed approach is that, because of the log-log scale, longer sampling times correspond to lower temperature values, where the low SNR requires longer time averaging. Such an argument is, however, mitigated by the nearly $1/f$ distribution of environmental thermal noise at low frequencies (Cielo et al., 1986a).

Another typical application of thermal inspection is the thickness evaluation of metallurgical coatings (Almond et al., 1985; Cielo and Dallaire, 1986). Such refractory metal or ceramic coatings are hot-sprayed on metallic substrates for corrosion or wear protection. The coating material has a thermal conductivity which is typically one order of magnitude smaller than the thermal conductivity of the substrate (Elchinger et al. 1979; Brandt, 1981). When optothermally testing with either pulsed or step excitation, the recorded thermograms will be of the kind shown in Figure 6-46. The departure of the thermal history curve from a straight line after a time period corresponding to the thermal propagation time through the coating thickness makes it possible to evaluate the coating thickness if its thermal diffusivity is known or vice versa. Furthermore, the thermal effusivity may be independently evaluated by comparison with the usually known effusivity of the substrate (Cielo, 1983a). An example of the correlation between the processed step-heating signal and the thickness of different ZrO_2 thermal barrier coatings is shown in Figure 6-47 (Cielo and Dallaire, 1986). For this application, the

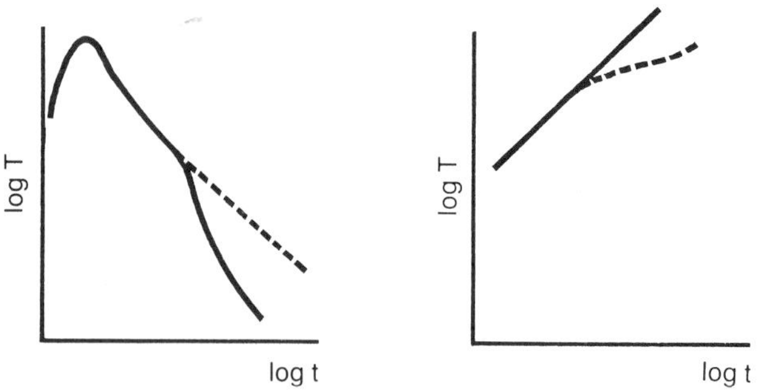

Figure 6-46 Typical thermogram obtained for either pulsed (left) or step (right) thermal excitation of a metallurgical coating on a metallic substrate.

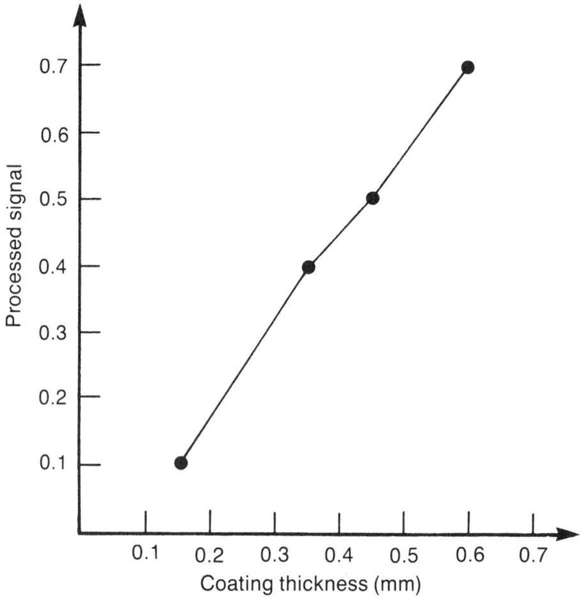

Figure 6-47 Processed signal obtained from step-heating measurements for the thickness evaluation of ZrO_2 thermal-barrier coatings on steel.

step-heating approach is more convenient because it requires a smaller dynamic range for the temperature excursion as Figure 6-46 shows.

Another beam modulation approach can be implemented by square-wave chopping the heating beam to obtain a periodic heat input, as shown in Figure 6-48 (see, e.g., Busse, 1982). The two signals shown are typical of a homogeneous material (continuous curve) and of a material with a shallow

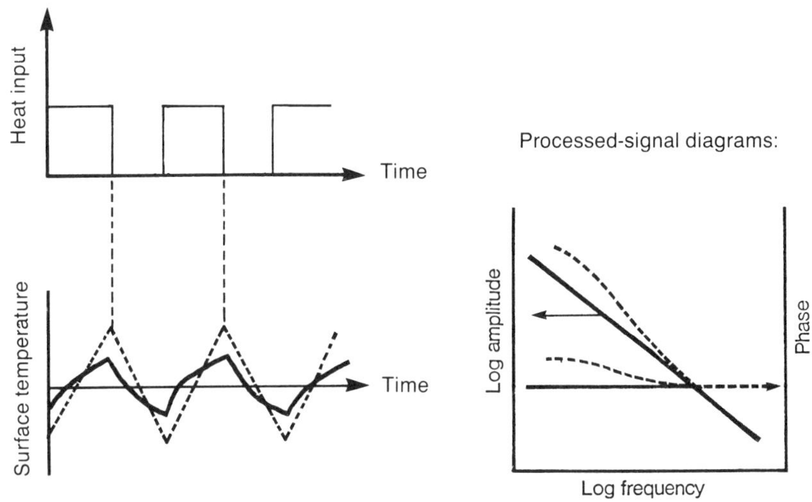

Figure 6-48 Schematic diagram of the surface temperature response of an opaque material irradiated by a chopped-laser source with (dotted lines) and without (continuous lines) a subsurface defect.

delamination (dotted curve). The curve obtained on the delamination has both a higher amplitude and a larger phase shift than the curve corresponding to a sound sample (the continuous curve has a phase shift of nearly 45° with respect to the square wave, while the dotted curve has a phase shift near to 90°). If we plot at different modulation frequencies the amplitude and phase of the recorded signal, we obtain curves of the kind shown on the lower right plot of Figure 6-48, which contain information similar to that of the thermogram in the transient-heating approach (the inverse of the modulation frequency is equivalent to the thermal propagation time). We can analyze the slope of the amplitude versus frequency plot, or the value of the phase shift, to obtain information on the presence and position of a subsurface defect.

The modulated-beam method allows signal averaging over a large number of cycles and thus provides a good SNR with limited heating power when the depth to be inspected is relatively small. For the industrial inspection of relatively thick materials, however, the measurement time tends to be impractically long if the modulated-beam approach is followed because of the need to extend the measurement over several heating periods. For the inspection of thermal-barrier coatings, a single phase measurement at a given modulation frequency may require typically 30 seconds, compared to 0.3 seconds in the transient case (Cielo and Dallaire, 1986). Even longer measurement times are required if a plot of the recorded signal versus frequency of the kind shown in Figure 6-48 is required. One limitation of the modulated-beam approach comes from the fact that only the phase shift of the first harmonic of the detected signal is recorded, while all the information contained in the higher order harmonics is wasted. Such an approach is justified when working at high frequencies, in the MHz range, when the signal level is very low and there is no time to record and superpose the signal obtained for each cycle. When the modulation frequency is in the KHz range or lower, i.e., for the nondestructive inspection of depths larger than some tens of micrometers, a transient-heating approach has the advantage of providing a depth-resolved signal within a single heating cycle. Intermediate approaches, such as a modulated-beam technique with recording of the higher order harmonics, or a pulsed approach where a transient-signal analysis is made on the signal obtained by averaging over several repetitive pulse excitation cycles (Tam and Coufal, 1983) may also be considered.

Thermal Characterization of Industrial Materials

Noncontact thermal techniques may be a valuable tool not only for the detection of subsurface features as mentioned above but also for the characterization of homogeneous industrial materials. As mentioned at the beginning of this section, laser-flash techniques are widely used to evaluate material properties such as phase composition or porosity of small samples. To use such techniques for continuous quality control on the production line, methods applicable *in situ* without any sample preparation should be developed. Ideally, such methods should require access from one side only of the processed material, and they should be noncontact and sufficiently rapid to be automatically implemented on products moving along the production line.

An example of a thermal characterization technique is shown in Figure 6-49 (Cielo et al., 1986a). The material is irradiated by a laser pulse resulting in a surface temperature increase of a few degrees centigrade. As can be seen from equation (6-7), the surface temperature level is inversely proportional to the square root of the thermal inertia $K\rho c$. Such a parameter can thus be evaluated from the infrared-monitored surface temperature, provided that the amount of absorbed laser energy is known. A highly reflective hemispherical cavity is used to ensure nearly 100% laser absorptivity and infrared

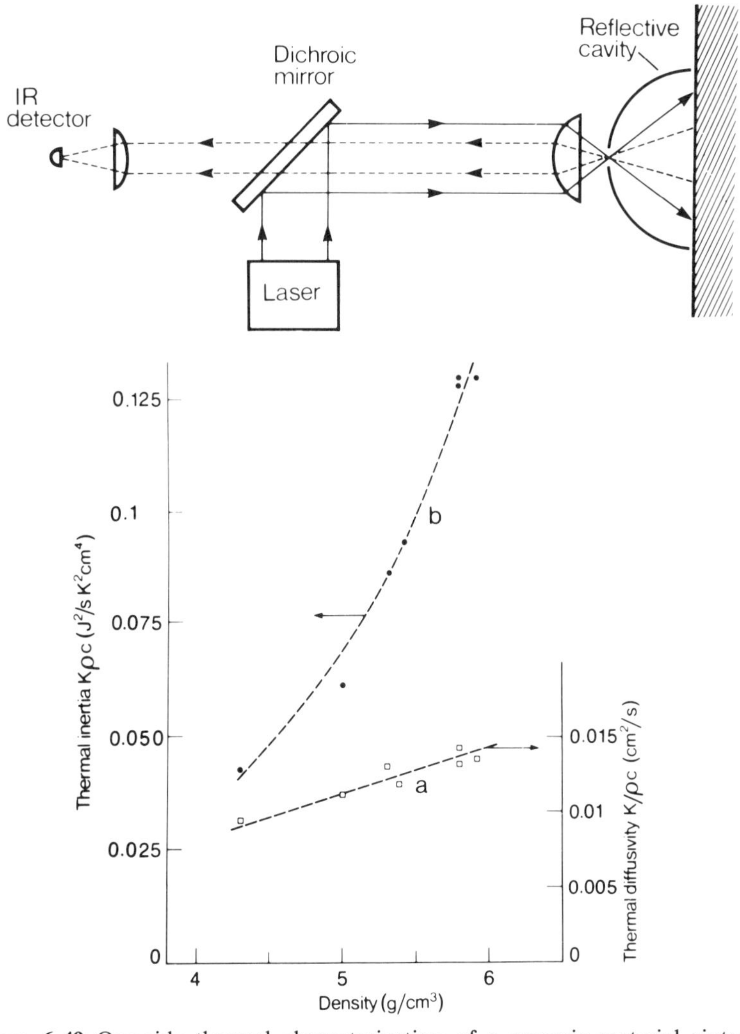

Figure 6-49 One-side thermal characterization of a ceramic material sintered at different temperatures. Top: experimental apparatus; bottom: measured thermal properties of different pellets sintered at temperatures ranging from 1250°C to 1550°C.

emissivity for the inspected material, thus enabling absolute measurements to be made on materials with unknown surface emissivity properties.

This method was applied to the nondestructive evaluation of a number of ZrO_2 pellets sintered at temperatures ranging from 1250°C to 1550°C, resulting in different levels of porosity and thus in different densities of the sintered material. The experimental results at the bottom of Figure 6-49 show that such a method, apart from its convenience in terms of *in situ* and one-side inspection capabilities, is actually more appropriate for porosity measurements as compared to the double-side laser-flash approach. The latter technique provides an evaluation of the thermal diffusivity $K/\rho c$. Because the conductivity K and the density ρ both increase with the degree of sintering, such a parameter is better evaluated through a measurement of the thermal inertia $K\rho c$.

Other material properties, such as the microstructural configuration or the phase composition, may be better characterized by thermal-diffusivity measurements. Again, one-side techniques are preferable for *in situ* quality control of bulk materials or of sheets of unkown thickness. Techniques of this kind using focused-laser heating and infrared monitoring of the three-dimensionally diverging thermal flow have been reported for the characterization of isotropic (Luukkala et al., 1982) or of anisotropic (Berrie et al., 1981) materials. A converging-flow approach has also been recently proposed to avoid surface overheating in such measurements (Cielo et al., 1986b). Finally, photothermal techniques have been shown to be applicable to thermal analysis measurements such as the monitoring of curing in epoxy resins (Busse and Eyerer, 1983).

6.4.2 *Thermal Imaging Techniques*

In the previous section a number of single-spot thermal techniques were described for the detection of subsurface defects or material characterization. Although material characterization can usually be performed at one, or a restricted number of, positions over the material surface, subsurface defect detection must usually be done over a given area. The choice of the proper surface scanning procedure is a question of considerable practical importance in thermal NDT because of the relatively slow propagation of the thermal wavefront through the inspected part, as mentioned above in relation with equation (6-8). This implies that, in contrast to thermoelastic or photoacoustic methods, point-by-point scanning may require an unacceptably long inspection time if relatively deep subsurface features are to be thermally imaged.

Different methods for surface scanning are shown in Figure 6-50. In the sequential scanning approach, the heating laser and the infrared-emission beams are collinearly scanned spot-by-spot over the inspected surface by a step-motor driven, angularly scanning mirror. As mentioned above, this approach may result in unacceptably large inspection times when inspecting

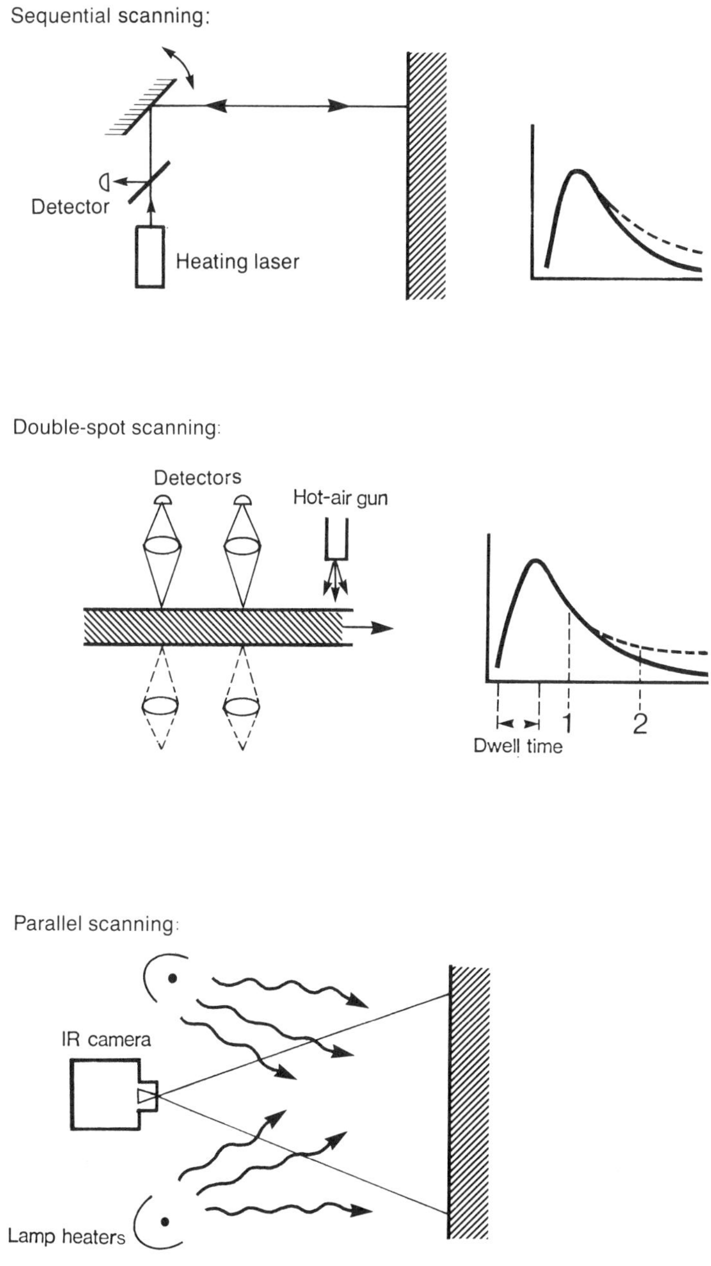

Figure 6-50 Different surface scanning approaches for thermal nondestructive detection of subsurface defects.

materials such as composite panels for which the thermal response time at each point may require more than ten seconds. Additional problems encountered with this approach include three-dimensional propagation effects which increase the dynamic range requirements for the temperature signal, as well as thermal cross-talk between adjacent spots.

Another approach (Green et al., 1984) uses a thermal source and two IR detectors aligned over a quickly moving material. The result is essentially a pulse-heating approach, where the thermogram is only sampled at times 1 and 2 during thermal decay (see diagram in Figure 6-50) and the value of the decay slope is related to the presence of subsurface defects. This method provides no information on the depth of the defect, because it neglects all the other points of the thermogram which could be easily accessed by a rapid scanning of the IR detector over the cooling surface.

The latter concept is retained in the parallel-scanning method, where an infrared camera produces an image of the temperature distribution over the surface heated by a large area heating source such as an infrared lamp. As a result, the full thermally imaged area is inspected during a single thermal-response time period, while the full thermal history information can be recorded for each pixel of the image. Three-dimensional heat flow as well as thermal cross-talk problems are avoided. Defects perpendicular to the surface, such as normal cracks, cannot, however, be imaged by this method unless a highly structured heat source is used.

One of the problems encountered in thermal inspection by parallel imaging is caused by the nonuniformity of the heat source. Some regions of the heated surface may appear as hot spots not because of subsurface defects, but simply because the heat input is higher in such a region. Figure 6-51 shows how standard image subtraction techniques can be used to counteract such a problem (Cielo et al., 1985c). A graphite-epoxy sheet is thermally inspected in transmission by heating the opposite side with an infrared lamp. The thermal image on the left shows a region without defects. The temperature distribution is highly nonuniform because of the nonuniformity of the heat input. The top right picture is a similar thermograph obtained over a 0.5 mm-deep, 1 cm-diameter subsurface delamination. The bottom thermograph shows the difference between the two images. It is clear from this example that reference subtraction techniques, where a reference from a sound area is kept in memory and subtracted from the thermographs obtained at different positions, is a powerful method to avoid heat source nonuniformity problems.

The method illustrated in Figure 6-51, which will be designated as

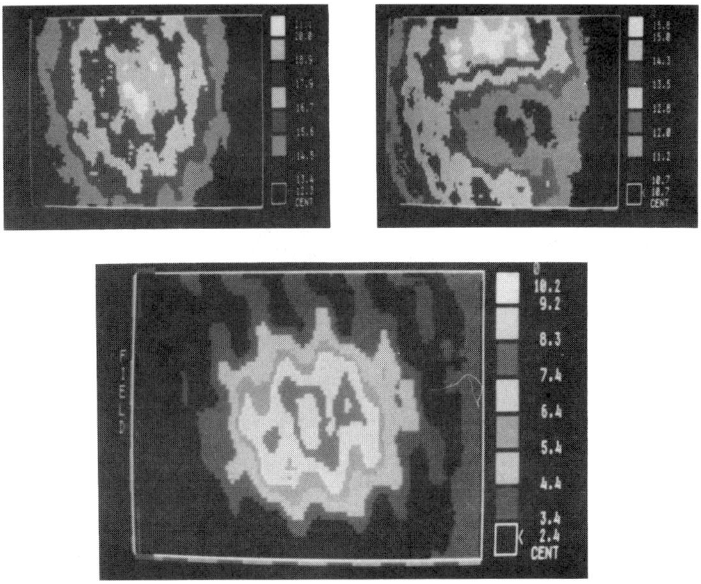

Figure 6-51 Thermal images obtained with a graphite-epoxy laminate heated on the opposite side by a tungsten-filament lamp. Top left thermograph: region without defects; top right: region with delaminations; bottom thermograph: difference between the two images, showing the position of the subsurface defect.

space-domain reference subtraction, cannot, however, compensate for the thermal image nonuniformity produced by surface absorptivity or emissivity variations over the inspected surface. The effect of surface emissivity variations over the surface of the woven graphite-epoxy sample can be seen as a superposed diagonal structure in the processed thermal image on the bottom of Figure 6-51. Such emissivity variations are different from sample to sample and cannot be eliminated by *a priori* knowledge about the inspection system.

Recent advances on quantitative thermal NDT techniques, such as the time resolved, emissivity-independent method illustrated in Figure 6-44, may offer a solution to the variable emissivity problem (Cielo et al., 1985c; Holmsten, 1986). The power of such an approach is demonstrated in Figure 6-52. The same locally-delaminated sample shown in Figure 6-51 was inspected under conditions of intense reflective noise (infrared reflection from a torch in the background). The two thermographs shown on the top of Figure 6-52 represent the apparent temperature distribution over the same sample region and were taken after three seconds (top right photo) and after

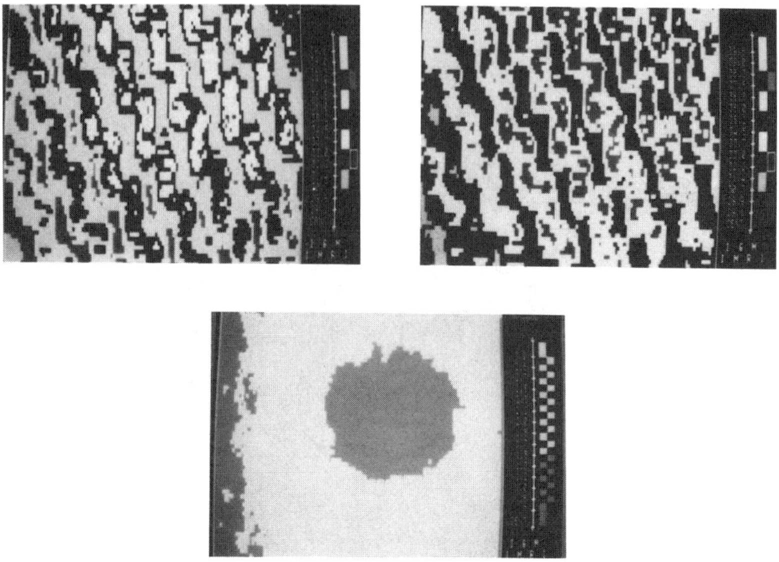

Figure 6-52 Time-domain reference subtraction technique. The delaminated sample of Figure 6-51 is observed under conditions of strong reflective noise after three seconds (top right) and five seconds (top left) from starting heating. The difference thermograph (bottom) shows clearly the position of the defect.

five seconds (top left) from the beginning of heating. The strong diagonal pattern is produced by the infrared radiation reflected by the variable-emissivity (and thus variable-reflectivity) surface of the woven composite. As such a surface noise is the same for the two images, it completely disappears from the difference image (bottom thermograph) which shows very clearly the shape of the subsurface delamination. This approach is called time-domain reference subtraction, because it is a parallel-imaging implementation of the time-resolved thermal techniques described in Section 6.4.1. Indeed, the processed image on the bottom of Figure 6-52 can be considered as a map of the thermal rate of change over the imaged area after heating. A number of variations to this basic approach can be conceived, such as the adoption of different heat modulation procedures, the evaluation of the logarithmic thermal decay derivatives at each imaged point to counteract multiplicative thermal noise or the comparison of the measured first- or second-derivative thermal decay map with similar maps recorded on sound samples of similar, though complex, geometry.

A different application of thermal imaging (Belgen, 1967; Mountain and Webber, 1978; Baker, 1986) is in the field of stress pattern analysis by thermal emission (SPATE). A structure is subject to cyclic loading at typical frequencies of 1 to 20,000 Hz, and the amount of adiabatic heat developed during one vibration cycle is synchronously detected to obtain a map of the stress field across the structure. This technique is different from vibro-thermography (Reifsnider and Williams, 1974) where heat development during several cycles is integrated to detect defects. In the SPATE approach, the heat differential between subsequent half cycles gives the distribution of the stress amplitude across the structure. A commerical instrument based on this principle, available from Omertron, can detect temperature differentials as small as 0.001°C, corresponding to 10^6 Pa in steel. Such a temperature sensitivity is, however, achieved at the expense of a relatively long scanning time, because the image is sequentially scanned with a dwell time on each pixel extending over several vibration cycles.

The SPATE technique can be applied to evaluate the stress distribution on mechanically loaded parts, similarly to the strain-distribution analysis obtained by holographic interferometry. The thermal approach has the advantage of being insensitive to ambient vibrations, because no interferometry is involved. The minimum detectable stress is, however, in the 10^6 Pa range, as compared to a stress sensitivity of the order of 10^4 Pa for classical holography (Klappert, 1976) and an even higher sensitivity for heterodyne holography or sequentially scanned focused-beam interferometry. The high temperature sensitivity reached by the SPATE camera represents an example of the capabilities offered by thermal imaging devices.

6.5 A Comparative Assessment

Different optical techniques for the nondestructive testing of materials and structures have been reviewed in this chapter. The fast rate of evolution of such techniques clearly demonstrates that optical methods have a potential extending much farther than the classical surface inspection line of applications. Although optical NDT techniques are still largely at the laboratory development stage, their implementation on the industrial floor appears to be possible in many cases, thanks in part to the increasing ruggedness and shrinking price of modern optoelectronic devices. Some unique characteristics of optical techniques, such as their noncontact nature and suitability to remote operation, actually make them a particularly convenient tool for operation in hostile industrial environments.

Different approaches have been described in this chapter, including methods based on the analysis of the mechanical, thermoelastic or vibrational generation of strain patterns, on the optical monitoring of photoacoustically-generated ultrasonic wave interaction with subsurface defects, as well as on the infrared probing of thermal flow. Both imaging and nonimaging techniques have been reviewed and discussed. A comparative assessment of such techniques can be given only with reference to specific NDT application fields, each of these techniques being appropriate in certain cases but not necessarily as a universal NDT tool. The main purpose of this discussion is thus to identify a number of relevant criteria which should be taken into consideration when choosing among the different optical techniques which are presently available.

Optical NDT techniques may first be classified as (1) strain-monitoring or (2) wave-propagation methods. The first category comprises holographic, speckle and focused-beam interferometric methods for the observation of mechanically or thermoelastically produced part deformation or vibration patterns. A lack of integrity of the inspected workpiece may be inferred from the presence of discontinuities in the recorded strain distribution or in the vibrational pattern as a consequence of the presence of subsurface defects. The second category, wave-propagation methods, includes optical monitoring of ultrasonic waves as well as thermal inspection techniques. In this case, defects can be seen when they affect the propagation of the ultrasonic or of the thermal wave.

As a general comment, strain monitoring methods, as compared to wave-propagation optical techniques or more classical nonoptical NDT techniques such as ultrasound, eddy current, X-ray or dye penetrant methods, have the advantage of providing information on the amount by which the structural integrity is affected by the presence of a defect, rather than simply on the presence of a fault. Not all flaws are critical, the criticality depending on the size, shape, nature and location of the flaw. Strain visualization techniques measure the strain behavior of the loaded structure, thus revealing flaws resulting in strain concentrations which substantially weaken the mechanical resistance of the structure. False rejection caused by cosmetic or noncritical defects can thus be avoided.

Another performance criterion is that of operation speed. As a general rule, parallel operation techniques, where a full image of the inspected part is processed simultaneously, are faster than sequential scanning methods.

This is certainly important in the choice of thermal-imaging over scanned-spot thermal techniques, as discussed in Section 6.4.2. However, when the single-spot signal response is relatively fast, as in interferometric vibration sensing and thermoelastic inspection techniques, the parallel processing capabilities offered by methods such as holographic or speckle techniques are no longer a prime consideration. If the single-spot observation time is of the order of one millisecond, as is normally the case in scanned thermoeleastic inspection, the length of time required to scan a given structure tends to be determined more by the available power of the pulsed heat source than by the mechanical scanning speed.

It should be observed that parallel techniques such as holographic and speckle interferometry are ingenious methods to produce pictorial information for quick evaluation by the human operator. In an automated inspection environment, however, obtaining pictorial information in the form of a fringe pattern which must subsequently be machine processed to extract the relevant information is not a necessary step. Point-by-point scanning for 3-D image acquisition by incoherent methods or for interferometric profilometry or displacement measurement may be as fast as parallel image evaluation in many automated inspection problems.

Other aspects of pulsed thermoelastic testing of layered materials as compared to thermally-stressed holography in terms of signal visibility, sensitivity to ambient vibrations and dependence on whole body structural resonances are more specifically analyzed in Section 6.3.2. Similarly, a comparative analysis of the practical merits of holograhic and speckle techniques can be found at the end of Section 6.2.2, while a discussion of different beam modulation approaches in thermal-wave testing is presented in Section 6.4.1.

An important criterion when comparing different techniques for subsurface defect detection is the minimum defect size that can be detected by the different approaches. Strain monitoring methods (holographic, speckle, thermoelastic) are practically limited to transversal defect dimensions of the order of or larger than the defect depth in order to obtain a surface displacement signal clearly discernible from the material's thermal expansion noise. Similar limitations, although based on different theoretical grounds (the diffusion nature of thermal waves) are valid for thermographic sensing of isotropic materials at the present state of signal processing capabilities. Laser-based ultrasonic inspection techniques, on the other hand, are not subject to this limitation. Pulse-echo signals can, in principle, be obtained from defects of size much smaller than their depth below the surface.

Another point to be considered is the cost and difficulty of implementation of the different techniques in day-to-day routine inspection by unskilled operators. Holographic techniques are relatively inexpensive but have long been considered as sophisticated techniques requiring highly skilled operators for the implementation and interpretation of the results. Automated-development thermoplastic film cameras may simplify such procedures in the future. ESPI techniques have also contributed to simplifying coherent strain observation techniques. The sensitivity of these techniques to environmental noise is, however, still a problem to be overcome when applying such methods in hostile industrial environments.

Focused-spot interferometric and photoacoustic methods are less sensitive to low frequency environmental perturbations, but require relatively complex and expensive interferometric probes, while thermoelastic and, in particular, laser-ultrasound techniques use pulsed heating laser sources of relatively high power which are subject to strict eye safety rules if they are to be installed on the industrial floor. Thermographic techniques are convenient in terms of cost and safety of the required heating source, while the detection process is relatively straightforward.

The choice among the different techniques is often based on the physical properties of the materials to be inspected. Highly polished metallic surfaces give poor results when thermographically inspected because of their small infrared emissivity, while graphite-epoxy materials are more conveniently inspected by such techniques because of the high emissivity and absorptivity of their surface. Similarly, the case-depth of surface-hardened steel may be conveniently probed by thermal characterization techniques because of the large variation in the thermal diffusivity of steel before and after hardening (Jaarinen and Luukkala, 1983), while the ultrasonic velocity change is limited to nearly 1% in such a case (Martin and Becker, 1980). As an opposite example, the evaluation of the coating thickness of galvanized steel (zinc-coated) is better performed by laser-ultrasound methods taking advantage of the difference in the ultrasonic velocity value by more than 20% between zinc and steel, while a thermal-wave technique would give poor results because of the very similar thermal effusivities of the two materials.

In conclusion, all the optical NDT techniques described in this chapter appear to have attractive features when applied to the appropriate NDT problem. Furthermore, they are all relatively new and unexplored approaches suggesting more intensive development work to determine the extent of their full potential.

Chapter 7

Optical Sensing for Process Control

Process control is an important requirement for most material processing and industrial manufacturing activities. Investment in on-line process analysis provides very high rates of return. It has been estimated, for example, that in a single modern petroleum refinery there may be more than $100 million worth of process control instrumentation (Hirschfield et al., 1984).

On-line sensors are a crucial component in the process control loop. The applications of optical modern technology in such an area are increasing at a fast pace (Rutledge, 1986; Kamal and Patterson, 1986). The requirements for ruggedness, unsupervised operation and reliability are very demanding in many production lines. This explains why several very powerful but rather complex analytical instruments, such as those based on nuclear magnetic resonance, electron-beam microprobing or Auger spectrometry, are mainly confined to the analytical laboratory. As Table 7-1 shows, most on-line instruments are based on rather simple and readily applicable technologies (Hirschfield et al., 1984).

Indirect information on process variables such as temperature, pressure or flow is by far the most widely spread type of data collected in automated plants because of the availability of cheap and rugged sensors. Many years of experience in small-scale pilot plants have taught process engineers the correlation between such indirect process variables and more direct data on

Table 7-1 On-line Instrument Utilization in Process Control

Widespread	Occasional	Potential
Temperature	Wet-chemical processor	Chemiluminescence
Pressure	Chemical tape systems	Microwave absorption
Flow	Specific-ion electrodes	Optical emission
Density	Fluorometry	X-ray fluorescence
Refractive index	Infrared absorption (scanning)	Raman
Ultraviolet-visible colorimetry	Mass spectrometers	Infrared emission
Infrared absorption (filter)	Ultraviolet spectrometry	Nuclear magnetic resonance
Gas chromatography	Gamma-ray absorption	Near-infrared reflectance
Indicator tags	Dielectrometry	Acoustic emission
pH electrodes	Magnetometry	Chemresistor
Acoustic velocity	Particle counters	Piezobalances
Amperometry	Viscosity	Acoustic attenuation
	Liquid chromatography	Turbulent flow
		Array chromatography
		Electrophoresis

the material properties of the manufactured product. Nevertheless, direct chemical composition sensors would be very valuable to process control engineering, and a variety of optical material analysis sensors are under intense development for on-line applications.

This chapter describes a number of process-parameter sensors based on optical technology. Optical sensors for direct material characterization will be the subject of the next chapter.

7.1 Fundamentals of Process Control

The need for process control (Di Stefano et al., 1967; Ogata, 1970; DeCarfort and Foulard, 1971; Schwarzenbach and Gill, 1978; Dorf, 1980; Considine, 1985) is increasing rapidly with the modern industrial trend toward automated production. More and more powerful control tools, such as computer-aided supervisory networks and flexible manufacturing structures, are stimulating a fast expansion as well as an extensive evolution of process control schemes. Under such conditions, requirements for on-line inspection devices are rapidly increasing. This section is an introduction to the basic notions of process control for nonspecialists with emphasis on the required sensor properties, such as response speed and proximity to the product transformation source. The following sections describe a number of optical-sensing techniques for process control and analyze the suitability of optical techniques to such requirements.

7.1.1 *Elements of Automatic Control*

Figure 7-1 shows an example of an industrial process to be controlled, a heat exchanger. The purpose of this process is to provide a given hot water output at a specified temperature by heating the cold water injected into a reservoir by a series of steam coils. The output of this system is the temperature of the output water, while the input is the rate of steam flow. The output temperature is continuously monitored by a thermometer to make sure it is within the required range. If this is not the case, the input parameter, steam flow, may be regulated through the steam valve in order to vary the injected amount of heat and thus approach the target temperature. If the steam valve is manually or automatically set at a conveniently predetermined level intended to assure the proper output temperature, we operate an open loop control. A closed loop, or feedback, control is one where the input variable setting depends on the value of the output parameter, here the sensed temperature.

An example of a closed loop control application for the system shown in Figure 7-1 is represented schematically in Figure 7-2. The detected temperature signal is transmitted to a bourdon-tube actuator which, in conjunction with a set-point knob, determines the vertical position of the plug in the pilot

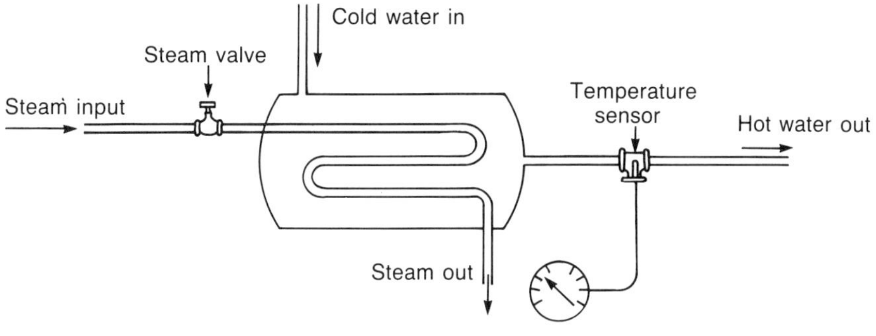

Figure 7-1 Typical heat exchange process

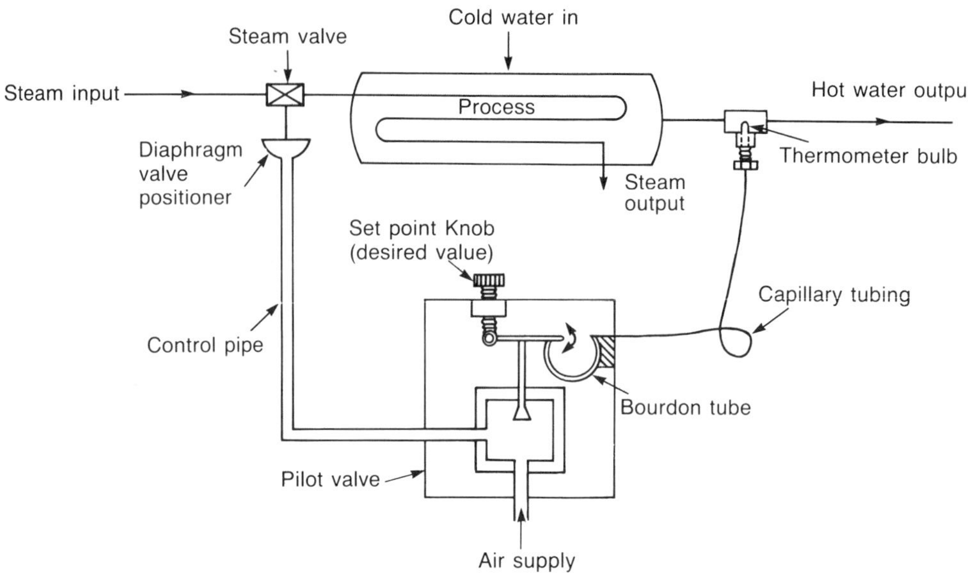

Figure 7-2 Example of a pneumatic feedback-control mechanism for the automatic control of the heat exchanger shown in Figure 7-1.

valve. This in turn determines the internal pressure in the control pipe and thus the setting of the steam valve whose vane position depends on the pressure applied to the diaphragm valve positioner. A sudden increase in the sensed temperature is thus automatically translated into a further occlusion of the pilot valve aperture and hence the steam valve flow. The corresponding decrease of the heat input reestablishes the normal hot water temperature level. The system is thereby capable of automatically reacting to environmental disturbances, such as a temperature change in the supplied cold water or variation of the steam pressure, to maintain the temperature value imposed by the set value of the knob.

Step Response of a Control Loop

An important property characterizing a control system is the output response to a step-function variation of the input variable. We shall assume here that the considered system, like the one shown in Figure 7-1, is self-regulating. In a self-regulating process, the output variable stabilizes at a new equilibrium level if the system is left untouched after a perturbation. In the system of Figure 7-1, for example, the water temperature reaches a higher level following sudden increase in steam input, corresponding to a new equilibrium between the energy input and the energy removed by the flowing water.

In a non-self-regulating system, such a spontaneous tendency to a new equilibrium does not exist, and the system is harder to control. An example is a tank into which a liquid is injected and from which the liquid is extracted by two constant flow pumps. Unless the inflow exactly equals the outflow, the tank will either empty or outflow. There is no tendency for this process to self-adjust to a new balanced condition when a step-function variation is applied to the input.

In a self-regulating system, the output-variable response to an input step is determined by the time lags affecting the process which may be of a capacitive, resistive or dead-time nature. The capacity of a process is its ability to store energy. In Figure 7-1, for example, the walls of the steam coils and the water in the tank can store thermal energy. The parts of a process that resist transfer of energy (or material) to or from the capacity constitute its resistance. In Figure 7-1 the walls of the steam coils and the insulating effect of the layers of steam and water on either side of them resist the transfer of energy between the steam in the coils and the water on the outside of the coils. As in an electrical RC circuit, the combined effect of a resistance and a capacitance produces exponential decay in the system toward the new equilibrium position after a perturbation.

Curve a in Figure 7-3 shows the typical response of a self-regulating system to a step-impulse of the input variable, such as a sudden drop of the steam input at time zero, if only single-capacitive and resistive time lags are present.

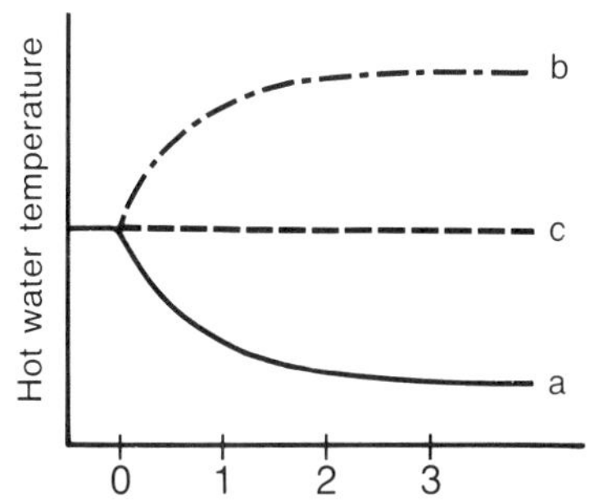

Figure 7-3 (a) Response of a system with resistive and capacitive time lags to a step variation of the input variable at time zero; (b) response to an exact correction; (c) feedback-compensated output.

The output starts changing with a nearly linear slope as soon as the perturbation is applied, so that the perturbation level can be immediately evaluated and a correction step impulse equal and opposite to the perturbation can in principle be applied by the control loop, producing a reaction of the kind shown in curve b. The feedback controlled output of such an ideal system will thus follow a perfectly compensated constant level as shown by curve c in Figure 7-3.

Most systems are affected by a certain amount of dead time. This is the time required to carry a change from one point to another in the process, such as the flow time of the water from the steam coils to the temperature detector in the system of Figure 7-1. A similar dead time may occur in the feedback control system. Figure 7-4 shows how dead time affects the operation of an automatic controller. Curve a is the response of the system to a step variation of the input variable at time zero. The reaction of the control loop cannot take place before the output perturbation takes place, while the applied correction, curve b, is affected by the same time lag as for curve a increased by the dead time of the feedback mechanism. The result, curve c, is an inevitable unbalance of the output variable for a time period of the order of the dead time.

The amount of dead time is thus an important problem in industrial controls. Not only does it produce output unbalance situations, but it is also a major cause of system destabilization, as we shall see later. To minimize such dead time, time delays in the feedback loop can first be reduced as much as possible. For example, in the system shown in Figure 7-2, the pneumatic temperature sensor and correction mechanism could be replaced by a

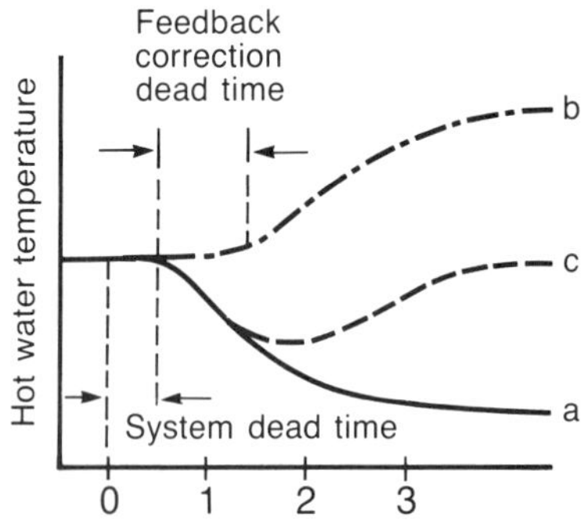

Figure 7-4 (a) Response of a system with dead-time lags to a step perturbation at time zero; (b) response to a correction applied as soon as the output perturbation is sensed; (c) feedback-compensated output.

thermoelectric sensor whose amplified output is directly applied to the vane of the steam valve for an immediate reaction time. Another improvement would be to place the temperature sensor as close as possible to the process transformation source, in this case the steam tubes. In many instances, such as in the continuous casting of steel products, such a proximity is difficult to achieve because of the extremely hostile environmental conditions in the vicinity of the process transformation source. This is where noncontact optical sensors are particularly useful for immediate and stable process control.

Control Actions

The control action on the input variable to correct for a deviation of the output may assume different forms. The simplest approach is the two-position control. One such application is shown in Figure 7-5 in relation to the system in Figure 7-1. In the on-off approach, the valve is either open or shut depending on whether the sensed temperature level is below or above the set point. This inevitably gives rise to a cycling of the output temperature, as Figure 7-5b shows.

To reduce the excursions of the output temperature, one can adopt a multiposition control where the valve opening may assume different discrete positions depending on the amount of detected temperature unbalance. An extension of such an approach is the continuous variation of the input variable as a function of the system output, the most widely used forms of which are the following:

(a) *Proportional control action.* In this scheme, the valve displacement with respect to the initial value (which under normal conditions would assume the proper temperature output) is proportional to the temperature deviation. Figure 7-6a shows a typical response of a proportional action control to a step variation of the load, such as a variation of the cold water temperature or flow rate or of the steam pressure in the system shown in Figure 7-1. As a consequence of the perturbation, the output temperature starts to change. The valve position changes accordingly, by an amount proportional to the temperature deviation, to counteract the temperature variation. The temperature excursion is reversed by the feedback control action and eventually stabilizes at a position slightly different from the initial set-point position which corresponded to the previous load condition. The presence of a residual offset is typical of proportional control.

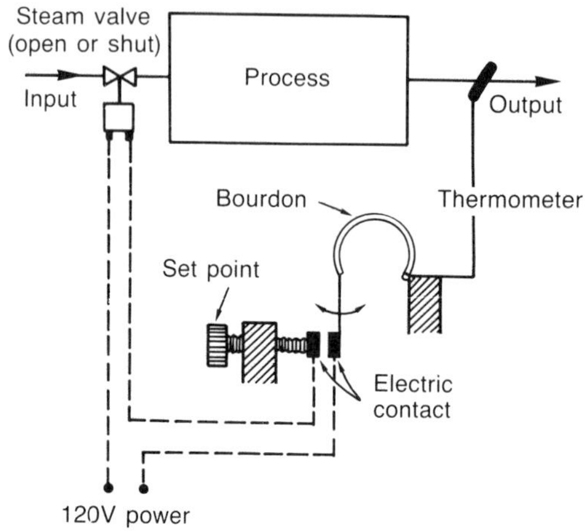

(a)

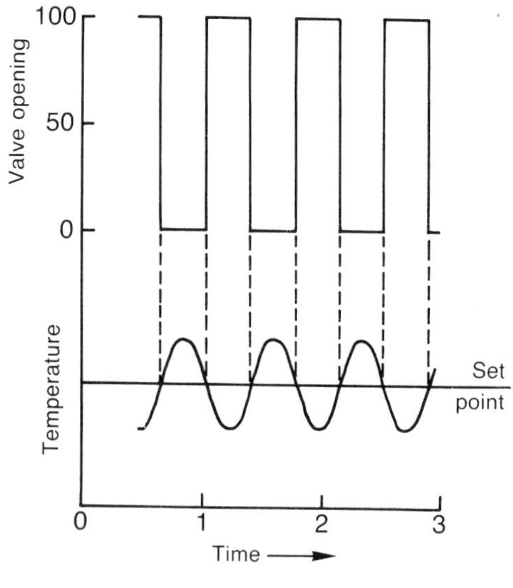

(b)

Figure 7-5 (a) Example of a two-position electrically operated control system; (b) time fluctuation of the input and output variables, showing the typical cycling behavior.

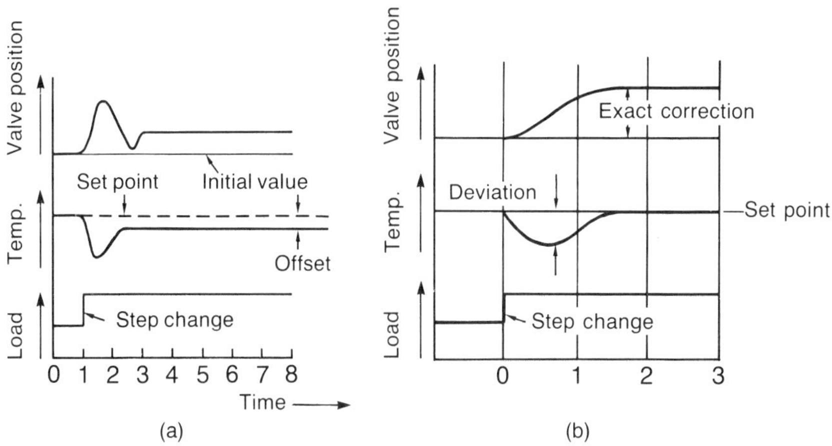

Figure 7-6 (a) Response of a process such as the one shown in Figure 7-1 to a step impulse perturbation under proportional control action: The temperatue tends to stabilize at an offset position as compared to the set point. (b) An example of integral control action: The temperature fluctuation is wider because of the slow corrective action, but the offset is eliminated.

(b) *Integral control action.* In this case, the valve position is linearly related to the integral of the temperature deviation, or in other terms, the rate of valve travel is proportional to the deviation. Figure 7-6b shows the response of such a system to a step perturbation. The reaction is slower than in the case of proportional action, but the correction to the valve position continues as long as the temperature differs from the set point, so that the offset is eliminated.

(c) *Derivative control action.* In this case the valve displacement is proportional to the temperature deviation. The reaction to a temperature deviation is very fast so that temperature excursion amplitudes are minimized. However, such a scheme is sensitive to high frequency, low amplitude noise, unless the input or output signals are low-pass filtered. As soon as the temperature signal has reached a constant value, no further corrective action is taken, even if the temperature level is far away from the set point. This approach is thus quite susceptible to offset errors. Derivative control is usually introduced in conjunction with proportional as well as integral control to speed up the reaction time, while the integral component eliminates offset errors. The step-response of a system of this kind is given in Figure 7-7, showing the different control components.

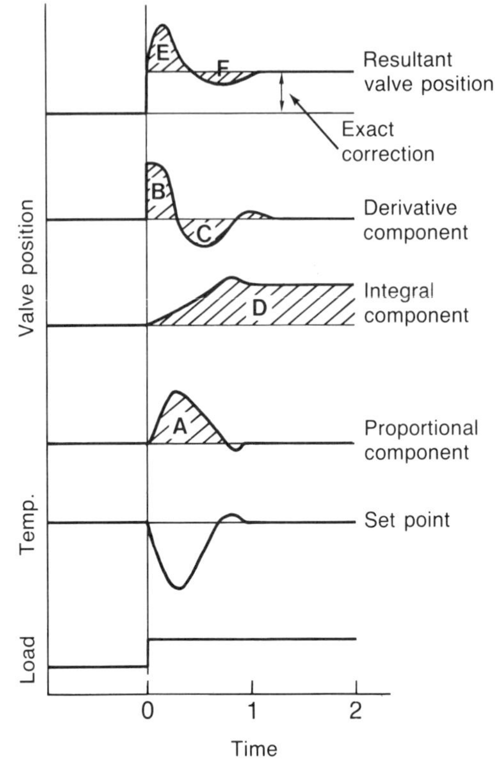

Figure 7-7 Step response of a proportional-integral-derivative (PID) control system showing the three components of the feedback reaction. The derivative component speeds up the reaction, thus reducing the temperature excursion amplitude. The integral component eliminates long term offsets.

7.1.2 *Stability of Control Loops*

A feedback loop is said to be stable when the output tends to a steady value without diverging after a disturbance. This is the case if the correction is always opposed to changes of the output variable. This means that, if the output of a system fluctuates sinusoidally, the correction applied by the control loop to the input should be phase-shifted by 180° with respect to the output. This is indeed the case for a proportional action control, as Figure 7-8 shows. The same figure also shows that the integral component is, by definition, phase-shifted by 270° with respect to the output variable, while the derivative component is phase-shifted by only 90°. The derivative and integral components tend to suppress each other at a given frequency.

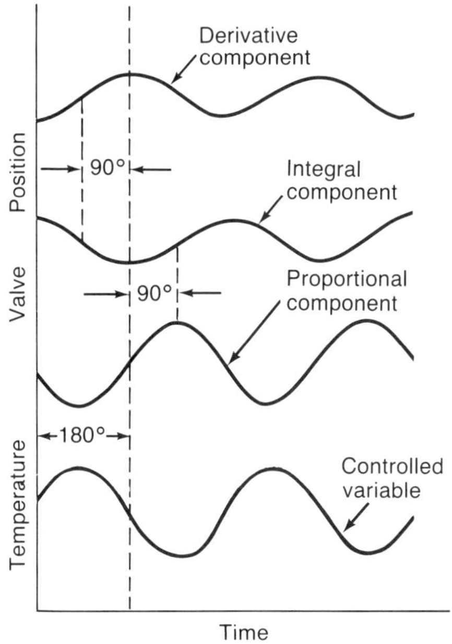

Figure 7-8 Response of the proportional, integral and derivative components of a feedback control to a sinusoidal fluctuation of the controlled variable.

However, the integral component normally dominates at low frequencies, while the derivative component prevails at high frequencies.

The presence of time lags in a feedback loop may produce instability if the delayed correction ends up being in phase with system fluctuation. This is illustrated in Figure 7-9, showing the response of a process and of the open-loop control system to a sinusoidal input of given frequency. If the system is linear, the process output is also sinusoidal with a phase shift θp depending on the time lags introduced by the process. The control system introduces an additional phase shift of 180° (assuming proportional control action), plus the phase shift θ_c due to the dead time of the control system. The total phase shift between the input variable and the output of the control system may thus very well reach 360° and add in phase with the input perturbation. If such a loop is closed and if the gain around the loop is higher than 1, the amplitude of the oscillation will diverge and the system will become unstable.

As time lags always exist in practical systems, a phase shift of 360° will always be reached for sufficiently high frequencies (the frequency corresponding to 360° phase shift is called the "crossover" frequency). For stable

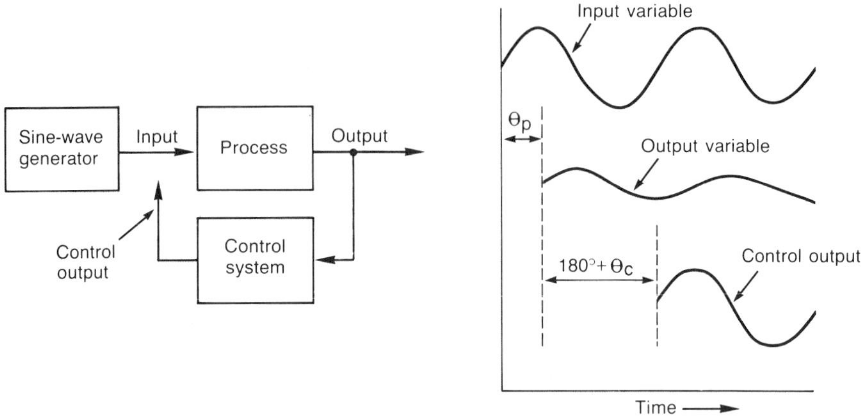

Figure 7-9 Response of a typical open-loop system to a sinusoidal input. The built-in 180° phase shift of the control system, when added to the phase shift θ_p of the process and θ_c of the control system, gives a 360° phase shift at the considered frequency. This system will thus be unstable if the total gain around the loop is larger than 1.

operation, the gain must be smaller than 1 for such frequencies. Curves of gain and phase shift versus frequency are thus the primary information required to characterize an industrial control system. Such curves can be obtained experimentally by applying different input frequencies in a configuration of the kind shown in Figure 7-9 or else by observing the system response to an applied transient perturbation.

Programmable controllers (PC), which are now widely used in industrial control systems, can be used to program the gain versus frequency response of a control system. The proportional, integral and derivative components of a PID (proprotional-integral-derivative) control can be tailored for maximum reaction speed and stable operation. It is important to mention that, as shown in Figure 7-8, the integral component introduces a longer phase shift in the response of the control system, thus worsening the stability performance. Its gain is, however, stronger at lower frequencies where a given time lag corresponds to a smaller phase shift. The inverse is true for the derivative control component.

An exact analysis of control systems can be performed once the transfer function of the process and the control loop are known. The control algorithm can thus be optimized for minimum output deviation and stable operation (see, e.g., Ogata, 1970 or Considine, 1985, p. 17–58 to 17–157). However, such a quantitative analysis requires a precise analytical model of the system, whose experimental evaluation may be exceedingly long, while new evaluations may be required after changes in the environmental conditions or in the operation parameters. Moreover, the gain is often nonlinear because of factors such as turbulence effects in the steam flow as the steam valve is opened or viscosity variations as the liquid temperature changes. Other problems which are often found in the practical implementation of industrial controls are related to cross-talk between control loops applied in cascade along a process: If two loops have similar natural frequencies of cycling, output changes from one loop may happen to resonate with the natural frequency of associated control loops producing what is called "harmonic cycling". Process control is therefore still much of an art, the process engineer proceeding in most cases by trial and error to finely tune the feedback system during process startup or by applying some empirical rules (see Considine, 1985, p. 17–41 to 17–47). Nevertheless, the increased availability of computer supervision in industrial processes makes quantitative process control analysis easier to perform and the application of sophisticated methods such as self-learning or adaptive control more realistic.

It can be concluded here that the performance and reliability of the sensing instrumentation is crucial to an effective process control procedure. Time lags in both system and control loop are the main problem to overcome for a prompt reaction and a stable feedback-loop operation. The information provided by the sensor should be as direct as possible and as little affected as possible by background noise and cross-talk. The next sections of this chapter review optical sensors for process control and discuss their suitability to process requirements.

7.2 On-line Optical Sensing of Temperature

A review of nonoptical temperature sensors was given in Section 2.5.3. This section describes infrared temperature sensors, which are based on the measurement of the infrared radiation emitted by hot materials. The main interest of infrared temperature sensors is their noncontact nature, which makes them attractive in an industrial environment for a number of reasons:

(a) Very high temperatures can be measured, such as the surface temperature of incandescent metals, by simply pointing the sensor toward the surface to be inspected.
(b) The temperature of rapidly moving objects can be monitored, such as rotating turbine blades, transmission belts or tire surfaces.

(c) The response speed can be very high, below 1 μs, because no thermal flow to the sensor is required as in the case of thermocouple or thermoresistive devices.
(d) The temperature of very small parts, such as microcirucit elements, can be readily measured because of the good spatial resolution of optically focused sensors and the absence of any contact which eliminates temperature perturbations introduced by contact sensors such as thermocouples.
(e) Remote or difficult-to-access parts can be temperature monitored provided that one can point to such parts through clear air or through infrared-transparent materials such as a sapphire window in the wall of a furnace.
(f) Finally, an infrared temperature reading can be performed very quickly because no accurate thermal bond must be made with the inspected surface, as is the case for thermal devices. Surface scanning is performed at rates of the order of 10^6 points/sec in modern infrared cameras.

The main problem with infrared temperature sensors is their low accuracy, mainly because of the usually unknown surface emissivity and sensitivity to reflections from hot bodies in the background.

7.2.1 *Infrared Thermal Sensors*

Infrared temperature sensors, sometimes called pyrometers, are either thermal or photoelectric detectors (see Section 3.3) whose sensitive area is imaged on the surface under inspection by infrared-transmitting optics. If the inspected diffuse-finish surface is wider than the image of the sensitive area, the radiated signal amplitude is independent of the viewing angle with respect to the surface. The surface of a blackbody at Kelvin temperature T emits infrared radiation following Planck's law (Wolfe and Zissis, 1978, Chapter 1)

$$W(\lambda, T) = \frac{C_1}{\lambda^5} \frac{1}{\exp(C_2/\lambda T) - 1}, \qquad (7\text{-}1)$$

where λ is the wavelength of the emitted radiation, $C_1 = 3.74.10^{-12}$ watts cm^2 and $C_2 = 1.44\ cm°K$. $W(\lambda, T)$ is the infrared emission power within a $2\pi^2$ steradian angular aperture per unit surface area of the radiating surface and per unit spectral bandwidth. A plot of $W(\lambda, T)$ expressed in watts cm^{-2} μm^{-1} is shown in Figure 7-10. It should be noted that, because of the log-log scale, most of the energy at a given temperature is emitted at wavelengths longer than the peak emission wavelength.

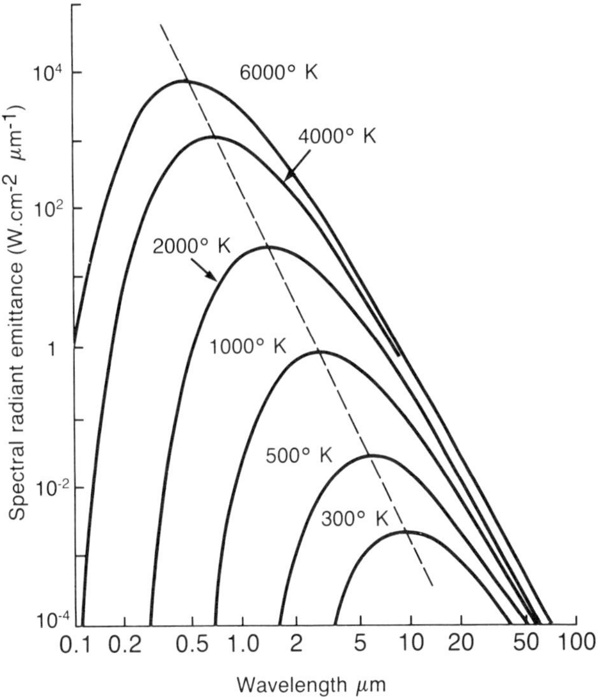

Figure 7-10 Blackbody spectral radiant emittance.

The total emission integrated over the full spectrum is

$$E = \sigma T^4 \tag{7-2}$$

where $\sigma = 5.67 \cdot 10^{-12}$ watts cm^{-2}K^{-4} is the Stefan–Baltzmann constant. The infrared emission is thus a rapidly increasing, strongly nonlinear function of the absolute temperature. By differentiating equation (7-1) with respect to λ, one obtains Wien's law for the peak wavelength λ_m of maximum infrared emission

$$\lambda_m = \frac{C}{T}, \tag{7-3}$$

where $C = 2898$ μm K. The peak emission shifts to smaller wavelengths with increasing object temperature, going from 10 μm at ambient temperature to 1 μm at 3000 K.

The above expressions are valid for a blackbody whose emissivity ε is equal to 1. Most materials have an emissivity smaller than 1, while polished metallic surfaces may have emissivities smaller than 0.05 at ambient temperature in the infrared. The expressions (7-1) and (7-2) must be multiplied by ε for nonblackbodies, while reflections from a warm background must be taken into consideration. For opaque materials, the surface reflectivity is equal to $1 - \varepsilon$, so we can write for the total power irradiated by a nonblackbody

$$W = \varepsilon\sigma T^4 + (1 - \varepsilon)\sigma T_A^4 \tag{7-4}$$

where T_A is the temperature of the ambient background, assumed uniform. Part of the background radiation comes from the detector package itself, unless it is cryogenically cooled, and from the room-temperature optics if it has a significant absorptivity (and thus emissivity) in a spectral band contained within the sensitivity range of the detector. Best results are obtained when pointing toward surfaces at temperatures T much higher than the ambient, in which case the second term of equation (1-4) can be neglected.

The emissivity value varies widely for different materials. Table 7-2 gives some orders of magnitude (Lawson and Sabey, 1970; a review of methods for emissivity measurement can be found in the handbook: *Thermophysical Properties of Matter* published by Y. S. Touloukian and coworkers of TPRC, Purdue University, Vol. 7, p. 29a–47a). The precise value of ε depends on the wavelength, temperature, emission angle (Sparrow and Cess, 1978, Section 2.2) and polarization for polished surfaces (Berry, 1980). The surface state is often difficult to quantify exactly in industrial conditions: As Table 7-1 shows, the emissivity of smooth metallic surfaces may vary by as much as an order of magnitude deending on the degree of surface oxidation, while the presence of high-emissivity surface layers such as oil is often unavoidable. Emissivity variations are thus the major problem in the quantitative evaluation of surface temperature by infrared detection.

However, variation of the emissivity by a certain factor does not imply an error in the estimated surface temperature by the same factor. The

Table 7-2 Emissivities of Typical Materials

Ag polished	0.03	Sand	0.75
Al polished	0.05	Cement	0.95
Cu polished	0.15	Brick	0.94
Fe polished	0.05	Wood	0.8–.09
Al anodized	0.3–0.5	Paper	~0.9
Cu oxidized	0.6		
Fe oxidized	0.6–0.9		

sensitivity of the temperature measurement to emissivity variations depends on the spectral region of detection. If the detector spectral sensitivity is limited to a relatively narrow bandwidth centered at λ_d, and λ_m is the wavelength of maximum thermal emittance, the detected radiation signal will be (Dreyfus, 1963; Lawson and Sabey, 1970; Kanstad and Nordal, 1986)

$$V(T) = \int_\Omega \int_\Gamma \int_\lambda \varepsilon(\lambda) W(\lambda, T) S(\lambda)\, d\lambda\, dA \frac{d\Omega}{2\pi^2} \approx KT^n \tag{7-5}$$

$$\text{with } n = \frac{5\lambda_m}{\lambda_d} \qquad \text{for } \lambda_d/\lambda_m \leqslant 2.5,$$

where Ω is the solid-angle aperture of the instrument, A is the object area imaged by the sensor optics on the sensitive surface of the detector, $S(\lambda)$ is the spectral sensitivity of the detector and K is a constant. As an example, suppose that the temperature of a body at 600°K in a cooler background has to be evaluated with an infrared detector centered at either (a) $\lambda_d = 5\,\mu$m or (b) $\lambda_d = 2\,\mu$m. Figure 7-10 shows that the signal level is higher by a factor of nearly 20 in case (a) compared to (b) (this is partially compensated by a higher detectivity at smaller wavelengths, see Figure 3-23). The sensitivity to emissivity errors is, however, much smaller in case (b): Equation (7-3) gives $\lambda_m = 5\,\mu$m, so that the n value is equal to 5 and to 12.5, respectively, for case (a) and (b), from equation (7-5). It follows that an error of 10% in the evaluation of the emissivity leads to an error in the estimated temperature of 2% in case (a) and of 0.8% in case (b). Other advantages of working at lower wavelengths are a reduced sensitivity to ambient-temperature background radiation and, quite importantly, the availability of rugged and sensitive infrared detectors which can operate without the need of cryogenic cooling.

The strong dependence of the n value on the relative values of λ_d and λ_m in equation (7-5) has made *double-wavelength infared sensors* attractive for the evaluation of temperature independent of emissivity (Barber, 1978). If two detectors centered at the adjacent wavelengths λ_1 and λ_2 are pointed toward the same surface through the same optics, the expression $V(T) \approx KT^n$ can be written for both signals, with different n values but with the same constant K if the emissivity can be assumed to be the same at λ_1 and λ_2. The ratio of these two expressions, combined with equation (7-3), gives the values of λ_m and T. This assumes, however, that the emissivity is the same in the two detection bands, which are chosen for this reason as close as possible to each other. Such an assumption is usually valid for polished metallic surfaces but not for oxidized or nonmetallic surfaces whose spectral emissivity curve is generally complex. Moreover, the rate of change of signal with temperature as well as the spectral bandwidth is smaller for ratio sensors than for single-band sensors (Peacock, 1979).

Finally, it should be mentioned that the above arguments assume perfect transmission of the infrared radiation through air. The infrared transmission through 1 m of air at 20°C and 55% relative humidity is shown in Figure 7-11. (Wolfe and Zissis, 1978, Chapter 5; Barber, 1978). As we can see, absorption by water and CO_2 molecules cannot be neglected even at such a close distance for certain spectral bands. However, the transmission is good within the "windows" of 2 to 2.5 μm, 3 to 5.5 μm and 8 to 14 μm. Single-, as well as double-wavelength detectors should be strictly limited to such windows to avoid dependence of the ratio signal over air-path length and relative humidity.

7.2.2 *On-line Industrial Applications*

Temperature monitoring is required in a wide range of industrial processes. Infrared pyrometers have long been recognized as valuable tools for on-line temperature monitoring, particularly in hostile environments. Examples

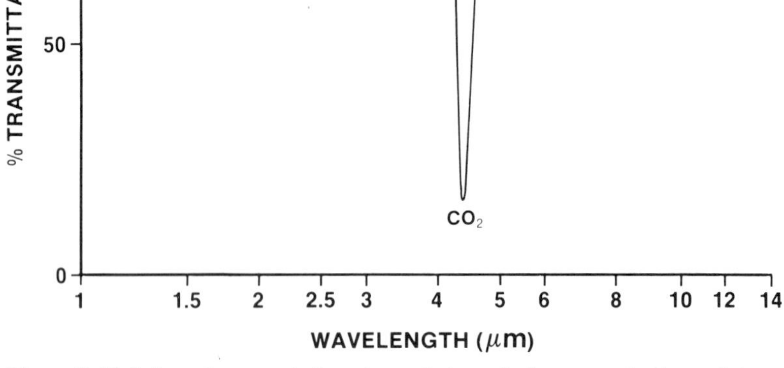

Figure 7-11 Infrared transmission through 1 m of air at standard conditions.

include the inspection of flames in combustion systems for intensity monitoring and optimization of burner stoichiometry (Godridge, 1983), the control of refractory lining wear in blast furnaces, stoves and ladles (Rogers, 1978), the temperature control of brazed joint repairs in heat exchangers (Haberlin and Campbell, 1983), thermographic quality control of hot-rolled steel sheets (Holmsten, 1986) as well as on-line temperature control of hot-formed products such as thermoplastic polymers (Bedrossian, 1987) and glass. In the last case, infrared wavelengths of different optical penetration within the material have been chosen to obtain the subsurface thermal gradient profile (Barber, 1980).

A typical application of infrared pyrometry is in the steelmaking process (Fulton, 1984). Figure 7-12 shows a continuous casting process with infrared temperature-monitoring units 1, 2 and 3 at typical locations along the hot-slab caster (Intrieri, 1977; Mairy and Ramelot, 1981). The infrared sensor is pointed directly, or through a fiber optic extension, to the slab surface which is typically at 900 to 1100°C. A descaling mechanical scraper is normally inserted above the inspected area to avoid signal fluctuations produced by emissivity variations and thermal-barrier effects in the presence of surface

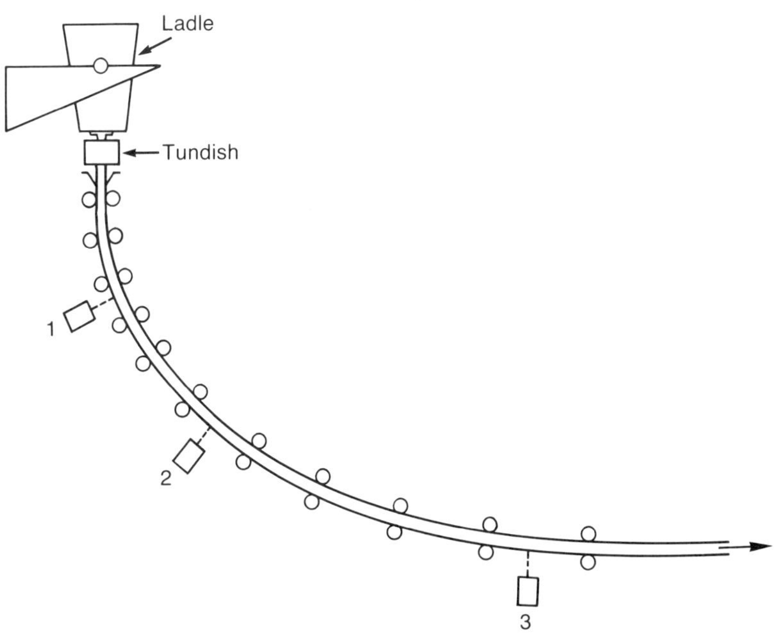

Figure 7-12 Schematic diagram of a continuous casting process with typical infrared temperature-sensing locations, 1, 2 and 3.

scale. Good temperature control can thus be achieved by feedback adjustment of the cooling settings to optimize the thermal profile of the strand as a function of casting conditions. Fiber optic sensors have also been installed between ladle and tundish to detect slag flow-out by inspecting the molten steel column (Tamura et al., 1985).

This application is an example of the possibilities offered by optical temperature sensors as opposed to contact devices. Because of their non-contact nature, optical sensors can be installed much higher on the production line to detect product conditions in otherwise unaccessible positions and to reduce time lags and feedback instabilities. If a temperature measurement on the hot slab was made further down the line by an intermittent-touching or rotating-wheel contact thermometer, both the dead-time due to the distant sensing position and the capacitive and resistive time delays required for thermal flow through the contact material would result in much longer time lags (see Section 7.1). Similar considerations are valid for the measurement of product temperature in furnaces or for the inspection of hot-rolled steel plates by single-spot pyrometers (Moore, 1972), photodiode arrays (Yamada et al., 1982) or fiber optic bundles (Yamada et al., 1984). The infrared thermometer can directly monitor the product temperature, rather than an indirect variable such as the temperature of the surrounding air or of a roller in contact with the material. Sensing errors and time delays are thus minimized.

Sensing Configurations

The installation of temperature sensors in high temperature or otherwise hostile industrial environments requires special skills to ensure reliable measurements are obtained with minimum maintenance during day-to-day operation. Cooling and air-purging supplies are often required to prevent instrument overheating and dirt accumulation over the viewing window. One of the most difficult problems to overcome when measuring target temperature in hot surroundings, such as when pointing the pyrometer to a target through an orifice within a furnace, is the unknown emissivity and reflectivity of the inspected surface. If the targeted surface is highly reflective, as in the case of a metal, and if the furnace wall temperature is of the order of or higher than the target temperature, the whole furnace can be considered as a blackbody cavity. In the case, the measured temperature is the average furnace temperature rather than the temperature of the target product.

A number of techniques can be used to reduce this kind of error. Figure 7-13 shows a configuration avoiding reflection errors when smooth-finish

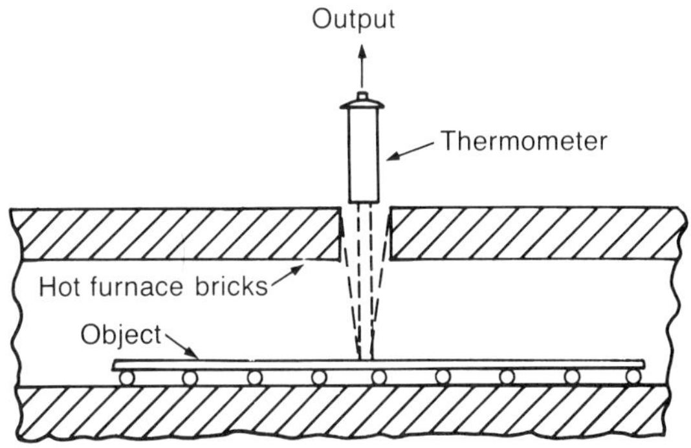

Figure 7-13 Technique for avoiding reflectivity errors when measuring the temperature of a specular surface in a hot background.

surfaces such as glass sheets in a tempering furnace are inspected for temperature. Sighting is performed at a normal incidence to the surface so that the reflected radiation from the hot surroundings can be neglected. Figure 7-14 shows another configuration which is often used when inspecting diffuse-surface objects in a furnace. A water-cooled (for sensor protection and also to avoid reflected radiation errors), blackened (to avoid multiple-reflection errors), wide shield is inserted into the furnace. For metal sheet materials, a ratio between shield diameter D and height H of 2 to 4 is sufficient to avoid most of the reflection-related error.

The methods shown in Figures 7-13 and 7-14 require particular settings which are not always conveniently installed in industrial furnaces. Figures

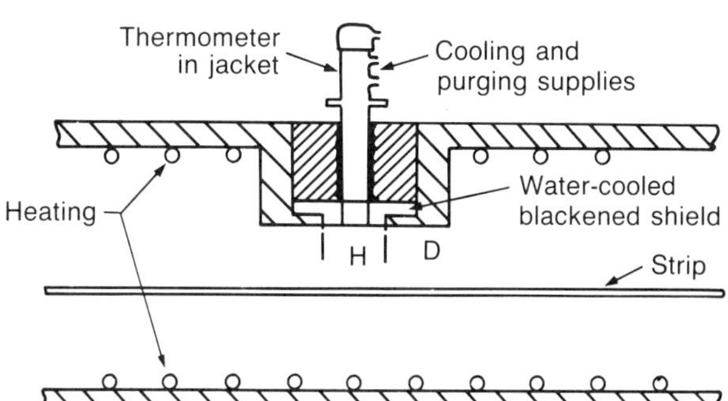

Figure 7-14 The shielded-target method to avoid reflectivity errors from diffuse surfaces.

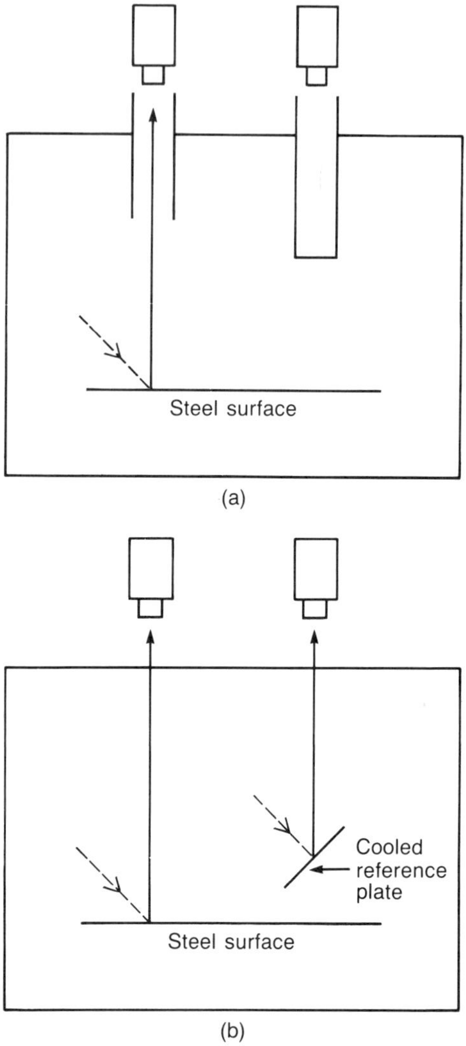

Figure 7-15 Dual-pyrometer reflectivity compensation methods for sensing metal temperature within a uniform high-temperature furnace: (a) closed-tube reference method; (b) cooled-reference method.

7-15 and 7-16 show some alternative configurations for sensing the temperature of metallic targets in high temperature furnaces with minimum perturbation of the furnace geometry (Roney, 1982). Figure 7-15 illustrates the use of a pyrometer pointed to a closed tube in thermal equilibrium with the furnace interior to sense the average furnace temperature. The closed-tube pyrometer may be replaced by a thermocouple if environmental conditions

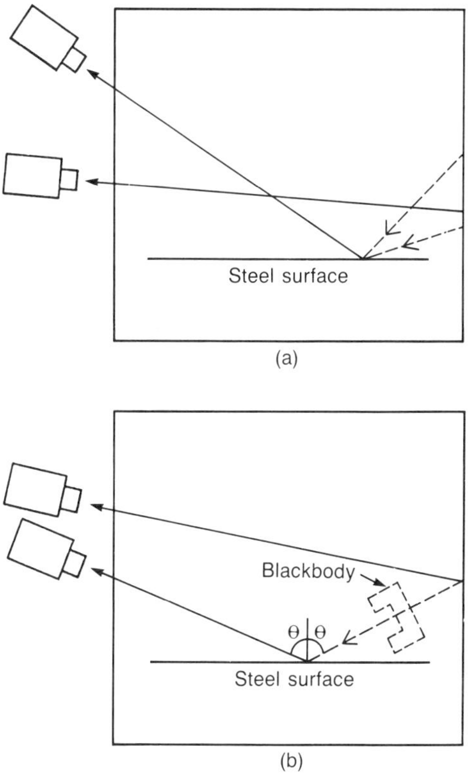

Figure 7-16 Reflectivity compensation in a nonuniform furnace for (a) a diffuse-surface sheet and (b) a specular surface. A blackbody cavity may be inserted in the path of the reflected beam to avoid directional-reflectance errors.

allow it. If the furnace interior is relatively uniform (no open flames or heating tubes whose radiation may be specularly reflected by the target), the closed-tube pyrometer gives the ambient temperature T_A of the furnace, which can be considered as a blackbody cavity. Such a value can be used to obtain the sheet temperature T if its emissivity ε is known by using equation (7-4). The approach shown in Figure 7-15b is similar, but the closed-end tube is now replaced by a cooled reference plate of emissivity similar to that of the steel sheet. The reference pyrometer will thus receive a signal comprising only the reflected component from the furnace interior. The higher signal obtained from the steel surface is related to the steel temperature, which can be calculated if its emissivity is known.

If the furnace is not at uniform temperature, configurations of the kind shown in Figure 7-16 can be resorted to. Figure 7-16a shows a dual-pyrometer configuration for reflectivity compensation from a diffuse-finish steel surface. The reference pyrometer senses the furnace wall temperature, which can be used to correct the signal of the steel-sighting pyrometer if the sighted points on the wall and on the steel surface are sufficiently close (otherwise the source of the steel-scattered signal may extend over a larger area than the sighted spot on the wall). If the steel surface is specularly smooth, the configuration of Figure 7-16b can be used where both the source of the reflected radiation and the spot inspected by the reference detector on the furnace wall coincide. Both of the configurations shown in Figure 7-16 are still affected by reflection errors if the area sighted by the reference pyrometer on the furnace wall is not perfectly black. Indeed, in a nonuniform furnace the amount of radiation reflected by such an area depends on the sighting angle. To avoid such an error, a blackbody cavity may be inserted on the reflected beam path as shown in Figure 7-16b.

All of the previous configurations show reflection compensation techniques for in-furnace temperature measurement. However, quantitative object-temperature evaluation still requires a knowledge of its surface emissivity. Such a parameter is often difficult to quantify, as mentioned earlier. In some cases, such as annealing furnaces where a hydrogen atmosphere is maintained to reduce the oxide coating, the steel-sheet emissivity actually decreases while the sheet advances along the furnace. Emissivity-independent methods, more powerful than the simple high n-value or the double-wavelength detection techniques described earlier, have thus been developed for such measurements. One such method uses a configuration of the kind shown in Figure 7-16b, without reference pyrometer and with a variable-temperature blackbody cavity along the reflected-light path (Kelsall, 1963; Iuchi and Kusaka, 1982). By recording the pyrometer output at two known temperatures of the blackbody, both the temperature and the angular emissivity of the steel surface can be determined. Such a method can, however, only be applied to specularly-reflecting surfaces or to surfaces whose angular scattering pattern is known and repetitive.

For surfaces of unknown topography, a reflecting cavity approach has been developed (Land, 1959; Iuchi and Kusaka, 1982). A schematic diagram of such a device is shown in Figure 7-17. A cylindrical cavity with a highly reflecting internal gold plating is held above the sheet surface to be inspected. A gold-plated rotating chopper is placed close to the open top of the cavity. A circular slit is present on the blades through which the pyrometer is pointed to the sheet surface. When the blade covers the cylinder top, the highly-reflecting cavity acts as a near-blackbody (Cielo, et al., 1985d; Cielo, et al.,

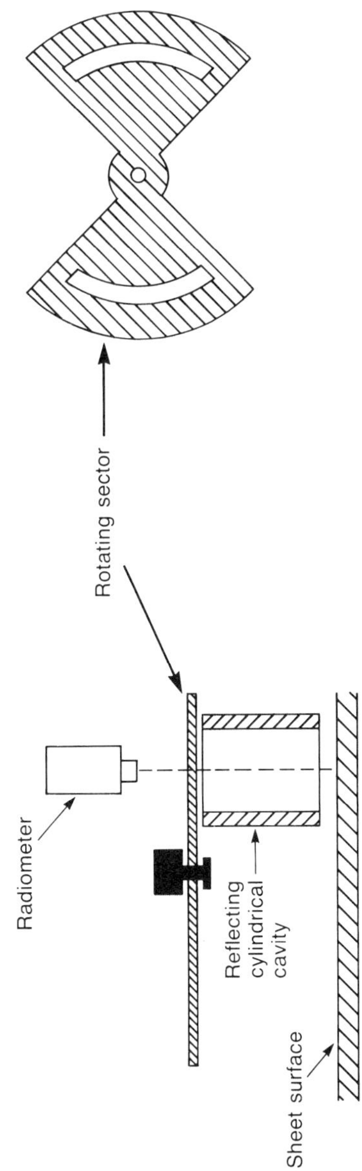

Figure 7-17 The reflecting-cavity method for the evaluation of both temperature and emissivity of a sheet surface.

1986a) so that the effective emissivity seen by the radiometer pointed to the slit is close to 1. When the upper opening of the cavity is not covered by the blade, no multiple reflection takes place and the detected signal is proportional to the real emissivity of the cavity. From these two measurements both the temperature and the emissivity of the sheet surface can be calculated. A similar device, using two hemispheric cavities — one reflecting and one blackened — and having no rotating parts, is under development at CRM, Liège, Belgium.

A similar approach, but not requiring the insertion of any object in the furnace, has been recently reported (Harada et al., 1985). The basic principle is shown in Figure 7-18. The emissivity-independent temperature reading is again based on the multiple reflections of the infrared beam in the wedge between the sheet and the outer surface of the roller. After contact, the two surfaces have a very similar temperature, which is measured effectively by pointing the pyrometer to such a near-blackbody cavity. As rollers are present in many locations along the production line, no extraneous mechanical system must be inserted in this case.

Finally, an example will be given of the capabilities of infrared temperature sensors for the inspection of fast-moving objects. One of the most striking

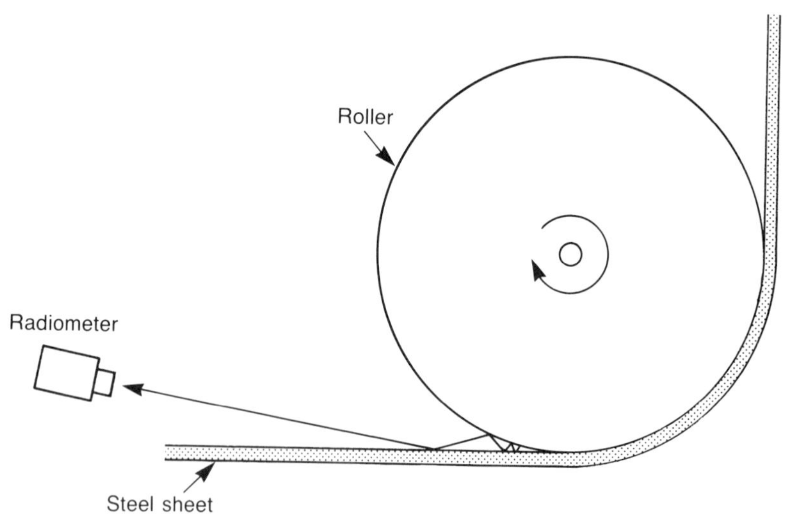

Figure 7-18 The wedge-pointing method for evaluating the temperature of a sheet metal independently of the emissivity.

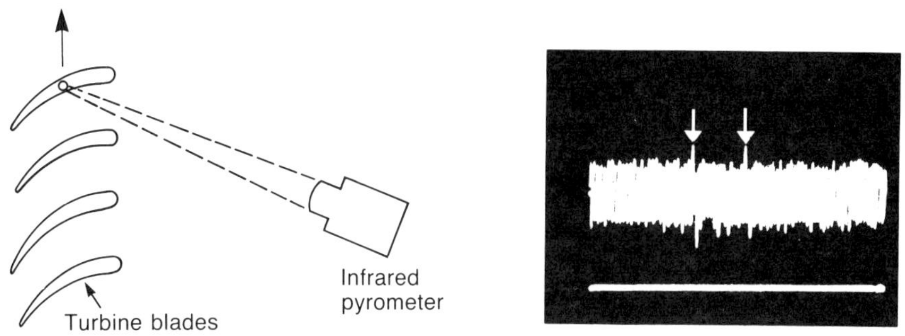

(a) (b)

Figure 7-19 Temperature inspection of rotating turbine blades: (a) sensing geometry; (b) sample of the detected signal, showing 35°C overheating signals from two faulty blades.

demonstrations of such capabilities is related to the temperature inspection of rotating turbine blades (Beynon, 1982; Smith, 1983; Davinson, 1985; Bedrossian, 1987). Similar applications to combustion flame monitoring will be described in Section 8.3.3. As Figure 7-19a shows, the pyrometer can be pointed to the tips of the turbine blades operating at temperatures of the order of 800°C within the rotor. Focused fiber optic probes are normally inserted in the turbine chamber to collect the infrared signal and transmit it to the outside detector. Typical performances are a 2 mm spot diameter on the blade tips and 1 μs response time. Both the average blade temperature and the presence of overheated blades due to faulty cooling represent important information for the prevention of catastrophic failures. Figure 7-19b shows a sample of the detected signal with two overheated blades clearly in evidence. Obviously, no such inspection requirements could be met with conventional nonoptical temperature sensors.

7.3 Optical Sensing of Flow and Speed

Sensing fluid flow is an important requirement of industrial process instrumentation. A review of nonoptical flow-sensing systems was given in Section 2.5.2. This section reviews optical systems for flow and speed sensing.

7.3.1 *Doppler Velocimetry Sensors*

Doppler velocimetry can be performed with electromagnetic (Skolnik, 1962) as well as with mechanical (Baumoel, 1980; Bard, 1983) waves. Similarly to the interferometric displacement-sensing techniques reviewed in Section 6.3.3., optical Doppler-velocimetry methods can be classified as either velocity-sensing or position-sensing techniques. In velocity-sensing techniques (Barker and Hollenbach, 1972; Cielo and Delisle, 1976a) the Doppler frequency shift of the beam reflected by the moving object is detected by a high-resolution filter such as an optical interferometer. Position-sensing velocimetry is performed, similarly to microwave velocimetry, by superposing the frequency-shifted beam with a coherent reference and recording the frequency of the resultant heterodyne beat signal. Because of its sensitivity to lower speed values, this second class of optical velocimetry has received a wider attention and will only be discussed in this section.

A large number of optical configurations have been suggested for optical Doppler velocimeters (Watrasiewicz and Rudd, 1976; Drain, 1980; Durst et al., 1981). Figure 7-20 shows a typical configuration for monitoring particle flow speed. A laser beam is split in two parts, which are focused in the middle of the particle stream to be monitored. In the beam-intersection region, stationary interference fringes of the kind shown in Figure 3-49 will form. As the particles cross such fringes, the forward-scattering signal collected by the detector will record a sinusoidally-varying signal whose frequency is proportional to the particle speed. The signal fluctuations can alternatively be interpreted as the beat signal between the two beams scattered by the particle. The beam coming from the top in Figure 7-20, whose vertical direction component is opposite to the particle movement, is scattered with a positive Doppler frequency shift, while the beam coming from the bottom suffers a frequency reduction after scattering. The superposition of the two scattered beams gives a heterodyne beat signal with frequency proportional to the Doppler shift and thus to the particle velocity.

Different versions of this basic configuration have been developed. A Bragg cell is often inserted along one of the projected beams to produce a signal at a convenient IF frequency which can more easily be demodulated by standard FM receivers (see Section 6.3.1). Another variation which is often implemented uses a double fringe pattern at perpendicular orientations to determine the two normal components of the flow speed. Figure 7-21 shows a commercial anemometer of this kind. Four beams are used in this case, two from the 488 nm and two from the 514 nm line emission of the Argon-ion laser source. Each of the emitted beams is split and focused by a system of

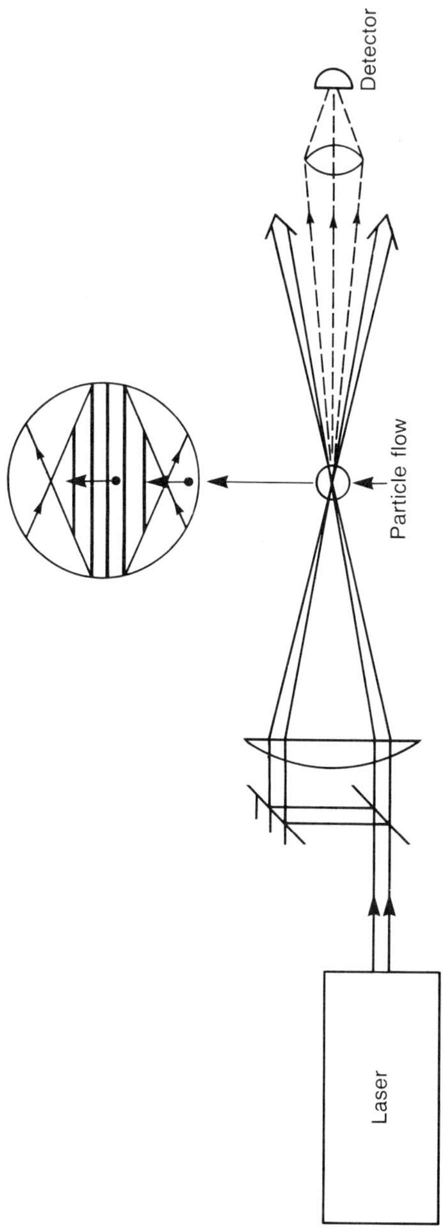

Figure 7-20 A typical laser-Doppler system for monitoring particle flow.

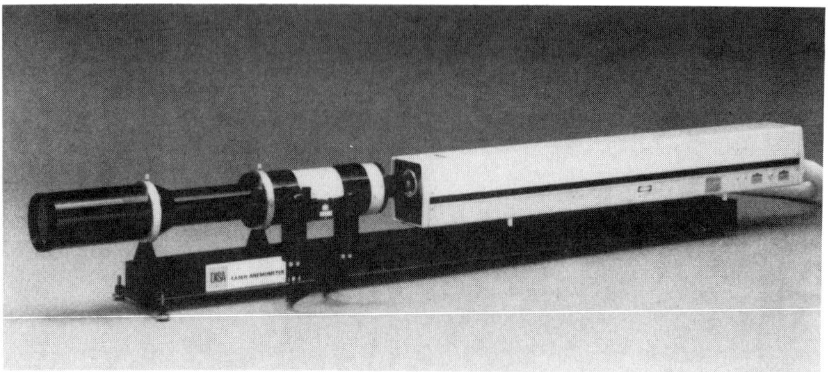

Figure 7-21 A commercial laser anemometer using two wavelengths to produce cross-intersecting fringes for the evaluation of the two normal components of the flow speed. (Courtesy of Dantec/DISA, Copenhagen.)

the kind shown in Figure 7-20, with the two systems oriented at right angle to each other to obtain two mutually perpendicular interference-fringe patterns. A dichroic mirror selects the 488 nm and the 514 nm components which are directed to different detectors to separately determine the flow speed components in the two directions normal to the crossed interference planes.

The main advantage of the laser-Doppler technique compared to the classical techniques described in Section 2.5.2 is its high accuracy and resolution, as well as the complete absence of inserted measuring devices which disturb the flow. The disadvantages are its higher cost, the requirement for particle feeding when inspecting homogeneous fluids and applicability only to relatively transparent fluids.

One of the main interests of laser-Doppler velocimetry is the possibility to finely map the velocity field distribution within a given volume by scanning the point of intersection of the two beams within such a volume. Measurements of laminar flow velocity distribution across a pipe (Durst et al., 1981, Section 13.6), of the spatial correlation of the transverse component of flow velocity (Drain, 1980, p. 209) and of different parameters in turbulent flows (Durst et al., 1981, Sections 13.10 to 13.23) have been reported.

The noncontact nature of laser-Doppler anemometry makes it possible to perform velocimetry analyses in situations where they would otherwise be impossible. One example is flame analysis, including industrial burners with turbulent recirculating regions (Baker et al., 1974) as well as flow studies around the sparking plug within one cylinder of an operating engine (Cole and Swords, 1979). Fiber optic microprobes for velocimetry studies within blood vessels have also been reported (Tanaka and Benedek, 1975).

7.3.2 *Other Optical Techniques for Fluid and Motion Sensing*

The structured-light pattern shown in Figure 7-20 can be obtained by incoherent-light methods, in which case no Doppler frequency shift takes place. For example, a white-light-illuminated Ronchi grid can be imaged on the flow volume to produce a square-wave light fluctuation along the particle path (Ballik and Chan, 1973). Alternatively, the grid can be placed in front of the detector shown in Figure 7-20 (Andermo et al., 1976). Such methods are appropriate for measurements with well-defined direction. For imperfectly defined particle paths, the grid period must be made very short to obtain good accuracy over the depth-of-field limited particle path within the structured-light volume.

Correlation Velocimetry

A variation of the grid imaging technique using a two-point inspection is the cross-correlation method, shown schematically in Figure 7-22 (Beck, 1981; Ziesemer, 1983). The material, such as the moving sheet shown here or a particle flow, moves below two focused detectors, which can sense either the ambient light reflected by the material or the natural emission from hot

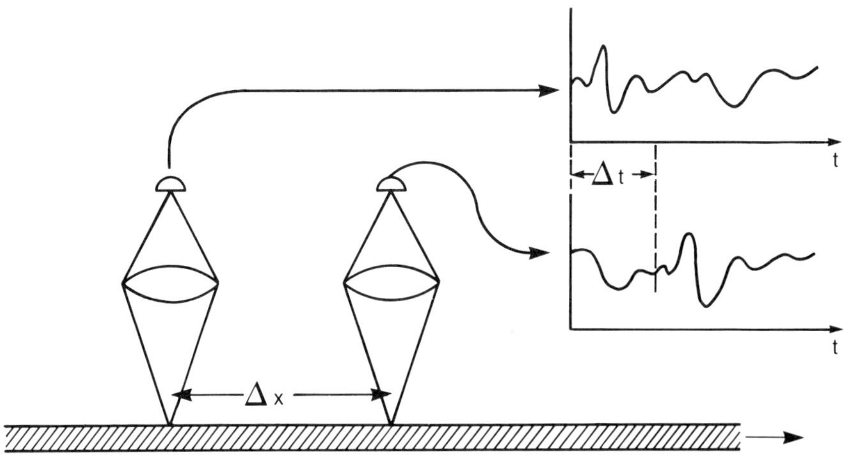

Figure 7-22 Schematic diagram of a cross-correlation velocimeter.

materials, or else use artificial illumination sources. Double-spot laser illumination at points sufficiently close to avoid speckle decorrelation (see Section 3.10) has been successfully implemented (Andermo et al., 1976; Bolon et al., 1983) for the inspection of materials, such as paper or metallic sheets, which have a uniform reflectivity under incoherent illumination.

The operation of a correlation velocimeter is as follows: If the two detectors have similar characteristics and are well aligned along the direction of sheet movement, the two detected signals will be similar but delayed by a time period $\Delta t = \Delta x/v$, where v is the sheet speed. By evaluating Δt from the cross-correlation function of the two signals (Bendat and Piersol, 1980), and knowing Δx, the velocity v can be obtained from this operation. Such a technique can be implemented with inexpensive incoherent sources such as LEDs, while the cross correlation can be evaluated with simple hardware devices.

Both the laser-Doppler anemometry and the correlation velocimetry technique require the presence of reflectivity fluctuations in the moving material to perform properly. When analyzing perfectly uniform materials such as transparent gases or liquids, artificial particle seeding is required to produce a detectable signal. A slightly different approach, thermally-excited correlation velocimetry (Zapka and Tam, 1982; Sell, 1985) can be applied on a homogeneous fluid without seeding.

A typical apparatus for such a technique is shown in Figure 7-23. A powerful pulsed laser is focused within the flowing stream through a suitable window. Localized heating, or even breakdown of the fluid, takes place at the focusing region if the fluid is moderately absorbing at the pump-laser

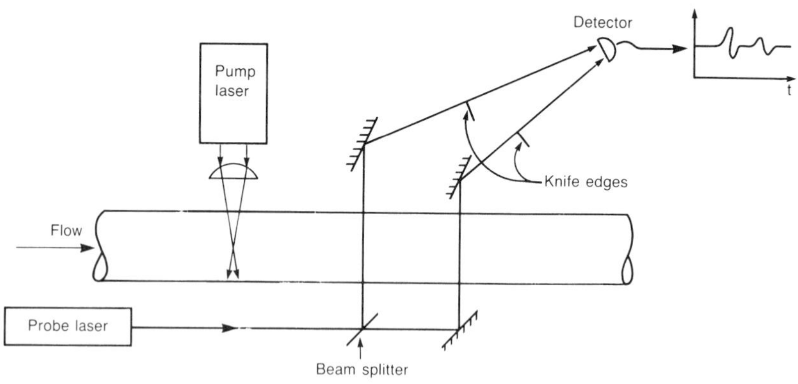

Figure 7-23 A typical apparatus for flow velocity measurement by local thermal excitation of the flowing fluid.

wavelength. Such a heated volume propagates down the pipe at the fluid speed and is successively detected by two probe beams which are deflected (by the optothermal beam-deflection effect, see Section 6.4.1) when the heated fluid reaches the beams. The time delay, and thus the fluid speed, are calculated by cross-correlating the two detected pulses. This approach is attractive, but requires a relatively powerful pump laser.

Particle Sizing in Fluid Flows

In laser-Doppler anemometry with artificial feeding of particles of given diameter, the visibility of the Doppler signal depends on the particle diameter. When a particle flows across the center of the interference fringe pattern (see Figure 7-20), a signal of the kind shown in Figure 7-24a is detected if the particle is smaller than the fringe spacing. If, however, the particle overlaps several fringes, a signal of the kind shown in Figure 7-24b will be seen. The visibility of the signal, i.e., the ratio of the periodic-modulation amplitude over the average amplitude, was therefore suggested as a parameter for the evaluation of the particle size in a stream (Farmer, 1972; Roberts, 1977). Particle concentration in dilute streams may also be evaluated by counting the number of Doppler signals recorded over a given time period. Instrumentation for simultaneous particle counting, sizing and velocity monitoring (Holve et al., 1981) is commercially available.

The modulation-visibility method for particle sizing assumes 100% modulation for particles smaller than the fringe spacing. This is not true if the particles do not intersect exactly the center of the fringe pattern. Moreover,

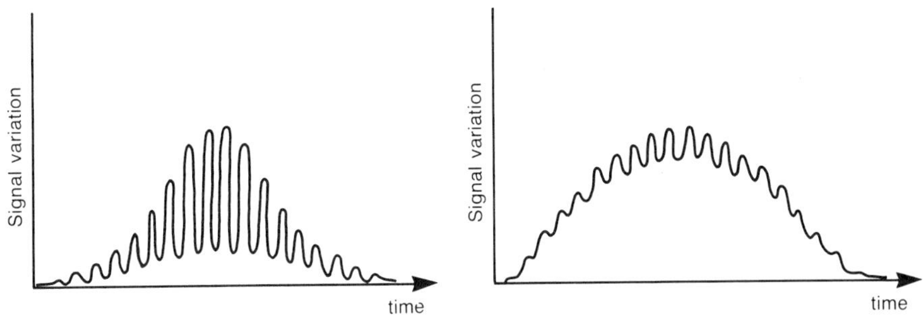

(a) (b)

Figure 7-24 Visibility of the laser-Doppler signal obtained with an apparatus of the kind shown in Figure 7-20 with flowing particles of diameter (a) smaller and (b) larger than the fringe spacing.

the intersecting laser beams must be perfectly coherent and have equal amplitude. The refractive index also has a strong influence on the visibility (Chu and Robinson, 1977). Some controversy thus exists over the validity of the modulation-visibility method as a quantitative approach for particle sizing (Durst and Heiber, 1977; Adrian and Orloff, 1977; Farmer, 1978). Further limitations of the modulation-visibility particle sizing technique are the requirement for smooth particle movement as well as a limited range of particle diameters and low concentration. Other optical particle-sizing methods based on turbidity or angular-scatter analysis which are independent of particle motion will be described in the next chapter.

Other Fluid Sensors

In many cases, the presence and density of particles in a flow must be determined qualitatively to obtain information such as the malfunction of a filtering unit, the degree of dispersion, etc. Such an analysis is usually performed visually by an operator through a sight window on the pipe wall. A simple electro-optic device to perform such an evaluation automatically on a continuous basis is shown in Figure 7-25a (Raab, 1984). In contrast with the anemometry approach, a single beam is projected within the stream and the scattered signal is angle-sighted by a detector. The frequency and amplitude of the scattered pulses indicate the density and relative size of the particles, while an unfocused beam can be used to evaluate the dispersion density.

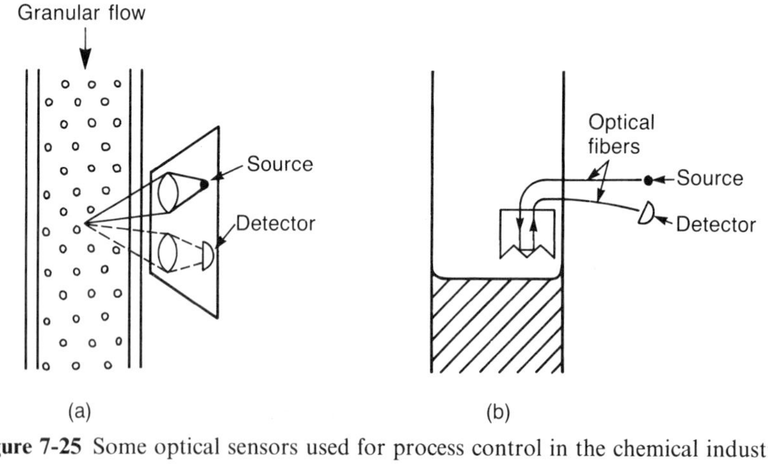

Figure 7-25 Some optical sensors used for process control in the chemical industry: (a) a flow-monitoring device; (b) a liquid-level sensor.

Figure 7-25b shows an optical liquid-level sensor. A retro-reflecting prism array is internally illuminated by a light beam carried by fiber optics. If the prism surface is in contact with air, the internal angle of incidence of nearly 45° is larger than the critical angle for total internal reflection, giving a strong reflected signal. When the liquid comes in contact with the surface, the critical angle becomes larger than the 45° angle of incidence, so that part of the light is transmitted within the liquid medium resulting in an abrupt decrease of the reflected-signal intensity. Other optical liquid-level sensors employ triangulation for noncontact inspection of high temperature materials such as molten glass (King and Merchant, 1982) or fiber optic probes, as described in the next section.

Motion Sensors

A number of proximity-sensing methods were described earlier, based on pneumatic (see Section 2.1) or electromagnetic (see Section 2.4.2) approaches. For a review of commercially available electromagnetic, ultrasonic and photoelectric sensors see the paper "Proximity Switches" in *Machine Design*, Apr.23, 1973, p. 33 or the publication *Inspection, Gaging and Testing* by SME, 1984 for a more up-to-date review. Proximity sensors may be realized optically by a triangulation device, as described in Section 4.1.4. For process control applications such as turbine blade tip clearance sensing (Davinson, 1984) or robot-hand installation (Rosen and Nitzan, 1977), rugged miniaturized devices are required. A frequent approach is the use of a transmitting-receiving fiber optic bundle (Catros and Espiau, 1980; Yanagi et al., 1980; Tokida, 1981; Cielo and Lapierre, 1982; Agrawal and Epstein, 1983; Krohn, 1984; Sridharan, 1985; Asher and Jackson, 1986). A sensing device of this kind is marketed by MTI, Latham, N.Y.

A schematic diagram of a fiber-bundle proximity sensor is shown in Figure 7-26. Light injected through one half of a bifurcated fiber optic bundle (the "transmitting" fibers) is reflected off the inspected surface back into the other part of the bundle (the "receiving" fibers). The detected signal level strongly depends on the probe-to-surface distance, as shown in Figure 7-27. Three curve shapes are given for three typical bundle configurations corresponding to a random, hemispherical or concentric distribution, respectively, of the transmitting (or receiving) optical fibers within the bundle. In all cases, the curve contains a first portion, the front slope, where the signal amplitude increases with the probe-to-surface distance. In this area, an increasing portion of the light transmitted by one fiber is reflected into the adjacent

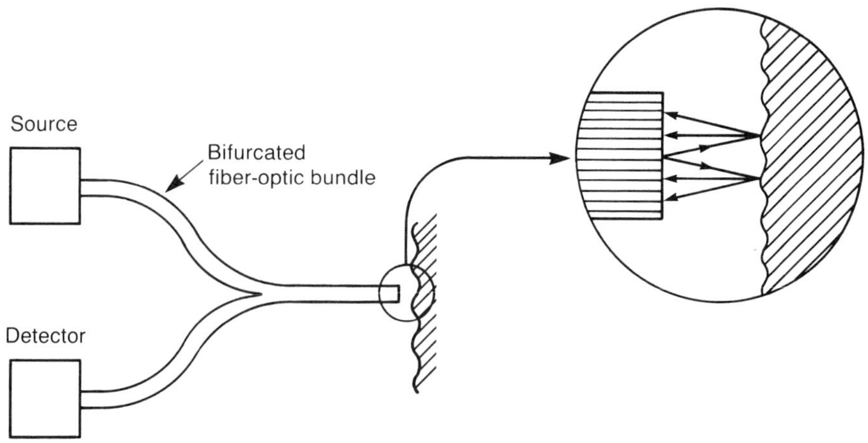

Figure 7-26 A fiber-bundle surface position sensor.

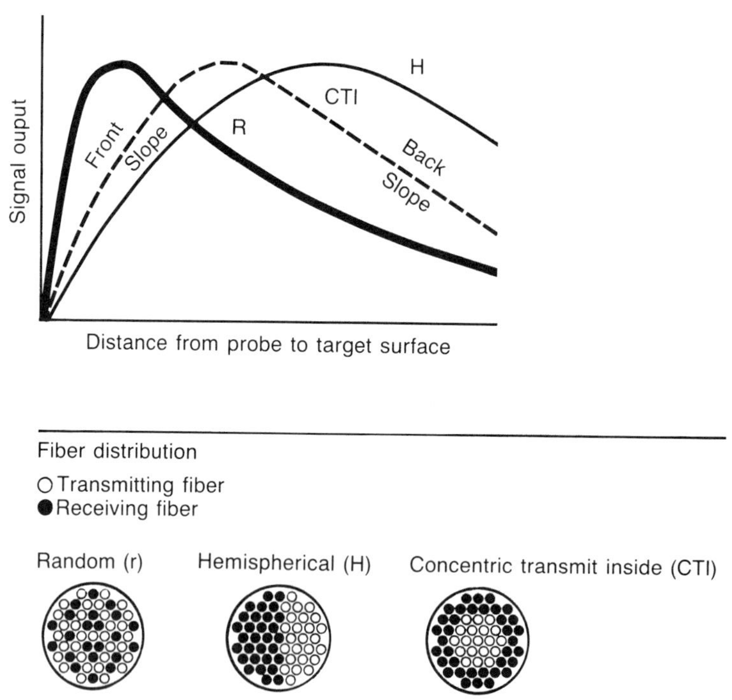

Figure 7-27 Possible fiber distribution and corresponding signal outputs for a typical commercial probe. (Courtesy of MTI, Latham, N.Y.)

receiving fibers (rather than being reflected within the same transmitting fiber for very close distances) as the probe-to-surface distance increases. For larger distances, the curve starts to decrease (back slope) with increasing distance when the reflected-beam diameter becomes larger than the probe diameter.

Typical probe-tip diameters range from 0.5 mm to 3 mm, and the probe-to-surface operating distances are in the 100 μm to 500 μm range for operation in the front-slope region and in the 0.5 mm to 5 mm range when operating in the back-slope region. Larger standoff distances are possible by lens-capped fiber-bundle probes (Asher and Jackson, 1986). Resolutions down to a few micrometers are possible for short-range probing of relative displacements In contrast to triangulation-based distance sensors, absolute distance measurements require a previous calibration, because the signal amplitude depends on the reflectivity and scattering characteristics of the surface (Pryor and North, 1985, p. 4). Typical applications include displacement, vibration, position or reflectivity sensing, as well as tachometer pulse generation.

7.4 Fiber Optic Sensors

As mentioned in Section 3.9, fiber optics has undergone enormous advances since the early 70s, to the point that fibers with attenuations below 1 dB/Km are now commercially available at very reasonable prices. Their suitability to long-range communication applications is now fully established, and new application areas are continuously emerging. One of these applications concerns industrial fiber optic sensors (Culshaw, 1984).

The total industrial sensor market in 1983 was $5 billion and is expected to grow 30% per year into the mid-1990s (Macfadyen, 1985). More than 80% of the transducer market for process control and monitoring is for pressure, temperature, flow and analysis sensors (Hingham and Medlock, 1984), all fields in which the potential of fiber optic transducers has been demonstrated. In spite of the strong competition from electronic sensors (Medlock, 1987), the total market for fiber optic sensors has been predicted by several market reports (Jones, 1985) to reach $400 million per annum in the mid-1990s.

Some optical sensors using optical fibers for the light transmission to and from the sensing location were described in the previous sections. This section reviews sensors where the optical fiber plays a role in the transduction mechanism.

7.4.1 *Phase Sensors*

The main interest in using fibers as optical sensors is that, because of their low attenuation, several kilometers of fiber may be reeled on to a compact loop to achieve a very long interaction length. If phase-sensitive interferometry methods (see Section 6.3) are used to detect faint perturbations of the fiber by the surrounding medium, a very high sensitivity can be achieved.

Pressure Sensors

Phase-sensitive fiber optic sensors were first developed as underwater acoustic sensors to detect low-amplitude sonar signals in the 1 to 1000 Hz range (Bucaro et al., 1977; Cole et al., 1977). A typical configuration is shown in Figure 7-28. A light beam, typically from a diode laser, is coupled into a single-mode optical fiber and subsequently split into roughly equal parts between a sensing and a reference fiber by a fiber coupler (single-mode fiber directional couplers have recently become commercially available). The

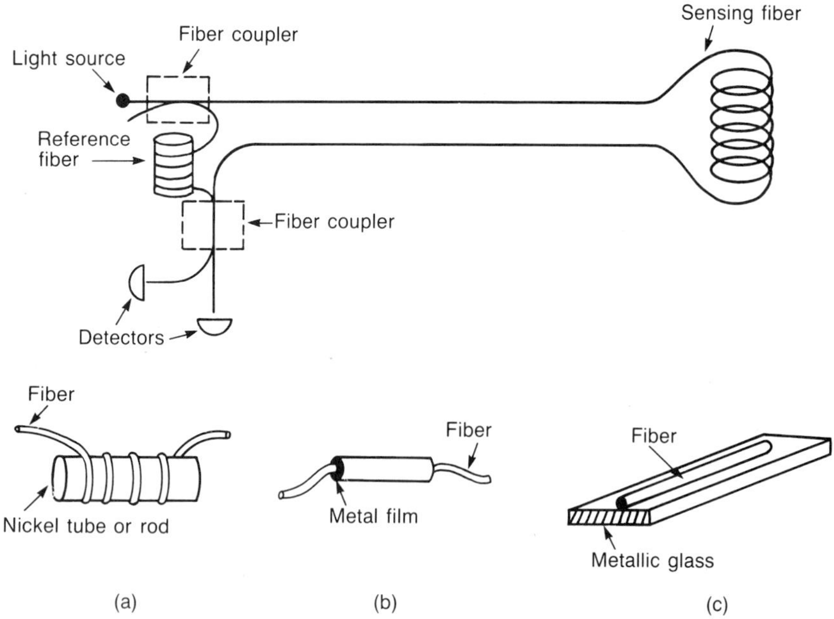

Figure 7-28 Phase-sensitive fiber optic sensing. Top: a Mach-Zehnder interferometric pressure sensor; bottom: some typical configurations for the sensing head of a magnetic fiber optic sensor: (a) mandrel sensor, (b) coated sensor and (c) stripline sensor.

sensing fiber is immersed in the liquid whose pressure fluctuations are to be probed, while the reference fiber is usually coiled around a piezoelectric cylinder to keep the output from the two fibers in quadrature for maximum sensitivity (see Section 6.3.1). After mixing at a second fiber coupler, the superposition of the reference and the probe beams can be detected at either of the two fiber outputs, or else the difference between the signals from the two detectors can be taken to eliminate common-mode noise (see Section 3.6.2). Polarization-maintaining single-mode fibers (normally obtained by applying a transverse stress to make the core slightly birefringent) are preferred to avoid polarization fading due to environmentally produced stress-dependent polarization rotation within the fibers (Stowe et al., 1982; Yamada et al., 1985).

When the sensing fiber is subject to a uniform pressure within the liquid, it undergoes a refractive index variation, Δn, and a length variation, ΔL, both of which produce a change of phase of the light beam propagating within the fiber

$$\Delta\Phi = k_0 n \Delta L + L k_0 \Delta n \tag{7-6}$$

where $k_0 = 2\pi/\lambda$ is the free-space propagation constant, n is the effective refractive index in the fiber and L is the sensing-fiber length. When the strain configuration is taken into account, equation (7-6) becomes (Cielo, 1979)

$$\Delta\Phi = k_0 \mathrm{n} S_1 L - \tfrac{1}{2} L k_0 n^3 \sum_{j=1}^{6} p_{2j} S_j \tag{7-7}$$

where $\mathbf{S} = (\varepsilon - 2\mu\varepsilon)[1, 1, 1, 0, 0, 0]$ is the strain vector for isotropic strain, and p is the photoelastic tensor in the standard contracted notation; ε is the strain and μ the Poisson ratio. The subscript 1 of the strain vector refers to the longitudinal direction, i.e., along the fiber's axis. When typical values are assumed for the mechanical properties of the silica fiber (Cielo, 1979; Cielo and Lapierre, 1982a), and with $\lambda = 0.63$ μm, equation (7-7) gives a phase sensitivity for the sensing fiber of the order of 10^{-10} rad/μPa m. As shown in Section 6.3.1, the shot-noise-limited ultimate sensitivity of a typical interferometer is of the order of 10^{-8} rad/Hz$^{1/2}$. This gives a pressure sensitivity of the order of 1 μPa in a 1 Hz bandwidth for a 100 m-long sensing fiber, better by nearly two orders of magnitude than a typical high-sensitivity piezoelectric hydrophone (Bucaro and Hickman, 1979).

As can be seen from equation (7-7), the phase shift produced on an isotropically compressed fiber coil is the sum of two components of opposite sign, a fiber-length contraction component which decreases the light-beam phase with an increase of the pressure and a refractive index component (due to the fiber compression), producing a phase increase of the light beam when the pressure is increased. These two components have similar magnitudes (Cielo, 1979), resulting in partial compensation of the two effects. Much higher pressure sensitivities can thus be obtained by embedding the sensing portion of the fiber in a compliant medium or by coiling it around a compliant cylinder to obtain a longitudinal, rather than isotropic, compression (McMahon and Cielo, 1979; Shajenko and Flatley, 1980). Sensitivities higher by two additional orders of value can be obtained by using a polymeric compliant medium.

Other interferometric configurations have been used. An increasingly popular scheme is the Febry-Pérot configuration (Cielo, 1979; Yoshino et al., 1982; Ohtsuka, 1982; Martinelli, 1982; Kist and Drope, 1983) In this case no separate reference fiber is needed, multiple reflections within the sensing-fiber cavity providing the required beam superposition. A cavity in the sensing fiber can be produced by introducing two fiber connectors or other discontinuities along the fiber. If the sensing cavity is deployed at a certain depth, such a configuration strongly reduces the environmental noise in a sonar hydrophone (Cielo, 1979; Cielo and McMahon, 1982).

A double-cavity may also be used for phase tuning as well as for reducing the laser source coherence-length requirements. As mentioned before, the coherence length of fiber-coupled laser diodes may be reduced to a few millimeters by feedback-induced line broadening (Miles et al., 1981; Dandridge and Miles, 1985). This is a serious problem if sensing-fiber lengths of hundreds of meters must be used. A double-cavity configuration only requires that the coherence length be longer than the difference between the lengths of the two cavities (Cielo, 1979).

Other Applications

Although the first application of phase-sensitive fiber optic sensors was for acoustic and pressure sensors, the same scheme can be used to sense other parameters. Temperature sensors can be obtained by inserting the sensing fiber of an interferometer of the kind shown in the top of Figure 7-28 in the medium whose temperature is to be measured. The thermomechanical expansion or contraction of the fiber is translated to phase variations which are detected by the sensitive interferometer. A similar application is the measurement of flow velocity by stretching the sensing fiber across the flowing stream in a pipe and detecting the phase modulation produced by vortex shedding (Lyle and Pitt, 1981; Barton and Saoudi, 1986).

Another possible application is in magnetic field sensing. The sensing fiber is here coiled around or imbedded in a magnetostrictive material, as shown in the bottom of Figure 7-28 (Giallorenzi et al., 1982). Variations in the surrounding magnetic field produce fiber stretching or contraction which are detected by the fiber interferometer.

Fiber optic gyroscopes have been developed where the relative phase of two counterpropagating beams in a coiled-fiber loop are sensed by interferometric means. When the loop is rotated around its axis, the relative phase of the two counterpropagating beams varies by an amount proportional to the rotational velocity because of the Sagnac effect. Predicted theoretical sensitivities reach values of the order of 10^{-6} degrees/(hour)$^{1/2}$ (Giallorenzi, 1982, Figure 21).

7.4.2 *Amplitude Sensors*

Although phase-sensitive fiber optic interferometers can reach very high sensitivities, they require a relatively complex optical setup and relatively expensive polarization-maintaining single-mode fibers. Cheaper sensors can be realized by simply monitoring the light intensity transmitted through the sensing fiber and comparing with a calibration table.

Several parameters can be measured by observing their effect on the transmissivity of the sensing fiber. Pressure can be sensed by inserting the fiber in a convoluted transducer of the kind shown in Figure 7-29 (Giallorenzi et al., 1982) to produce pressure-dependent microbend losses (see Figure 3-71 in Section 3.9). When pressure is applied on the plate, part of the light intensity is coupled from the fiber core to the cladding, where it is eliminated by mode-stripping devices (essentially a portion of the bare fiber whose cladding is coated with a black paint or in contact with a matching-index liquid or epoxy). An alternative approach is the detection of polarization rotation produced by the unidirectional transverse stress (Cielo and Lapierre, 1982a).

Luminescent-Emission Temperature Sensors

Temperature sensors can be obtained by inserting a photoluminescent crystal in front of the fiber tip as shown in Figure 7-30a (Ovren et al., 1985). The sensor structure features an encapsulated crystal of GaAs, on which crystal layers of different compositions have been processed by means of epitaxial crystal growth techniques. The photoluminescence takes place in a thin (typically 3 μm-thick) layer of GaAs surrounded by AlGaAs layers. The

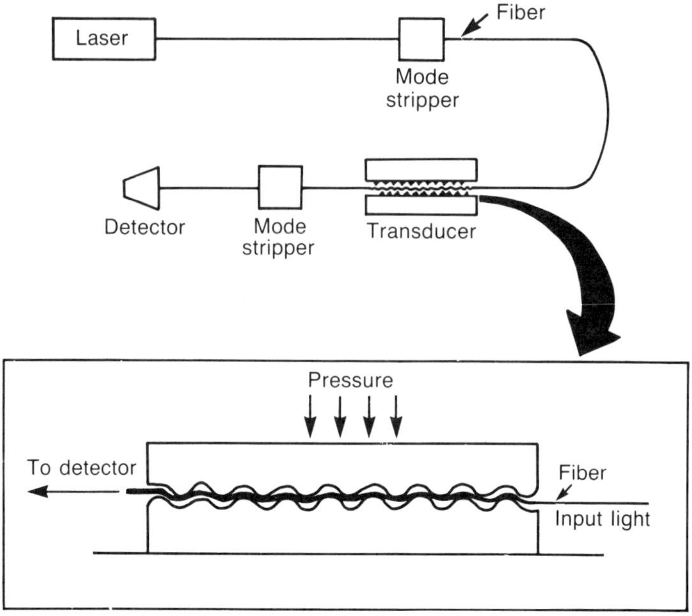

Figure 7-29 Pressure-sensitive optical sensor based on intensity variations generated by externally produced microbend losses.

crystal is excited by the nearly 750 nm emission from a LED source and emits photoluminescent radiation with peaks centered on 860 nm up to 920 nm as the temperature is varied from 0 to 200°C. The photoluminescent signal is split by optical filters with edge wavelength at 900 nm and detected by two detectors. Figure 7-30b shows typical outputs from the low-wavelength, the high-wavelength and the ratio signal as a function of temperature. Sensitivities of the order of 0.1°C are obtained. A similar device, using a 130 nm-thick phosphor layer on top of the fiber (Wickersheim and Alves, 1979) is now marketed by Luxtron, Mountain View, CA. Such a tiny sensor can be introduced in small recesses for applications such as temperature control during microwave heating, monitoring polymer curing, or biochemical temperature measurement. The induced surface perturbation is minimum because of the low thermal conductivity of the fiber compared to thermocouple wires, and maximum temperatures of 260°C can be reached.

Higher temperatures, in the 1000°C range and more, can be probed by fiber-blackbody sensors (Diles, 1983). As shown in Figure 7-31, such sensors contain a fiber microcavity which is produced by sputtering a thin iridium film on the end of a 0.25 to 1.25 mm diameter sapphire fiber and subsequently

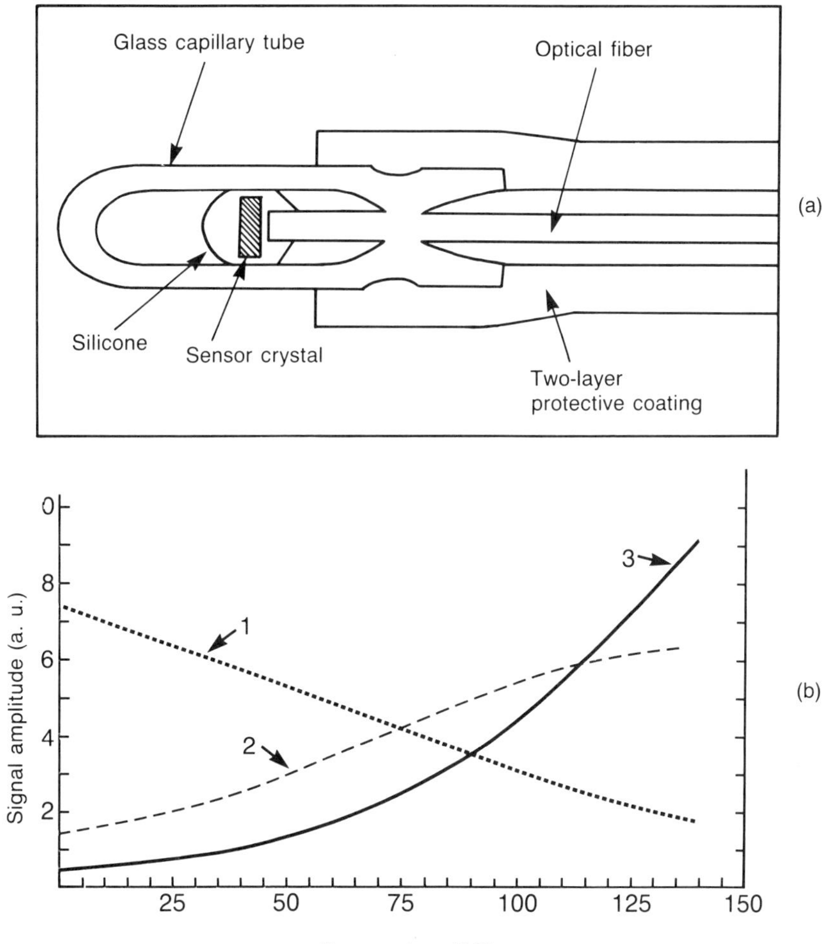

Figure 7-30 (a) Sensing head of a fiber optic photoluminescence-based temperature sensor; (b) Relative signal output from the low wavelength detector (curve 1), high wavelength detector (curve 2) and ratio signal (curve 3). (Courtesy of ASEA Research and Innovation.)

coating with an aluminum oxide film for protection. When introduced into the hot medium, the infrared radiance from the cavity is transmitted through the fiber to two sensors, typically at 0.6 and 0.7 μm, for double-wavelength detection (see Section 7.2.1). Accuracies of 0.05% in a 0–100 KHz bandwidth have been reported.

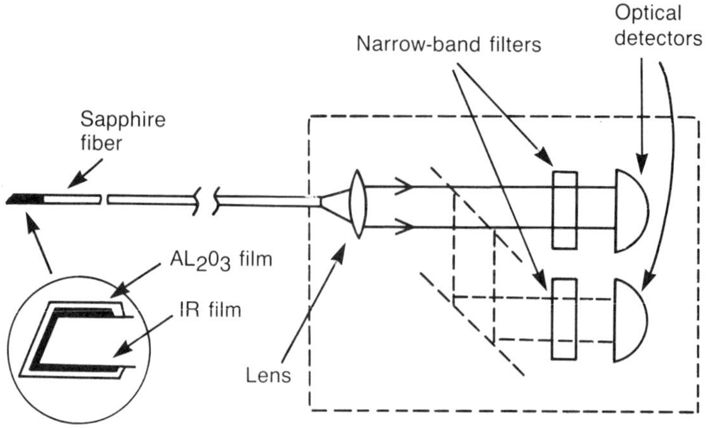

Figure 7-31 Blackbody fiber optic sensor for temperature measurement.

Chemical Analyzers

Process analyzers are required for many material characterization tasks in the chemical industry, as further discussed in Section 8-3. A number of noncontact optics-based approaches for product evaluation will be described in the next chapter. Contact fiber optic chemical sensors are increasingly reported for such applications (Harmer, 1984; Seitz, 1984; Snook, 1985; Maggioli, 1985). An example is fiber optic sensing of chemical pH level (Peterson et al., 1980; Sarti and Leitz, 1982; Kirkbright et al., 1984). A typical design is shown in Figure 7-32. An indicator dye such as bromothymol blue is immobilized on a support matrix. The analyte diffuses through a corrosion-resistant membrane such as PTFE and reacts with the dye, causing changes in the attenuation of specific reflectance bands. A fiber optic bundle addresses the sensor with a light beam corresponding in

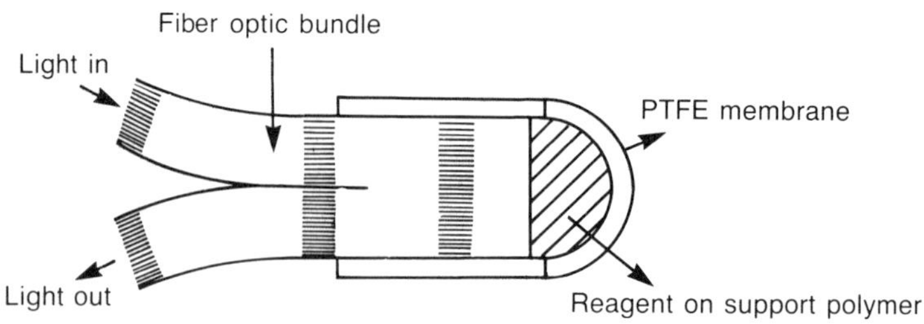

Figure 7-32 Sensing head of a pH-probing fiber optic reflectance sensor.

wavelength to one of these reflectance bands. The reflected signal is compared to the reflected intensity at a neutral wavelength where the dye is essentially transparent. Several other fiber optic sensors have been developed for monitoring other properties, such as the O_2, glucose, methane or ammonia concentrations, as well as for remote fluorimetry, nephelometry or oil pollution detection (Seitz, 1984; Harmer, 1984).

While the fiber optic analyzers described above use fibers for the transmission to and from the sensing volume, totally fiber optic analyzers, where the spectrum of the light beam carried by the fiber is sensitive to the properties of the surrounding medium through attenuated-total-reflection spectroscopy (Harrick, 1967; Hutchinson, 1981) can be conceived. In this case the light beam is always contained within the fiber, but it is subject to a wavelength dependent attenuation in the absorbing medium surrounding the fiber because of the extension of the evanescent wave (see Section 3.9) in such a medium if the fiber cladding is sufficiently thin (Chabay, 1982; Lew and Depeursinge, 1984). Applications to *in situ* quality control of glass-reinforced plastics (Black and Johnson, 1980; Farrar and Ashbee, 1983; see also Section 8.2.2), to polymer cure monitoring (Fanconi et al., 1986), as well as to biosensors (Owen, 1985) can be considered.

Many further applications have been proposed for fiber optic sensors (Busurin et al., 1985). Examples are the detection of acceleration by bonding the fiber to an inert mass (Rines, 1981; Ovren et al., 1985), liquid-level and smoke detectors (Baumbick and Alexander, 1980; Jones, 1985), tachometry (Macfayden, 1985), as well as monitoring of composite breakage by embedding a fiber optic matrix during structure assembly and detecting fiber integrity by light transmission (Crane et al., 1982; Jen et al., 1984). However, most of these configurations are still at the proposal or laboratory demonstration stage.

7.5 An Example of an Optically Controlled Process: Robotic Welding

Robotics is an emerging technology which is being called on to play a major role in productivity improvements in the manufacturing industry. A robot may be defined as a reprogrammable manipulator which can be quickly retrained to execute a wide variety of manufacturing tasks as compared to classical automation (Halivi, 1980; Engelberger, 1980; Hunt, 1983; Ray, 1985). The basic concept of robotics is flexible automation, where the same automated machine can be used to process small batches of products with different properties or geometry instead of designing a different production line for each batch.

Robotic manufacturing is economically advantageous for intermediate production levels, as shown in Figure 7.33. This is a typical industrial cost diagram which can be used to determine the economic viability of either manual, robotic or rigidly automated production depending on the volume of a given product. Manual production has the lowest initial cost (the intersection of the cost curve with the zero-production ordinate), whereas rigid automation gives the lowest marginal costs (the slope of the curve). Manual production is thus least expensive for the production of a non-repetitive product, as may be found in the construction industry, while rigid automation is jutisified when a large volume of a repetitive product is required, as in the case of steel sheet or plastic film manufacturing. Robotics is economically advantageous for a wide range of intermediate production volumes for which flexible automation results in the lowest total cost. The extent of such an intermediary range is expected to expand as the robot performance will improve and prices will decrease in the near future.

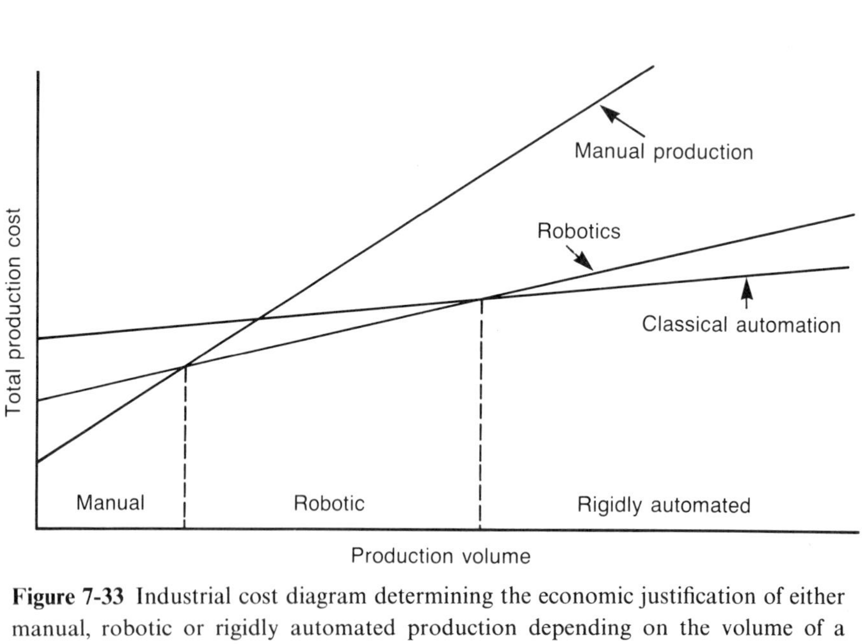

Figure 7-33 Industrial cost diagram determining the economic justification of either manual, robotic or rigidly automated production depending on the volume of a repetitive product.

7.5.1 *Fundamentals of Robotic Arc Welding*

As much as 40% of all industrial robots are being used for welding tasks (Ross, 1984). Such a trend reflects the need for increased productivity and improved quality of the welded products. Manual welding must be limited to short periods because of the required setup time, operator discomfort and safety considerations. Consequently, manual arc-welding time is generally less than 30% of the total operator working time (Boillot et al., 1985). This compares to a welding time fraction which may reach 85% for a welding robot, yielding an increase in productivity by a factor of 2.8, assuming that the same welding speed is followed in the two cases. Higher welding speeds may eventually be possible in robotic welding as advanced techniques like deep-penetration laser welding (Bass, 1983) and on-line automated monitoring devices become affordable.

Although a substantial fraction of spot welding is performed by robots in the automotive industry, it has been estimated that only 8% of arc-welding activities use robots (Ross, 1984). Such a low fraction is partly due to the high precision continuously required during arc welding. Figure 7-34 shows a typical setup for welding two metal plates along one side. Both plates must be symmetrically melted by the arc power to provide a good metallurgical bond with the filler metal.

The quality of robotic welding is now limited by the difficulty of adapting the robot to the welding process variables, which include unwelded part,

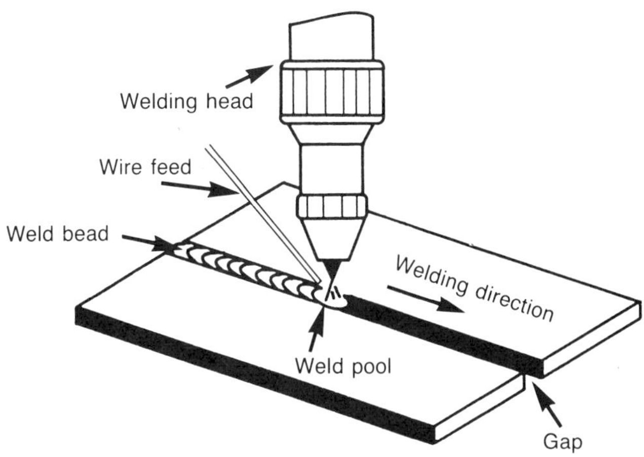

Figure 7-34 Schematic view of a GTAW setup for butt-welding two metal plates.

weld machine, metallurgical and welding head positioning variables. Unwelded part variables such as joint mismatch, heat balance, root opening position and width have strong effects on the welding process. Examples of weld machine variables are random fluctuations in current and voltage, arc fluctuations related to plate topographical factors or arc blow, gas composition and flow rate. Metallurgical variables are mainly related to surface contamination and to the visco-dynamic, thermal and interface properties of the base and filler metal. Welding head positioning variables result largely from an inability to maintain tolerances in part dimensions, part position in the fixturing equipment and part distortions during welding. Most of these variables are treated as stochastic parameters in present-day robot welding machines because their value is not known during the welding process.

Sensing systems are being developed to provide adequate in-process information on the physical state of the weld area for a partially or fully adaptive robot operation. Such sensors would track variations in dynamic weld conditions, providing real-time information to be immediately fed back to the machine controller. This information would update the process status model in the monitoring unit. After comparison with the stored data base, the unit would produce the appropriate corrective actions.

An important parameter to be continuously monitored during arc welding is the position of the welding torch relative to the gap between the two plates to be welded. A misalignment as small as 1 mm between the welding head and the gap results in a strongly asymmetric thermal distribution in the workpiece, as shown in Figure 7-35 (Boillot et al., 1985). This figure shows the result of a finite-difference simulation performed to evaluate the temperature distribution within the plates under different conditions. Square-groove carbon steel plates, 5 mm-thick with a 2 mm-wide root opening (which corresponds to the air-filled gap in front of the weld pool in Figure 7-36), were considered in this case for a welding speed of 2.6 mm/s. If the torch is well aligned, Figure 7-35a, the temperature distribution is symmetric so that a good weld is produced if the heat input is properly chosen. However, if the torch is misplaced by as little as 1 mm, a strongly asymmetric temperature distribution will result (figure 7-35b). In this case, the metal plate on the right side of the gap is not completely melted. The molten filler thus penetrates within the gap, but its adherence to the right-side plate will be inadequate because no metallurgical bond is formed.

Figure 7-36 shows a metallographic micrograph of the section of a weld bead produced with a well-centered and an off-center torch, respectively. The lack of adherence between the right plate and the root of the bead is apparent is the misaligned case. The strength of such a weld to bending stress is orders

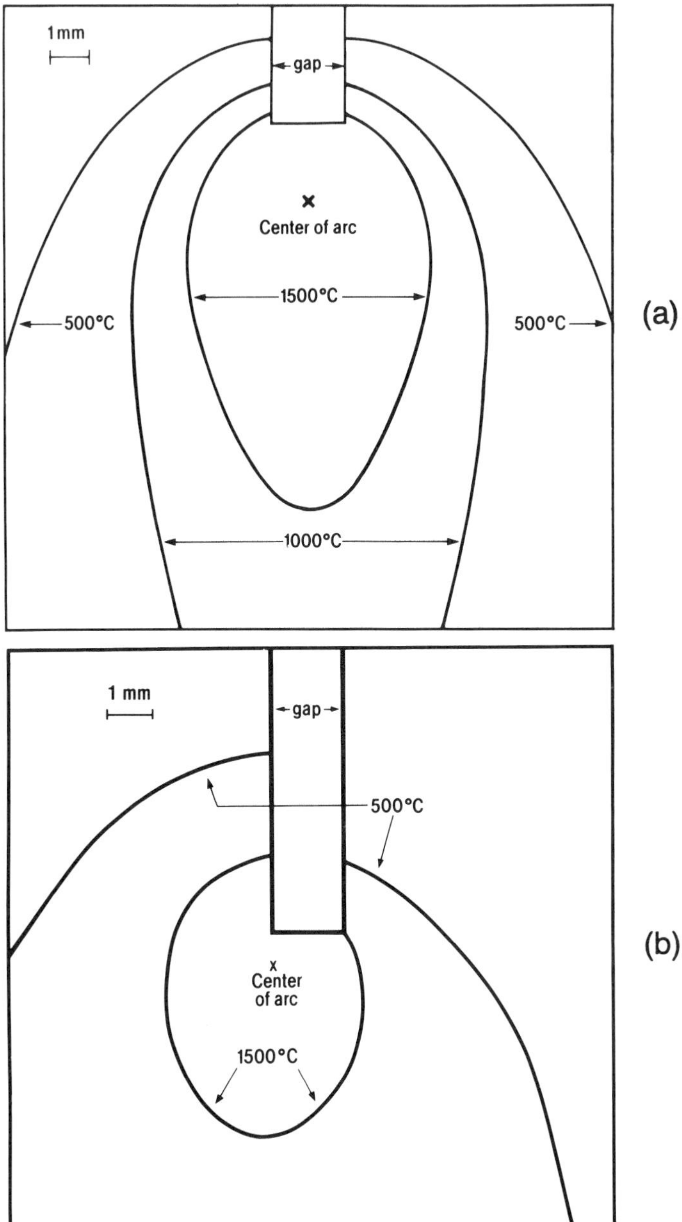

Figure 7-35 Finite-difference simulation of the stationary surface temperature distribution in the metal plates being welded (carbon-steel, welding speed 2.6 mm/s; the 1500°C isotherm corresponds to the boundaries of the molten weld-pool) for a well-centered torch (a) and for a torch laterally displaced by 1 mm with respect to the plate-to-plate gap (b).

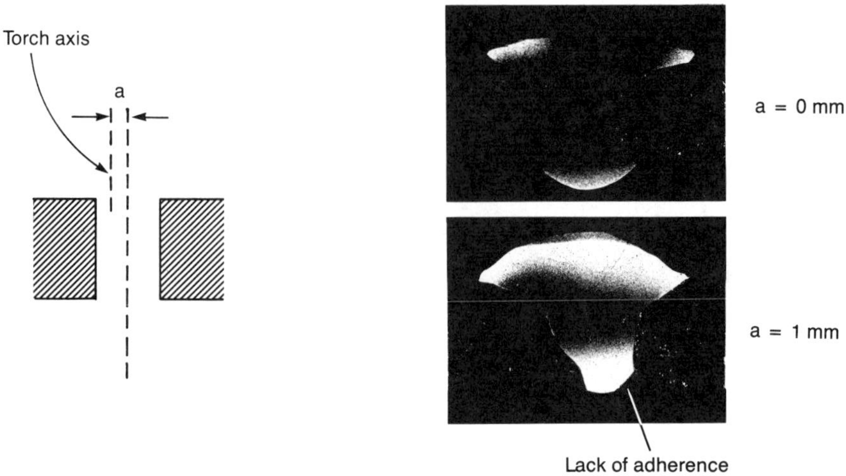

Figure 7-36 Metallographic micrograph of a weld cross-section obtained with a well-aligned or misaligned welding torch. (Courtesy of the IMRI welding group.)

of magnitude smaller than a well-centered weld. The positioning precision of a welding robot is of the order of ± 0.2 mm. Such precision is, however, not sufficient to assure proper torch positioning: In practical conditions, the gap position may vary appreciably with respect to the theoretical profile because of lack of straightness of the plate edges, inaccurate fixturing or thermal deformations during welding.

Although the examples shown above correspond to plate-to-plate butt welding, a variety of workpiece geometries may be submitted to a given welding robot. Figure 7-37 shows another typical configuration where the roots between two plates at 90° to each other must be joined. In this case, the path to be followed is the intersection between the two surface planes. Similar considerations are valid concerning the accuracy required to position the torch on the V-shaped edge. In many cases, the path to be followed is not rectilinear. The precise acquisition of the path geometry in present-day robots is usually done by a semiautomated procedure by which the operator manually positions the welding head at the beginning and at the end of the path, as well as at all points where there is a change in the welding direction, and inserts the corresponding xyz coordinates in the robot memory. This will not account for variations in edge straightness or dynamic deformations, and it must be repeated for every workpiece of a repetitive batch if fixturing inaccuracies are to be compensated. Consequently, an artificial vision system which could determine in real time the position of the edge to follow relative

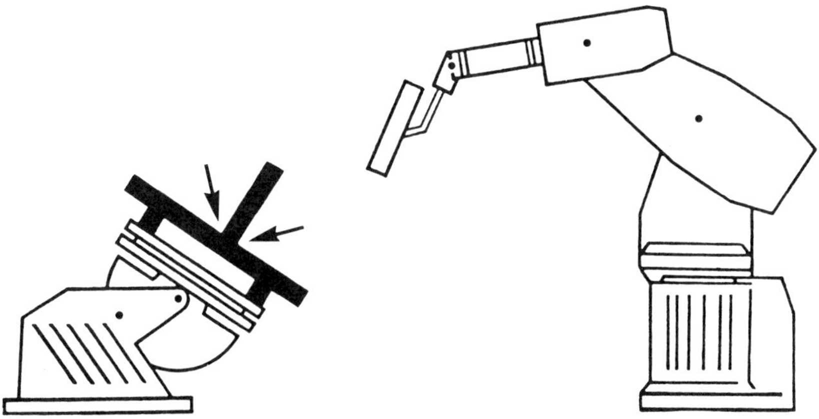

Figure 7-37 A typical configuration for robotic welding. The arrows indicate the edges to be welded.

to the welding head would be a highly desirable tool for weld-quality assurance and for productivity maximization.

7.5.2 *Active Optical Sensors for Robotic Welding Monitoring*

Optical sensing systems are under extensive development for adaptive welding applications. The main advantages of the optical approach are noncontact operation, versatility and the availability of a large amount of information from the spatial as well as the spectral features of the optical output. Optical sensing systems can be classified into active or passive, depending on whether the observed image is produced by selective illumination with an artificial light source such as a laser or whether the welding scene is observed without artificial illumination. Passive systems can be divided into reflective or emissive systems, depending on whether the observed visual pattern is produced by reflection of the ambient light from the inspected object or whether the observed light pattern is produced by direct thermal emission from the hot surface.

Most of the active sensing systems use pattern projection and triangulation techniques of the kind described in Chapter 5 to determine the surface profile in front of the welding pool for robot arm guiding and arc control purposes (Bamba et al., 1982; Clocksin et al., 1982; Cook, 1983; Oomen and Verbeek, 1983; Dufour and Bégin, 1983; Dufour and Cielo, 1984; Clocksin et al., 1985;

Beranek et al., 1986) Either a stationary laminar light beam or a mechanically-scanned focused light beam is projected on the workpiece surface in front of the welding torch. As described in Section 5.3, such a light pattern is viewed at an angle to obtain the three-dimensional profile of the part to be welded.

A typical optical light-beam projection and angle-viewing system for adaptive welding is shown in Figure 7-38 (Dufour and Cielo, 1984). A light pattern is projected in front of the welding torch and is viewed at an angle by a video camera. The shape of the pattern as seen by the camera provides information about the surface topography in front of the welding pool, including gap width and position with respect to the torch as well as plate orientation and mismatch. The projection unit is a matrix of optical fibers, each individually fed by a diode laser, which is imaged on the surface through a proper combination of spherical and cylindrical lenses in order to obtain a smooth line profile (Dufour and Cielo, 1984). The surface geometry analysis algorithm is based on the measurement of the area and of the first moments of the illuminated pattern inside the computer-programmable window (Dufour and Bégin, 1983). In an actual sensor-equipped welding system, both the projection unit and the camera are solidly attached to the welding head for compactness and also to provide a coherent reference system between the torch and the camera. A close-up of the IMRI prototype ready to weld along the edge of a V-shaped groove is shown in Figure 7-39. It includes, from left to right, the CCD camera, the projection unit and the welding torch.

A typical image seen by the camera when the laminar light beam is projected across the intersection between the two tilted surfaces is shown in Figure 7-40.

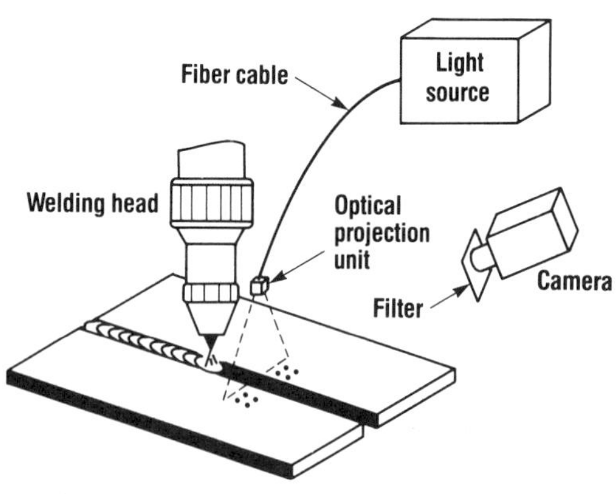

Figure 7-38 Schematic view of an active optical inspection system for real-time control of an adaptive welding process.

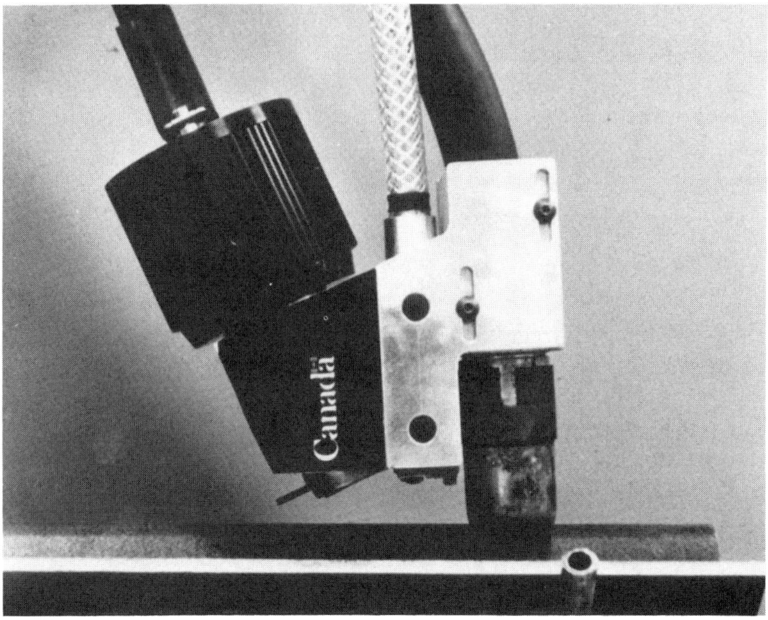

Figure 7-39 Close-up of a welding head with incorporated optical sensing system for torch guiding along the edge of a V-shaped workpiece. (Courtesy of M. Dufour, IMRI.)

Figure 7-40a shows a raw image of the light pattern seen at an angle by the camera. The V-shaped projected pattern can be seen in the center of the camera field, while the irregular dark area on the left corresponds to the light emission, either direct or reflected by the workpiece surface, from the welding arc. Although the arc-emission noise can be substantially reduced by using an interference filter matched to the emission spectrum of the inert gas arc shield (Dufour and Cielo, 1984; Lenef and Gardner, 1985), such a noise cannot be completely eliminated because of the very high brightness level of the arc. The arc-emission pattern can, however, be effectively eliminated by proper application of digital filtering algorithms of the kind described in Section 5.2. Figure 7.40b shows a typical single-line camera output obtained along a horizontally-directed line in Figure 7-40a. Although the arc-emission noise, on the left side of the signal, has a substantial amplitude, it spreads over a wider area than the narrow peak detected at time *Tnp* and corresponding to the cross section of the projected beam. Signal differentiation through a convenient convolutional operator (see Section 5.2) thus results in a clean filtered pattern profile as shown in Figure 7-40c.

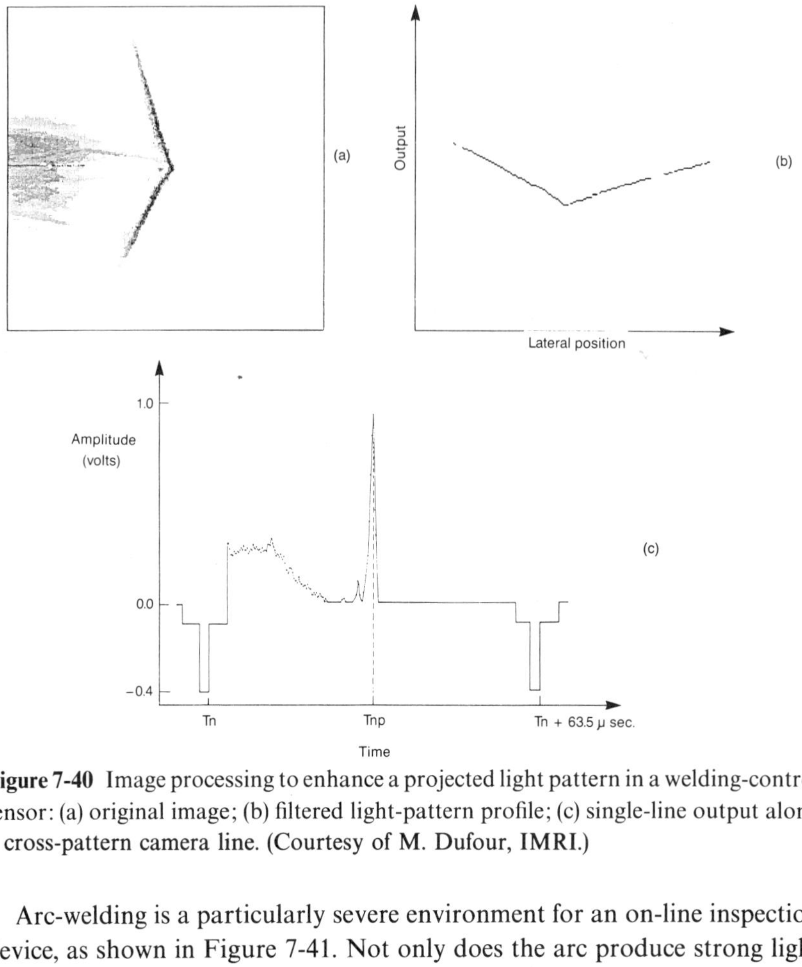

Figure 7-40 Image processing to enhance a projected light pattern in a welding-control sensor: (a) original image; (b) filtered light-pattern profile; (c) single-line output along a cross-pattern camera line. (Courtesy of M. Dufour, IMRI.)

Arc-welding is a particularly severe environment for an on-line inspection device, as shown in Figure 7-41. Not only does the arc produce strong light emission which interferes with the camera image, but it also produces a considerable amount of heat, fumes and an abundance of sparks, particularly in certain welding modes such as the short-circuit GMAW. Such projected molten particles tend to accumulate on the optical windows, which must be repolished or substituted at regular intervals. In order to decrease the frequency of these operations, a robust camera-assembly system has been developed at IMRI, as shown in Figure 7-42. The high image brightness makes it possible to decrease the entrance pupil of the camera to nearly 1 mm^2. A telecentric imaging system (see Section 3.8) can thus be used with the input pinhole in the front focal plane of the camera lens. The amount of sparks entering the camera enclosure is thus minimized, while a moderate

Figure 7-41 A welding robot equipped with the IMRI vision system is performing a weld in the GMAW short-circuit mode (Coutesy of M. Dufour, IMRI, and J. P. Boillot, Servo-Robot).

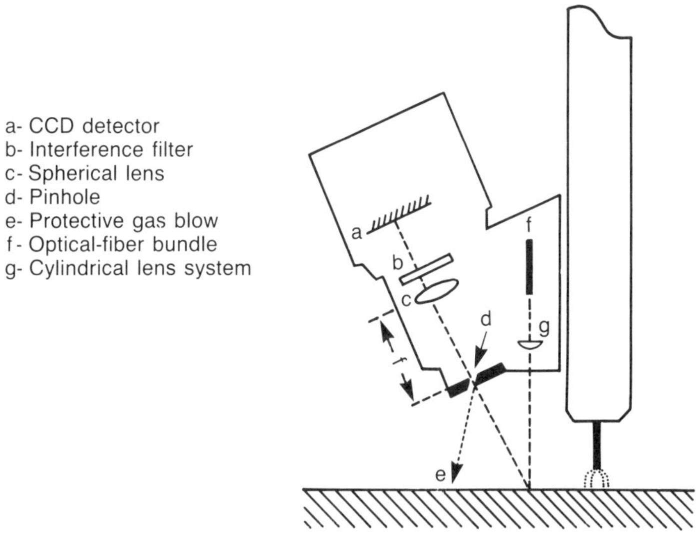

Figure 7-42 Schematic diagram of the telecentric optical sensor system used at IMRI for protection against sparks.

inert-gas flow keeps the fumes out of the optics and reduces the speed and temperature of the sparks entering through the pinhole. The telecentric design has the additional advantage of producing a nearly collimated beam between the lens and the camera retina, so that narrower band-pass interference filters can be used for better projected-beam discrimination (see Section 3.6.1). Synchronous detection (see Section 3.6) may also be used to substantially reduce the arc-emission noise (Dufour et al., 1986).

Flying sparks also produce highly luminous streaks across the camera field of view. As they are very narrow, they cannot be eliminated by spatial filtering techniques. A promising approach now under development at IMRI makes use of a double-wavelength differential inspection scheme to eliminate arc and projected-spark noise by discriminating the wide-spectral-bandwidth noise from the single-wavelength projected laser light.

7.5.3 *Passive Optical Inspection*

As mentioned above, passive systems are optical inspection devices which do not employ artificial illumination and include reflective and emissive sensors. Reflective systems are based on sensing the light reflected by the workpiece in order to obtain information on the welding parameters. The light source may be the ambient light or the light emission by the welding arc. The position and the width of the open gap in front of the weld pool may usually be clearly determined from an analysis of the bidimensional image which can be recorded by a camera situated in front of the torch. When no gap is present, however, as in lap-welding or when a V-shaped edge is followed, such a two-dimensional imaging approach is totally inadequate as compared to the three-dimensional pattern-projection active approach.

Passive techniques based on the detection of infrared thermal emission from the hot workpiece are attractive for a number of reasons. The possibility to inspect in real time the temperature of the welded materials is very attractive, because the temperature distribution around the weld pool provides important information on the welding status. Factors such as the pool shape, absolute temperature and symmetry of the temperature distribution are directly related to several welding process variables, such as joint mismatch, root opening and plate thickness fluctuations, arc blow, or the visco-dynamic and thermal conductivity properties of the plates.

In some cases, thermal sensing may substitute active projection techniques for robot guiding. This can be done by sensing the temperature distribution in front of the welding pool using an IR camera (Chin et al., 1983; Wang and Chin, 1986) or a fiber optic device of the kind shown in Figure 7-43 (Boillot et al., 1985). A bundle of 8 to 20 optical fibers aligned side-by-side in a linear array is inserted in the welding torch and the fiber ends are imaged by a lens on a linear array in front of the welding pool. The infrared radiation collected by each fiber is transmitted to a detector array whose output provides a linear scan of the temperature in front of the pool. If the torch is well-aligned with respect to the gap (Figure 7-35a), the temperature of the two plates in front of the welding pool is the same and a symmetric thermal distribution of the kind shown in Figure 7-44 is obtained. If, on the other hand, the torch is misplaced (Figure 7-35b), the temperature of the two plates is unbalanced resulting in an asymmetric thermal scan. Such information is fed back to the torch-positioning mechanism for a prompt correction of the welding path. Irregular paths of the kind shown in Figure 7-45 were followed using such a sensor. It should be noticed that the correction is made on the basis of the actual weld pool position and not on the position of the torch with respect to the gap as in the case of an active-projection sensor: Welding errors produced by pool displacements resulting from magnetic lateral

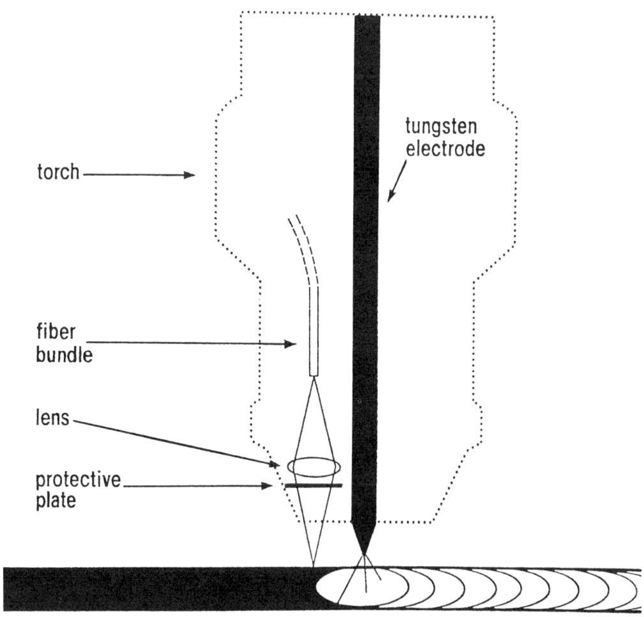

Figure 7-43 Schematic view of a fiber optic thermographic sensor inserted in a GTAW welding torch to monitor the temperature distribution in front of the weld pool.

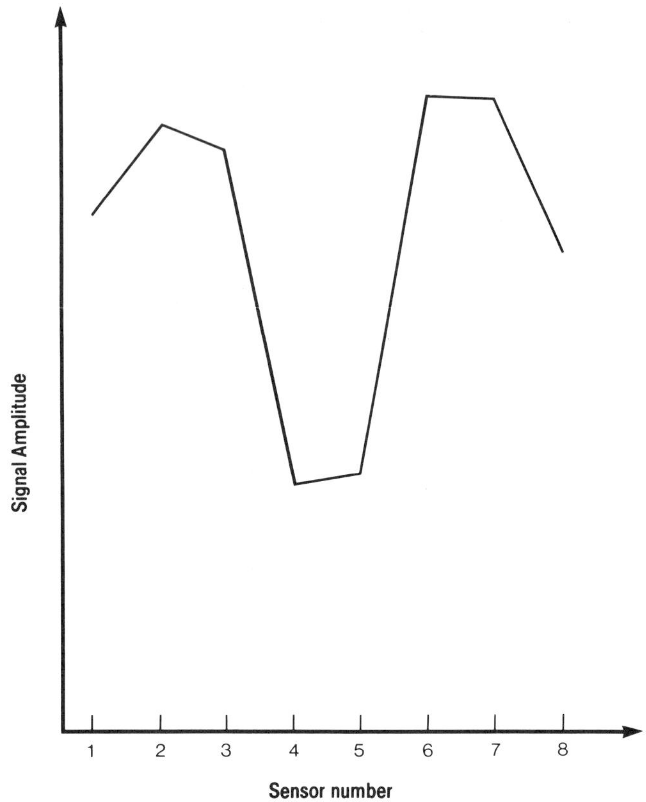

Figure 7-44 Typical signal obtained from the fiber array imaged in front of the welding pool.

blowing of the arc or from the presence of asymmetric thermal flow in the piece (as produced, e.g., by an asymmetric geometry) are naturally avoided by such an approach. Once again, the basic advantage of optical noncontact sensors consists of making it possible to monitor the temperature very close to the process source. This results in faster and more direct information, making the control task easier, as discussed in Section 7.1.

Time lags in the control loop, and thus loop instability, are minimized by pointing the sensors as closely as possible to the weld pool. A numerical-modelling analysis has shown that in a typical arc welding situation (carbon-steel workpiece, welding speed 2.6 mm/s), the thermal time lag is of the order of one second if the fiber optic sensor is pointed at a position situated 5 mm in front of the arc, while it reaches three seconds if the sensed area is 8 mm

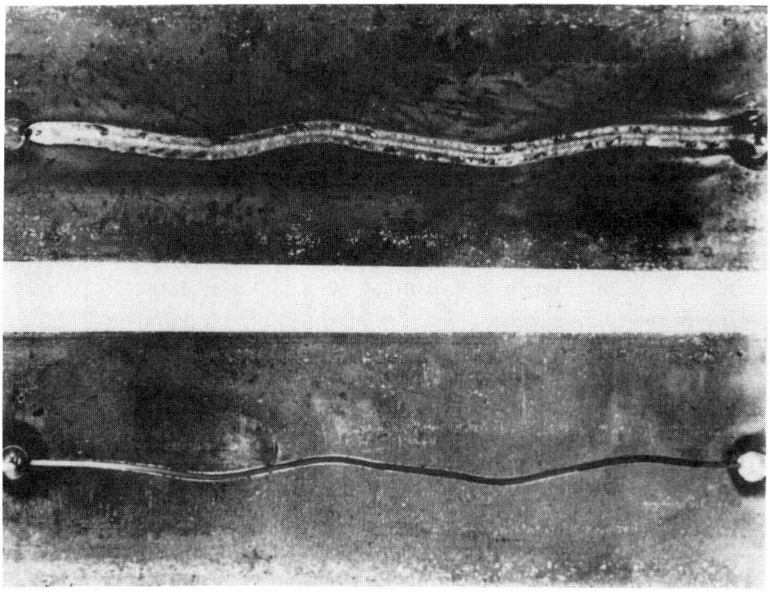

Figure 7-45 Example of an irregular welding path followed by a GMAW head guided by a mechanically scanned fiber optic thermographic sensor before (top) and after (bottom) welding.

from the arc (Boillot et al., 1985). Short-wavelength sensing close to the welding pool is thus preferable to long-wavelength monitoring of the lower temperature distribution far from the arc (Chin et al., 1983) if the environmental noise represented by the arc emission can be overcome. One method to avoid such an interference in pulsed arc-welding is to collect the sensing information during the periods of arc shut-off (Niepold and Bruemmer, 1982). For continuous-welding processes, noise from reflected arc light can be reduced by using some of the techniques described in Section 7.2, such as by properly choosing the viewing direction in order to avoid specular reflections from the smooth molten pool, filtering the infrared radiation in order to avoid the strong emission lines of the arc gas or using a double-wavelength detection in order to compensate for emissivity variations and to recognize the high-temperature arc emission component.

When no gap exists between the two plates or when the welding speed is too high for the thermal diffusion to heat the plates in front of the torch at a detectable level, thermal sensing is not effective for the robot-guiding function. A thermal sensor may, however, be a useful complement to a light-projection guiding system for a full control of the welding parameters.

One of the parameters which can be monitored by infrared thermal sensing is the temperature distribution, and thus the rate of cooling of the solidified bead after welding (Lukens and Morris, 1982). Such a cooling rate may substantially affect the microstructure of the welded volume and hence its mechanical properties such as hardness and brittleness. Even more important is the possibility to evaluate, by passive thermal-imaging sensing, the size of the weld pool. The pool size must be sufficiently large to produce full melting of the plate edge but not too large to avoid a collapse of the molten volume. Such an evaluation can simply be made by measuring the width of the area whose brightness exceeds a given threshold from a video-camera image or by using a photodiode array (Rider, 1975) or an infrared camera (Wang and Chin, 1986). However, the arc-emission interference as well as the strong variation of the emissivity near the welding pool (the emissivity may decrease by as much as a factor of five between the oxidized surface of the plates and the molten weld pool surface) make such a measurement difficult. The above mentioned spatial and spectral filtering techniques should be effective in filtering out the precise weld pool size.

One of the important parameters which can be estimated from a measurement of the weld pool size is the depth of penetration of the weld pool between the plates. As mentioned above, complete melting of the plates across their full thickness must take place to obtain good resistance of the welded interface to shear stress. In a first approximation, it can be assumed that the weld pool has a hemispherical cross section so that its penetration depth is roughly equal to half its diameter on the upper surface. Better estimation of the weld pool volume can be obtained by measuring the pool solidification time during pulsed welding (Fihey et al., 1983).

In conclusion, optical techniques have a strong potential as high-performance, rugged and reliabile sensors for industrial process control. Some of these techniques, such as infrared pyrometry, are already widely used in the production environment, while others, such as fiber optic, Doppler-velocimetry and robot-guiding sensors, are in a phase of intense development. Such efforts will certainly be pursued in the future, stimulated both by the availability of high performance and cost-effective optoelectronic components and by the increasing need for on-line sensors in the modern computerized factory.

Chapter 8

Optical Techniques for Material Analysis

As mentioned in Chapter 4, optical inspection techniques include morphological and nonmorphological technologies depending on whether the part texture or shape is analyzed or else whether some material properties other than its surface morphology are evaluated. Chapters 4 and 5 described a number of optical techniques belonging to the first category, which correspond more closely to the human vision function. Chapter 6 was devoted to the inspection of product properties which, while not directly related to the exterior surface configuration, are related to the internal shape or to the mechanical integrity of the part, such as the presence of deep cracks or ply-to-ply delaminations. The techniques described in Chapter 7, such as infrared thermometry or fiber optic sensing methods, make use of optical beams for purposes completely different from the analysis of the external shape of the inspected objects.

Along this line, the interaction of light with matter may provide information on the intrinsic properties of the material composing the industrial product, such as its chemical composition or its microstructural state. As mentioned in the last chapter, on-line material analysis provides more direct information than temperature, flow or pressure sensing for process control purposes. The propose of this chapter is to give some examples of optical techniques for the analysis of industrial materials. The examples are chosen in the fields of spectroscopic and of light-scattering methods, which are particularly powerful while being directly applicable to in-process analysis. Other optical analysis techniques which are less relevant to industrial inspection, such as ellipsometry (Azzam and Bashara, 1977) or combustion diagnostics (Lapp and Penney, 1974; Croesley, 1980; Schmieder and Kerstein, 1980), are described in more specialized reviews.

8.1 Spectroscopic Techniques

Spectroscopy refers to all those techniques where a relative comparison of the light intensity at different optical wavelengths is performed after the light beam has interacted with the analyzed material. Although largely a chemical laboratory tool at the present time, its applications to the production line are increasing rapidly. We have seen in Section 5.1.3 how a spectroscopic technique can be used to determine the thickness of a transparent film. In this section the powerful capabilities of spectroscopy for material analysis will be described and examples of in-process quality control applications will be given.

8.1.1 *Overview of the Most Common Optical Spectroscopy Methods*

Spectroscopic techniques (Straughan and Walker, 1976; Smith, 1979) may first be classified as either absorption or emission methods. *Emission spectroscopy* is used to analyze the spectrum of light emitted by materials whose atoms or molecules undergo a transition from an excited state to a lower energy level. An example is flame spectroscopy, where the atom is raised to an excited or ionized state by the thermal energy from the flame and decays to lower energy levels emitting photons of characteristic wavelengths.

In *absorption spectroscopy*, which is more widely used, the spectrum of the incident beam is observed before and after interaction with the analyzed material. Transitions from an unexcited to an excited energy level take place with absorption of photons of a specific wavelength, producing absorption peaks in the light-beam spectrum after interaction with the material. Some instruments, such as the flame spectrometer shown in Figure 8-1, may provide both an emission and an absorption spectrum (Olsen, 1975, Chapter 5).

The width of the absorption or emission lines depends on the kind of material which is inspected. Low-pressure gases give sharp line spectra, the

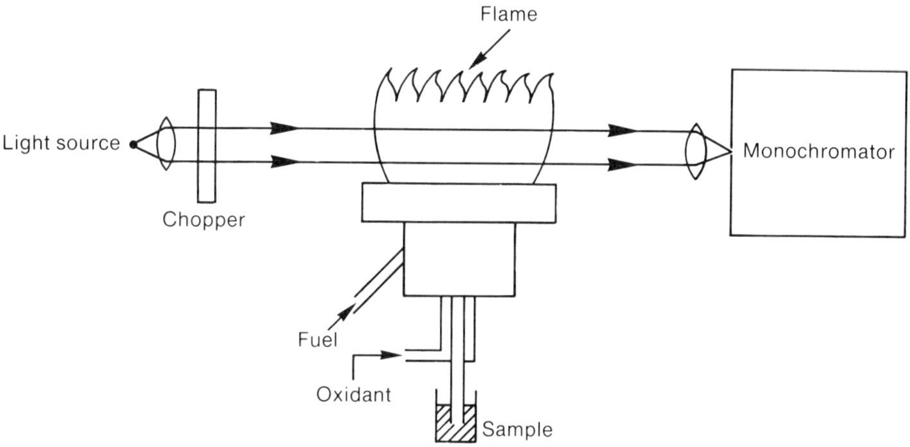

Figure 8-1 Schematic diagram of an emission/absorption spectrometer. The flame-emission spectrum is recorded by switching off the light source.

width of each line being of the order of 10^8 Hz as determined principally by Doppler and collision broadening of the line spectrum. Spectra of liquid and solid samples present much wider peaks as a result of the strong interactions between the closely-packed atoms and molecules in the condensed matter. Such line broadening decreases the amount of information which can be obtained from the spectrum of a liquid or solid material. However, the information contained in such spectra is often sufficient to identify and quantitatively evaluate the relative proportions of a given compound in the inspected material, while the decreased spectral sharpness relaxes the resolution requirements for the spectroscopic equipment. As a result, simple and inexpensive spectrometers can often be used for industrial inspection purposes, and in some cases detection of light at two specific wavelengths using two bandpass filters is sufficient for in-process analytical purposes.

Another distinction is often made between atomic and molecular spectroscopy, depending on whether the transition energy levels correspond to the quantum states of an atom or of a molecule. Atomic transitions typically take place in the ultraviolet, visible and near-infrared regions and correspond to the passage of an electron from one quantum state to another. Molecular spectroscopy is somewhat broader in scope, since in addition to electronic transitions there is also the possibility of transitions between different vibrational and rotational quantum states. The latter tend to occur in the infrared and microwave regions. From an instrumental point of view, visible and near-infrared spectroscopy is easier to apply in an industrial environment because of the availability of high-intensity thermal sources and effective photo-detectors, as discussed in Sections 3.1 and 3.3. *Near-IR molecular spectroscopy* is thus receiving increasing attention for industrial inspection purposes (Wetzel, 1983). As a general rule, the resonance frequencies of the molecules are higher for lighter atomic groups and strong bonds. Near-IR spectra are thus observed mainly for groups such as C−H, O−H or N−H which are mainly found in organic compounds. This is a major field of industrial inspection, covering industrial materials such as polymers, textiles, pulp and paper, pharmaceutical or food-related products. Morever, near-IR absorption peaks in the 0.7 to 2.5 μm wavelength range are often overtones or cumbinations of fundamentals in the mid-IR range. As a consequence, the absorption is weak requiring interaction lengths often of the order of millimeters in a solid sample for good visibility. Such an inspection can thus naturally be performed *in situ* on sheetlike products (in transmission) or on translucent materials (by diffuse reflectance). This is in contrast to the need for elaborate sample preparation techniques including fine grinding, diluting in a solvent and production of a thin layer such as an oil mull or a film, which are normally required in mid-IR spectroscopy. (Kendall, 1966, Chapter 4).

Spectroscopic Instrumentation

From an optical-instrumentation point of view, one can distinguish spectroscopic techniques based on nondispersive analysis, wavelength scanning or Fourier-transform infrared (FTIR) spectroscopy. Nondispersive analyzers use two or more bandpass optical filters to compare the material absorbance at certain characteristic wavelengths. Examples of nondispersive analyzer architectures are shown in Figure 5-12. Another method is to use a tunable laser, such as a dye laser, as a variable-wavelength light source (laser spectroscopy). Advantages of the nondispersive approach are speed and simplicity of the required apparatus, leading to low instrumentation costs and minimum signal-processing requirements. Disadvantages are low flexibility, requiring a different filter for each spectral range to be probed, limited analytical power and difficulty to compensate for environmental perturbations such as spectrally-selective beam attenuation by fumes or vapor in the atmosphere.

Wavelength scanning instrumentation normally uses dispersive optical components, mainly gratings, to scan through the optical spectrum, as shown in Figure 8-2a. A large fraction of the optical intensity emitted from the light

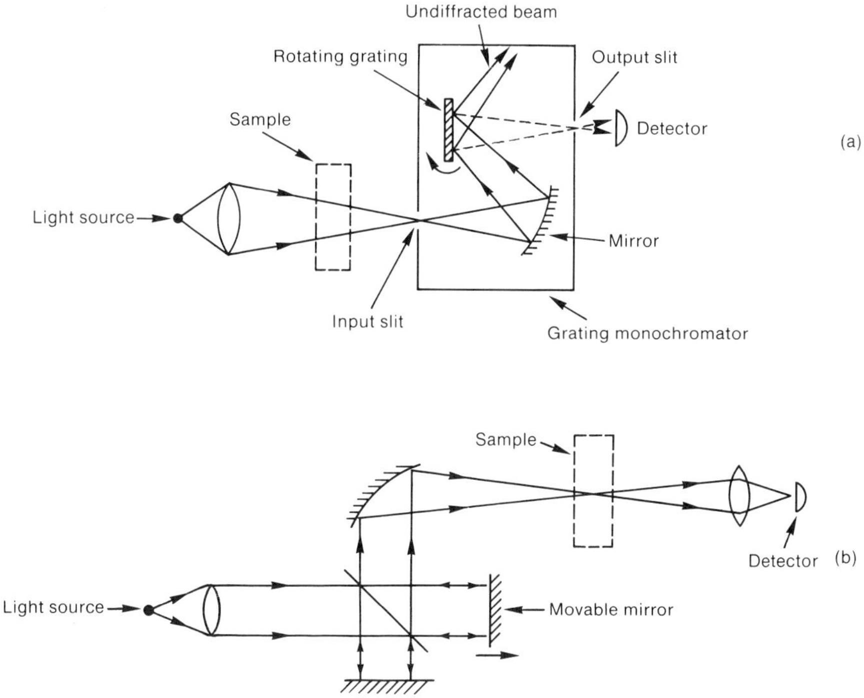

Figure 8-2 Typical configurations for (a) a wavelength-scanning spectrometer and (b) an FTIR spectrometer.

source is wasted in this case for two reasons, namely: (1) the input slit normally selects a small area of the light source image, and (2) only a small fraction of the diffracted beam is collected at a given time by the output slit. Both sources of optical power loss are less severe if a reduced spectral resolution is requested, in which case the width of the monochromator slits can be increased. A way to eliminate the second source of intensity losses is to use a detector array rather than a single detector with angular scanning. In such a case, the whole spectrum is continuously recorded in parallel, while the requirement for a mechanical scanning system is eliminated. Rugged and fast spectrometers for industrial tasks using such a design are commercially available for near-IR spectroscopy, where convenient detector arrays are more widely found.

The use of FTIR spectrometers (Bell, 1972; Griffiths and de Haseth, 1986) avoids both of the above mentioned sources of optical intensity losses. As shown in Figure 8-2, the monochromator is in this case replaced by a

Michelson interferometer, one of whose mirrors is linearly scanned to obtain the detected interferogram which is Fourier transformed to produce the spectrum. The interferometer receives information about the entire spectral range during an entire scan, while the wavelength-scanning spectrometer analyzes only a narrow band at a given time (Fellgett advantage). Furthermore, the interferometer can be operated with a large circular source at the input with no strong limitation of the collimation-dependent resolution, rather than requiring a narrow slit at the entrance pupil as is the case for the monochromator (Jacquinot advantage). The high light collection efficiency makes the S/N ratio of FTIR spectrometers particularly attractive for high resolution spectra, although the performances of practical devices are well below the theoretical predictions (Griffiths and de Haseth, 1986, p. 275).

The requirement for a very good continuous alignment of the linearly scanned mirror makes such an instrument rather delicate and expensive. FTIR spectrometers have thus been mainly limited until now to laboratory use, although rugged, miniaturized versions of the kind shown in Figure 8-3 are under development (Truett et al., 1984). In this case, the variation in the interferometer path-difference is performed by rotation a refractor plate.

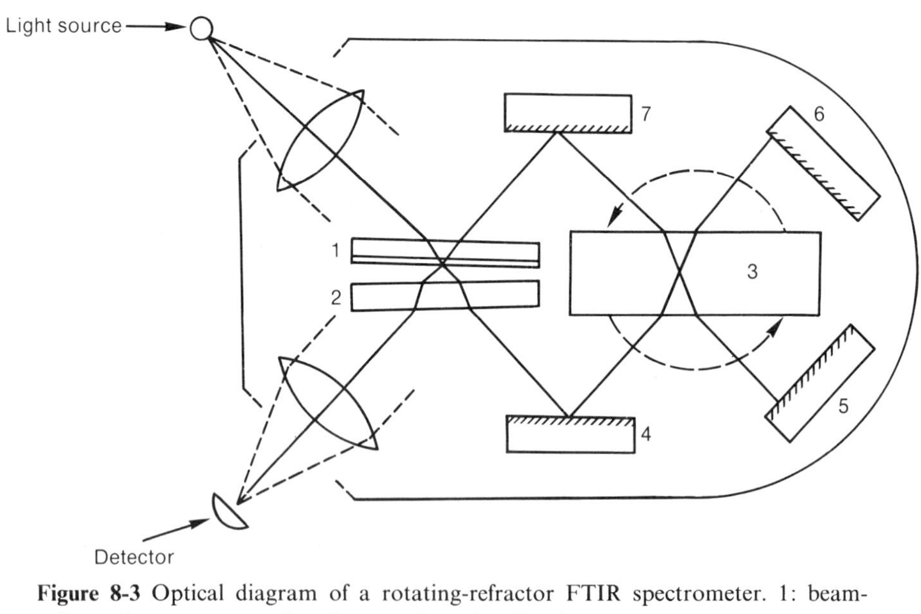

Figure 8-3 Optical diagram of a rotating-refractor FTIR spectrometer. 1: beam-splitter; 2: compensator; 3: refractor plate; 4 to 7: mirrors.

Alignment needs are thus greatly relaxed, while the symmetric design makes the instrument less sensitive to ambient vibrations. The optical head measures just 10 × 10 × 15 cm, and the sample compartment may be situated anywhere between the light source and the beam-splitter or between the beam-splitter and the detector. Vibration- and temperature-shielded FTIR spectrometers are increasingly being used in hostile industrial environments for quality and process control purposes (Hewitt et al., 1985), sometimes in combination with other chemical analysis tecniques such as gas and liquid chromatography (Compton et al., 1985). The ability to obtain full-spectrum information coupled to the data-processing capabilities of the built-in computer are making the FTIR approach attractive for multicomponent analysis (La Butti et al., 1980; Roth et al., 1985) compared with the more rigid discrete-wavelength process analyzers which are described in Section 8.3.

8.1.2 *Transmission and Reflection Spectroscopy*

In the laboratory, spectral measurements are usually made in transmission, if possible. Gases and liquids are enclosed in cells with transparent windows. Solids are measured in the form of a thin film or as a finely ground powder dispersed in an oil mull or pressed in a pellet with an nonabsorbing matrix. The sample is irradiated by a light source and the spectral distribution of the transmitted beam is analyzed by a suitable spectrometer as shown in Figure 8-2.

For industrial applications, such as the *in situ* analysis of massive solid materials or the characterization of thin surface layers, reflection spectroscopy is a convenient approach (Wendlandt and Hecht, 1966; Kortüm, 1969; Harrick, 1977). Reflection spectroscopy may assume different forms, as Figure 8-4 shows.

A diagram of a reflection-absorption spectroscopy configuration (see, e.g., Golden, 1985) is shown in Figure 8-4a. A parallel-polarized light beam is incident on a film-coated metallic substrate at a near-grazing angle (see Section 8.2.1). The highly reflective substrate does not affect the spectral distribution of the reflected beam. Thus the spectral distribution is determined only by the absorption peaks of the film through which the beam is doubly transmitted. The recorded spectrum is very similar to a transmission spectrum of the free-standing film.

Internal-reflection spectroscopy (Harrick, 1979) is mainly useful for the inspection of high-absorbance materials. A typical configuration is shown in Figure 8-4b. A light beam is multiply reflected inside a transparent prism coated with the material to be analyzed. Fluid-cell instruments where the

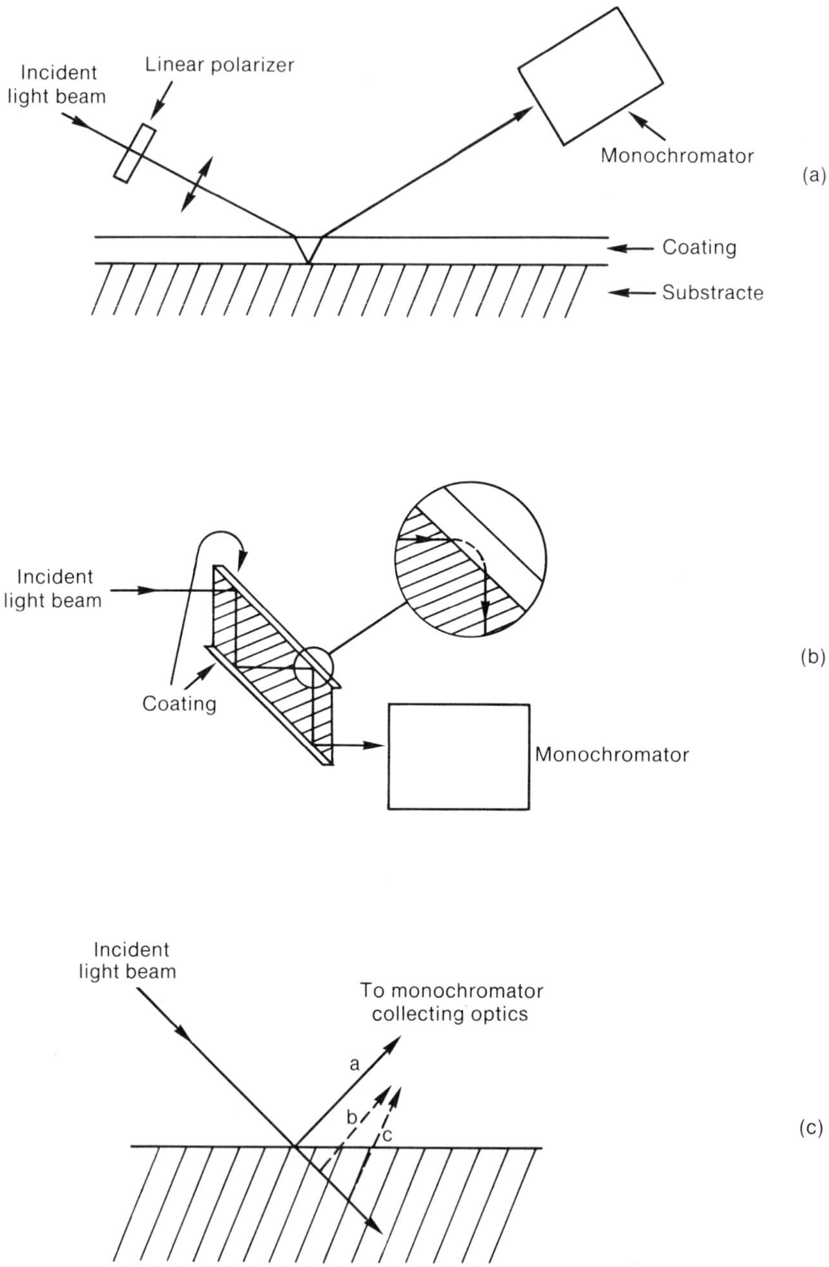

Figure 8-4 Different forms of reflection spectroscopy: (a) reflection-absorption spectroscopy; (b) internal-reflection spectroscopy; (c) diffuse-reflectance spectroscopy.

prism surface is in contact with the gas or liquid to be examined are also available. As shown in the insert, the beam is incident on the prism's internal surface at an angle larger than the critical angle for total internal reflection, so that it is reflected back into the prism. An evanescent wave penetrates, however, by a fraction of a wavelength within the material to be inspected. The spectral distribution of the beam detected after a number of reflections within the prism corresponds to the absorption spectrum of the medium in contact with the transparent prism. This technique may be convenient for a rapid qualitative analysis of industrial materials. As an example, a polymeric resin may be analyzed for composition, degree of curing or moisture content by simply pressing it against the external bare surface of the prism. This is much quicker than removing a sample of the resin, diluting it in a neutral solvent and analyzing the sample by transmission spectroscopy. Only the near-surface medium can, however, be analyzed by the internal-reflection technique. A high sensitivity internal-reflection probe can be obtained by replacing the prism, where only a limited number of reflections may take place, with an optical fiber of appropriate spectral transmissivity and sufficiently high refractive index. The fiber may be immersed in the fluid or viscous medium to be analyzed (Lew et al., 1984; Newby et al., 1984; Owen, 1985) or may be incorporated in a polymeric material before curing for an *in situ* analysis during polymerization or during the service life.

Figure 8-4c shows schematically the beam path within a translucent material in the case of diffuse-reflectance spectroscopy (Wendlandt and Hecht, 1966; Kortüm, 1969; Wetzel, 1983). The diffuse-reflectance approach is quite convenient because it can be rapidly applied to solid materials with little or no sample preparation requirements. In the case of highly absorbing powdered materials, better spectra are obtained if the sample is diluted by mixing with particles of a highly reflective and spectrally neutral compound such as potassium bromide. Examples of industrial applications are moisture monitoring in detergent powder, cosmetics, corn, coal, paper or textile products (Wetzel, 1983) or *in situ* evaluation of the degree of curing of graphite-epoxy prepregs (Cole et al., 1985). Diffuse-reflectance spectra may, however, differ appreciably from absorption spectra, since the relative intensity of the spectral peaks may vary considerably depending on the degree of finish of the inspected surface, the polarization and angle of incidence of the probing beam, as well as the local degree of material porosity. Because of these complications, caution must be exercised in applying the technique for quantitative analysis.

The features of diffuse-reflectance spectra may be understood with reference to Figure 8-4c. If the probing light beam is incident on a translucent or porous material, part of the light is reflected at the surface (beam a in Figure 8-4c). The surface reflectivity R of a given medium is governed by the laws of Fresnel and depends on the angle of incidence and the polarization of the beam, as well as on the index of refraction n and the absorption coefficient k of the material. For normal incidence we can write (Wendlandt and Hecht, 1966; Cole et al., 1986)

$$R = \frac{(n-1)^2 + (k/4\pi\bar{\nu})^2}{(n+1)^2 + (k/4\pi\bar{\nu})^2} \tag{8-1}$$

where $\bar{\nu}$ is the frequency in wave numbers. The portion of the beam which is not reflected at the surface penetrates within the material and may eventually be scattered back toward the spectrometer collecting optics by discontinuities or reflecting impurities within the material (beams b and c in Figure 8-4c). As in the case of Figure 8-4a, such a bulk-reflected beam has been selectively absorbed by the material encountered along its path, so that its spectrum is essentially similar to an absorption spectrum. The magnitude of the bulk-backscattered reflectivity R_∞ for an infinitely thick scattering sample is determined by the absorption coefficient k and the scattering coefficient s of the material following the Kubelka-Munk expression (Wendlandt and Hecht, 1966; Kortüm, 1969; Cole et al., 1986)

$$\frac{k}{s} = \frac{(1-R_\infty)^2}{2R_\infty}, \tag{8-2}$$

which implies that $R_\infty = 1$ if the material has a scattering coefficient much larger than the absorption coefficient (such as white paper), while $R_\infty = 0$ if the material has a large k/s ratio (such as a highly absorbing material or else a moderately absorbing but highly homogeneous material with very few backscattering impurities).

Both the surface-reflected and the bulk-reflected beams are collected by the spectrometer, the relative intensities of the two components being dependent on the material condition as well as on the beam polarization and direction with respect to the detector angular position. The intensity of the surface-reflected component may be reduced by using parallel polarization at a Brewster incidence and by avoiding specular reflections. Such a component cannot, however, be eliminated when a matt-finish surface is inspected. On the other hand, it should be realized that the bulk-reflected light depends on the amount of Fresnel reflections at the interface of the grains within a porous material, so that Fresnel reflectivity and bulk absorptivity both affect the backscattered spectrum.

As equation (8-1) indicates, the spectral distribution of the Fresnel-reflected component may differ substantially from the bulk-reflected light. For a typical transparent material such as glass or plastics, for which $n \approx 1.5$ and $k = 0$, we have $R \approx 4\%$, while in the region of an absorption band k passes through a maximum and n varies according to a dispersion-type curve, as shown in Figure 8-5 (see, e.g., Born and Wolf, 1970, Chapter 2, or Harrick, 1977). The surface reflectivity may then, paradoxically, increase substantially at frequencies where the sample strongly absorbs (reststrahlen effect). In the limiting case of very large k, R approaches 1 from equation (8-1). The surface reflectivity component may thus have a different spectral distribution than the bulk reflectivity, which presents negative peaks at the absorption frequencies. As a consequence, a diffuse-reflectance spectrum may be used to identify a material from the position of the spectral peaks, but if front-surface reflection effects causes distortion of the peak shapes it becomes difficult to obtain accurate quantitative information, as exemplified in Figure 8-6 (Cole et al., 1985).

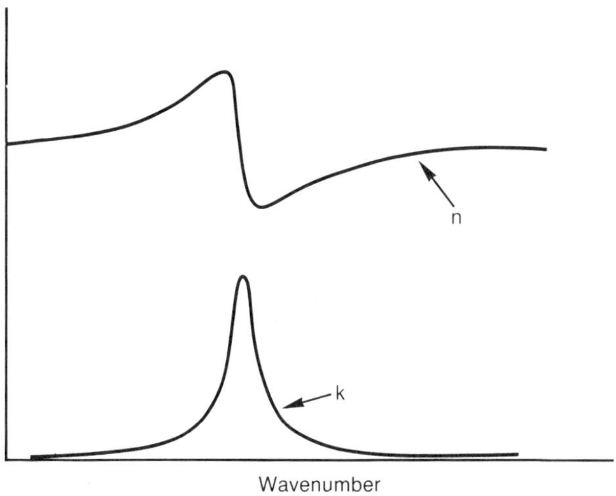

Figure 8-5 Schematic representation of the variations of the real refractive index n and of the absorption coefficient k of a material near an absorption band.

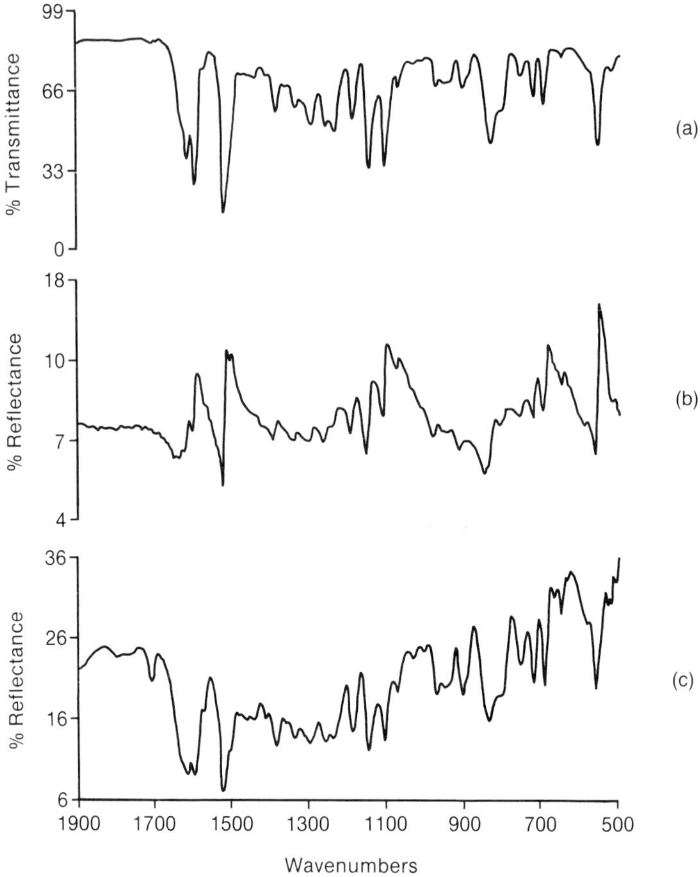

Figure 8-6 Infrared spectroscopy of a carbon-fiber-reinforced epoxy material: (a) transmission spectrum of a thin film of epoxy resin; (b) front-surface-reflectance spectrum of a smooth resin film; (c) diffuse-reflectance spectrum of the carbon/epoxy prepreg. The epoxy is Narmco 5208.

8.1.3 *Raman and luminescence Spectroscopy*

Raman spectroscopy (Rosenbaum et al., 1960; Olsen, 1975, Chapter 6) has become popular since the advent of the laser. The classical molecular-absorption infrared spectroscopy and the Raman technique provide similar information on the vibrational or rotational levels of the analyzed material but are based on different physical mechanisms. Infrared spectroscopy is based on the absorption of the incident photon by the molecule or radical, which is raised to an excited energy level as a consequence of such an absorption. In Raman spectroscopy, the incident photon is scattered by the molecule which may be induced by the collision to undergo a rotational or vibrational change. As a result, the scattered frequency differs from the frequency of the incident wave by an amount corresponding to the rotational or vibrational energy. The selection rules which determine the visibility of a transition line in the recorded spectrum are accordingly different for the two spectroscopic techniques. The probability of Raman transitions depends on the molecular polarizability, while in infrared spectroscopy only transitions which induce a change in the molecular dipole moment are allowed.

Figure 8-7 shows a typical configuration illustrating the formation of a Raman spectrum. An intense monochromatic beam of fequency ν_0 is incident on the medium whose molecules are characterized by a given energy-level diagram, Figure 8-7a. Most of the incident power is undeviated, while a small fraction, typically 0.1%, is scattered in all directions. The radiation scattered along the observation direction contains a large proportion of elastically scattered photons at the same frequency δ as the incident beam, but in addition contains frequencies $\delta_0 - \delta_1$ or $\delta_0 + \delta_1$, corresponding to inelastic scattering from a molecule which undergoes an increase or a decrease, respectively, of its rotational or vibrational energy level. As Figure 8-7a shows, the Stokes transitions which correspond to a decrease in the scattered light frequency raise the molecule to a higher excited level, while the anti-Stokes lines take place when the initially-excited molecule makes a transition to a lower excitation level as a result of the interaction with the incident photon. Since at ordinary temperatures the population of molecules at the ground level is always greater than in excited levels, the intensities of the anti-Stokes lines will always be weaker than those of the Stokes lines, as Figure 8-7b shows.

Because only a small fraction of the incident light beam is scattered, while only a small fraction of the scattered light is shifted in frequency, Raman scattering accounts for less than 10^{-7} of the incident light intensity, even considering the full $4\pi^2$ steradians scattering aperture. The intensity of the Raman signal is thus extremely low, which explains why such a technique has become popular only after the development of high-power lasers. Typical lasers used in Raman spectroscopy are argon lasers which can deliver an intensity of a few watts in a single line at 488 or 514.5 μm. The shifted wavelengths can thus be observed in the neighborhood of such lines. Special care must be used to avoid spurious reflections of the undeviated laser beam.

On the other hand, Raman spectroscopy has the advantage over infrared

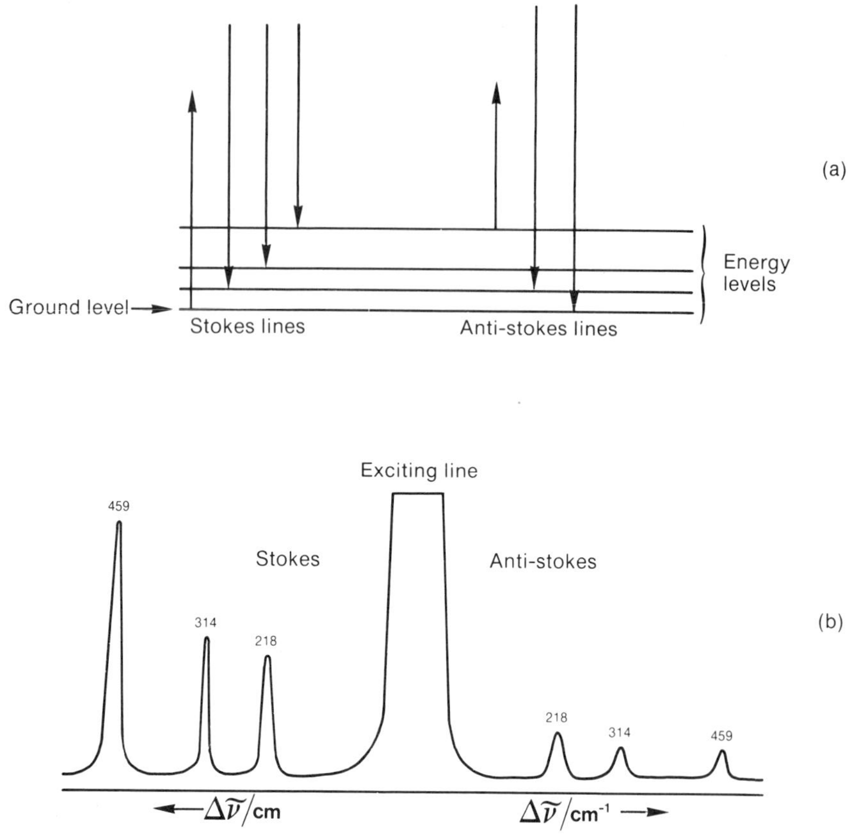

Figure 8-7 (a): Schematic diagram illustrating the typical energy transitions in Raman scattering; (b) Raman spectrum of CCl_4 showing Stokes and anti-Stokes lines.

spectroscopy of shifting the operating frequency to the visible range. Consequently, water in the analyzed sample is not a major problem, whereas its opacity in the infrared region prevents its use as a solvent in infrared spectroscopy. Cells and optical windows may be made out of glass, as opposed to the requirement for more expensive and less rugged infrared materials for infrared spectroscopy (see Section 3.7.2). Finally, optical detection takes place in the visible, where sensitive photoemissive or photoconductive devices are available (see Section 3.3).

The short-wavelength laser beam which is used in Raman spectroscopy can be focused to a very small spot for a microscopic spectral analysis of a sample surface. Imaging microprobes can thus be developed where a 1 μm-diameter laser spot is scanned over the sample surface to produce a map of a component distribution or to detect impurities in semiconductor materials (Dhamelincourt et al., 1979).

Another interesting feature of Raman spectroscopy is the possibility of studying low-frequency transitions, in the 100 cm^{-1} range or less, without having to resort to far-infrared spectroscopy, where light sources are inadequate and detectors have a low sensitivity. Low-frequency data may provide information on low-energy bonds such as intermolecular interactions in partially-crystalline solids. An important industrial application is the evaluation of the degree of chain-orientation and crystallinity in polymeric yarns (Bulkin et al., 1985). Moreover, the heat contents, free energy, entropy and heat capacity can be calculated from fundamental vibrational frequencies, bond distances and angles by means of statistical mechanics. Partially-oriented yarn material under different annealing conditions was recently characterized by Raman probing of the vibrational density of states through an analysis of the phonon distribution within the solid (Bulkin et al., 1986).

The very low signal level to be probed in Raman spectroscopy makes the industrial applications of such a technique a difficult task. Nevertheless, the applicability of Raman spectroscopy in particularly hostile environmental conditions, such as for the *in situ* characterization of thermal-barrier coatings under attack by molten sodium vanadate at 700°C (Hamilton and Nagelberg, 1984) or coating monitoring for offshore structures (Dao et al., 1986) shows that the industrial potential of this technique must not be underestimated.

Fluorescence and Phosphorescence Spectroscopy

In contrast with Raman spectroscopy, where the incident photon only plays the role of stimulating a molecular transition under conditions of inelastic scattering, luminescence spectroscopy (Ewing, 1969, Chapter 4; Olsen, 1975, Chapter 2) requires the actual excitation of the molecule to a higher energy level, from which a decay takes place after a relatively long time period. Consequently, the wavelength of the incident beam must correspond to an absorption band of the molecule in luminescence spectroscopy, while in Raman scattering the choice of the incident laser wavelength is relatively unimportant.

The molecular mechanism for luminescence spectroscopy is illustrated in Figure 8-8. The absorption of an incident photon raises the molecule from the ground electronic-vibrational state to an excited electronic-vibrational state. When a molecule is in a stable excited state, as often happens with

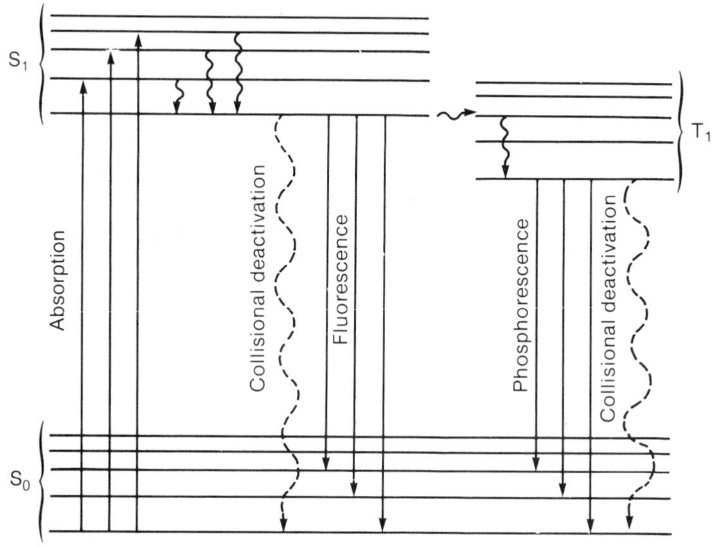

Figure 8-8 Energy-level diagram showing the energy transitions involved in luminescence spectroscopy (S_0 = singlet ground state; S_1 = singlet first-excited state; T_1 = triplet first-excited state).

large organic molecules, the excess vibrational energy will be dissipated by collisions while the molecule remains in the excited state; the molecule may then decay radiatively from the fundamental vibrational state of the excited electronic level to the ground level after a decay half-time which is usually in the 10^{-4} to 10^{-8} seconds range (fluorescent decay). Alternatively, the molecule may shift to a metastable triplet state from which it may decay after a time delay which is usually much longer, in the 10^{-4} to 20 seconds range (phosphorescence), provided that the molecule has not undergone a collisional deactivation in the meantime.

It is clear from the energy-level diagram of Figure 8-8 that the decay-photon energy is normally smaller than the energy of the incident photon. Consequently, reemission takes place at wavelengths which are equal to or longer than the exciting wavelength. Also, absorption and emission spectra show an approximate mirror symmetry. Figure 8-9 shows typical spectra of an organic molecule such as anthracene. Fluorescence is normally observed by exciting the molecule with a near-ultraviolet laser or mercury-discharge beam at a specific wavelength, while the emission spectrum is analyzed with a spectrometer at an angle from the exciting-beam direction. As the molecular excitation and radiative emission take place at different times, no correlation

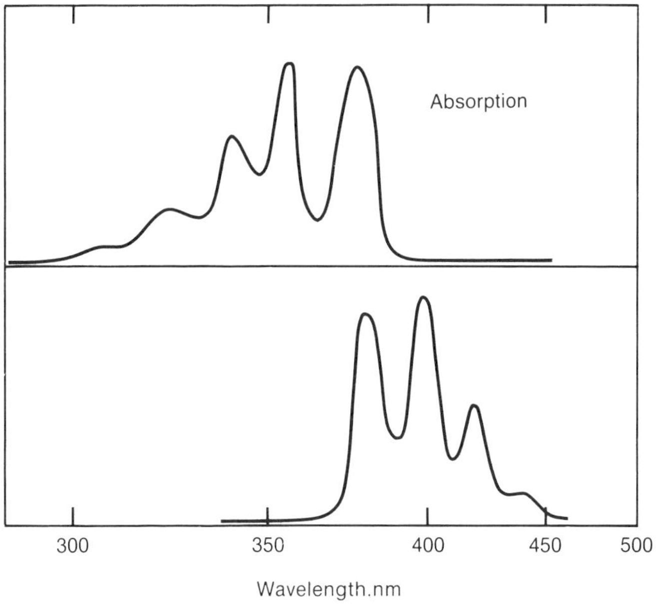

Figure 8-9 Typical absorption and emission spectra of a fluorescent molecule.

exists between the angular directions of the pumping beam and of the emitted radiation.

Luminescence spectroscopy has the advantage over absorption spectroscopy of a higher sensitivity, mainly because in absorption spectroscopy the detector must distinguish a small difference between two large signals, each of which is affected by a large superimposed detector noise, while in luminescence the detector must distinguish a small emission signal against a black background. The spectral selectivity may be enhanced by studying both the absorption curve at a fixed emission wavelength and vice versa or else by applying time-resolved techniques. Some practical problems exist, however, in the quantitative evaluation of luminescent spectra, the most important of which is the possibility of a radiationless collisional deactivation of the excited molecule before the fluorescent or phosphorescent photon emission may take place (see Figure 8-8). Such a mechanism, which depends on a number of factors such as the sample composition or the ambient temperature, must sometimes be minimized by methods such as the inclusion of the material in a rigid medium to reduce collisonal processes. Indeed, the most common technique for observing phosphorescence is to freeze the phosphorescent molecule in special glass which is frozen at liquid-nitrogen temperature (Olsen, 1975, p. 386).

8.2 Spectroscopy of Industrial Materials

Some examples of spectroscopic material characterization of particular interest for inspection on the industrial floor will now be given. This section describes some applications which are still in the exploratory or developmental phase, while the next section will cover some examples of spectroscopic process analyzers which are more established as on-line sensors for industrial quality control or process monitoring. The emphasis is put on general-purpose spectroscopic devices in this section, while a number of specialized sensors for monitoring a specific material property will be described in the next section.

8.2.1 *Analysis of Surface Layers*

Surface quality is important in a number of industrial applications. The presence of oxides or of organic contaminants may adversely affect the spot weldability and the paintability of metallic sheets (Coduti, 1980; Oles and Perfetti, 1980) or decrease the adherence of coatings or of adhesive layers. Although a number of powerful methods are available for the laboratory chemical analysis of surface composition, such as Auger or ESCA spectroscopy (Slane et al., 1978), such techniques are not adequate for on-line automated inspection of the surface state. Optical spectroscopy is an attractive approach for such inspection tasks because of its nondestructive nature and remote-operation capability. A number of applictions where the sensitivity of optical spectroscopic techniques appears to be adequate will now be described.

Measure of Zinc-Phosphate Primer Coatings

An important step in the metal-finishing industry for coating a wide variety of products is the phosphating process (Burns and Bradley, 1959, p. 545). A crystalline coating of zinc phosphate, in the 1 μm thickness range, is produced by exposing the surface to an aqueous mixture of phosphoric acid, zinc phosphate and oxidizing agents. This coating provides a base for impregnation with paints, lacquers or enamels. while protecting the metal against corrosion by improving paint adhesion, restricting moisture penetration and inhibiting the spread of underfilm corrosion. Quality control of the measurable parameters of the phosphate coating, such as thickness, uniformity, porosity and adhesion is an important concern in the metal-finishing industry, particularly for automotive applications.

The suitability of reflection-absorption spectroscopy (Figure 8-4a) for the evaluation of phosphate-coating weight has recently been reported (Cheevers, 1978; Mansour, 1983). The spectra of the light reflected by the coated surface under different coating-weight conditions are shown in Figure 8-10. The three absorption peaks in the 8.5 to 12 μm area are due to the phosphate coating. By taking the ratio of the spectral reflectivity at 9.1 and 8.3 μm, the film absorbance could be calculated using equation (5-1). The absorbance shows a satisfactory linear correlation with the coating weight in the 0.7 to 2 μm film-thickness range. The conditions for this spectroscopic application are relatively favorable, the 8 to 11 μm spectral range being free from atmospheric absorption (see Figure 7-11), while relatively inexpensive thermal sources and sensitive detectors are available in this region (see Sections 3.1.2 and 3.3).

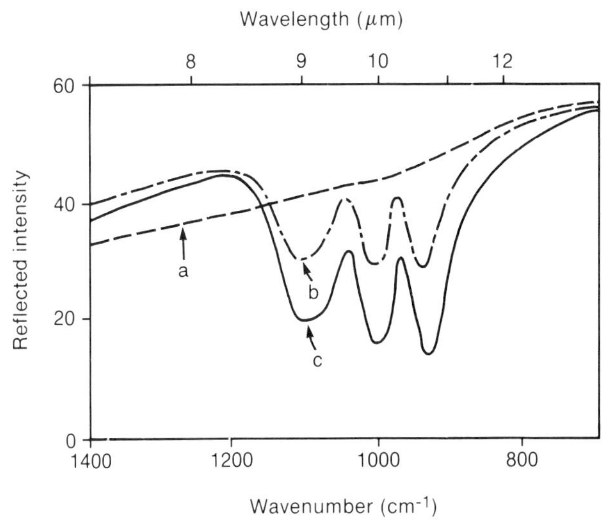

Figure 8-10 Spectral distribution of light reflected from zinc-phosphated panels: (a) base panel; (b) and (c): panels with different coating thickness.

Spectroscopic Measurement of Surface Cleanliness

Another field of possible applicability of spectroscopic methods to surface inspection is the evaluation of surface cleanliness. As mentioned above, surface organic deposits such as oils or grease may have detrimental effects on coating adhesion. Moreover, rolling-mill companies have an interest in determining the cleanliness of the metal sheets in the cold-rolling process before the bell furnace. An excessive amount of oil which is carbonized in the furnace results in a large amount of carbon on the steel surface, which reduces the phosphatability and decreases the efficiency of the anti-corrosion treatments. A fluorescence approach was recently used by a Swedish research group (Galle et al., 1979; Herder et al., 1982) for a remote evaluation of the oil-film thickness at the surface of the steel sheet.

Most oils fluoresce substantially under ultraviolet illumination. Figure 8-11a shows the fluoresence spectra in the visible range of some typical roll oils under ultraviolet excitation (Herder et al., 1982). The excitation is conveniently provided by a compact nigrogen laser giving 10 ns, 4 mJ pulses at a wavelength of 337 nm. Figure 8-11b shows the fluorescent intensity at 420 nm as a function of the oil film thickness. An approximately linear relationship between the emission intensity and the oil thickness is obtained providing a quantitative means for film-thickness evaluation. For other oils, the fluorescent emission under 337 nm excitation saturates as the film thickness increases, showing that the exciting UV light penetrates less than 20 μm within the oil film. A linear relationship within the 0 to 20 μm film-thickness range of interest could, however, be obtained by using a longer excitation wavelength, such as a nitrogen-laser-pumped dye laser at 480 nm. On the other hand, the 337 nm excitation wavelength may be used to evaluate much thinner oil films, down to 0.03 μm-thick. A different approach using a spectral scan in the visible range for the measurement of surface cleanliness on pickled steel sheet was also reported recently (Torao et al., 1986).

Diffuse-reflectance was recently considered for the detection of contamination by mold-release agent on the surface of epoxy-based composites (Cole et al., 1986). Such a substance is applied to the mold surface to facilitate removal. Transfer of release agent to the surface of the reinforced-plastics part can interfere with subsequent processing such as painting or adhesive joining. Diffuse-reflectance FTIR spectroscopy was shown to be effective for release-agent quantification with a detection limit of 200 μg/cm^2.

A similar experiment has recently been presented for the detection of contaminants within narrow tubes (Ottesen, 1986). Figure 8-12a shows the

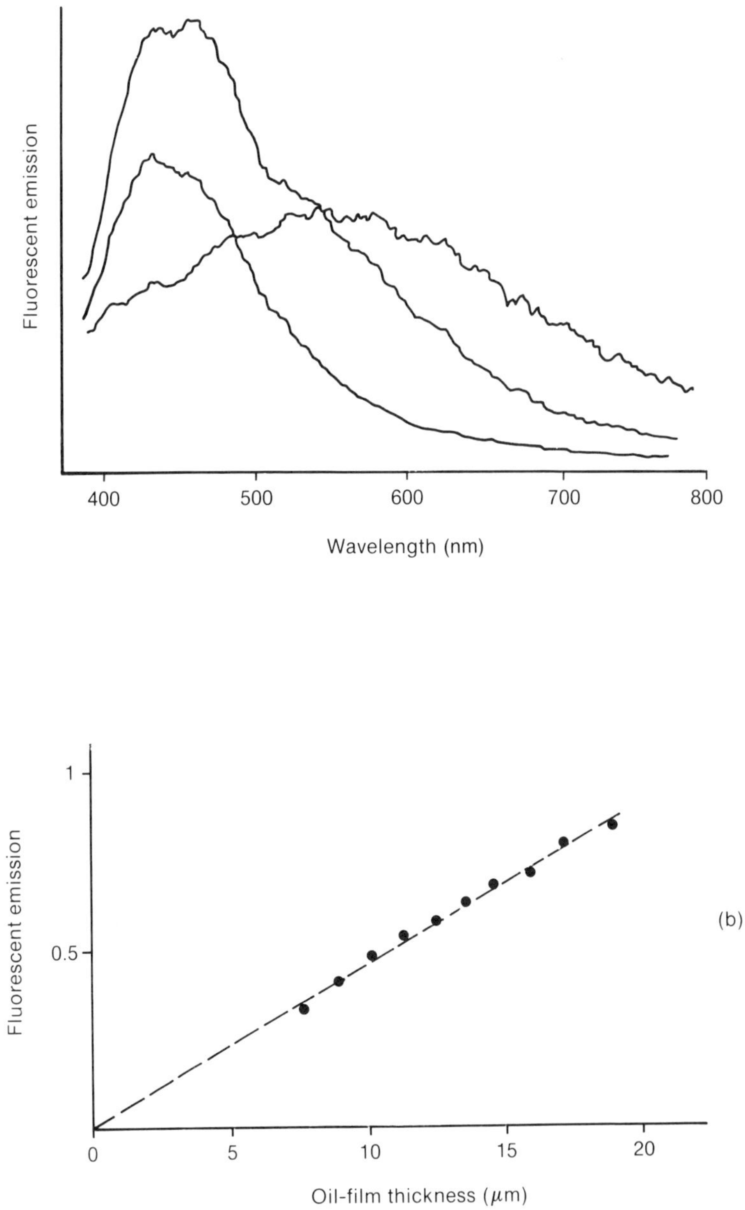

Figure 8-11 Fluorescence emission under ultraviolet excitation from a grease-oil film on a metallic surface; (a) fluorescence spectra of some typical roll oils; (b) emission intensity as a function of oil thickness.

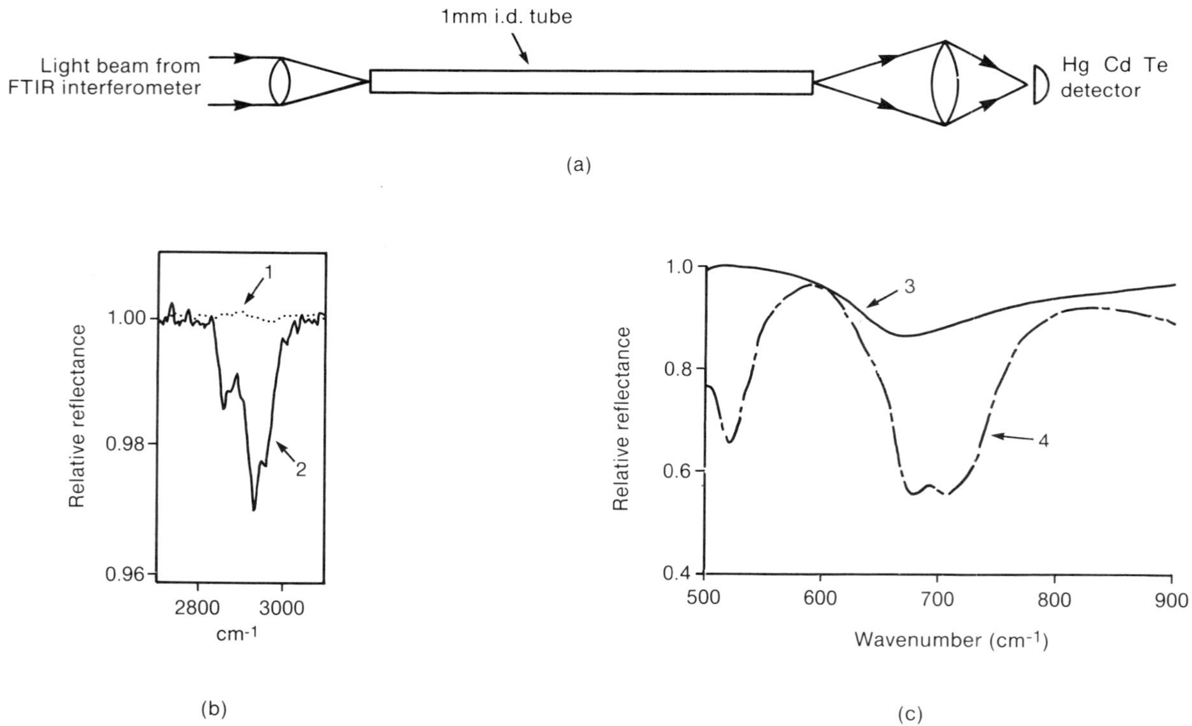

Figure 8-12 Spectroscopic detection of contaminants within a tube by reflection spectroscopy; (a) experimental setup; (b) near-IR spectrum of a clean (1) and an as-machined (2) tube; (c) mid-IR spectrum of a tube oxidized in air for an hour at 500°C (3) and at 700°C (4).

experimental apparatus. The light beam from the interferometer of an FTIR spectrometer is focused within one end of a narrow-bore stainless-steel tube used to contain high pressure gases. Surface contaminants have been observed to affect the integrity of pinch weldments in these tubes. The light beam is multiply reflected within the tube and is collected at the end by the wide-aperture detector optics. Figure 8-12b gives the near-IR spectrum of a clean (1) and of an as-machined (2) tube, showing evidence of absorption by organic material (machining oil) in the 2900 cm^{-1} range. Figure 8-12c shows the spectrum of a tube which was oxidized by exposure in air during an hour at 500°C (3) and at 700°C (4). Satisfactory agreement was obtained with theoretical calculations using the recorded values of the real refractive index and of the absorption coefficient as a function of the wavelength for the contaminating substances. This is an interesting application of more fundamental grazing-incidence FTIR spectroscopic studies of metallic-oxide films of different compositions in the 3 to 25 μm thickness range (Ottesen and Nagelberg, 1980).

Measure of very Thin Layers

The sensitivity of reflection-absorption spectroscopy to thin layers of contaminants on metal surfaces may be increased by a proper choice of the light beam polarization and angle of incidence (Golden, 1985). At near-normal incidence, the incident and reflected waves form a standing wave with a node at the surface of the metal, i.e., at the metal surface the amplitude of the standing wave is zero. If the electric field is zero, it cannot interact with an adsorbed layer of thickness which is small with respect to the wavelength. It can, however, be shown (Greenler, 1966) that the amplitude of the standing wave can be substantially increased by operating at near-grazing incidence and with a polarization parallel to the plane of incidence. Figure 8-13 shows how the visibility of the absorbance peaks of a thin film of silicone grease on aluminum can be enhanced by using parallel-polarized light at a grazing incidence (Krishnan and Ferraro, 1982). For the inspection of dielectric films of thickness larger than the wavelength, the optimum conditions are, again, a parallel polarization but an angle of incidence close to the Brewster angle. In such conditions, the light beam penetrates within the film without front-surface reflection and without interference fringes, providing a good-quality absorption-type spectrum.

Using parallel-polarized light, polymeric films of thickness down to 9 nm could be sensed by FTIR reflection-absorption spectroscopy (Ishitani et al.,

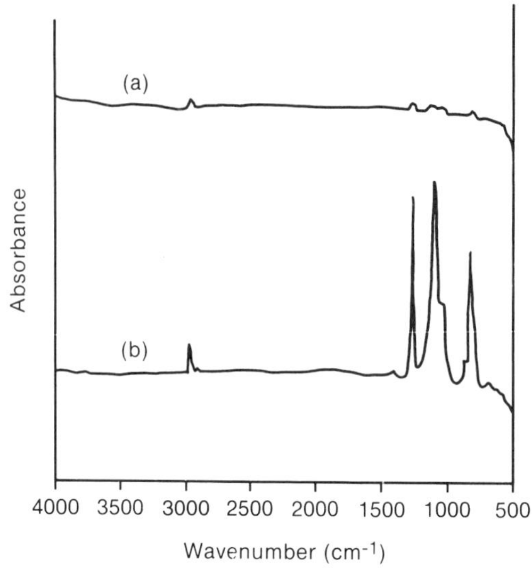

Figure 8-13 FTIR reflection-absorption spectrum of a thin film of silicone grease on an aluminum foil, using (a) a perpendicular-polarized incident beam and (b) a parallel-polarized beam. Both spectra are plotted on the same scale.

1982). Figure 8-14 shows three spectra displayed with different vertical scales for the three indicated film thicknesses (the film thickness was measured by ellipsometry). The absorbance peaks in the 2900 and 1400 cm^{-1} areas are clearly visible in the case of the 26 nm-thick film, while being barely discernible for the 9 nm-thick film.

An interesting approach for high accuracy thin-film monitoring makes use of a polarization-modulated incident beam. Figure 8-15 shows the spectrum obtained by such a technique for a very thin film (only 6 monolayers) of cadmium arachidate absorbed on silver (Golden, 1985). The light beam emerging from the interferometer is polarization-modulated from the parallel to the perpendicular direction by a photoelastic modulator (such as a KDP crystal, see, e.g., Yariv, 1976) at a frequency of the order of 100 KHz. The AC component $I_p - I_s$ of the signal is synchronously detected and divided by the average signal $I_p + I_s$ corresponding to the signal obtained with a randomly-polarized incident beam. Such a ratio signal is free from noise arising from absorption by impurities in the atmosphere or from variations in the surface reflectivity of the substrate as well as from low-frequency electronic noise. As mentioned in Section 3.6.3, synchronous detection is also an effective means for discriminating against environmental noise, making

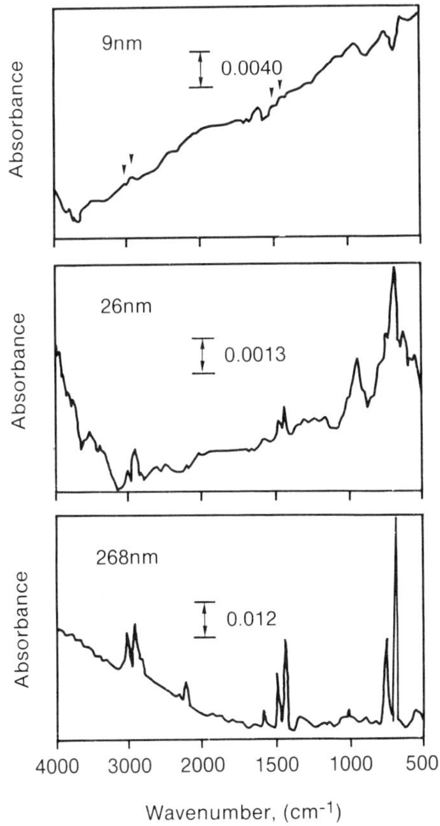

Figure 8-14 FTIR reflection-absorption spectra of thin films of poly(acrylonitrile co-styrene) adsorbed on aluminum.

such an approach convenient for industrial applications where unwanted ambient-light reflections or strong IR emission from high-temperature parts are present.

8.2.2 *Quality Control of Epoxy-based Products*

Spectroscopic techniques are widely used for the investigation of many properties of polymeric materials. Examples are component-proportion quantification, determination of the degree of crystallinity before and after annealing, and monitoring of degradation by oxidation processes as well as by hydrolysis consequent to moisture infiltration (Koenig and Antoon, 1978; May et al., 1978; Liao and Koenig, 1984; Chalmers and MacKenzie, 1985).

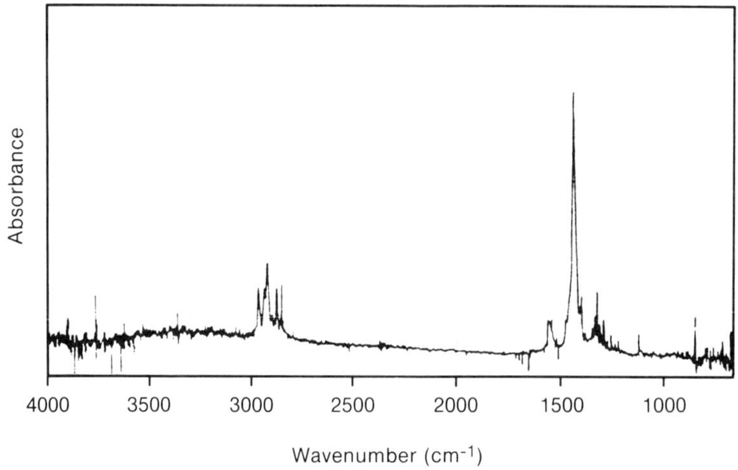

Figure 8-15 Polarization-modulated FTIR reflection-absorption spectrum of six monolayers of cadmium arachidate adsorbed on silver.

The use of Fourier-transform spectroscopic equipment in conjunction with appropriate signal-processing algorithms such as spectrum subtraction and least-square fitting makes it possible to identify and to quantify subtle differences between spectra of different products or of the same product before and after processing.

One of the application fields where the quality-assurance industrial needs are particularly strong is the nondestructive characterization of epoxy-based structural composites. Considerable growth has been experienced in recent years, particularly in aerospace materials based on a combination of high-strength reinforcing fibers, such as graphite or kevlar, with a polymeric matrix, usually epoxy. Such parts are often made from prepreg (fiber reinforcement impregnated with uncured resin) by laying down several layers in a mold and heating under pressure to crosslink the epoxy. Because of the importance of safety considerations in the aerospace industry, as well as of the relatively recent introduction of such materials whose properties are not so constant and widely documented as they are for metals, quality control is an important requirement for the acceptance of composites as structural components. Careful inspection is performed (1) on the raw-material, i.e., on the prepreg sheets as received by the aerospace company, which must be checked for composition and pre-cure level, (2) on the maufactured part after curing, which is tested to verify that the proper degree of curing has been achieved and to spot possible disbonds, and (3) on the finished product after a given service life to verify whether any damage has occurred as a result of fatigue, impacts, etc. or whether the integrity of the structural materials has been affected by environmental agents such as moisture or ultraviolet exposure. Nondestructive testing methods for the assurance of part integrity after assembly have been discussed in detail in Chapter 6. This section describes some infrared spectroscopy techniques for quality control of the prepregs and for the evaluation of the degree of cure and moisture infiltration.

Prepregs are shipped to the manufacturing company in refrigerated containers under strictly controlled pre-cure conditions. The degree of pre-cure must be sufficiently advanced to give the prepreg the required consistence for easy manipulation and to avoid excessive flow of the epoxy matrix during pressure curing. On the other hand, excessive pre-cure would reduce the prepreg chemical bondability and mechanical flexibility, leading to severe disbond problems. Infrared spectroscopy is one of the most effective techniques for the characterization of the degree of cure of epoxy resins, the other standard techniques being liquid chromatography and thermal analysis (May et al., 1979; Liao and Koenig, 1984; Cole and Noël, 1986). Infrared spectroscopy has the advantages of being potentially nondestructive, powerful and useful for both qualitative and quantitative analyses of the material properties.

Figure 8-16 shows the transmission spectra of a typical epoxy resin (Narmco 5208) before and after curing (Cole and Noël, 1986). A number of changes can be observed in the spectral distribution as a consequence of the curing reaction. The N−H stretching bands (3500–3200 cm^{-1}) have been largely replaced by a broad OH band centered around 3400 cm^{-1}. In the C−H stretching region (3100 − 2800 cm^{-1}) the band shape has changed because of the disappearance of the epoxy C−H peak at 2996 cm^{-1}. The epoxy peaks at 1000–900 cm^{-1} have also practically disappeared. Other slight modifications in the relative intensity of the different absorbance peaks can be noticed, as well as the appearance of the new peaks at 1722 and 1667 cm^{-1}, which are, however, due to carbonyl groups formed because of degradation by oxidation of the resin.

The epoxy peaks between 973 and 906 cm^{-1}, which are due to absorption by unreacted epoxy rings in uncured epoxy, are a particularly reliable index for the quantitative evaluation of the degree of cure. The area under such peaks, as indicated in Figure 8-17, may be taken as an index of the amount of curing after normalization by reference with the amplitude of a constant-absorbance peak, such as the one at 1517 cm^{-1}, to compensate for thickness-variation effects. Such an index is plotted in Figure 8-18 together with the

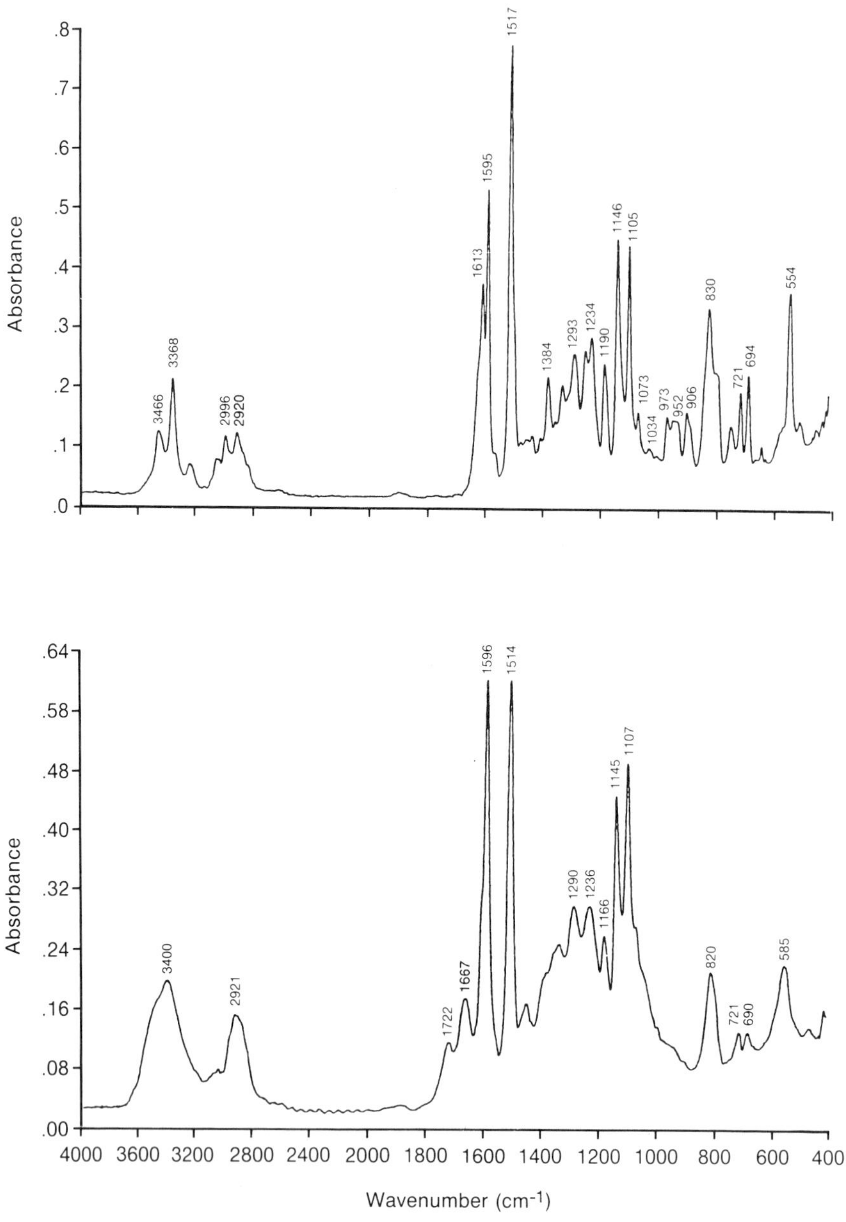

Figure 8-16 Transmission FTIR spectra of Narmco 5208 epoxy resin before (top) and after (bottom) curing.

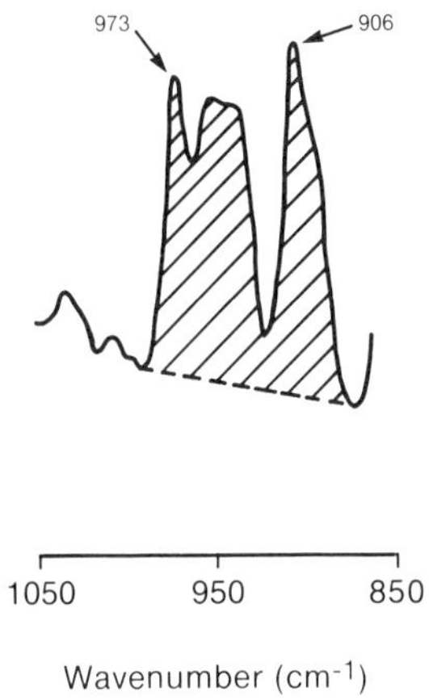

Figure 8-17 Absorbance peaks produced by unreacted epoxy rings in uncured epoxy. The shaded area may be taken as an index to determine the degree of cure.

temperature history to which the epoxy film was subjected during curing. As we can see, curing starts at the first temperature step of 135°C and rapidly accelerates when the temperature is increased to 177°C. By the end of the third hour, only about 15% of the epoxy groups remain. Over the next two hours, there is further gradual reaction, but at the end of the recommended cure cycle (five hours) there are still some unreacted epoxy groups (about 7% of the inital amount).

Epoxy resins may also be characterized by near-IR analysis (Wetzel, 1983; Cole, 1984). Absorption coefficients tend to be much weaker than in the mid-IR, so thicker samples must be used. As mentioned previously, the advantages of working in this region are the availability of strong infrared sources and sensitive detectors as well as the possibility to use inexpensive glass optics. Figure 8-19 shows the near-IR spectra of some epoxy films (Cole, 1984). The disappearance of the epoxy-ring peak at 4524 cm^{-1} (2.21 μm) may be taken as an index of the degree of cure. Also shown in Figure 8-19 is the spectrum of the same cured film after absorption of 5% by weight of water as a result of immersion in boiling water for five hours to saturate the cured resin with moisture (Cole and Noël, 1986). The appearance of the water-absorption band at nearly 5200 cm^{-1} (1.92 μm) shows that the amount of absorbed water can be evaluated by near-IR spectroscopy.

Transmission spectroscopy is a powerful laboratory tool, but it is often not conveniently applicable to on-line inspection. To obtain the previously

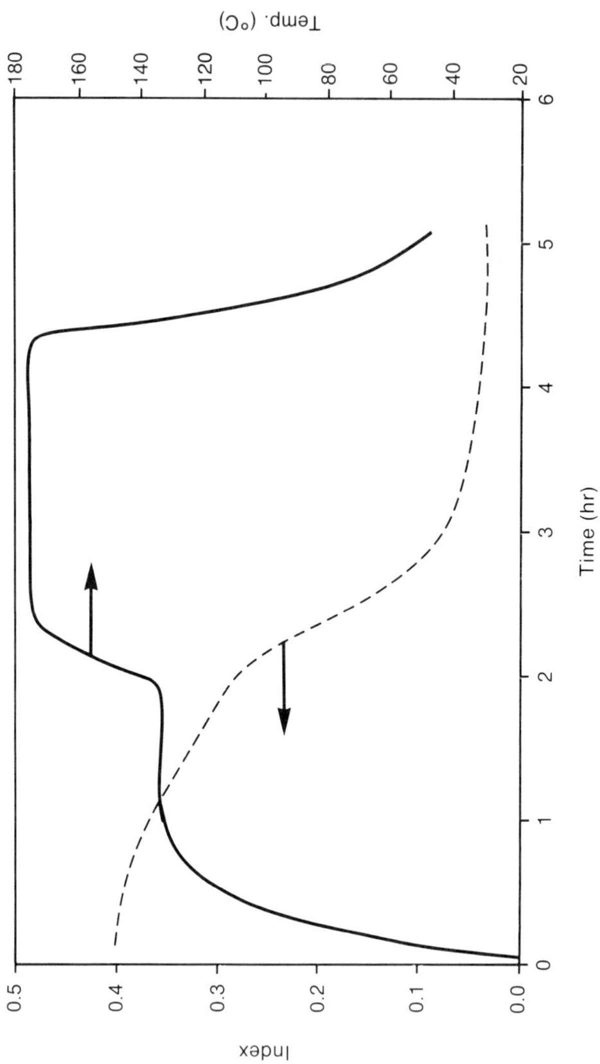

Figure 8-18 Evolution of the epoxy index of cure (extent of the shaded spectral region shown in the previous figure) as the epoxy film is cured by exposure to the indicated temperature program.

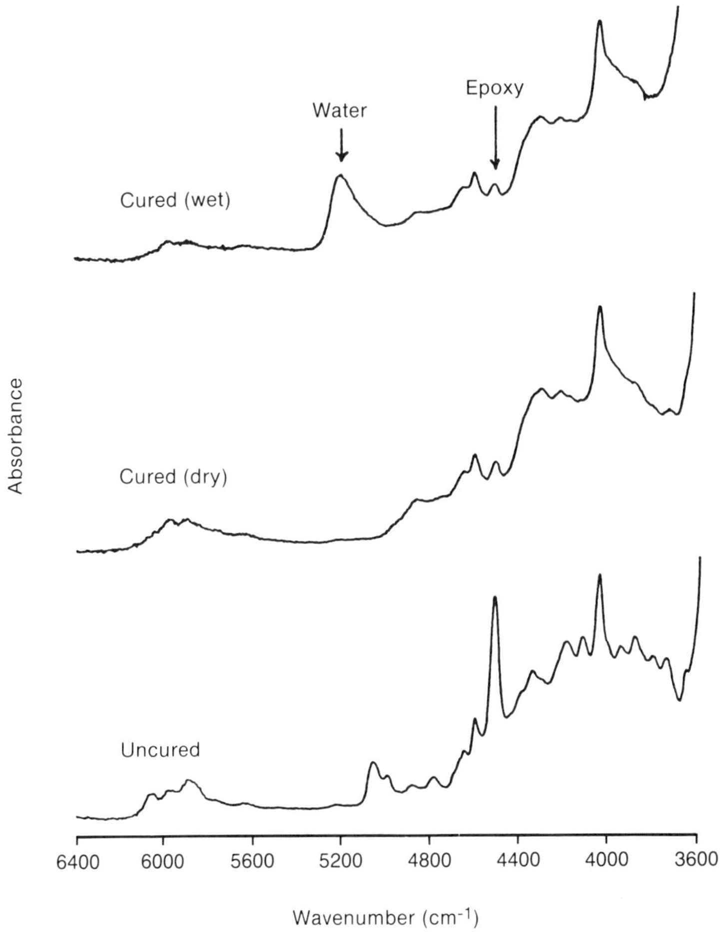

Figure 8-19 Near-IR transmission spectra of 0.2 mm-thick films of Narmco 5208 epoxy resin uncured (bottom), cured (middle) and cured but saturated with 5% water (top curve).

shown spectra, samples of the epoxy resin were separated from the opaque carbon-fiber structure with a solvent, a few drops of the solution were placed on an IR-transparent window, and the solvent was evaporated at a moderate temperature to produce a cast film to be analyzed in transmission. A diffuse-reflectance FTIR spectroscopy investigation was recently performed to analyze the possibility of *in situ* and nondestructive spectrosopic evaluation of the degree of cure in as-received prepregs and cured panels (Cole et al.,

1985). A high-aperture elliptical half-mirror was used to focus the infrared beam onto the surface of the sample, and a similar half-mirror collected a substantial fraction of the diffusely-reflected light. No sample preparation was required other than cutting a piece small enough to fit inside the diffuse reflectance accessory. With a properly modified optical configuration, large samples could be anlayzed without the need of cutting a piece (Cole and Noël, 1986).

Figure 8-20 shows some diffuse-reflectance spectra of uncured (left) and cured (right) carbon-epoxy composites. In spite of the strong attenuation, the different features of the resin spectra are still visible when the plane of incidence is parallel to the graphite fibers (upper sectra). The plane of incidence is here defined as the plane containing the axes of the wide-aperture beams which are incident on and reflected from the diffuse surface. The disappearance of the epoxy rings in the 900–1000 cm^{-1} region can again be noticed after curing. Figure 8-20 shows, however, that the spectra may be unacceptably deformed when the plane of incidence is perpendicular to the fiber direction, in which case the attenuation due to fiber absorption is very important. Moreover, it was observed that because of the hard-to-predict surface reflectance effect (see Section 8.1.2), the measured value of the epoxy-curing index was affected by errors of the order of 10% to 20%, as compared to a reproducibility of 2% in transmission measurements (Cole et al., 1985). It is thus felt that diffuse-reflectance spectroscopy is a convenient technique for qualitative inspection, but it is of limited value for quantitative analysis of composites.

An alternative approach for polymer cure monitoring with an imbedded fiber-optic probe is based on a fluorescence analysis from viscosity-dependent dye pigments dispersed in the resin (Fanconi et al., 1986; Birnbaum et al., 1986). The dye molecules are excited by the evanescent wave at the fiber-resin interface. The fluorescence spectrum is related to the viscosity changes in the gel to glass-solid range during polymerization. Some resins exhibit viscosity-dependent fluorescence without the need for a dye additive (Levy, 1986).

8.2.3 *Laser-induced Breakdown Spectroscopy*

We have seen in Section 8.1.1 that atomic-emission spectroscopy is widely used for material analysis. Laser-induced breakdown spectroscopy (LIBS) is an attractive spectrochemical analysis technique for rapid *in situ* characterization of materials such as molten metal (Runge et al., 1966), fuel in flames (Schmieder and Kerstein, 1980), gasified coal or other aerosols (Loree et al.,

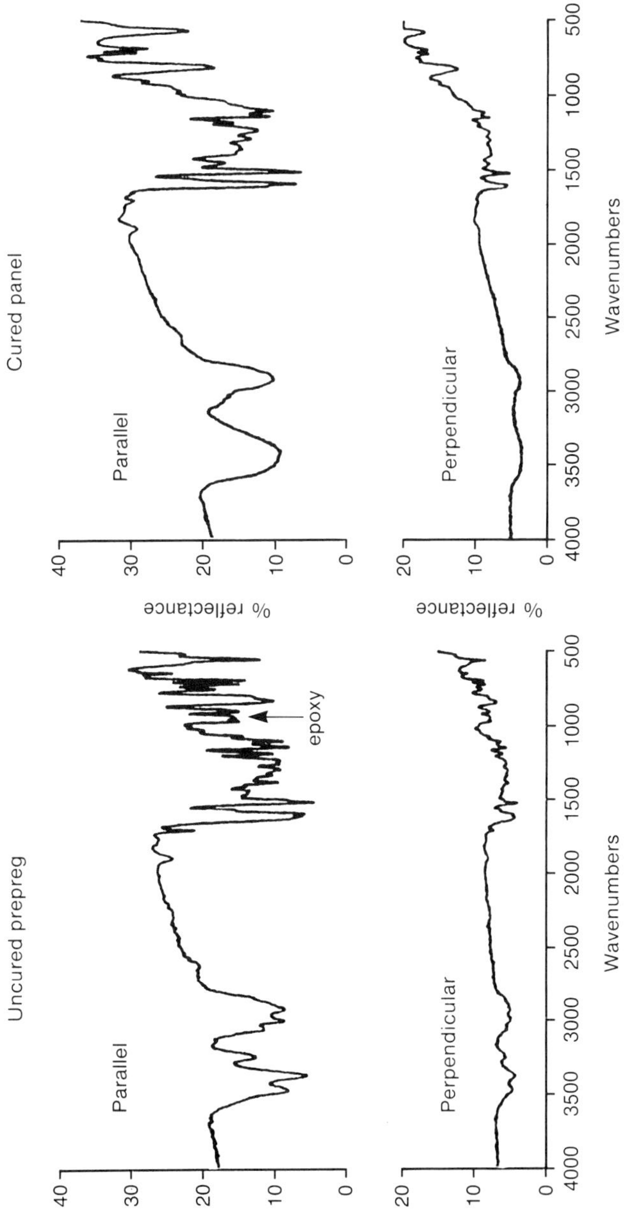

Figure 8-20 Diffuse reflectance IR spectra of Narmco 5208/WC3000 prepreg and cured laminate, showing the effect of carbon-fiber orientation. Left: uncured prepreg; right: cured panel; top curves: plane of incidence parallel to the predominant carbon-fiber direction; bottom curves: plane of incidence perpendicular to the fibers.

1982). Such a technique consists of producing a spark breakdown by irradiating the material to be analyzed with a high-power focused laser pulse and observing the characteristic spectral emission from the spark to determine the material composition. The laser probe has a number of advantages as compared to more classical electric-discharge emission spectroscopy probes using DC arcs, AC arcs or AC sparks (Olsen, 1975, Section 4-2). It is a physically noninvasive probe, requiring only line-of-sight access through a window. Compared to electric-spark sources, the plasma is clean (no contribution of electrode material and its spectrum) and versatile (the sample may be in almost any physical state).

In situ Analysis of Gases and Aerosols

Figure 8-21 shows a typical apparatus for remote LIBS probing in hostile environments such as a coal gasifier (Loree and Radziemski, 1981; Loree et al., 1982). A pulsed laser such as a Q-switched Nd : YAG capable of giving 10–100 MW/cm^2 focused power densities during the 10–20 ns pulse duration is focused on the periphery of the gasified-coal stream(the stream is rather opaque; one-side configurations of the kind shown in Figure 3-33 in Chapter 3 of this book should also be possible). In the focal region, the electric field of the beam is strong enough to start avalanche ionization of any material located there, producing a localized plasma of electrons and atomic species (excited ions and atoms). The molecules are dissociated in the plasma, so

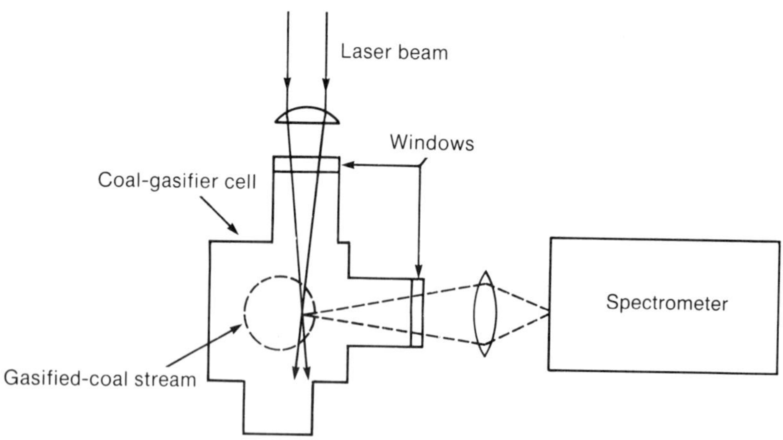

Figure 8-21 Schematic components of a laser-induced breakdown spectroscopy probe used for detection of Na and K in the product stream of a coal gasifier.

that only atomic species can effectively be traced by this method, similar to other atomic-emission spectroscopic techniques. The atomic species radiate with decreasing intensity for many microseconds after plasma initiation, and such a radiation is probed by a spectrometer whose entrance slit is imaged on the small plasma volume. A grating spectrometer with a staring array of 1024 photodiodes may be used to obtain the full spectrum in a single shot, eventually averaging the signal over several hundred shots for more reliable data and quantitative analysis. Alternatively, a mechanically scanned spectrometer with a single photomultiplier detector can be used to scan the spectrum.

Figure 8-22 shows a typical spectrum obtained during the coal-gasifier tests. The corrosion of gasifier turbine blades is related to the amount of alkali metals in the product stream. A real-time sensor detecting the presence of sodium and potassium either in the gas or in particulates entrained in the gas stream is thus of interest for on-line process monitoring. Sodium and potassium emission lines are visible in the spectrum shown in Figure 8-22. The sodium doublet at 589 nm is particularly prominent and may be used to detect sodium impurities down to a level of 10 ppb (Loree et al., 1982).

A further development of the LIBS technique makes use of the time-resolved spectral features of the laser-induced plasma to improve the

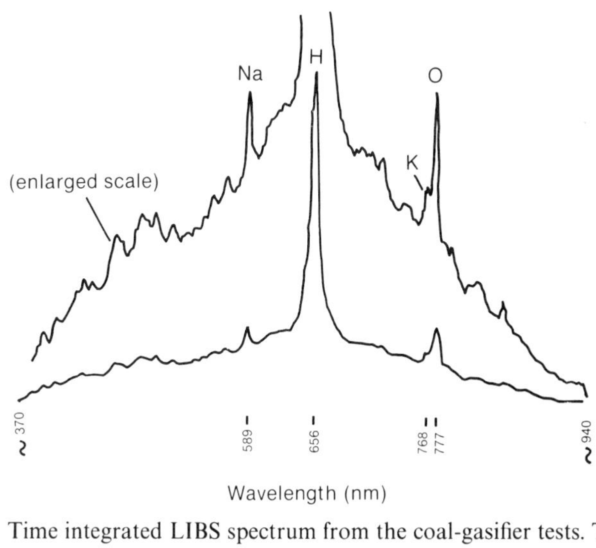

Figure 8-22 Time integrated LIBS spectrum from the coal-gasifier tests. The Na and K emission lines can be seen superimposed over the background plasma emission.

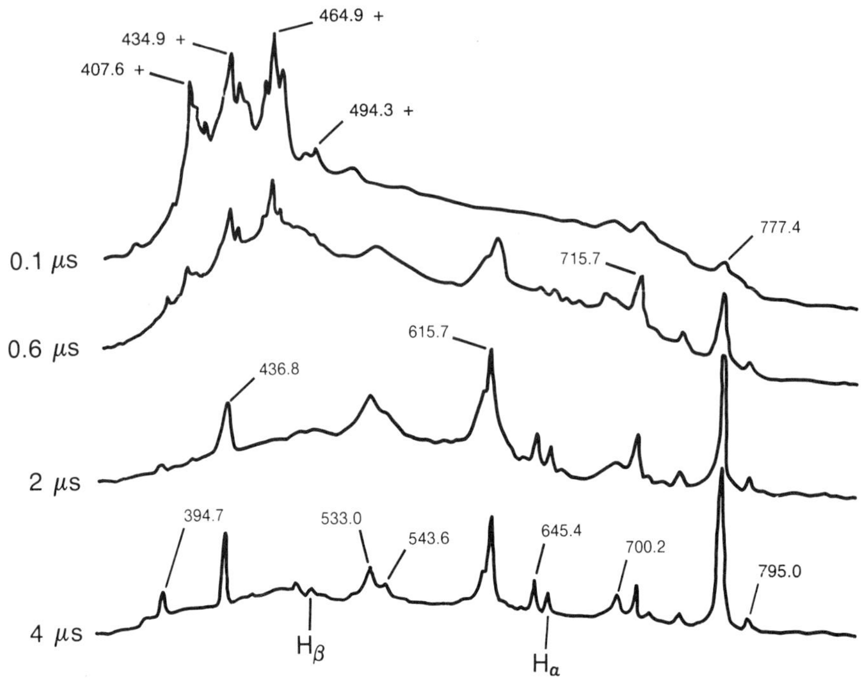

Figure 8-23 Evolution of a LIBS spectrum (in a 1-atm oxygen gas) at increasing delay times after plasma initiation by the 20 ns laser pulse. As the time delay increases, the continuum background and ionic lines (marked +) both decrease, while emission from neutral atoms becomes dominant.

signal-to-noise ratio (Radziemski and Loree, 1981; Loree et al., 1982; Radziemski et al., 1983). As Figure 8-23 shows, a background continuum from the plasma at several tens of thousands of degrees C, with strong ion-emission lines, makes a large contribution in the first 100 ns after the laser pulse. As time progresses, the intensity of the background emission decreases and shifts to longer wavelengths, while singly ionized oxygen lines are replaced by neutral atomic emission. Since the detector gain was increased for each successive scan, the ionic lines would be predominant in a time-integrated LIBS spectrum. The obvious improvement in signal-to-noise ratio at late times is strikingly shown by the visibility of a pair of hydrogen lines from a minor impurity in the oxygen gas. Time-delayed spectra can be obtained by using a time-gated silicon-diode array or a gated photomultiplier. Other applications of laser-induced breakdown spectroscopy of fluids include the detection of beryllium, chlorine and fluorine in air or liquids (Radziemski et al., 1983; Cremers and Radziemski, 1983; Cremers et al., 1984), the mapping of the fuel/air ratio in a diffusion flame, as well as the performance of elemental abundance analysis on a sample of shale oil vapor (Schmieder and Kerstein, 1980).

Surface Analysis of High-Temperature Metals

The noncontact nature of the laser-induced breakdown spectroscopy approach makes it attractive for the *in situ* analysis of high temperature materials, for which contact techniques must be excluded. The possibility of using a laser pulse to produce the plasma spark required for atomic emission spectroscopy dates back to the early years after the development of lasers (Runge et al., 1966). This idea is being revived (see *Photonics Spectra*, Sept. 1983, p. 20) by the above-mentioned success of the LIBS technique in the remote analysis of fluids.

Nickel- and chromium-based alloys were spectrochemically analyzed at high temperature by bombarding the metal surface with a focused ruby laser giving typically 50 mJ, 20 ns pulses (Runge et al, 1966). The laser pulse produced a shallow crater, 0.5 mm in diameter, on the surface of the metal sample which was induction heated to reach the melting temperature. The highly luminous, 3 mm-diameter spark was spectrally analyzed by a prism spectrometer to monitor the relative intensities of the iron, nickel and chromium characteristic emission lines. The ratio of the line emission pairs 341 nm (Ni)/376 nm (Fe) and 396 nm (Cr)/376 nm (Fe) were measured for steel samples with different nickel and chromium concentrations. Figure 8-24 shows the log intensity ratios plotted against the nickel and chromium percentages. The relationships are straight lines over the covered concentration range for both the solid and molten states. Similarly encouraging results have been obtained at lower temperatures by using alternative discharge-creating mechanisms, such as luminescent-discharge lamps or a hybrid laser-excited, electrically sustained plasma (Bojic et al., 1970; Hobson and Ward, 1975; Berneron, 1978; see also *Laser Focus*, Apr. 1981, p. 30).

Rapid in-process analysis of molten metal would be quite valuable for a number of tasks such as determining iron composition during the blast furnace cast, analyzing the steel during the refining process or monitoring sulfur removal. Yearly savings possible by such measurements may be estimated to hundreds of M$ (Achey, 1982). The main problem appears to be related to the presence of oxide or slag patches which normally float at the surface of the liquid metal.

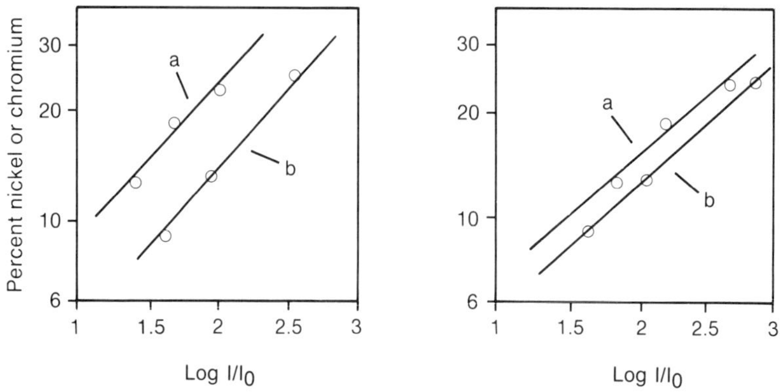

Figure 8-24 Results obtained in the spectrochemical analysis of high temperature steel by laser-induced breakdown spectroscopy. Left: solid metal; right: molten metal; curves a; chromium-based alloy; curves b: nickel-based alloy.

8.3 Nondispersive Process Analyzers

Dedicated instrumentation for on-line composition monitoring has been in use for decades in the chemical industry (Olsen, 1975; Considine, 1985; Yeh, 1986). An overview of commercial analyzers based on a variety of physical principles can be found in Clevett (1986). Nondispersive spectroscopic instruments typically use fixed-filter devices of the kind shown in Figure 5-12 and are normally more rugged and less expensive than dispersive or FTIR spectrometers, at the expense of a much reduced versatility. This section briefly reviews such systems and some typical industrial applications. The emphasis is on sensors for in-process compositional analysis; description of other wavelength-sensing instrumentation which is used for other purposes, such as for on-line control of product color, can be found in more specialized textbooks (Judd and Wyszecki, 1975; Chamberlin and Chamberlin, 1980; McCamy, 1986). Specific papers also deal with color inspection on selected products such as paper, textile and plastics (Cook, 1980; Kishner, 1980) and steel sheets (Torao et al., 1986).

8.3.1 *Basic Principles*

The basic principle of a nondispersive device for in-process material analysis is very similar to that of spectrosocopic material characterization: The attenuation of the transmitted beam due to absorption at a characteristic wavelength is related through the Beer-Lambert law to the amount of the absorbing substance present in the sample. We have seen the first form of this law in Section 5.1.3, stating that the absorbance of the light beam by a sheet of a given material is proportional to the sheet thickness. This law was formulated by Lambert in the seventeenth century. Beer's law, stated a century later, deals with a sample of fixed thickness and variable concentration of the absorbing species, and can be written (see, e.g., Olsen, 1975, Section 1-5)

$$\log\left(\frac{I_0}{I}\right) = aCd \tag{8-3}$$

where a is the absorptivity and C is the concentration of the absorbing species, d is the sample thickness, I is the light intensity transmitted through the sample and I_0 is the initial light intensity. The product aCd is called absorbance, or optical density, of the sample.

In a nondispersive process analyzer, a double-wavelength device of the kind shown in Figure 5-12 can be used to detect the intensity I at the wavelength of absorption for the analyzed substance, as well as the intensity I_0 at a nonabsorbed wavelength. The absorbance is then measured independently of fluctuations in the light source power or of spectrally neutral dirt accumulation along the common light path. An alternative differential approach, which is very popular for the detection of minor impurities in a gas or liquid, is the Luft-type balanced condenser microphone system shown in Figure 8-25 (see, e.g., Considine, 1985, p. 6.74). A light beam is split in two parts, one being transmitted through the gas or liquid continuously sampled from the process to be monitored and the other beam being transmitted

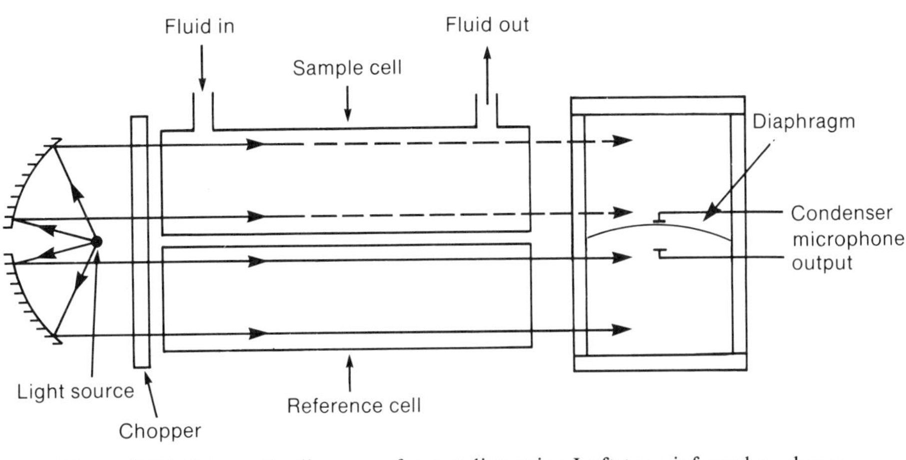

Figure 8-25 Schematic diagram of a nondispersive Luft-type infrared analyzer.

through a reference cell filled with a nonabsorbing gas such as nitrogen. The two transmitted beams are finally absorbed within the two halves of the detecting cell, which is filled with a highly absorbing gas and is divided by a diaphragm whose displacement is monitored by a condenser microphone. When an absorbing substance is present in the sample cell, the two chopped light beams reaching the detecting cell have unbalanced intensities, resulting in an AC signal whose amplitude is related to the concentration of the absorbing substance. More recent detecting cells use flow sensors (sensing the flow of gas through an orifice from one half to the other half of the detecting cell, see, e.g., Nakano, 1980) to eliminate one of the major causes of detector failure, i.e., failure of the thin diaphragm. Photodetectors are increasingly used instead of the detecting cell beams, because they are more rugged and have more powerful signal processing capabilities.

Tunable diode lasers may also be used instead of a filtered incoherent source (Wall et al., 1980; Grisar, 1983; Forrest, 1986). Their still relatively high price and limited output and spectral range tend, however, to limit their applicability to situations where their very narrow spectral bandwidth is required. Finally, optical fibers are being increasingly used as a convenient extension line for remote sensing in the near-ultraviolet, visible and near-infrared range (Fitch and Gargus, 1985) as well as for Raman spectroscopy (Nguyen and Plaza, 1986). This approach makes it possible to use relatively complex and cumbersome spectroscopic instrumentation for on-line applications by using the fibers as a remote probe. The future availability of optical fibers transmitting in the mid-IR (see Section 3-9) will possibly extend the applicability of infrared process analysis to several cases where the hostile environmental conditions prevent their use at the present time.

Ultraviolet and Infrared Analyzers

The spectral range of detection depends on the substance to be monitored. Ultraviolet absorption analyzers are useful for quantifying halogen elements or aromatic compounds such as benzene, naphthalene, anthracene and related compounds. The presence of such substances is monitored, e.g., in wastewater streams for pollution control. Sulphur-containing compounds are also strong ultraviolet absorbers. Ultraviolet analysis can be performed in air down to approximately 200 nm, below which limit the air starts to absorb, requiring vacuum spectrometers (Olsen, 1975, Section 2-5). The ultraviolet spectra of molecules are generally composed of broad vibrational and rotational bands, whose fine structure is normally not resolved, except for

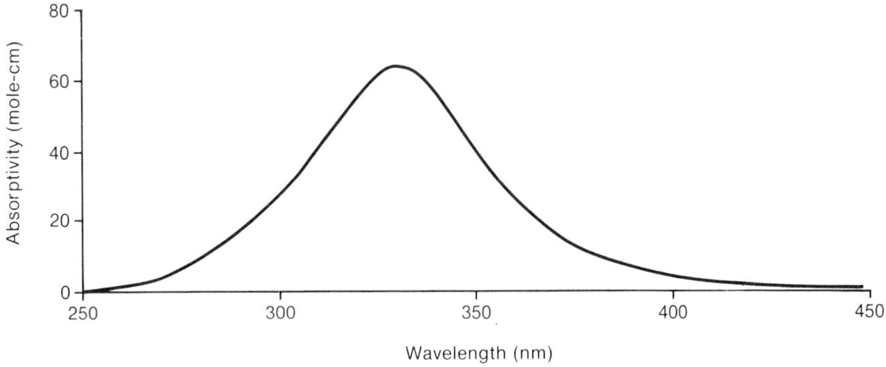

Figure 8-26 Spectral absorptivity of chlorine in the ultraviolet range.

low-pressure gas. Figure 8-26 shows as an example the spectral absorptivity of the chlorine molecule. The absorptivity spectra of several substances overlap considerably in the ultraviolet range, so that this approach provides reduced selectivity when analyzing multicomponent fluids. Ultraviolet analysis has, however, some attractive characteristics which explain its relatively broad use. One of the main advantages is the low absorptivity of water, a ubiquitous substance, in the ultraviolet, while its high infrared absorbance prevents many *in situ* infrared-transmission analyses. Also, relatively inexpensive quartz cells and optical elements can be used in the ultraviolet.

Infared process analyzers are a convenient tool for process monitoring. The more selective spectral-absorptivity characteristics in the infared are more effective in providing unique "fingerprints" of a small quantity of a given substance in a multicomponent mixture, particularly if a multiple analysis based on a circular variable filter wheel (LaButti et al., 1980) is performed. Near-infrared (0.7 to 3 μm) analyzers suffer little attenuation from water, while a mid-infrared (3 to 15 μm) analysis normally requires very thin transmission cells or internal-reflection elements (LaButti et al, 1980). Examples of substances which can be effectively monitored by infrared analyzers are CO and CO_2 in ammonia plant synthesis gas, moisture in paper and in cereals (where fat and protein contents can also be evaluated in the near-IR), as well as a large range of gases whose low density allows the use of longer transmission cells.

Range of Validity of Beer's Law

The accuracy of process analyzers for a quantitative evaluation of the concentration of the absorbing substance depends on the validity of Beer's law as stated in equation (8-3). Under practical conditions, it is found that the range of validity of this equation is limited. Indeed, if the experimental values of the absorbance obtained by varying the absorbing substance concentration are plotted as a function of the concentration, the linear range of such a curve is normally limited (see, e.g., Olsen, 1975, Section 1-5).

A fundamental cause of nonlinearity in the absorbance versus concentration curve is due to the dependence of the absorptivity on the refractive index, which is assumed constant in equation (8-3). Since the refractive index of the tested material varies with the concentration, so does the absorptivity. This cause of error is, however, negligible at small concentrations, and its effect is normally much smaller as compared to other sources of nonlinearity. Such nonlinearities are mainly due to violations of the Beer's law assumptions, in particular that: The attenuation of the transmitted beam is due only to absorption; the incident light beam is monochromatic; and the absorbing species act independently of one another.

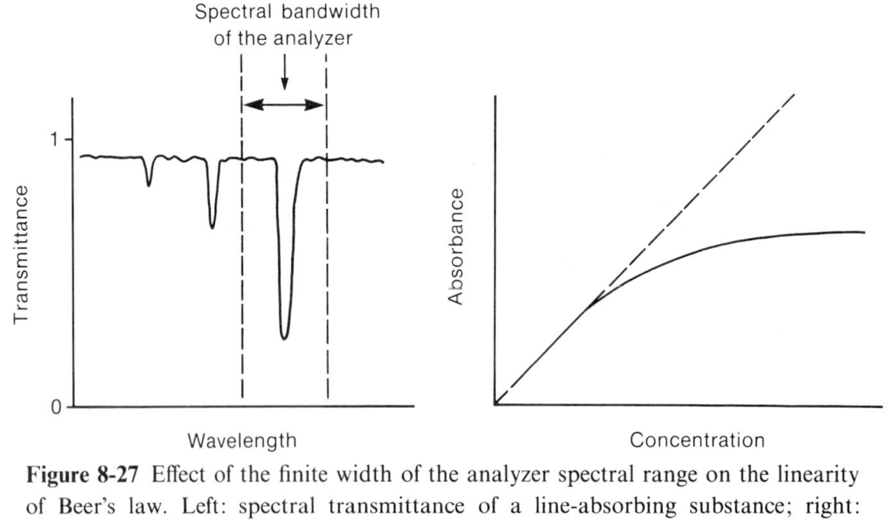

Figure 8-27 Effect of the finite width of the analyzer spectral range on the linearity of Beer's law. Left: spectral transmittance of a line-absorbing substance; right: resultant nonlinearity in the absorbance versus concentration curve.

The first assumption implicit in the formulation of equation (8-3), namely that only absorption is present, is usually violated by the presence of other sources of attenuation, such as scattering and photoluminescence. Since the scattered or reemitted radiation is distributed over a wide angular aperture, the amount of such attenuation which may be attribted to absorption is dependent on the angular aperture of the collecting optics. Another cause for the violation of Beer's law is illustrated in Figure 8-27. If the absorbing substance has a narrow line spectrum, while the spectral bandwidth accepted by the analyzer is broader than the width of such lines, the absorbed fraction of the transmitted light will never exceed the ratio between the linewidth and the spectral bandwidth of the analyzer. The absorbance versus concentration curve correspondingly saturates, limiting the extent of the linear region. Finally, errors in the strict application of Beer's law may arise from the coalescence of the absorbing species at high concentration or to the partial reaction of such species with other compounds in the solution. A quantitative evaluation of the absorbing species concentration can thus be made only with limited accuracy, usually within a few percent at best, unless a specific nonlinear calibration curve is recorded for each substance to be analyzed.

8.3.2 *In-process Material Analysis*

Spectroscopic Fluid Analyzers

Infrared nondispersive gas analyzers are widely used for process control (Jutila, 1979; Nakano, 1980; Durham, 1983; Saltzman and Dell, 1984; Sonner et al., 1984; Cardis and Lastrapes, 1984; Fisher, 1984; Lord and Maier, 1984; Fitch and Gargus, 1985; Wilks, 1985). Table 8-1 lists typical gases and concentration ranges which can normally be monitored by nondispersive IR analyzers (Durham, 1983; Considine, 1985, Chapter 6). Table 8-2 gives some typical specifications of an NDIR process analyzer (Nakano, 1980; Considine, 1985, Chapter 6). Fiber-optic sensors for fluorescence or near-IR absorption spectroscopy are increasingly used for pH or compositional analysis (Hirschfield, 1986).

Ultraviolet analyzers are used for monitoring halogen elements or aromatic compounds (Fitch and Gargus, 1985), with typical sensitivities down in the parts per million range, and have response times generally shorter than a second (Jutila, 1979). Fiber-optic extension probes may also be used for on-line UV-absorption monitoring (Saturday, 1983). Besides the chemical and petrochemical industries, other sectors are increasingly considering the opportunity to use such noninvasive and relatively inexpensive equipment

Table 8-1 Typical Gases and Concentrations Measured by NDIR Analyzers

Gas	Range (full-scale concentration	
	Minimum ppm	Maximum, %
Ammonia	1000	100
Butane	300	100
Carbon monoxide	500	100
Carbon dioxide	200	100
Ethylene	1000	100
Ethane	500	100
Ethylene oxide	1500	100
Hexane	400	5
Methane	400	100
Nitrous oxide	150	100
Propane	400	100
Sulfur dioxide	400	100
Water vapor	1500	10

Table 8-2 Typical Specifications for an NDIR Process Analyzer

Repeatability ±1% of fullscale	Noise level Less than 1% of full scale
Zero drift ±1% of full scale per 24 hours	Warm-up 15 minutes
Span drift ±1% of full scale per 24 hours	Internal span check Electrically activated
Linearity ±0.5% of full scale fitted to theoretical curve	Temperature range 0-50°C (32–122°F)
	Ambient humidity Up to 95% relative humidity

for continuous process monitoring. One example is the measurement of lignin in the black liquor of the pulp and paper industry, for which ultraviolet (Bentley, 1986), infrared (Sugden, 1986) or fluorescence spectroscopy methods (Horvath and Semerjian, 1986) are currently being taken into consideration. Chlorine dioxide residual monitoring in kraft pulp bleaching is another recently reported application (Gilbert, 1986).

Turbidimetric, Nephelometric and Refractometric Analyzers

Turbidity is a measure of the loss in transmitted light intensity caused by scattering from inhomogeneities within the inspected medium. On-line instrumentation based on turbidity measurements has been in use for a long time in industry (Krigman, 1984; Considine, 1985, p. 6.97). The applications are often qualitative — for instance, to detect breakthrough of a filter or to indicate change in feedstock composition. Most applications involve measurement of solids in suspension in liquid media such as drinking water or wastewater, although evaluations of gas turbidity for pollution monitoring or detection of inclusions such as air bubbles in clear solid materials are also performed. More quantitative applications of turbidimetric methods will be discussed in Section 8.4.

Turbidity is normally monitored by measuring the attenuation suffered by the light beam after transmission through a given length of the medium. Double-beam methods of the kind described in Section 8.3.1 may also be used for improved accuracy and reliability against source-power fluctuations or dirt accumulation. Calibration is usually performed by substituting the inspected medium with a reference solution of known turbidity. A number of instruments, called nephelometers, detect the amount of light scattered at 90° from the direction of illumination, rather than the attenuation of the transmitted light. Such instruments give a better precision when monitoring very low turbidity media. The units which are most often used are the so-called nephelos units, with 100 units corresponding to a scattering loss of 2% of the total light passing through the sample (Considine, 1985, p. 6.95).

The main on-line industrial applications of turbidity or haze sensors are in the beverage, chemical and pharmaceutical industries. One example is the evaluation of fat and protein concentration in food processing applications. Accuracies of 0.005% have been reported in such measurements by using a He–Ne laser beam at two different angles (Atutov et al., 1981). Another application is the monitoring of oil in water using fiber-optic sensors (Extance et al., 1983) or double-wavelength techniques (Frenkel et al., 1982). By using double-angle monitoring devices, oil concentration can be evaluated with an accuracy of 5% within different ranges from 0 to 10,000 ppm (Colman and Thew, 1983). Turbidity is also used to monitor fibers in liquids such as asbestos slurries (Shiuh, 1972) or paper pulps (see next section).

Another type of inexpensive process analyzer based on simple geometrical optics is represented by refractometry sensors (Mooney, 1984; Clevett, 1986). Such sensors are based on the measurement of the refractive index of the material, usually a liquid, by methods such as the evaluation of the refraction angle through a prism-shaped container or the detection of the critical angle for total internal reflection. Similarly to internal-reflection spectroscopy (see Section 8.1.2), critical angle sensors have the advantage of being applicable to opaque liquids. Their sensitivity is also quite satisfactory, reaching $2 \cdot 10^{-4}$ of a refractive index unit. An even higher precision, in the 10^{-6} range, can be reached with differential refractometry configurations, where the refractive index of the probed material is compared to that of a reference fluid (Mooney, 1984). Such instrumentation is quite appropriate for on-line analysis in the beverage industry (Stackhouse, 1966).

Moisture Sensing

Near-IR analysis for moisture sensing in paper is one of the most popular fields of application for infrared process analyzers (Tyler, 1972; Brewster and Robinson, 1974; LaButti et al., 1980; Pugh, 1980). Infrared sensors for light-grade, low-moisture sheets and microwave sensors for heavy-grade, high-moisture paper are gradually replacing, in this field, the previously used direct or AC current sensors using either conductance or capacitance techniques (Chu and Balls, 1979). The intermediate frequency range of the far-IR lasers is also being considered for such measurements (Boulay et al., 1984).

A number of other products can be inspected for moisture by infrared, including polymers and composites (Ryan et al., 1978; Levy et al., 1979; Boulay et al., 1985), as well as a number of organic materials (Wetzel, 1983; Mooney, 1984). Table 8-3 gives a number of food-related products which can be analyzed for moisture by infrared sensors tegether with their typical moisture range (Mooney, 1984).

The main advantages of infrared moisture-sensing devices, as compared, e.g., to electric-impedance resonator cells, are their remote-inspection capabilities, reliability and relative insensitivity to basis weight (sheet thickness) because of the double-wavelength normalization (see Section 8.3.1). Infrared moisture-sensing devices are particularly convenient to install when they operate from one side only of the material to be inspected. Such devices can be used for surface-moisture sensing of thick or even opaque materials, such as roasted coffee or metallic ores.

Figure 8-28 shows the diagram of a typical infrared moisture analyzer operating in reflection (Tyler, 1972). A wide-spectral-bandwidth light beam from a thermal source is projected on the material to be analyzed and the backscattered light is detected by a PbS photodiode. A rotating wheel contains

Table 8-3 Some Typical Ranges of Moisture Measurement for Solid Products

Material	Moisture range
Acetate flake	0–2%
Cereals	5–10%
Coffee, roasted	0–4%
Clay, ceramic	0–1%
Detergents	0–15%
Fertilizers	0–5% & 0–10%
Milk, dry powdered	0–5%
Metallic ores	0–15%
Nylon chips	0–3%
PVC, granular	0–1%
Potato, dried	3–5% & 5–15%
Peat	0–80%
Rubber, crumb	0–2%
Sugar beet pulp	0–15%
Sand	0–10%
Santacure	0–0.5%
Seaweed	20–80%
Vegetables, dried	0–5% & 0–10%

bandpass filters at suitable center wavelengths to select two spectral regions, typically at the water absorption band near 1.94 μm (see Figure 7-11 or 8-19) and at a neutral reference point such as 1.81 μm (Chu and Balls, 1979). The projected light beam is backscattered after a number of internal reflections within the translucent material (see Section 8.1.2). The reflectivity R_∞ will be attenuated at the water-absorbed wavelength as compared to the reference wavelength following the Kubelka-Munk equation (8-2). The ratio of the signals detected at the absorption and reference wavelengths may be related to the amount of absorbed water through an appropriate calibration curve.

It is important to note that only the surface moisture can be evaluated by a backscattering analyzer, the probed depth below the surface corresponding to the penetration distance of the light beam. An experiment performed to evaluate such a penetration depth in paper is illustrated in Figure 8-29 (Pugh, 1980). A thin (21 g/m^2) paper sheet with a 48% water content is sealed in an impermeable bag and placed at various positions within a 12-layer stack of paper sheets at about 6% moisture. As the graph show, the probed

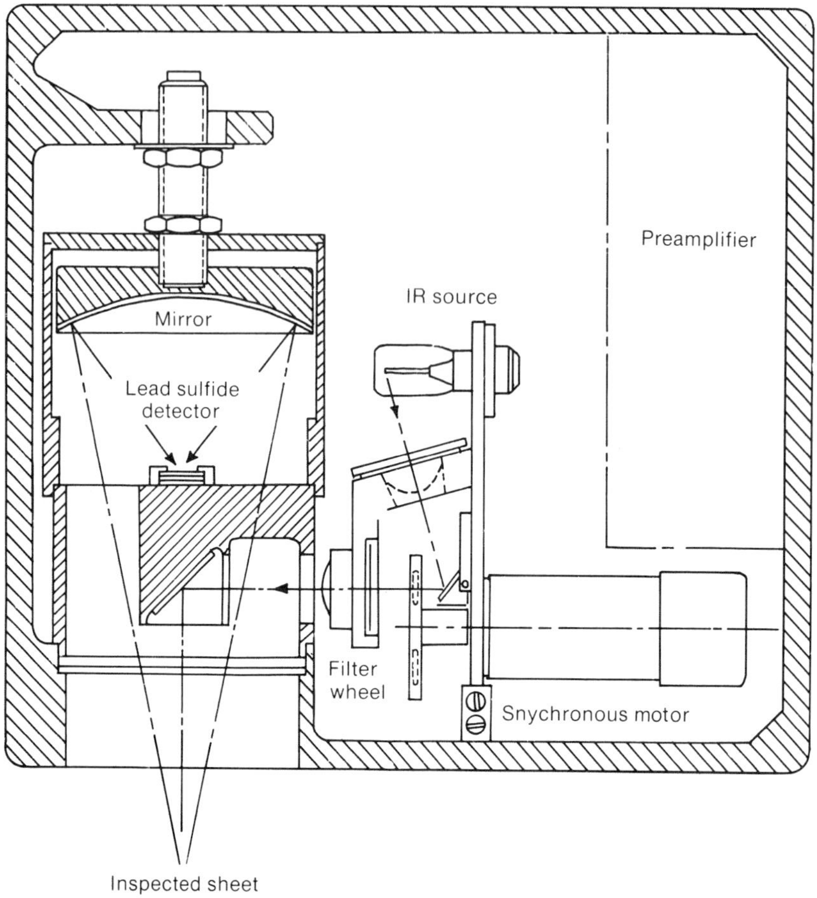

Figure 8-28 Diagram of a typical backscattering moisture analyzer.

depth does not exceed two or three sheets, corresponding to a penetration depth of the order of 100 μm in paper. Such a penetration is larger for materials with a lower scattering coefficient than paper.

8.3.3 *Combustion Monitoring*

Control of the combustion process is an important requirement for energy efficiency and pollution minimization (Stokely, 1983). The greatest amout of heat generation in a flame occurs when the amount of supplied oxygen is just enough to completely burn the fuel (stoichiometric combustion). If there is insufficient air to oxidize the fuel, unburned combustibles will result, causing

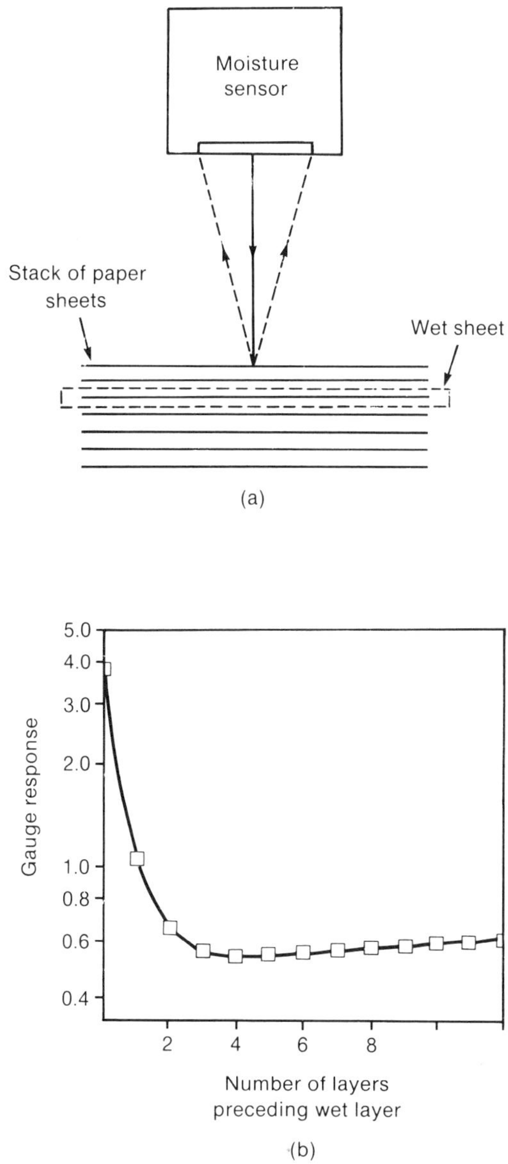

Figure 8-29 Experimental setup to evaluate the depth of penetration of a back-scattering moisture sensor; (b) decrease in sensitivity as a function of the depth of the wet sheet down the stack.

a waste of fuel and the production of unsafe levels of CO and hydrocarbons. On the other hand, excess air insures complete combustion but decreases flame temperature and efficiency, because part of the heat energy will be lost to warm the extra nitrogen and unused oxygen in the air flow.

The air/fuel ratio in industrial burners is normally adjusted at a fixed level which would assure stoichiometric combustion conditions. Partial clogging of the fuel channel or of the air intake, as well as variations in fuel composition require, however, a continuous on-line control of the combustion conditions for optimum efficiency. Techniques for direct control of the combustion process include flame temperature monitoring by Coherent Antistokes Raman Scattering (CARS: see, e.g., Crosley, 1980, p. 207–231, or Schooley, 1982, p. 575–709), direct spectroscopic flame analysis (Olsen, 1975, Chapter 5; Astheimer, 1983) optical inspection of flame geometry (Morey, 1986), absorption spectroscopy of the exhaust gas (Gilbert, 1980; Shafer and Smith, 1982) or diffusion-broadening spectroscopy (Weill et al., 1986).

Spectroscopic analyzers for combustion monitoring are based on the detection of the relative amounts of O_2, CO_2 and CO gas present in the external portion of the flame or in the exhaust gas by spectral filtering on the appropriate bands. Characteristic spectral bands from O_2 and CO_2 in the emission spectrum of the flame have been shown to be correlated to the fuel/air ratio in the individual burners of an industrial boiler facility (Astheimer, 1983). In such an investigation, an infared pyrometer with a series of filters in the 2.6 μm wavelength range was directly pointed toward the flame periphery through a viewing window in the furnace. CO measurement is, however, considered to be the most sensitive parameter for suspension-fired combustion control, since unburned fuel is sampled directly. The amount of CO relative to the amount of CO_2 generated by a full combustion can be monitored by sensing the transmissivity of the exhaust gas at 4.4 μm and 4.6 μm, corresponding, respectively, to the absorption bands of the CO_2 and CO molecules, as Figure 8-30 shows (Durham, 1983).

A schematic diagram of a CO/CO_2 analyzer for *in situ* combustion monitoring is shown in Figure 8-31 (Shafer and Smith, 1982). The sensing head is made from two parallel stainless steel tubes welded to a stainless steel corner-cube reflector. Infrared radiation beamed down one tube is reflected at the far end and returns back up the second tube. Each tube has narrow metallic-wool filters mounted over slots through the tubing along each side to allow the flue gas to diffuse within the sensor but to avoid direct accumulation on the optical components. The filters are periodically and automatically cleaned by an air-purge burst. Evaluation of the amount of

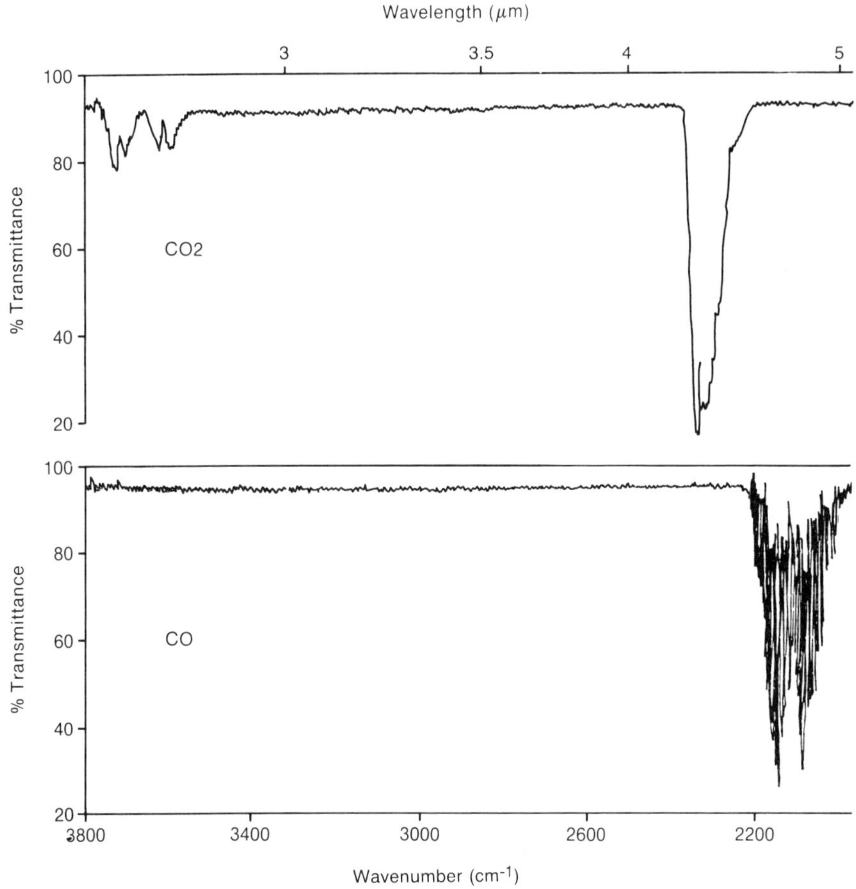

Figure 8-30 Absorption spectra of CO_2 and CO.

CO/CO_2 present is made either by a rotating filter wheel or else by a double-beam technique in which a cell filled with the appropriate gas and a reference cell are alternatively placed in the IR beam by a rotating wheel. The temperature of the sensor head is monitored by a thermocouple to allow the built-in microprocessor to correct the gas absorption data which vary as a function of temperature in a complicated way, as well as to take into account the temperature-dependence of the gas density.

A different approach was recently reported for flame diagnostics inside a gas turbine combustor by direct video-imaging of the flame configuration (Morey, 1986). Figure 8-32 shows a schematic diagram of the combustor rig

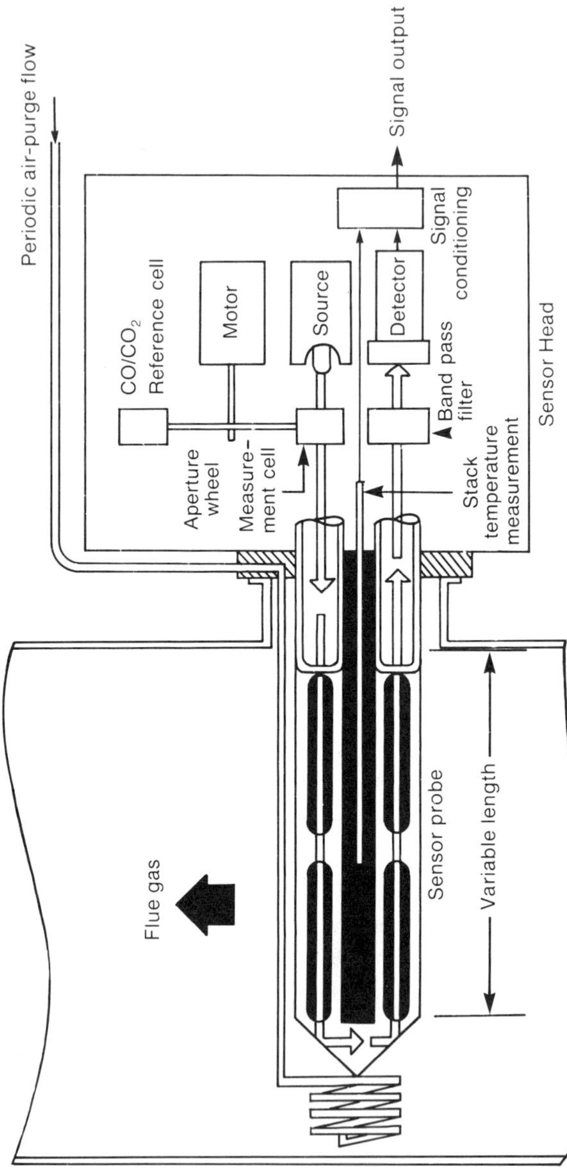

Figure 8-31 Schematic diagram of a CO/CO_2 flue-gas analyzer for combustion efficiency optimization.

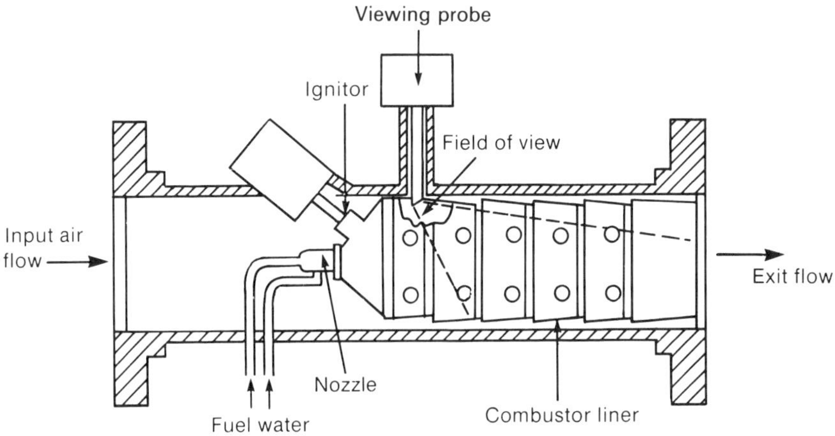

Figure 8-32 Schematic view of a combustor rig with mounted imaging probe for inside inspection of a gas turbine combustor.

used in the tests. The flame image is collected by a lens-capped, 3.2 mm diameter fiber-optic image conduit which transfers the image outside the combustor case. At this point, the image is transferred to a 5 m-long flexible image guide that carries the image to a remote-located video camera. The probe head is water cooled and air purged. Sensors are also provided to detect variations in emission wavelengths as well as flicker frequencies of the emitted-light signal.

Flame data, along with other engine and sensor data taken over long periods of time and under different engine conditions, can give signatures to help diagnose developing, impending or suddenly occurring problems in the engine hot-gas path. An example of two digitized flame-image contours is shown in Figure 8-33. The two flame configurations illustrate an anomalous shift of the flame position at low fuel rate. The fiber-optic image sensor collects a 120° field-of-view image of the flame from the combustor exit to a point several centimeters in front of the fuel nozzle. At a critical fuel rate, the flame flips from a location away from the nozzle back towards the nozzle, and some combustion takes place in the liner dome area around the fuel nozzle, which is located just below the bottom limit of the plots. The flame pattern parameters which are calculated are the X and Y centroid that define the location of the flame, the second-order X and Y moments that define the size of the flame, the cross moment which gives the orientation and the total optical signals which is related to the rate of fuel combustion. A number of simulated process malfunctions such as air-flow nonuniformities, fuel

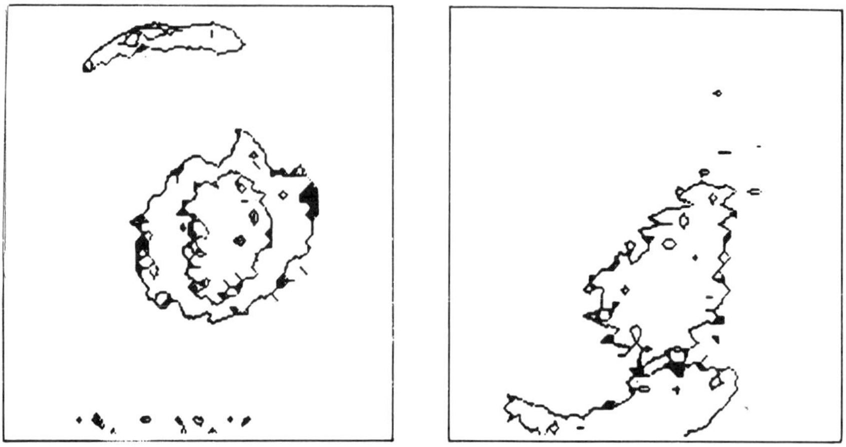

Figure 8-33 Examples of flame-image configurations collected by the imaging probe shown in the previous figure before (left) and after (right) a characteristic flame-position transition at fuel rate 5% below full load.

nozzle changes, linear damage, as well as fuel, water injection and air-flow changes could be related to characteristic modifications of the sensed signal.

8.4 Materials Characterization by Light-Scattering Methods

We have seen in Section 8.3.2 that turbidity caused by scattering from particles in suspension in a liquid is often monitored by on-line process analyzers to collect qualitative data related to process malfunctions or to the end of a batch of feedstock. In this section some examples will be given of optical techniques for a quantitative particle-size characterization of the processed material using light-scattering methods. A large number of optical methods have been investigated for particle sizing as applied to either industrial quality control or to chemical analysis, ranging from computerized micro-imaging of relatively large particles to time-dependent scattering from Brownian motion of sub-micron particles (Barth, 1984).

8.4.1 *Light Scattering by an Aggregate of Particles*

The diffraction of light is one of the fundamental theories of wave optics (Born and Wolf, 1970). Diffraction by spherical particles whose diameter is of the order of magnitude of light wavelengths has been studied in particular by G. Mie at the beginning of this century. The Mie theory provides the basis for a number of techniques of small-particle sizing by an analysis of the scattered-light pattern (Oster, 1948, Van de Hulst, 1957; Kerker, 1969; Barth, 1984).

When a parallel, monochromatic light beam is incident on an opaque, spherical particle, the polar diagram of the scattered light (relative light intensity as a function of the corresponding scattering angle) has the typical shapes shown in Figure 8-34 for different ratios of the particle diameter d and the light wavelength λ. For relatively large particle diameters with respect to the light wavelength, a typical Fraunhofer angular pattern is observed in the far field with a narrow central lobe and a number of secondary peaks within an angular range, which gradually compresses as the ratio d/λ increases. The secondary lobes are averaged out if a particle aggregate with a finite particle-diameter distribution range is inspected or if an incoherent light beam with a finite spectral bandwidth is used. Also, some distortion of the scattered light pattern may be observed when transparent particles are inspected because of the presence of light refracted through the particles. Nevertheless, the angular halfwidth $\Delta\theta$ of the central peak is strongly correlated to the average particle diameter, providing a rapid means to evaluate particle size over large statistical ensembles, as we shall see in the next section.

If the particle diameter is small compared to the wavelength, the central scattering lobe gradually widens and the secondary lobes disappear, as shown in the right part of Figure 8-34. In the limit of very small scattering particles

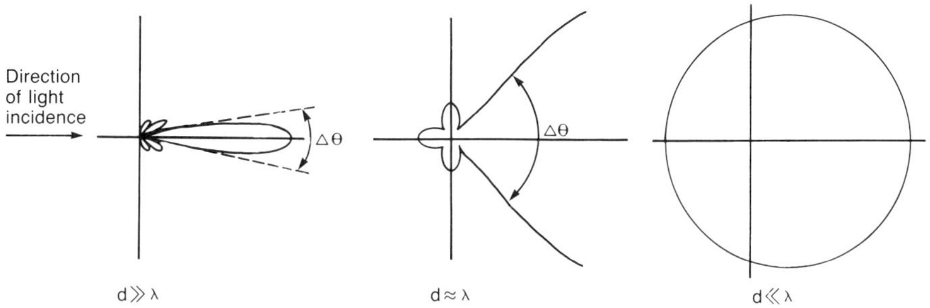

Figure 8-34 Polar diagrams of the light scattered by a spherical particle for different values of the particle diameter d with respect to the light wavelength λ. The angular width $\Delta\theta$ of the central lobe increases as the ratio d/λ decreases. The actual light intensity scale decreases from left to right.

(Rayleigh scattering), the polar diagram becomes a circle centered at the origin. By measuring the asymmetry of such an angular pattern, the size of particles smaller than the wavelength may be evaluated. As an example, the average length of tobacco mosaic virus particles in the range of 100 to 300 nm could be evaluated by taking the ratio of 546 nm wavelength light scattered at 45° and at 135° from the forward direction (Oster, 1948, p. 342).

The angular pattern of the scattered light is, however, essentially limited to the inspection of relatively thin or very dilute samples, for which multiple scattering is negligible. When a substantial amount of multiple scattering is present (i.e., when the light scattered by a particle is normally scattered by several other particles before reaching the detector), a cosine-distributed Lambertian angular pattern is obtained independently of the particle size. One can consider that multiple scattering can no longer be neglected at extinction values (i.e., attenuation of the undeviated light beam) of 10% to 30% (Van de Hulst, 1957; Schroeder and Rosolowski, 1981). At such relatively low scattering levels, noise from extraneous-particle inclusions or surface roughness in sheet samples is normally a major source of angular pattern perturbation. As an example, such a noise source must be eliminated in the case of polarized-light inspection of birefringent spherulitic particles in semicrystalline polymer films (Samuels, 1971; Stein, 1978) by placing a thin film sample between flat glass plates using a matching-index liquid to avoid reflections at the polymer-film surface. Such a requirement is obviously not convenient for on-line industrial applications.

Rather than monitoring the angular-scattering pattern, one may evaluate the average particle size by other light-scattering methods. One approach using the visibility of Doppler-velocimetry signals was mentioned in Section 7.3.2 for the analysis of a stream of moving particles. An experimentally simpler approach, which can as well be applied to a still distribution of particles, consists of measuring the material turbidity at given light wavelengths (Azzopardi, 1979; Durst et al., 1981). The basic principles of such a method are valid even in the presence of substantial multiple scattering (see, e.g., Kortüm, 1969, p. 199–207). It can be shown (Hodkinson, 1966; Azzopardi, 1979; Durst et al., 1981) that the light intensity transmitted through a transparent medium containing disperse spherical particles is given by the following expression

$$I = I_0 \exp(-NSEt), \tag{8-4}$$

where I_0 is the transmitted intensity if no particles are present, N is the number of particles per unit volume, S is the projected particle area in a plane normal to the beam direction, E is the extinction coefficient and t is the film thickness. The variation of the extinction coefficient as a function of the adimensional parameter $(4\pi d/\lambda)(m-1)$ is shown in Figure 8-35, d being the particle diameter and m the ratio between the refractive indices of the particle and of the surrounding medium. The extinction ratio becomes very small for values of the particle diameter much smaller than the wavelength approaching, in the limit of very small particles, the Rayleigh scattering regime where (Hodkinson, 1966)

$$E = \frac{8}{3}\left(\frac{\pi d}{\lambda}\right)^4 \left(\frac{m^2-1}{m^2+2}\right)^4. \tag{8-5}$$

At relatively large values of the particle diameter, the extinction-coefficient curve exhibits damped oscillations, tending to a value of 2 for large diameters. The fluctuations are more damped in the case of a polydisperse distribution, where the particles have diameters distributed around the mean value d. If $E = 2$ in equation (8-4) and if the thickness and the second-phase volume are constant, then $N \propto d^{-3}$, $S \propto d^2$ and equation (8-4) becomes

$$I = I_0 \exp\left(\frac{-K}{d}\right) \tag{8-6}$$

where K is a constant. The transmissivity increases in this case as the particle

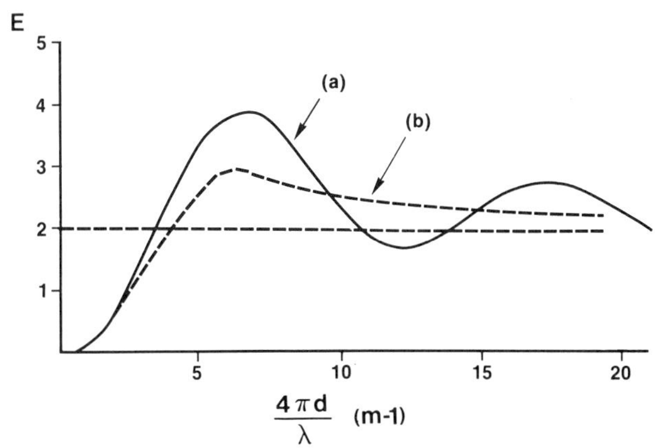

Figure 8-35 Variation of the light-extinction coefficient E in a transparent medium containing dispersed spherical particles as a function of the particle mean diameter d and relative refractive index m for (a) a monodisperse and (b) a polydisperse distribution.

diameter increases, because the light beam encounters a smaller number of first-to-second-phase interfaces across the film thickness.

Equation (8-6) can be used to evaluate the average particle diameter in a polymer blend from transmissivity measurements across a film of known thickness. Similar techniques have been used to characterize the porosity of very thick ceramic materials by one-side imaging techniques (Cielo et al., 1986e). Our derivation assumes, however, that (1) the relative proportion of the two phases is known, and that (2) all of the second phase is contained in particles with diameters large with respect to the wavelength. This second assumption cannot always be made because of the very wide range of particle diameters which are often found in practice.

A more powerful approach for particle-diameter evaluation in polydisperse blends is based on spectral analysis. If the light wavelength is varied, the spectral attenuation will be large for wavelengths much smaller than the particle diameter and small for wavelengths much larger than the particle diameter, as shown in Figure (8-35). By measuring the relative spectral distribution of the light transmitted through the inspected material, one can thus obtain particle-size information irrespective of the film thickness or of the relative amount of the two separate phases. The application of such an approach to the characterization of an immiscible polymer blend will be discussed in the next section.

8.4.2 *Industrial Inspection of Quasi-Spherical Particles*

Angular-Pattern Methods for Particles Larger than the Wavelength

Light-scattering devices for on-line material characterization are widely used in industry. Simple turbidity meters for a qualitative evaluation of the amount of particles in suspension in a liquid slurry were reviewed in Section 8.3.2. Particle-size distributions in the 1 to 100 μm-diameter range can be obtained by spatial filtering systems of the kind shown in Figure 8-36. A parallel light beam is used to illuminate a medium containing dispersed particles, either still or moving in a plane normal to the direction of the light beam. The spatial Fourier transform (Goodman, 1968) of the light scattered by the sheet of particles is obtained at the focal plane of a focusing lens. Either a spatial filter in the form of a rotating mask with spiral-shaped aperture (Wertheimer and Wilcock, 1976) or a detector array (Thompson, 1978; Frock and Muly, 1980) are inserted in the Fourier plane. In the first

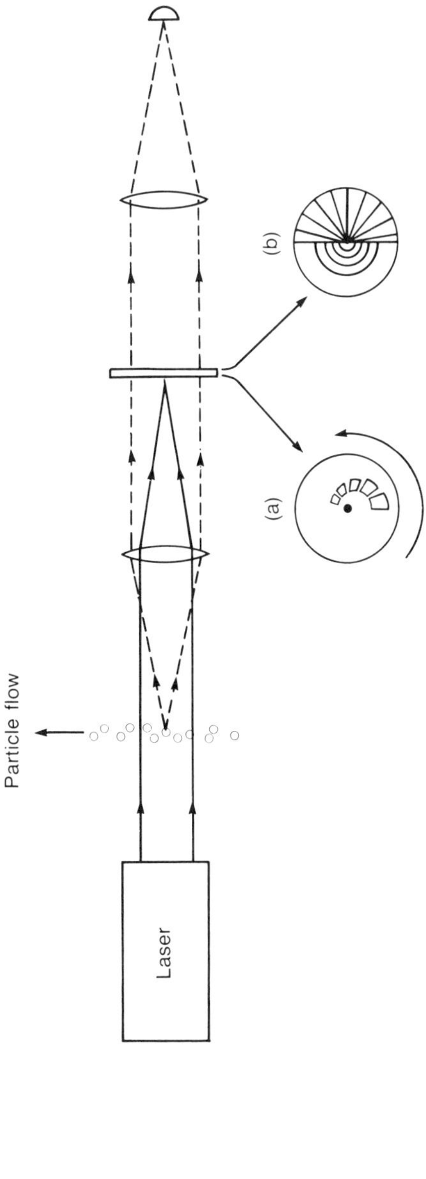

Figure 8-36 A typical optical-correlator apparatus for particle size evaluation by an analysis of the diffraction pattern using (a) a rotating spatial filter or (b) a radial detector array.

case, a signal is detected whose amplitude is a sampling of the light intensity distribution at increasing diffraction angles as the mask is rotated by 180°. In the second case, a radial array of detectors can be used to sample the diffracted light intensity in the Fourier plane. Segmented radial-array detectors containing 32 semi-annular ring elements and 32 wedge-shaped elements are available (Mead et al., 1975) for a simultaneous readout of the full diffraction pattern. A quadrant sensing device is also incorporated in the center to allow for alignment. The total area of the radial elements increases from the center to the periphery of the sensor to compensate for the lower intensity level of the light scattered at higher angles.

A wide number of powder materials can be characterized by such a technique. Examples are plastic powder, asbestos, cement, abrasives, ceramics, mineral ores, metal powder and chemical precipitates. Instrumentation is commercially available both for laboratory analyses of particle emulsions in a test-cell sample and for on-line characterization by on-stream sample extraction. Figure 8-37 shows a sampling pipe directly inserted in the liquid

Figure 8-37 On-line monitoring of a suspended-particle slurry by a light-scattering instrument. The sampling pipe for continuous on-stream sampling is visible in the photo. (Courtesy of Leeds and Northrup/Microtrac, Largo, FL.)

slurry for continuous monitoring of the liquid product. Dry samples can also be obtained in an air stream at typical rates of 10 to 30 grams per minute.

One of the advantages of this light-scattering approach, as compared, for example, to a manual or automated particle-by-particle microscope evaluation, is the intrinsic speed of the parallel operation. All of the particles within the illuminated area are simultaneously inspected by the optical-correlator approach so that statistically significant samples are quickly analyzed. Multiple scattering is, however, not allowed with such an approach so that relatively thin cells or dilute slurries must be used.

Turbidity Spectra

When the material cannot be diluted, as in the case of solid-state mixtures, or when smaller particles, of the order of magnitude of the light wavelength, are to be inspected, spectral turbidity methods are preferred. As mentioned above, this approach consists of measuring the relative transmission through the particulate medium in different spectral regions and inferring from such measurements the average particle size with reference to curves of the kind shown in Figure 8-35. One of the main applications of quantitative turbidity spectrometry is in the inspection of latex particles (Maron et al., 1963; Kiparissides et al., 1980). The particle size and particle size distribution are important specifications for a commercial latex. In addition, particle size distributions affect production rate and molecular weight development, making such a measurement useful during emulsion polymerization.

Another industrial inspection field where light-scattering methods can be applied is the quality control of polymer films (Samuels, 1971; Stein, 1978; Koberstein et al., 1979). Applications to the crystallinity analysis of thin films by polarized-light inspection have been reported (Tilgner, 1976; Bingxin et al., 1983). Turbidity measurements at a given wavelength as the temperature is varied through a lower-critical-solution-temperature range have also been associated with polymer polydispersity properties (Varennes et al., 1984). An application of turbidity spectroscopy to particle size evaluation in polymer blends will now be described.

Polymer blends are physical mixtures of structurally different homo- or copolymers. The study of blends is currently one of the fastest growing areas in polymer research (Paul and Barlow, 1980; Robson, 1984). The great majority of polymer blends are immiscible from the thermodynamic standpoint and may be conceptualized in terms of relatively large deformable domains (usually of the order of microns) in a sea or matrix. The morphology of a blend may be changed dramatically depending on the processing conditions or material composition employed. Such changes are significant since the size and shape of the minor component is very important to the final performance of the product, for example: 1) the size of the disperse phase has a significant effect on the impact properties of blends such as high impact polystyrene and 2) transition from a spherical to a fibrillar morphology has a dramatic effect on the strength and modulus of the final material. It is evident, therefore, that it is important to develop instruments which can rapidly analyze a polymer blend with respect to particle size and shape.

An investigation was recently performed (Cielo et al., 1986d) to evaluate the applicability of turbidity spectrometry to particle size analysis in 10%

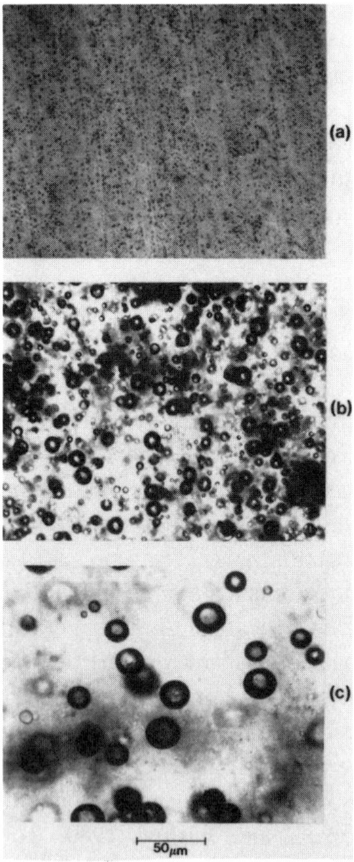

Figure 8-38 Micrographs illustrating blend morphology of polycarbonate particles in a polypropylene matrix for a polyblend material processed under different conditions: (a) blend 1, (b) blend 2, (c) blend 3.

polycarbonate/90% polypropylene blends. Three films were processed under different conditions, resulting in different sizes of the PC particles in the PP matrix, as shown in Figure 8-38. The volume average diameter of the roughly spherical PC particles, as determined by microscopic analysis, was 18.9 μm, 9.8 μm and 1.6 μm, respectively, for blends 1 to 3.

Light transmitted through the polyblend films was analyzed by an FTIR spectrometer to evaluate the feasibility of using the spectral turbidity approach for a control of the particle size in extruded or cast polyblend films. Figure 8-39 shows the spectra of two films obtained with pure polypropylene

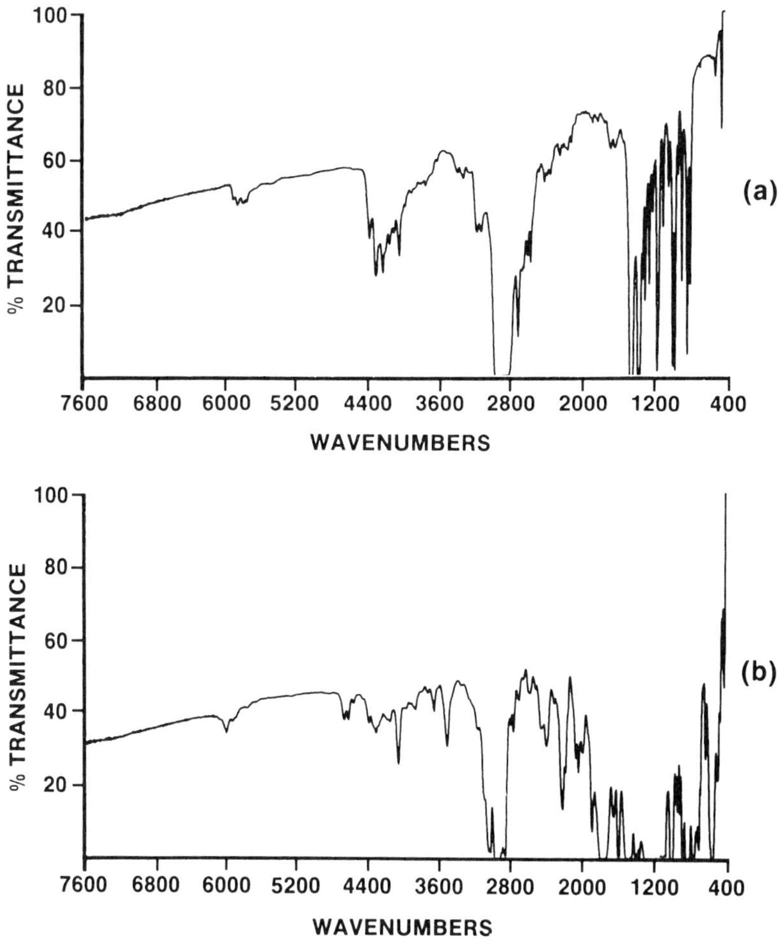

Figure 8-39 Transmission spectra of (a) a 150 μm-thick polypropylene film and (b) a 200 μm-thick polycarbonate film.

and pure polycarbonate, respectively. As we can see, except for the gradual attenuation due to surface scattering at short wavelengths and the absorption bands near 6000 cm^{-1}, 4400 cm^{-1} and 2900 cm^{-1}, the transmittance is relatively high in the 1800 to 7600 cm^{-1} range.

When the spectra of some polyblend films are considered, Figure 8-40, a much stronger attenuation is observed at low wavelengths because of scattering by the dispersed polycarbonate particles (the refractive indices at standard conditions of polycarbonate, 1.59, and polypropylene, 1.49, are sufficiently different to produce significant reflections at the particle-to-matrix interfaces). As we can see, the attenuation range is gradually displaced toward longer wavelengths as the average particle diameter increases, corresponding to a displacement of the cutting edge of the extinction-ratio curve, Figure 8-35. Such spectra can thus be used to obtain particle-size information.

The spectra obtained with the medium-size and with the large-size particles, curves (b) and (c) in Figure 8-40, require some additional discussion. As we can see, the attenuation introduced by the coarse-particle film, curve (c), is 35% stronger than for curve (b) in the low-absorption window around 2400 cm^{-1}. Conversely, it is the medium-particle film, curve (b), which introduces an attenuation 30% larger than the coarse-particle film in the 7000 cm^{-1} range, also corresponding to a low-absorptivity region. This can

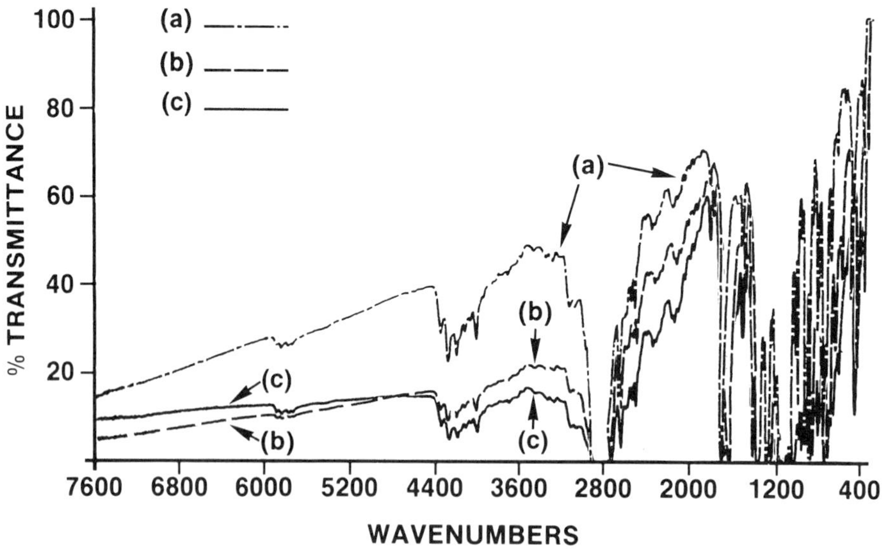

Figure 8-40 Comparative transmission spectra for the 90% polypropylene, 10% polycarbonate polymer blends of Figure 8-38.

be understood if we consider that both films have the same relative amount of polycarbonate, 10%, only the particle size being different. Consequently, in the low-wavelength region, the coarse-particle film introduces lower scattering losses because of equation (8-6), while in the high-wavelength region, the extinction level increases with an increase of the particle diameter, as shown in Figure 8-35. Although a precise determination of the particle size distribution requires a previous knowledge of the particle's shape and concentration (Kiparissides et al., 1980), a simple ratio between the attenuation at two or more wavelengths thus appears to be a convenient and practical tool for the characterization of processed polymer blends in an industrial environment.

Finally, it will be mentioned that larger particulates in composite materials, such as fiberglass or mica fillers of typical sizes from 10 μm to 1 mm, may be optically monitored using far-IR lasers (Boulay et al., 1986).

8.4.3 *Fiber Monitoring in the Pulp and Paper Industry*

Pulp monitoring is an important requirement for quality and process control in pulp and paper mills. The relative concentration of fines (diameter smaller than 10 μm, length smaller than 1 mm), fibers (diameter between 10 and 80 μm, length between 1 and 5 mm) and shives (diameter larger than 80 μm, length between 1 and 20 mm) in the pulp effluent must be evaluated on line for quality control and for regulation of the thermomechanical and chemical pulping machines. The average fiber contents can be evaluated by turbidity or consistency measurements (see Section 2.5.2 for nonoptical consistency-measuring methods; a review of optical transmission or backscattering methods can be found in Keitaaniemi and Kujansivu, 1986), but the evaluation of the average particle size is a more difficult problem, which is normally addressed by manual periodic sampling followed by microscopic visual inspection or else by mechanically screening with different mesh sizes.

Automated optical methods have long been needed for a continuous on-line control of the pulp characteristics. Shading or coherent optical approaches of the kind described in Section 5.1.1 cannot be applied to the inspection of a multitude of randomly distributed fibers in a suspension. Angular-pattern sensing methods of the kind described in the last section for determining the average size of the fibers from an inspection of the Fraunhofer diffraction pattern have been considered (Powers and Somerford, 1979; Somerford and Watson, 1983). Such an approach is, however, difficult to apply in the case of transparent fibers because of the presence of an important light component which is transmitted through the irregularly shaped fibers. This makes it necessary to isolate subsequent diffraction patterns of single fibers, leading to measurement times of the order of five to ten minutes in order to collect statistically significant samples. On the other hand, only information on the fiber diameters may be obtained by this approach; the more relevant fiber length cannot be measured because it is far too large with respect to the light wavelength. The latter argument also excludes spectral turbidity methods for fiber sizing in pulp mills, while far-IR methods (Boulay et al., 1986a) are limited to the inspection of dry samples because of the strong absorption by water.

Fiber Counting Methods

A light-scattering approach for on-line fiber monitoring in the pulp effluent is shown schematically in Figure 8-41 (Hill and Eriksson, 1976; Fladda et al., 1980; Johnsson and Hill, 1983). In the method illustrated in Figure 8-41a, a light beam is focused through a window in the middle of the pipe carrying the pulp slurry. When a suspended particle intersects the beam at the focal point, an appreciable amount of light is scattered out of the original cone of light and detected by a suitably positioned optical sensor. This gives rise to spikes exceeding a given threshold in the recorded signal. The relative width of such spikes depends on the particle size, longer fibers giving longer pulses for a given pulp-flow speed. By taking the ratio of the AC signal amplitude over the DC signal (which depends on the total particle concentration), the average particle size can thus be roughly evaluated.

More precise measurements are obtained by the technique shown in Figure 8-41b. In this case, a part of the pulp is deviated from the main flow and, after dilution to less than 0.1% of solid material, it is made to pass through a capillary tube, typically 1 mm in transversal size with transparent windows. When a fiber passes through the tube in front of the detectors, the decrease in the signal level registered by the sensors is a measure of the fiber diameter. Moreover, the length of the signal gives an evaluation of the fiber length for a given flow speed. More sophisticated techniques using crossed polarizers for increased signal-to-noise level, as well as line-array detectors for fiber-length evaluation independently of the flow speed, have also been proposed (Tiikkaja et al., 1983). Typical fiber-length distributions for hardwood and softwood fibers, as well as for softwood of different age, as obtained with the latter type of sensor, are shown in Figure 8-42 (Piirainen, 1986).

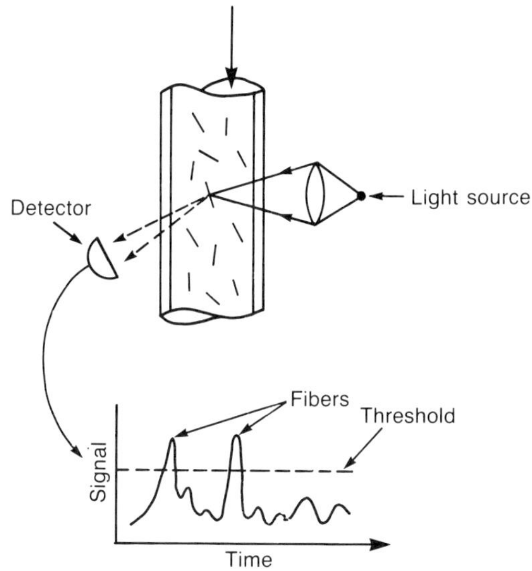

(a)

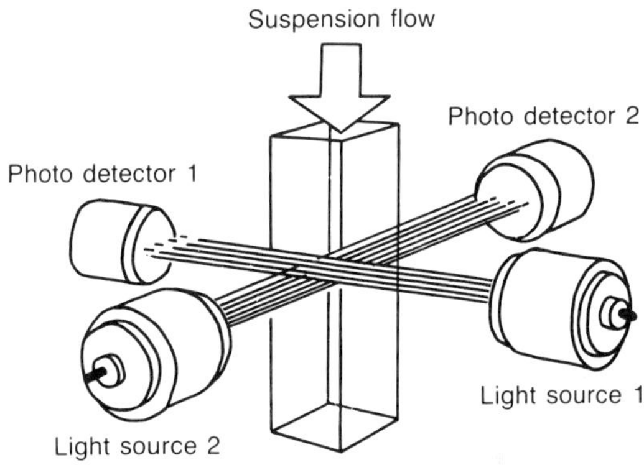

(b)

Figure 8-41 Two methods for fiber monitoring in a paper pulp effluent: (a) measurement of the scattered component of a light beam focused in the center of the stream when intercepted by a fiber; (b) single-fiber length and diameter monitoring by analyzing a diluted sample of the slurry through a capillary tube.

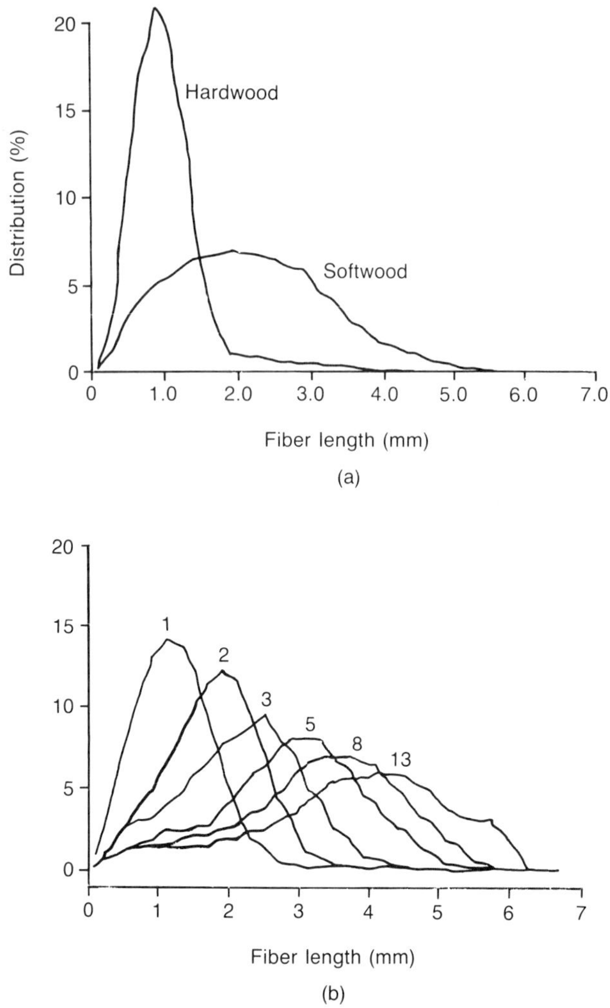

Figure 8-42 Length-weighted fiber length distribution (a) of typical hardwood and softwood fibers and (b) of different annual rings of a typical softwood (pine).

Optoacoustic Method

One of the limitations of the fiber-counting methods is related to the need for the deviation of part of the slurry through a narrow pipe, with occlusion problems. Moreover, the measurement speed is limited by the time required to collect a statistically significant sample for the calculation of a meaningful average. To avoid such problems, a different approach using an optoacoustical technique has recently been proposed (Dion et al., 1982). The basic principle of the optoacoustic technique is shown in Figure 8-43. The pulp slurry is introduced in a resonant cell between a PZT ultrasonic transducer driven at a typical frequency of 500 KHz and an acoustic reflector. When the ultrasonic transducer is powered, a stationary wave sets up in the cell, producing a number of nodal planes parallel to the xz-plane. The fiber particles tend to migrate and to align parallel to such planes in a time-period of a few seconds. The speed of fiber reorientation is faster for fibers of small dimensions, which encounter lower viscous resistance. By monitoring the average fiber orientation from light-scattering measurments with an optical detector conveniently positioned in the yz-plane of Figure 8-43a, the average fiber length can thus be evaluated. Figure 8-43b shows some typical signals

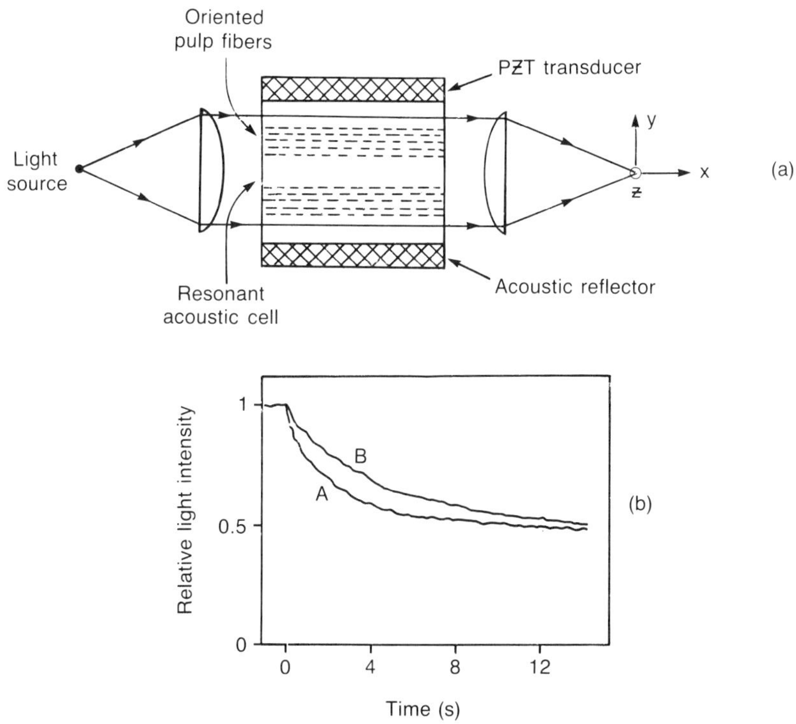

Figure 8-43 (a) The optoacoustic approach for fiber-length evaluation in a paper-pulp emulsion by ultrasonic stationary-wave fiber migration and light-scattering analysis of fiber orientation; (b) Typical results obtained with 200-mesh (curve A) and 100-mesh (curve B) fibers. The fiber-reorientation time constant is smaller for shorter fibers.

obtained with 200-mesh fibers (of shorter length) and with 100-mesh fibers (of longer length) in the slurry, with the optical detector situated off-center along the z-axis in Figure 8-43a. When the fibers reorient after application of the ultrasonic wave, light is preferentially scattered along the y-direction, so that the light intensity along the z-direction decreases. The time-constant of the signal is smaller for shorter fibers, thus providing an evaluation of the average fiber length. The method allows quick evaluation of a pulp volume large enough to provide a statistically significant sample, while occlusion problems in capillary tubes are avoided. This technique is now under intense development (Dion et al., 1986).

8.5 Conclusion

The broad variety of optical techniques which have been described in this textbook, as applied to a number of industrial inspection tasks, should have provided the reader with a feeling for the potential of electro-optical technology in this area of increasing industrial concern. The industrial practitioner may find in certain sections of the book some specific suggestions on how to solve his own sensing requirements. The many examples of optics-based on-line devices should convince the nonspecialist that this area of modern technology is now available as a practical and cost-effective tool to meet the challenges of today's computerized production. As to the electro-optical instrumentation engineer, he should find the cross-comparison of different optical and nonoptical methods helpful in the choice of the best approach for a specific requirement. Such examples should also stimulate the search for creative new applications of this versatile discipline.

Techniques involving geometrical metrology and surface morphology, which are mainly covered by Chapters 3 and 4, are probably those which will experience the largest rate of growth in the immediate future. Such devices are closely associated with the visual function of human operators so that their applications in the industrial environment are often self-suggesting. Although single-sensor devices based on increasingly robust and inexpensive optoelectronic components will certainly find increasing uses in a variety of specific applications, an explosive growth is expected by many economic analyzers in the fields of two dimensional and three-dimensional imaging (see, e.g., *Report on Electro-optical Inspection Systems* by Frost and Sullivan, London, 1986; see also *Machine Vision Markets*, MIRC, Palo Alto, CA, 1986). The availability of increasingly powerful and inexpensive computer image processors is mainly responsible for this optimism, although some caution is advisable against premature overselling (see, e.g. *Sensor Review*, October 1986, p. 204).

The various optical techniques described in Chapters 6 and 7 which are oriented toward specific sectors of the industrial inspection market should be the object of increasing efforts of transfer from the research laboratory to industrial practice in the years to come. The bright future forecasted for some of these devices, such as fiber optic sensors and robot vision systems, is spurring a hectic research and development activity in these areas. As to the optical techniques for material characterization, reviewed in Chapter 8, they represent a good example of a field where a higher degree of communication between high-level research activities and specific industrial sensor implementation would be very fruitful. Such an interaction would be mutually beneficial, both in terms of the increased performance of process analyzers and of the added motivation and practical significance of the research activities in these fields.

It should be mentioned at this point that the overview of optical inspection techniques given in this book, although being sufficiently broad to provide an ample perspective, is not intended to be exhaustive. Some additional disciplines, such as photoacoustic and photothermal techniques for material analysis or spectroscopic on-line surface scanning to assure constant color characteristics in the outcoming product, stand at the intersection of technologies described in different chapters of this book. Other applications of optical technology to industrial instrumentation have not been mentioned in cases where optics plays an auxiliary role in nonoptical devices. Examples are gas or liquid chromatography with IR or UV readout and thermal extensometry with optical displacement sensing devices.

In conclusion, electro-optics is certainly one of the faster growing disciplines of applied physics and it can be safely stated that its applications to industrial instrumentation are going to experience a substantial and sustained development in the years to come. Optical inspection represents one of the major fields of application of this young technology, and its adoption may take place at a very fast pace in the near future, particularly if it is generally perceived as a simple and reliable tool to meet the international challenge towards an increased industrial productivity.

References

Abramovich, A. (1985) "Advanced vision technology for printed circuit inspection." *Vision'85 Conf. Proc.*, p. 5–168. MVA/SME, Dearborne, MI.

Abramson, N. (1981) *The Making and Evaluation of Holograms.* Academic Press, New York.

Achey, F.A, (1982) "Rapid in-process analysis of molten steel." *Report of the AISI Sensor Development Unit 5-2.* The American Iron and Steel Inst., Pittsburg.

Adrian, R.J. and Orloff, K.L. (1977) "Laser anemometer signals: Visibility characteristics and application to particle sizing." *Appl. Opt.* **16**, 677.

Aggarwal, S.K. and Johnston, R.H. (1985) "The effect of temperature on the accuracy of microwave moisture measurements on sandstone cores." *IEEE Trans. Instr. Meas. IM*-**34**, 21.

Agin, G.J. and Binford, T.O. (1973) "Computer description of curved objects." *Proc. Int. Conf. on Artificial Intelligence,* p. 629. Stanford University, Aug. 20–23.

Agin, G.J. and Duda, R.O. (1977) "SRI vision research for advanced industrial automation." *Proc. 2nd USA/Japan Computer Conf.*, p. 113.

Agrawal, A. and Epstein, M. (1983) "Robot eye-in-hand using fiber optics." In: *Intelligent Robots: RoViSeC 3,* D.P. Casasent and E.L. Hall, editors, vol. **449**, p. 48. SPIE, Bellingham, WA.

Ahlers, R.J. (1985) "White-light method: A new sensor for the optical evaluation of rough surfaces." *Opt. Eng.* **24**, 423.

Aindow, A.M., Dewhurst, R.J. and Palmer, S.B. (1982) "Laser-generation of directional surface acoustic wave pulses in metals." *Opt. Commun.* **42**, 116.

Aindow, A.M., Dewhurst, R.J., Palmer, S.B. and Scruby, C.B. (1984) "Laser-based

NDT techniques for the ultrasonic characterization of subsurface flaws." *NDT Int.* **17**, 329.

Akutsu, S., Mashino, Y., Watanabe, Y. and Fujita, M. (1976) "Automatic surface-flaw inspection system of cold strip." *Int. Meet. on Iron and Steel Making, paper 6.2.2.* Eur. Comm. for Steel and Coal, Düsseldorf.

Alguard, M.J. and Lucas, J.M. (1981) "On-line roughness and printability." *Proc. Coat. Conf. Techn. Assoc. Pulp and Paper Ind.*, Houston, May 17–21, p. 135.

Allan, R. (1985) "Nonvision sensors." *Electron. Des.* **33**, 103.

Almond, D.P., Patel, P.M., Pickup, T.M. and Reiter, H. (1985) "An evaluation of the suitability of thermal wave interferometry for the testing of plasma-sprayed coatings." *NDT Int.* **18**.

Altschuler, M.D., Altschuler, B.R. and Taboada, J. (1981) "Laser electro-optic system for rapid 3-D topographic mapping of surfaces." *Opt. Eng.* **20**, 953.

Ameri, S., Ash, E.A., Neuman, V. and Petts, C.R. (1981) "Photo-displacement imaging." *Electron. Lett.* **17**, 337.

Andermo, I., Mörkoch, G. and Sjölund, A. (1976) "Velocity measurement by the correlation method: Accurate, contactless, nonmoving parts." *Teknisk Tidskrift*, **106** (3), 18.

Anderson, J.G., (1980) "Paper/board moisture measurement by microwave loss." In: *Instrumentation and Automation in the Paper, Rubber, Plastics and Polymerization Industries*, A. Van Cauwenberghe, editor, p. 75. Pergamon Press, Oxford.

Arai, S., Shibata, I., Ogawa, K. and Ogawa, S. (1983) "A new servo method which increases track-following range of a one-beam optical head." In: *Optical Disk Systems and Applications*, E.V. La Budde, editor, vol. **421**, p. 143. SPIE Bellingham, WA.

Arai, T., Kikuchi, M., Sakuragi, S., Saito, M. and Takizawa, M. (1985) "CO laser power delivery by As_2O_3 IR glass fiber with Teflon cladding." In: *Optical Fibers in Medicine and Biology*," A. Katzir and A.A. Lamole, editors, vol. **576,** p. 24. SPIE, Bellingham, WA.

Archbold, E. and Ennos, A.E. (1972) "Displacement measurement from double-exposure laser photographs." *Opt. Acta* **19**, 253.

Arecchi, F.T., Bertani, D., and Ciliberto, S. (1979) "A fast versatile optical profilometer." *Opt. Comm.* **31**, 263.

Asakura, T. (1978) "Surface roughness measurement." In *Speckle Metrology*, R.K. Erf, editor. Academic Press, New York.

Asano, Y., (1985) "On-line laser sensor for surface roughness of cold-rolled steel sheets." *Trans. ISI Japan*, **25**, 261.

Asher, J.A. and Jackson, R.L. (1986) "Noncontacting optical measurement and inspection systems." In: *Optical Techniques for Industrial Inspection*, P. Cielo, editor, vol. **665**, p. 139. SPIE, Bellingham, WA.

Ashkin, A., Dziedzie, J.M. and Stolen, R.H. (1981) "Outer diameter measurement of low birefringence optical fibers by a new resonant backscatter technique." *Appl. Opt.* **20**, 2299.

Aspnes, D.E., Theeten, J.B. and Hottier, F. (1979) "Investigation of effective medium models of microscopic surface roughness by spectroscopic ellipsometry." *Phys. Rev.* **B20**, 3292.

Astheimer, R.W. (1983) "Study of boiler optimization by means of spectral emission analysis." In: *Electro-optical Instrumentation for Industrial Applications*, F.M. Zweibaum and R.A. Carella, editors, vol. **411**, p. 62. SPIE, Bellingham, WA.

Attali, S.F. and Cohen, F.S. (1986) "Surface inspection based on stochastic modelling." In *Optical Techniques for Industrial Inspection*, P. Cielo, editor, vol. **665**, p. 46. SPIE, Bellingham, WA.

Atutov, S.N., Bednarzhevskii, S.S., Maltsev, V.P., Saprykin, E.G., Smirtnov, G.I. and Soloboev, V.E. (1981) "Two-parameter laser turbidity meter." *Avtometriya*, May/June 1981, p. 119.

Azzam, R.M.A., and Bashara, N.M. (1977) *Ellipsometry and Polarized Light*. North-Holland, Amsterdam.

Azzopardi, B.J., (1979) "Measurement of drop sizes." *Int. J. Heat and Mass Transfer*, **22**, 1245.

Bahr, A.J. (1982) *Microwave Nondestructive Testing Methods*. Gordon and Breach, New York.

Baier, S.W. (1981) *Thickness testing of Electroplated and Related coatings*, Portcullis Press, U.K.

Bailey, S.J. (1984) "Pressure sensors and transmitters affected by technological change." *Control Eng.* **31**, 97.

Baird, J.D. (1983) "Recent applications of unique noninvasive Doppler transducers for flow measurement." *Proc. ISA Symp., Productivity through Control Technology*, p. 201. The Instrument Society of America, Research Triangle Park, N.C.

Bajcsy, R. and Lieberman, L. (1976) "Texture gradient as a depth case." *Computer Vision, Graphics and Image Processing* **5**, 52.

Baker, L.P. (1976) "Some scanning techniques in automatic inspection." In: *The Engineering Uses of Coherent Optics*, E.R. Robertson, editor, p. 713. Cambridge University Press, London.

Baker, L. (1986) "Optical inspection in industry" In: *Optical Techniques for Industrial Inspection*, P. Cielo, editor, vol. **665**, p. 2. SPIE, Bellingham, WA.

Baker, R.J., Hutchinson, P. and Whitelaw, J.H. (1974) "Velocity measurements in the recirculating region of an industrial burner flame by laser anemometry with light frequency shifting." *Combustion and Flame* **23**, 57.

Balageas, D.L., Krapez, J.C. and Cielo, P. (1986) "Pulsed photothermal modelling of layered materials." *J. Appl. Phys.* **59**, 348.

Ballard, D.H. and Brown, C.M. (1982) *Computer Vision*. Prentice-Hall, Englewood Cliffs.

Ballik, E.A. and Chan, J.H. (1973) "Fringe image technique for the measurement of flow velocities." *Appl. Opt.* **12**, 2607.

Bamba, T., Marygama, H., Ohno, E. and Shiga, Y. (1982) "Welding sensor uses structured light and 2-D photocell" *Sensor Rev.*, **2**, 29.

Bamba, T., Muruyama, H. Kodaira, N. and Tsuda, E. (1984) "A visual seam tracking system for arc-welding robots." *Proc. 14th Int. Symp. Industrial Robots*, N. Martensson, editor, p. 365, IFS/North Holland.

Barber, R. (1978) "Noncontact temperature measurement of metal surfaces in the open" *Proc. ISA-78 Conf. Advances in Instrumentation*, p. 412. The Instrument Society of America, Research Triangle Park, N.C.

Barber, R. (1980) "Review of IR thermometry for the glass industry and the interpretation of their readings." *J. Noncryst. Sol.* **38**, 903.

Barker, A.J. and Brook, R.A. (1978) "A design study of an automatic system for on-line detection and classification of surface defects on cold-rolled steel strip." *Opt. Acta*, **25**, 1187.

Barker, L.M. and Hollenbach, R.E. (1972) "Laser interferometer for measuring high velocities of any reflecting surface" *J. Appl. Phys.* **43**, 4669.

Barth, H.G., ed. (1984) *Modern Methods of Particle Size Analysis.* J. Wiley & Sons, New York.

Barton, J.R. and Kusenberger, F.N. (1982) "Computer-aided nondestructive inspection of ball and bearing components." In *Automated Inspection and Product Control*, B. Batchelor, editor, p. 297. IFS, U.K.

Barton, J.S. and Saoudi, M. (1986) "A fiber-optic vortex flowmeter." *J. Phys. E* **19**, 64.

Bass, M. (1983) *Laser Materials Processing.* North-Holland.

Bassani, R. DiPasquale, E. and Vitali, C. (1984) "Probablistic model for metal-polymer friction." *Wear* **95**, 177.

Batchelor, B.G. and Williams, G.A. (1979) "Defect detection on the internal surfaces of hydraulic cylinders for motor vehicles." In: *Imaging Applications for Automated Industrial Inspection and Assembly*," R.P. Kruger, editor, vol. **182**, p. 65. SPIE, Bellingham, WA.

Batchelor, B.G., Hill, D.A. and Hodgson, D.C. (1985) *Automated Visual Inspection.* IFS/North-Holland, U.K.

Baumbick, R.J. and Alexander, J. (1980) "Fiber optics sense process variables." *Control Eng.*, March 1980.

Baumoel, J. (1980) "Ultrasonic flowmetering comes of age in the 80's" *Proc. ISA Conf., Advances in Instrumentation*, vol. **35**, p. 465. The Instrument Society of America, Research Triangle Park, N.C.

Baxendall, P.J. (1968) "Noise in transistor circuits — part 2" *Wireless World*, p. 454, Dec. 1968.

Beck, M.S. (1981) "Correlation in instruments. Cross-correlation flowmeters." *J. Phys. E Sci. Intrum.* **14**, 7.

Beckmann, P. and Spizzichino, A., (1963) *The Scattering of Electromagnetic Waves from Rough Surfaces.* Pergamon Press, Oxford.

Bedrossian, J. (1987) "Industrial applications of fiber optics in IR thermal imaging" in: *Thermosense IX*, R.P. Madding, editor, vol. **780**, p. 90. SPIE, Bellingham, WA.

Beheim, G. and Fritsch, K. (1986) "Range finding using frequency-modulated laser diode" *Appl. Opt.* **25**, 1439.

Beiser, L. (1985) "Laser scanning and recording: Developments and trends" *Laser Focus* **21**, 88 (February 1985).

Belgen, M.H. (1967) "Structural stress measurement with an infrared radiometer". *ISA Trans.* **6**, 49.

Bell, A.G. (1981) "Upon the production of sound by radiant energy." *Phil. Mag.* **11**, 510.

Bell, R.J. (1972) *Introductory Fourier Transform Spectroscopy*. Academic Press, New York.

Bendat, J.S. and Piersol, A.G. (1980) *Engineering Applications of Correlation and Spectral Analysis.* Wiley, New York.

Benedict, R.P. (1969) *Fundamentals of Temperature, Pressure and Flow Measurement.* J. Wiley and Sons, New York.

Benford, J.R. (1965) "Microscope objectives." In: *Applied Optics and Optical Engineering*, R. Kingslake, editor, vol. **3**, p. 146. Academic Press, New York.

Benford, J.R. and Rosenberger, H.E. (1967) "Microscopes" In: *Applied Optics and Optical Engineering*, R. Kingslake, editor, vol. **4**, p. 31. Academic Press, New York.

Bennet, H.E. (1978) "Scattering characteristics of optical materials." *Opt. Eng.* **17**, 480 (1978).

Bennet, J.M. and Dancy, J.H. (1981) "Stylus profiling instrument for measuring statistical properties of smooth optical surfaces." *Appl. Opt.* **20**, 1985.

Bennet, J.M. (1985) "Comparison of techniques for measuring the roughness of optical surfaces." *Opt. Eng.* **24**, 380.

Bentley, R.G. (1986) " An optical approach to the measurement of the lignin content of Kraft pulps, part A: Using UV measurements." In: *Optical Techniques for Industrial Inspection*, P. Cielo, editor, vol. **665**, p. 265. SPIE, Bellingham, WA.

Beranek, B., Boillot, J.P. and Ferrie, F.P. (1986) "Laser sensor for adaptive welding" In: *Optical Techniques for Industrial Inspection*, P. Cielo, editor, vol. **665**, p. 195. SPIE, Bellingham, WA.

Berger, D.M. (1975) "Detecting film flaws in coatings." *Chemical Engineering*, March 1975, p. 79.

Berneron, R. (1978) " L'analyse des surfaces métalliques par spectrométrie d'émission à décharge luminescente." *Spectr. Acta*, **33B**, 665.

Berrie, M.A., Puttick, K.E., Rider, J.G., Rudman, M. and Whitehead, R.D. (1981) "Thermal probe analysis of orientation in polymers and composites." *Plastics and Rubber Proc. Appl.* **1**, 129.

Berry, K.H. (1980) "Who needs measurements of emissivity?" *Proc. SPIE, vol.* **234**, p. 56.

Besl, P., Delp, E. and Jain, R. (1985) "Automatic visual solder joint inspection." *Vision '85 Conf. Proc.* p. 5–54. MVA/SME, Dearborne, MI.

Betz, C.E. (1973) *Principles of Magnetic Particle Testing*. Magnaflux, Chicago.

Betz, C.E. (1979) *Principles of Penetrants*. Magnaflux, Chicago.

Beuter, K. and Weiss, R. (1985) "Sound pattern recognition supports automatic inspection." *Sensor Review*, **5**, 13.

Beynon, T.G.R. (1982) "Radiation thermometry applied to the development and control of gas turbine engines" In: *Temperature, its Measurement and Control in Science and Industry*. J.F. Schooley, editor, p. 471. Am. Inst. of Physics, New York.

Bible, R.E. and Bible, R.E. Jr (1979) "Automated inspection using electro-optics" *Optics in Quality Assurance II*, H.K. Kasdan, editor, vol. **170**, p. 78. SPIE, Bellingham, WA.

Bickel, G., Häusler, G. and Maul, M. (1985) "Triangulation with expanded range of depth." *Opt. Eng.* **24**, 975.

Binford, T.O. (1979) "Visual perception by computer" *Proc. IEEE Conf. on Systems and Control*. Miami, December 1979.

Bingxin, Y., Qinglin, Y., Haoyu, Z. and Chunroug, S. (1983) "JSY-390 laser light scattering spectrometer." *Proc. Miconex '83 Conf.*, Shanghai Apr. 12–16, p. 183. The Instrument Society of America, Research Triangle Park, N.C.

Birchenough, D.R. and Marshall, M.J. (1972) "Magnetic and electromagnetic methods of coating thickness measurements." *Meas. Control* **5**, 509.

Birk, J., et al. (1979) "Image feature extraction using diameter-limited gradient direction histograms." *IEEE Trans. PAMI*, **1**, 228.

Birnbaum, G. and Vest, C.M. (1983) "Holographic nondestructive evaluation: Status and future." In: *International Advances in NDT*. W.J. McGonnagle, editor, vol. **9**, p. 257. Gordon and Breach, New York.

Birnbaum, G. and White, G.S. (1984) "Laser techniques in NDE" In: *Research Techniques in NDT*, R.S. Sharpe, editor, vol. **7**, p. 259. Academic Press, London.

Birnbaum, G., Nyyssonen, D., Vest, C.M. and Vorburger, T.V. (1986) "Optical NDE at the National Bureau of Standards" In: *Holographic NDT*, C.M. Vest, editor, vol. **604**. SPIE, Bellingham, WA.

Bittle, R.R. and Taylor, R.E. (1984) "Step-heating technique for thermal diffusivity measurements of large-grained heterogeneous materials." *J. Amer. Ceram. Soc.* **67**, 186.

Black, G.T. and Johnson, R.A. (1980) "Fiber-optic moisture sensor for composite structures." U.S. Patent 4, 221, 962.

Blais, F. and Rioux, M. (1986) "BIRIS: a simple 3-D sensor" In *Optics, Illumination and Image Sensing for Machine Vision*, D.J. Svetkoff editor, vol. **728**, p. 235. SPIE, Bellingham, WA.

Bloom, A.L. (1965) "Noise in lasers and laser detectors." *Spectra-Physics Laser Techn. Bull.*, No. **4**. Spectra-Physics, Mountain View, CA

Blumenthal, J. (1984) "Improving productivity through mass flow measurement and control." *Proc. Pacific Cascade Instrumentation Conf.*, p. 163. The Instrument Society of America.

Boccara, A.C., Fournier, D. and Badoz, J. (1980) "Thermo-optical spectroscopy: Detection by the mirage effect." *Appl. Phys. Lett.* **36**, 136.

Boillot, J.P., Cielo, P., Michel, C., Lessard, M., Fafard, P. and Villemure, D. (1985) "Adaptive welding by fiber-optic thermographic sensing: An analysis of thermal and instrumental considerations." *Welding Journ.*, **64**, 209-s.

Bojic, M., Bourdieu, J.M., Jorre, D. and Barbier, G. (1970) "Etude au laboratorie de l'analyse spectrographique directe sur fontes et aciers liquides." *Rev. Met.* **67**, 237.

Bolon, Ph., Moisan, E. and Sabater, J. (1983) "Jet-speed measurement by correlation of optical signals." In *Instrumentation and Automation in the Paper, Rubber, Plastics and Polymerization Industries*, A.R. Van Cauwenberghe, editor, p. 51. IFAC/Pergamon Press, Oxford.

Bondarenko, A.N., Drobot, Y.B. and Kruglov, S.V. (1976) "Optical excitation and detection of nanosecond acoustic pulses in NDT" *Sov. J. NDT* **12**, 655.

Born, M. and Wolf, E. (1970) *Principles of Optics*. Pergamon Press, Oxford.

Bosisio, R.G. and Giroux, M. (1970) "Paper sheet moisture measurements by microwave perturbation techniques." *Tappi*, **53**, 458.

Botsco, R.J. (1968) "Nondestructive testing of plastics with microwaves." In: *Plastic Design and Processing* (reprinted from November and December 1968 issues, Lake Publishing Corp., Libertyville, III.).

Boulay, R., Drouin, B., Gagnon, R. and Izatt, J.R. (1984) "Newsprint moisture and basis weight measurement using a submillimetre laser." *Journ. Pulp & Paper Science* **10**, J98.

Boulay, R., Drouin, B., Gagon, R., Bélanger, P.A. and Cielo, P. (1985) "Far-IR moisturing in polymeric composites". *Polym. Comm.* **26**, 69.

Boulay, R., Dubé, P., Bélanger, P.A., Vu-Khanh, T. and Cielo, P. (1986) "Analysis of filler concentration and orientation in reinforced polymers far-IR techniques." In: *Optical Techniques for Industrial Inspection*, P. Cielo, editor, vol. **665**, p. 352. SPIE, Bellingham, WA.

Boulay, R., Drouin, B., Gagnon, R. and Bernard, P. (1986a) "Paper fibre orientation measurement with a submillimetre laser." *J. Pulp and Paper Science*, **12**, J26.

Bowers, J.E. (1982) "Fiber-optical sensor for surface acoustic waves." *Appl. Phys. Lett.* **4**, 231.

Bowers, J.E., Jungerman, R.L., Khuri-Yakub, B.T. and Kino, G.S. (1983) "An all fiber-optic sensor for surface acoustic wave measurements." *J. Lightwave Tech.* **LT-1**, 429.

Bowyer, A., Graham, D. and Henry, G. (1985) "The measurement of 3-D features using laser triangulation." In: *Automated Inspection and Product Control*, P. A. McKeown, editor, p. 313. IFS, U.K.

Boyer, H.E., editor (1976) *Metals Handbook*, vol. 11. American Society for Metals, Metals Park, OH.

Boyer, H.E. and Gall, T.L. (1985) *Metals Handbook*. American Society for Metals, Metals Park, OH.

Brammer, J.A. and Percival, C.M. (1970) "Elevated-temperature elastic moduli of 2024 aluminum obtained by a laser-pulse technique." *Exp. Mech.* **10**, 245.

Brandt, R. (1981) "Thermal diffusivity measurements on plasma-sprayed CaO-stabilized ZrO_2." *High Temp.-High Press.* **13**, 79.

Bretschi, J. (1981) *Automated Inspection Systems for Industry*. IFS, U.K.

Brewster, D.B. and Robinson, W.I. (1974) "Papermaking sensors: The key to successful computer control." *Pulp & Paper*, March 1974, p. 70.

Brillouin, L. (1962) *Science and Information Theory.* Academic Press, New York.

Brittle, R. R. and Taylor R. E. (1984) "Step-heating technique for thermal diffusivity measurements of large-grained heterogeneous materials." *J. Amer. Cer. Soc.* **67**, 186.

Brodmann, R. and Thurn, G. (1984) "An optical instrument for measuring the surface roughness in production control." *Annals of the CIRP*, **33**, 403.

Brodmann, R., Gerstorfer, O. and Thurn, G. (1985) "Optical roughness measuring instrument for fine-machined surfaces." *Opt. Eng.* **24**, 408.

Brody, H. (1982) 'Holography creates new breed of optical components." *High Technology*, **4**, 68 (July/August 1982).

Brown, G.M. (1974) "Pneumatic tire inspection." In: *Holographic NDT*, R.K. Erf, editor, p. 355. Academic Press, N.Y.

Brown, G.M. and Wales, R.R. (1983) "Vibration analysis of automotive structures using holographic interferometry." In: *Industrial Applications of Laser Technology*, W.F. Fagan, editor, vol. **398**, p. 82. SPIE, Bellingham, WA.

Bruins, D.E., Garland, C.W. and Greytak, T.J. (1975) "New interferometric method for piezoelectric measurements." *Rev. Sci. Instrum.* **46**, 1167.

Bruning, J.H. (1978) "Fringe-scanning interferometers" In; *Optical Shop Testing*, D. Malacara, editor, p. 409. John Wiley and Sons, New York.

Brunk, W. (1982) "Geometric control by industrial robots." *Proc. 2nd Int. Conf. on Robot Vision and Sensory Controls*, Stuttgart, Nov. 2–4, p. 223.

Bucaro, J.A., Dardy, H.D. and Carome, E. (1977) "Fiber-optic hydrophone." *J. Acoust. Soc. Am.* **62**, 1302.

Bucaro, J.A. and Hickman, T.R. (1979) "Measurement of sensitivity of optical fibers for acoustic detection." *Appl. Opt.* **18**, 938.

Buchanan, J.G. and Thurston, R.C.A. (1956) "The measurement of crack depths by the direct-current conduction method." *Nondestructive Testing*, **14**, 36.

Buchhave, P. (1975) "Laser Doppler vibration measurements using variable frequency shift." *Disa Information Nr. 18.* Dantec/Disa, Copenhagen.

Buckley, S. (1985) "Fast listening finds defects." *Sensor Review* **5**, 153.

Buerger, M.J. (1960) *Crystal-Structure Analysis.* Wiley, New York.

Bulkin, B.J., Lewin, M. and DeBlase, F.J. (1985) "Conformational change, chain orientation and crystallinity in polyethylene terephtalate yarns: Raman spectroscopic study". *Macromolecules*, **18**, 2587.

Bulkin, B.J., DeBlase, F.J. and Lewin, M. (1986) "Raman spectroscopy applied to polymers: Low-frequency spectra of polyester fibers." In: *Optical Techniques for Industrial Inspection*, P. Cielo, editor, vol. **665**, p. 234. SPIE, Bellingham, WA.

Burk, D.L. and Simcic, N.F. (1978) "Mill applications using self-scanning diode arrays." *Iron and Steel Eng.*, July 1978, p. 43.

Burk, G. (1984) "Infrared sensors control thickness of film-coextrusion plies." *Modern Plastics*, January 1984, p. 84.

Burke, H.K. and Michon, G.J. (1976) "Charge injection imaging: Operating techniques and performance characteristics" *IEEE Trans. El. Dev.* **ED23**, 189.

Burns, R.M. and Bradley, W.W. (1959) *Protective Coatings for Metals.* Reinhold, New York.

Busse, G. (1982) "Optoacoustic and photothermal material inspection techniques." *Appl. Opt.* **21**, 107.

Busse, G. and Eyerer, P. (1983) "Thermal wave remote and nondestructive inspection of polymers." *Appl. Phys. Lett.* **43**, 355.

Busse, G. (1985) "Imaging with optically generated thermal waves." *IEEE Trans. Son. Ultrason.* **SU-32**, 355.

Busurin, V.I., Semenov, A.S. and Udalov, N.P. (1985) "Optical and fiber-optic sensors (a review)." *Sov. J. Quant. Electron.* **15**, 595.

Butters, J.N. and Leendertz, J.A. (1971) "Holographic and video techniques applied to engineering measurements." *J. Inst. Meas. and Control*, **4**, 349.

Butters, J.N. (1977) "Applications of ESPI to NDT." *Opt. Laser Tech.* **9**, 117.

Cachier, G. (1971) "Laser excitation of microwave sound in solids." *J. Acoust. Soc. Am.* **49**, 974.

Calder, C.A. and Wilcox, W.W. (1978) "High-temperature noncontact material testing." In: *Characterization of Materials for Service at Elevated Temperatures*, G.V. Smith, editor, p. 169. ASME Series MPC-7.

Calder, C.A. and Wilcox, W.W. (1979) "Acoustic velocity measurement across the diameter of a liquid metal column." *Exp. Mech.* **19**. 171.

Calder, C.A. and Wilcox, W.W. (1980) "Noncontact material testing using laser energy deposition and interferometry." *Mater. Eval.* **38**, 86.

Campbell, J.M. and McLachlan, E.H. (1979) "Holographic NDT" *Brit. J. NDT*, **21**, 71.

Campus, A., Matey, G. and Legoupil, H. (1985) "Laser-based determination of the cleanliness of high-voltage insulation compounds." *NDT Int.* **18**, 199.

Cantor, C. (1979) "Electro-optical sensor for measuring road surface texture." In: *Smart Sensors*, D.F. Barbe, editor, vol. **178**, p. 127. SPIE, Bellingham, WA.

Cardis, T.M. and Lastrapes, T.M. (1984) "Continuous monitoring of sodium hydroxide by IR analysis." In: *Instrumentation and Control in the 80's*, Des Plaines, Ill., May 1–3. The instrument Society of America, Research Triangle Park, N.C.

Carlisle, S.S. (1980) "On-line fault detection and pattern recognition on moving webs." In: *Instrumentation and Automation in the Paper*, Rubber, Plastics and Polymerization Industries, A. Van Cauwenberghe, editor, p. 533. IFAC/Pergamon Press, Oxford.

Carlson, F.E. and Clark, C.N. (1965) "Light sources for optical devices." In: *Applied Optics and Optical Engineering*, R. Kingslake, editor. Academic Press, New York.

Carslaw, H.S. and Jaeger, T.C. (1959) *Conduction of Heat in Solids.* Oxford Univeristy Press, Oxford.

Castleman, K.R. (1979) *Digital Image Processing.* Prentice-Hall, Englewood Cliffs.

Catros, J.-Y. and Espiau, B., (1980) "Utilisation de capteurs optiques de proximité en robotique industrielle." *Nouvel Automatisme*, **47**.

Cawley, P. (1985) "The operation of NDT instruments based on the impedance method." *Composite Structures*, **3**, 215.

Chabey, I. (1982) "Optical waveguides." *Anal. Chem.* **54**, 1071A.

Chalmers, J.M. and MacKenzie, M.W. (1985) "Some industrial applications of FTIR diffuse reflectance spectroscopy." *Appl. Spectr.* **39**, 634.

Chamberlin, G.J. and Chamberlin, D.G. (1980) *Colour: Its Measurement, Computation and Application.* Heyden, London.

Chang, B.J. and Winick, K.A. "Holographic optical elements." In: *Advances in Laser Scanning Technology*, L. Beiser, editor, **vol. 299**, p. 157. SPIE, Bellingham, WA.

Chaskelis, H.H. and Clark, A.V. (1980) "Ultrasonic nondestructive bond evaluation: An analysis of the problem." *Mater. Eval.* **38**, 20.

Cheevers, G. D. (1978) "Nondestructive determination of zinc phosphate coating weights using specular reflectance infrared absorbance." *Journ. Coating Technol.* **50**, 78.

Chesmond, C.J. (1984) *Control System Technology.* Edurard Arnold, London.

Chi, Z.C. (1986) "Recognition for 3-D surface shape using multiple distance feature." In: *Optical Techniques for Industrial Inspection*, P. Cielo, editor, vol. **665**, p. 217. SPIE, Bellingham WA.

Chiang, F.P. (1975) "A shadow moiré method with two discrete sensitivities." *Exp. Mech.*, p. 382.

Chijiima, M. and Morii, T. (1980) "Measurement of vibrations in audio components using a laser vibrometer." *Disa Information* **Nr. 27**. Dantec/Disa, Copenhagen.

Chin, B.A., Goodling, J.S. and Madsen, N.H. (1983) "Infrared thermography shows promise for sensors in robotic welding." *Robotics Today*, **5**, 85.

Chin, R.T. and Harlow, C.A. (1982) "Automated visual inspection: A survey" *IEEE Trans. P.A.M.I.*, **4**, 557.

Chottiner, J., Sanjana, Z.N., Kodani, M.R., Lengel, K.W. and Rosenblatt, G.B. "Monitoring cure of autoclave-molded parts by dielectric analysis." *Polymer Composites*, **3**, 59.

Chu, F.Y. and Balls, B.W. (1979) "Moisture measurement: Infrared vs. microwave." *Pulp & Paper Canada*, **80**, 71.

Chu, W.P., Robinson, D.M. and Goad, J.H. (1972) "Holographic nondestructive testing with impact excitation." *Appl. Opt.* **11**, 1644.

Chu, W.P. and Robinson, D.M. (1977) "Scattering from a moving spherical particle by two crossed coherent plane waves." *Appl. Opt.* **16**, 619.

Church, E.L. and Zavada, J.M. (1976) "Effects of surface microroughness in ellipsometry." *J. Opt. Soc. Am.* **66**, 1136A.

Church, E.L., Jenkinson, H.A. and Zavada, J.M. (1977) "Measurement of the finish of diamond-turned metal surfaces by differential light scattering." *Opt. Eng.* **16**, 370.

Church, E.L. (1979) "The measurement of surface texture and topography by differential light scattering." *Wear* **57**, 93.

Church, E.L. and Takacs, P.Z. (1985) "Survey of the finish characteristics of machined optical surfaces." *Opt. Eng.* **24**, 396.

Cielo, P. and Delisle, C. (1975) "Filtres à réflexion interne totale." *Appl. Opt.* **14**, 562.

Cielo, P. and Delisle, C. (1975a) "Quelques applications de la réflexion frustrée en optique instrumentale." *J. Can. Phys.* **53**, 1743.

Cielo, P. and Delisle, C. (1976) "Velocimétrie Doppler par modulation spectrale." *Can. J. Phys.* **54**, 1199.

Cielo, P. and Delisle, C. (1976a) "Multiplexage en communication optique par interférométrie à grande différence de marche en lumière blanche." *J. Can Phys.* **54**, 2322.

Cielo, P., Arsenault H.H., and Delisle, C., (1977) "Analyse de modifications superficielles par photographie speckle." *45e Congrès ACFAS*, Trois-Rivières, Qué.

Cielo, P. and Westwood, W.D. (1978) "Printing by modulation of ink viscosity." U.S. Patent 4,164,745, filed on May 8, 1978, issued on Aug. 14, 1979.

Cielo, P. (1979) "Fiber-optic hydrophone: Improved strain configuration and environmental noise protection." *Appl. Opt.* **18**, 2933.

Cielo, P. (1981) "Optical detection of acoustic waves for the characterization of materials with unpolished surfaces." *Proc. 102nd ASA Meeting — J. Acoust. Soc. Am.* vol. **70**, Suppl. 1, p. 546.

Cielo, P. and Lapierre, J. (1982) " Micro-optical spatial filtering for optical inspection." *Conf. on Lasers and Electro-optics* (CLEO '82), Phoenix, Apr. 14–16.

Cielo, P. and Lapierre, J. (1982a) "Fiber-optic ultrasound sensing for the evaluation of materials." *Appl. Opt.* **21**, 572.

Cielo, P. and McMahon, G.W. (1982) "Stable fiber-optic hydrophone." *Can. Pat.* 1,124,384, issued on May 25, 1982; U.S. Pat. 4,525,818, issued on June 25, 1985.

Cielo, P. (1983) "Optical inspection by internal reflection and spatial filtering." *Opt. Laser Techn.* **15**, 145.

Cielo, P. (1983a) "Analysis of pulsed thermal inspection." *Proc. 14th Symp. NDE*, D.W. Moore and G.A. Matzkanin, editors, p. 418. SRI/NTIAC, San Antonio, TX.

Cielo, P., Dufour, M. and Vaudreuil, G. (1983) "Interim report on optical roughness sensing for a wire-coating extruder." *Report 83-331-102*. NRCC-IMRI, Boucherville, Québec.

Cielo, P., Arsenault, P., Bruneau, L. and Lapierre, J. (1983a) "Inspection interférométrique de la déformation d'une valve hydraulique". *Report 83-216-01* Industrial Materials Research Institute, Boucherville, Québec.

Cielo, P. (1984) "Pulsed photothermal evaluation of layered materials", *J. Appl. Phys.* **56**, 230.

Cielo, P. and Vaudreuil, G. (1984) "Interim report on the optical roughness sensor for a wire-coating extruder." *Report 84-404-01*. NRCC-IMRI, Boucherville, Qué.

Cielo, P., Rousset, G. and Bertrand, L. (1984) "Nondestructive interferometric detection of unbonded layers" *Opt. Laser Eng.* **5**, 231.

Cielo, P. (1985) "Thermoelastic testing of stratified materials." In: *International Advances in NDT*, W.J. McGonnagle, editor, vol. **11**, p. 175–217.

Cielo, P. and Bussière, J. (1985) "Efficient laser generation of surface acoustic waves." U.S. Patent 4,541,280.

Cielo, P. and Vaudreuil, G. (1985) "Optical device for the detection of surface roughness on extruded wires." *Report 85-404-01*. NRCC-IMRI, Boucherville, Québec.

Cielo, P., Cole, K. and Vaudreuil, G. (1985) "Optical detectability of surface scale on hot-rolled steel sheets." *Report 85-404-05*. NRCC-IMRI, Boucherville, Québec.

Cielo, P., Nadeau, F. and Lamontagne, M. (1985a) "Laser generation of convergent acoustic waves for materials inspection." *Ultrasonics*, **23**, 55.

Cielo, P., Maldague, X., Rousset, G. and Jen, C.K. (1985b) "Thermo-elastic inspection of layered materials: Dynamic analysis." *Mater. Eval.* **43**, 1111.

Cielo, P., Lewak, R. and Balageas, D.L. (1985c) "Thermal sensing for industrial quality control." In *Thermosense VIII*, H. Kaplan, editor, vol. **581**, p. 47. SPIE, Bellingham, WA.

Cielo, P. and Dallaire, S. (1985d) "Laser surface coupler." U.S. Patent 4,480,168, issued 11 June 1985.

Cielo, P. and Dallaire, S. (1986) "Optothermal NDE of thermal-barrier coatings.". *ASM Materials Week*, Toronto, Oct. 12–17, 1985. (To be published in *Journ. Mat. En. Syst.*, Vol. **9(1)**, March 1986).

Cielo, P., Jen, C.K. and Maldague, X. (1986) "The converging SAW technique: Analysis and applications to NDE." *Can. J. Phys.* **64**, 1325.

Cielo, P., Dallaire, S., Lamonde, G. and Johar, S. (1986a) "Measurement of thermal inertia by the reflective-cavity method." *Can. J. Phys* **64**, 1217.

Cielo, P., Utracki, L.A. and Lamontagne, M. (1986b) "Thermal diffusivity measurements by the converging-thermal-wave techniques." *Can. J. Phys.* **64**, 1172.

Cielo, P., Cole, K. and Lamontagne, M. (1986c) *Report IGM 86-404-01* Industrial Materials Research Institute, Boucherville, Qué.

Cielo, P., Cole, K. and Favis, B. (1986d) "Optical inspection for industrial quality and process control." *6th Int. Conf. on Instrumentation and Automation in the Paper, Rubber, Plastics and Polymerization Industries*, Akron, OH, Oct. 27–29. The International Federation for Automation and Controls.

Cielo, P., Maldague, X., Johar, S. and Lauzon, P. (1986e) "Some laser-based techniques for the characterization of sintered ceramics." *Mater. Eval.* **44**, 770.

Cielo, P., Rousset, G. and Bertrand, L. (1986f) "Photoacoustic and photothermal evaluation of stratified materials." *Appl. Opt.* **25**, 1327.

Clark, G.L. (1955) *Applied X-rays*, 4th edition. McGraw-Hill, New York.

Clarke, G.M. and Bedford, J. (1978) "Laser scanning for defects, dimensions and surface finish." *Proc. 3rd Int. Conf. on Automatic Inspection and Product Control*, Nottingham, p. 119. IFS, U.K.

Clarke, G.M. and Thomas, T.R. (1979) "Roughness measurement with a laser-scanning analyzer." *Wear* **57**, 107.

Clevett, K.J. (1986) *Process Analyzer Technology*. Wiley, New York.

Clifton, R.J. (1970) "Analysis of the laser velocity interferometer." *J. Appl. Phys.* **41**, 5335.

Clocksin, W.F., Davey, P.G., Morgan, C.G. and Vilder, A.R. (1982) "Progress in visual feedback for robot arc-welding of thin-sheet steel." *2nd Int. Conf. on Robot Vision and Sensory Controls*, Stuttgert, Nov. 2–4.

Clocksin, W.F., Browley, J.S.E., Davey, P.G., Vidler, A.R. and Morgan, C.G. (1985) "An implementation of model-based visual feedback for robot arc welding of thin sheet steel." *Int. J. Robotics Res.* **4**, 13.

Close, D.H. (1979) "Optically recorded holographic optical elements." In: *Handbook of Optical Holography*, H.J. Caulfield, editor, p. 573–586. Academic Press, New York.

Coduti, P.L. (1980) "Effect of residual carbon on the paintability of steel strip.". *Metal Finishing* **7–8**, 51.

Cohen, D.K., Gee, W.H., Ludeke, M. and Lewkowicz, J. (1984) "Automatic focus control: The astigmatic lens approach." *Appl. Opt.* **23**, 565.

Cohen, R.L. and West, K.W. (1983) "Characterization of metals and alloys by electrical resistivity measurements." *Mater. Eval.* **41**, 1074.

Cohen, R.L. and West, K.W. (1983a) "NDT of spot-weld strength by electrical conductivity measurements." *Qual Test-2 Conf. Proc.*, Dallas Oct. 25–27. ASNT, Columbus, OH.

Colandene, B.T. (1964) "Program costs vs reliability." *10th Annual Symposium on Reliability and Quality Control, IEEE*, p. 386.

Cole, J.B. and Swords, M.D. (1979) "Laser-Doppler anemometry measurements in an engine." *Appl. Opt.* **18**, 1539.

Cole, J.H., Johnson, R.L. and Bhuta, P.B. (1977) "Fiber-optic detection of sound." *J. Acoust. Soc. Am* **62**, 1136.

Cole, J.H. (1980) "Low-frequency laser noise of several commercial lasers." *Appl. Opt.* **19**, 1023.

Cole, K.C. (1984) "Caractérisation de la résine époxyde Narmco 5208 dans le domaine proche-infrarouge." *Internal report 84-101-02.* Industrial Materials Research Institute, Boucherville, Qúebec.

Cole, K., Noël, D. and Hechler, J.J. (1985) "Characterization of epoxy-graphite composites by diffuse-reflectance FTIR." In: *Fourier and Computerized Infrared Spectroscopy*, J.G. Grasseeli and D.G. Cameron, editors, vol. **553**, p. 114. SPIE, Bellingham, WA.

Cole, K.C., Lehto, C. and Yuhasz, M. (1986) "Detection of mould release agent contamination of the surface of epoxy-based composites by diffuse-reflectance FTIR spectroscopy." In: *Optical Techniques for Industrial Inspection*, P. Cielo, editor, vol. **665**, p. 243. SPIE, Bellingham, WA.

Cole, K.C. and Noel, D. (1986) "Physicochemical characterization of high-performance fibre-reinforced organic-matrix composites." *Report IGM 86–101, –307, –321 and –424.* Industrial Materials Research Institute, Boucherville, Québec.

Coleman, D.A. and Thew, M.T. (1983) "The use of on-line optical devices for measuring oil in water content." *Proc. Int. Conf. on Optical Techniques in Process Control*, The Hague , June 14–16, p. 97. BHRA, Cranfield, U.K.

Compton, D.A.C., Markelov, M. Mittleman, M.L. and Grasselly, J.G. (1985) "The use of a small, low-cost FTIR spectrometer as a mobile analyzer for fluid streams." *Appl. Spectr.* **39**, 909.

Connors, R.W. and Harlow, C.A. (1980) "A theoretical comparison of texture algorithms." *IEEE Trans. Pattern Anal. Machine Intell.* **2**, 204.

Conrady, A.E. (1960) *Applied Optics and Optical Design*, Volumes 1 and 2. Dover, New York.

Considine, D.M. (1985) *Process Instruments and Control Handbook*. McGraw-Hill, New York.

Cook, A.J. (1980) "Advances in on-line appearance measurement." In: *Instrumentation and Automation in the Paper, Rubber Plastics and Polymerization Industries*, A. Van Cauwenberghe, editor, p. 67. IFAC/Pergamon Press, Oxford.

Cook, C.E. and Bernfeld, M. (1967) *Radar Signals*. Academic Press, New York.

Cook, G.E. (1983) "Robotic arc welding: Research in sensory feedback control." *IEEE Trans.*, vol. **IE-30**, p. 252.

Corey, H. S. (1980) "Laser surface finish gauge." In: *Advances in Laser Engineering and Applications*, M.L. Stitch, editor, vol. **247**, p. 173. SPIE, Bellingham, Wa.

Corey, H.S. (1986) "Automatic, transparent, thin-film-thickness inspection system for complex shapes." In: *Optical Techniques for Industrial Inspection*, P. Cielo, editor, vol. **665**, p. 128. SPIE, Bellingham, WA.

Corman, S. (1983) "Optical inspection technology for multi-layer printed wiring boards." *Test & Measurement Techn.*, May 1983.

Cornish, D.K., Jepson, G. and Smurthwaite, M.J. (1981) *Sampling Systems for Process Analyzers*. Sira/Page Bros., U.K.

Cowan, R.D. (1961) "Proposed method of measuring thermal diffusivity at high temperatures." *J. Appl. Phys.* **32**, 1363.

Cowhig, W.T. (1969) "Automatic inspection of textile yarns and fabrics." *Sira Conf. Autom. Insp.*, Eastbourne, May 1969.

Crane, R. (1969) "New developments in interferometry. 5. Interference phase measurement". *Appl. Opt.* **8**, 538.

Crane, R.M., Macander, A.B. and Gagorik, J. (1982) "Fiber optic damage assessment system for fiber reinforced composite structures." *Technical Report*, D.W. Taylor Naval Ship R & D Center, Bethesda, MD.

Cremers, D.A. and Radziemski, L.J. (1983) "Detection of chlorine and fluoride in air by laser-induced breakdown spectrometry." *Anal. Chem.* **55**, 1252.

Cremers, D.A., Radziemski, L.J. and Loree, T.R. (1984) "Spectrochemical analysis of liquids using the laser spark.". *Appl. Spectr.* **38**, 721.

Crosley, D.R., editor (1980) *Laser Probes for Combustion Chemistry*. American Chemical Society, Washington, D.C.

Culshaw, B. (1984) *Optical Fibre Sensing and Signal Processing*. Peter Peregrinus Ltd., London.

Dahne, C. and Lanzl, F. (1980) "A microscope-focusing technique for a digital image analysis system." *Optik* **55**, 437.

Dakin, J.P. and Withers, P.B. (1986) "The reduction of semiconductor laser phase noise for sensor applications." *Opt. Acta* **33**, 489.

Dändliker, R., Ineichen, B. and Mottier, F.M. (1973) "High resolution holographic interferometry by electronic phase measurement." *Opt. Comm.* **9**, 412.

Dändliker, R. (1980) "Heterodyne holographic interferometry." In: *Progress in Optics*, E. Wolf, editor, **Vol. 17**, p. 1, North-Holland, Amsterdam.

Dändliker, R. and Willemin, J.F. (1981) "Measuring microvibrations by heterodyne speckle interferometry." *Opt. Lett.* **6**, 165.

Dändliker, R. and Thalmann, R. (1985) "Heterodyne and quasi-heterodyne holographic interferometry.". *Opt. Eng.* **24**, 824.

Dandridge, A., Tveten, A.B., Miles, R.O. and Giallorenzi, T.G. (1980) "Laser noise in fiber-optic interferometer systems." *Appl. Phys. Lett.* **37**, 526.

Dandridge, A. and Tveten, A.B. (1981) "Noise reduction in fiber-optic interferometer systems." *Appl. Opt.* **20**, 2337.

Dandridge, A. (1982) "Noise in diode lasers". In: *Fiber-Optic Rotation Sensors and Related Technologies*, S. Ezekiel and H.J. Arditty, editors, p. 245. Springer-Verlag, Berlin.

Dandridge, A. and Tveten, A.B. (1983) "Properties of diode lasers with intensity noise control.". *Appl. Opt.* **22**, 310.

Dandridge, A. and Miles, R.O. (1985) "Coherent length and noise properties of GaAlAs broad-band sources." *J. Lightwave Techn.* **LT-3**, 338.

Danielson, G.L. (1980) "Aircraft quality assurance using close-range photogrammetry." In: *Optics in Metrology and Quality Assurance*, H.K. Kasdan, editor, vol. 220, p. 89. SPIE, Bellingham, WA.

Dao, N.Q., Prod'homme, M., Plaza, P. and Joyeux, M. (1986) "Étude in-situ et à distance par spectrométrie Raman associée aux fibres optiques, du dépôt sur acier en milieu marin." *C.R. Acad. Sc. Paris* **t. 302**, Ser. 2, p. 313.

Davenport, W. B. and Root, W. L. (1958) *An Introduction to the Theory of Random Signals and Noise.* McGraw-Hill, New York.

Davenport, W.B. (1970) *Probability and Random Processes.* McGraw-Hill, New York.

Davinson, I. (1984) "The applications of fibre optics in gas turbine engine instrumentation." In: *Fibre Optics '84*, J.M. Tait, editor, vol. 468, p. 190. SPIE, Bellingham, WA.

DeCarfort, F. and Foulard, C. (1971) *Asservissements Linéares Continus.* Dunod, Paris.

DeKeijzer, A. and Degroot, R.J. (1984) "Laser-based arc welding sensor monitors weld preparation profile." *Sensor Review*, January 1984, p. 8.

Deferrari, H.A, Darby, R.A. and Andrews, F.A. (1967) "Vibrational displacement and mode-shape measurement by laser interferometer." *J. Acoust. Soc. Am.* **42**, 982.

Deiniger, W.D. and Kupperman, D.S. (1980) "Infrared techniques for the evaluation of silicon carbide heat-exchanger tubing." *J. Test. Eval.* **8**, 48.

De La Rue, R.M., Humphryes, R.F., Mason, I.M. and Ash, E.A. (1972) "Acoustic-surface-wave amplitude and phase measurements using laser probes." *Proc. IEEE* **119**, 117.

Denstman, H. (1980) "State-of-the-art optics: Automated image focusing." *Ind. Phot.* **29**, 33.

Depew, C.A. and Weir, R.D. (1971) "Surface roughness determination by the measurement of reflectance." *Appl. Opt.* **10**, 969.

Detrio, J.A. and Miner, S.M. (1985) "Standardized total integrated scatter measurements of optical surfaces." *Opt. Eng.* **24**, 419.

DeVriendt, A. (1982) *La Transmission de la Chaleur*, vols. 1 and 2. G. Morin, Chicoutimi.

Devyatykh, G.G. and Dianov, E.M. (1984) "Research in KRS-5 and chalcogenide glass fibers." In: *IR Optical Materials and Fibers*, P. Klocek, editor, vol. **484**, p. 104. SPIE, Bellingham, WA.

Dewhurst, R.J. and Swift, K.J. (1983) "A laser electro-optic device for the orientation of mass-produced components." *Optics and Lasers in Engineering* **4**, 203.

Dewhurst, R.J. and Swift, K.J. (1985) "Laser-scattering aids inspection and orientation." *Sensor Review* **5**, 214.

Dhamelincourt, P., Wallert, F., Leclercq, M., Nguyen, A.T. and Landon, D.O. (1979) "Laser Raman molecule microprobe." *Anal. Chem.* **51**, 414A.

Dickson, L.D. and Sincerbox, G.T. (1981) "Optics and holography in the IBM supermarket scanner." In: *Advances in Laser Scanning Technology*, L. Beiser, editor, vol. **299**, p. 163. SPIE, Bellingham, WA.

Dickson, M.S. and Harkness, D. (1969) "Surveying and tracking instruments." In: *Applied Optics and Optical Engineering*, R. Kingslake, editor, vol. **5**, p. 231. Academic Press, New York.

Diles, R.R. (1983) "High-temperature optical fiber thermometer." *J. Appl. Phys.* **54**, 1198.

Dion, J.L., Malutta, A. and Cielo, P. (1982) "Ultrasonic inspection of fiber suspensions." *J. Acoust. Soc. Am.* **72**, 1524.

Dion, J.L., Garceau, J.J. and Morissette, J.C. (1986) "Acousto-optical evaluation of fiber size in wood pulp." In: *Optical Techniques for Industrial Inspection*, P. Cielo, editor, vol. **665**, p. 361. SPIE, Bellingham, WA.

DiStafano, J.J., Stubberud, A.R. and Williamsen, I.J. (1967) *Feedback and Control Systems*. McGraw-Hill, New York.

Dorf, R.C. (1980) *Modern Control Systems*. Addison-Wesley, Reading, MA.

Dover, W.D. and Collins, R. (1980) "Recent advances in the detection and sizing of cracks using AC field measurements." *British Journal of NDT*, p. 291.

Drain, L.E., Speake, J.H. and Moss, B.C. (1977) "Displacement and vibration measurement by laser interferometry." *Proc. 1st Eur. Congr. on Optics. applied to Metrology*, vol. **136**, p. 52, SPIE, Bellingham, WA.

Drain, L.E. (1980) *The Laser Doppler Technique*. John Wiley, Chichester, U.K.

Drain, L.E. and Moss, B.C. (1983) "Applications of laser interferometry to process control." *Proc. Int. Conf. on Optical Techniques in Process Control*, p. 279. BHRA, Cranfield, Bedford, U.K.

Dremel, W., Häusler, G. and Maul, M. (1986) "Triangulation with large dynamical range." In: *Optical Techniques for Industrial Inspection*, P. Cielo, editor, vol. **665**, p. 182. SPIE, Bellingham, WA.

Drewery, J.W. (1985) "A flexible approach to computer-aided quality management." *Proc. 7th Int. Conf. on Automated Inspection and Product Control*, P.A. McKeown, editor, p. 83. IFS/North-Holland.

Dreyfus, M.G. (1963) "Spectral variation of blackbody radiation" *Appl. Opt.* **2**, 1113.

Drouet, M.G. and Nadeau, F. (1982) "Acoustic measurement of the arc voltage applicable to arc welding and arc furnaces." *Journal of Physics E* **15**, 268.

Duda, R. and Hart, P. (1973) *Pattern Classification and Scene Analysis*. Wiley, New York.

Duda, R.O., Nitzan, D. and Barret, P. (1979) "Use of range and reflectance data to find planar surface regions." *IEEE Trans. P.A.M.I.* **1**, 259.

Dufour, M. and Bégin, G. (1983) "Adaptive robotic welding using a rapid image pre-processor." *Proc. 3rd Int. Conf. on Robot Vision and Sensory Controls*, Cambridge, Nov. 7–10, p. 482.

Dufour, M. and Cielo, P. (1984) "Optical inspection for adaptive welding." *Appl. Opt.* **23**, 1132.

Dufour, M., Maldague, X. and Cielo, P. (1986) "Environmental-noise analysis in active-vision systems for adaptive welding." In: *Optical Techniques for Industrial Inspection*, P. Cielo, editor, vol. **665**, p. 321. SPIE, Bellingham, WA.

Duley, W.W. (1976) *CO_2 Lasers, Effects and Applications*. Academic Press, New York.

Dumont-Fillon, J., Lacroix, M. and Pinard, J. (1980) "Détection automatique des défauts de surface sur tôles". *Rev. Metall.*, **77**, 597.

Duncan, J.P., Dean, D.P. and Pate, G.C. (1980) "Moiré contourography and computer-aided replication of human anatomy." *Eng. Med.* **9**, 29.

Dupuy, O. (1964) "Méthode de pointés longitudinaux de grande précision. Application à la détermination des états de surface." *Rev. Opt.* **43**, 217.

Durham, R.M. (1983) "Process gas analyzers using nondispersive infrared." In: *Electro-optical Instrumentation for Industrial Applications*, F.M. Zweibaum and R.A. Carella, editors, vol. **441**, p. 72. SPIE, Bellingham, WA.

Durst, F. and Eliasson, B. (1975) "Properties of laser-Doppler signals and their exploitation for particle size measurements." *Proc. LDA-75 Symposium*, Technical University of Denmark.

Durst, F., Melling, A. and Whitelaw, J.H. (1981) *Principles and Practice of Laser-Doppler Anemometry*. Academic Press, London.

Dyos, G.T. and Smith, S.A. (1986) "Two new methods for noncontact temperature measurement." In: *Optical Techniques for Industrial Inspection*, P. Cielo, editor, vol. **665**, p. 96. SPIE, Bellingham, WA.

Earman, A.M. (1982) "Optical focus servo for optical disk mass data storage system application." In: *Optical Disk Technology*, R.A. Sprague, editor, vol. **329**, p. 89. SPIE, Bellingham, WA.

Eastman, J.M. and Zavislan, J.M. (1983) "A new optical surface microprofiling instrument." In: *Precision Surface Metrology*, J.C. Wyant, editor, vol. **429**, p. 56, SPIE, Bellingham, WA.

Eberhardt, F.J. and Andrews, F.A. (1970) "Laser heterodyne system for measurement and analysis of vibration." *J. Acoust. Soc. Am.* **48**, 603.

Eby, J.E. and Levin, R.E. (1979) "Incoherent light sources". In *Applied Optics and Optical Engineering*, R.R. Shannon and J.C. Wyant, editors, vol. **7**, p. 1. Academic Press, New York.

Edgar, R.F. and Hindle, P.H. (1980) "Developments in infrared gauging in the paper and plastics industries." In: *Instrumentation and Automation in the Paper, Rubber, Plastics and Polymerization Industries"*, A. Van Cauwenberghe, editor, p. 85. IFAC/Pergamon Press.

Edgar, R.F. and Hindle, P.H. (1986) "Infrared absorption gauging: The future." In: *Optical techniques for Industrial Inspection*, P. Cielo, editor, vol. **665**, p. 101. SPIE, Bellingham, WA.

Egan, W.G. and Hilgeman, T.W. (1979) *Optical Properties of Inhomogeneous Materials*. Academic Press, New York.

Ejiri, M., Uno, T., Mese, M. and Ikeda, S. (1973) "A process for detecting defects in complicated patterns." *Computer Graphics and Image Processing*, 1973, p. 326.

Elchinger, M.F., Martin, C. and Fauchais, P. (1979) "Measure of the diffusivity of ceramic materials either sintered or sprayed by means of a flash method with a positive or negative pulse." *Rev. Int. Hautes Temper. Refract.* **16**, 317.

Elliott, J.E. and Mickelson, L. (1979) "Optical transducer and focusing system." U.S. Patent No. 4,152,586, issued on May 1, 1979.

Elson, J.M. and Bennet, J.M. (1979) "Relation between the angular dependence of scattering and the statistical properties of optical surfaces." *J. Opt. Soc. Am.* **69**, 31.

Elson, J.M., Bennet, H.E. and Bennet, J.M. (1979) "Scattering from optical surfaces." In: *Applied Optics and Optical Engineering*, R.R. Shannon and J.C. Wyant, editors, vol. **7**, p. 191. Academic Press, New York.

Emmony, P., Godfrey, M. and White, R. (1983) In: *Review of Progress in Quauntita ive NDE*, D.O. Thompson and D.E. Chimenti, editors, vol. **2A**, p. 533. Plenum Press, New York.

Engelberger, J.F. (1980) *Robotics in Practice*. IFS, U.K.

Ennos, A.E. (1975) "Speckle interferometry". In: *Laser Speckle and Related Phenomena*, J.C. Dainty, editor, p. 203–253. Springer-Verlag, Berlin.

Ennos, A.E. (1976) "Optical fingerprinting with the laser." In: *The Engineering Uses of Coherent Optics*, E.R. Robertson, editor, p. 240. Cambridge University Press, London.

Ennos, A.E. and Virdee, M.S. (1982) "Application of reflection holography to deformation measurement problems." *Exp. Mech.* **22**, 202.

Ennos, A.E. and Virdee, M.S. (1983) "Precision measurement of surface form by autocollimation." In: *Industrial Applications of Laser Technology*, W.F. Fagan, editor, vol. **398**, p. 252. SPIE, Bellingham, WA.

Erf, R.K. (1974) *Holographic Nondestructive Testing*. Academic Press, New York.

Erf, R.K., Gagosz, R.M. and Waters, J.P. (1976) "Holography in an industrial environment." In: *The Engineering uses of Coherent Optics*, E.R. Robertson, editor, p. 23. Cambridge University Press, London.

Erf, R.K., editor (1978) *Speckle Metrology*. Academic Press, New York.

Evtikhiev, N. N., Snezhko, Y.A., Tychinskii, V.P., Levinson, G.R. and Zakharov, V.P. (1977) "Laser interferometer profilograph." *Sov. J. Quantum Electron* **7**, 36.

Ewing, G.W. (1969) *Instrumental Methods of Chemical Analysis*. McGraw-Hill, New York.

Extance, P., Pitt, G.D., Scott, B.J. and Verrels, M.V. (1983) "Intelligent turbidity monitoring using fiber optics." *1st Int. Conf. on Optical Fibre Sensors*. IEEE, London.

Fainman, Y., Lenz, E. and Shamir, J. (1982) "Optical profilometer: A new method for high sensitivity and wide dynamic range." *Appl. Opt.* 21, 3200.

Fairlie, M., Akkerman, J.G. and Timsit, R. S. (1986) "Optical techniques for the surface evaluation of bright aluminum sheet." In: *Optical Techniques for Industrial Inspection*, P. Cielo, editor, vol. **665**, p. 32. SPIE, Bellingham, WA.

Fairand, B.P. and Clauer, A.H. (1976) "Effect of water and paint coatings on the magnitude of laser-generated shocks." *Opt. Commun.* **18**, 588.

Fanconi, B., Wang, F., Hunston, D. and Mopsik, F. (1986) "Cure monitoring for polymer matrix composites." In: *Materials Characterization for Systems Performance and Reliability*, J.W. McCauley and V. Weiss, editors. Plenum Press, New York.

Farmer, W.M. (1972) "Measurement of particle size, number density, and velocity using a laser interferometer" *Appl. Opt.* **11**, 2603.

Farmer, W.M. (1978) "Measurement of particle size and concentrations using LDV techniques." *Proc. Dynamic Flow Conf.*, Marseille, p.373.

Farrar, N.R. and Ashbee, K.H.G. (1983) "Glass-reinforced plastics as optical waveguides — a method for NDE of GRP." *NDT Int.* **16**, 13.

Favis, B.D., Fisa, B. and Sanschagrin, B. (1984) "Factors affecting the surface texture in extruded mica-polypropylene composites." *Polym. Comp.* **5**, 299.

Favre, F. and Le Guen, D. (1980) "High-frequency stability of laser-diode for heterodyne communication systems." *Electron Lett.* **16**, 709.

Feigenbaum, A.V. (1983) *Total Quality Control*. McGraw-Hill, New York.

Felske, A. and Happe, A. (1976) "Double-pulse holography as a diagnostic method in the automotive industry." In *The Engineering Uses of Coherent Optics*, E.R. Robertson, editor, p. 595. Cambridge University Press, London.

Ferrière, D. and Jaillard, J.R. (1983) "New apparatus for surface detection at the pickling line of Usinor's Montataire." In: *On-line Inspection of Steel Products*, p. 51. The International Iron and Steel Institute, Brussels.

Ferrière, D. and Savarit, P. (1985) "Contrôle optique automatisé des bandes d'acier décapées." *Revue de Métallurgie*. **82**, 281.

Fihey, J.L., Cielo, P. and Bégin, G. (1983) "On-line weld penetration measurement using an infrared sensor." *Proc. Int. Conf. on Welding in Energy-Related Projects*, Toronto, Sept. 20–21. The Welding Institute of Canada, Oakville, Ontario.

Fisher, A.D. and Warde, C. (1979) "Simple closed loop system for real-time optical phase measurement." *Opt. Lett.* **4**, 131.

Fisher, D.A. (1985) "Quality control in the manufacture and retreading of aircraft tyres." *Brit. Journ. NDT* **27**, 31.

Fisher, J.F. (1984) "Reduction of the water interference in the near-IR measurement of alcohols." In: *Instrumentation and Control in the 80's*, Des Plaines, Ill., May 1–3. The Instrument Society of America, Research Triangle Park, N.C.

Fisli, T.S. (1983) "Multifunction document processor." In: *Advances in Laser Scanning and Recording*, L. Beiser, editor, vol. **396**, p. 22. SPIE, Bellingham, WA.

Fitch, P. and Gargus, A.G. (1985) "Remote UV-VIS-NIR spectroscopy using fiber optic chemical sensing." *Am. Lab.* **17**, 64 (Dec. 1985).

Fladda, G., Petterson, T., Eriksson, L. and Tidstam, G. (1980) "A new optical method for measuring suspended solids in pulp and paper mill effluents." In *Instrumentation and Automation in the Paper, Rubber, Plastics and Polymerization Industries*, A.R. Van Cauwenberghe, editor, p. 19. IFAC/Pergamon Press, Oxford.

Fleischer, J.M. (1973) "Light scanning and printing system." U.S. Patent 3,750,189.

Florian, R., Pelzl, J. Rosenberg, M., Vargas, H. and Wernhardt, R. (1978) "Photoacoustic detection of phase transitions". *Phys. Stat. Sol.* (a) **48**, K35.

Forrest, G.T. (1986) "Compact tunable IR diode lasers as sources for process monitoring." *Proc. SPIE*, vol. 621.

Fourney, M.E. (1978) "Scattered light speckle interferometry." In *Speckle Metrology*, R.K. Erf, editor, p. 281. Academic Press, New York.

Fox, J.D., Khuri-Yakub, B.T. and Kino, G.S. (1983) "High-frequency acoustic wave measurements in air." *Proceedings of the 1983 Ultrasonic Symposium*, p. 581. IEEE.

Françon, M. (1966) *Optical Interferometry*. Academic Press, New York.

Frenkel, M., Shwartz, R. and Garti, N. (1982) "Turbidity measurements as a technique for evaluation of water-in-oil emulsion stability." *Journal of Dispersion Science and Technology* **3**, 195.

Frock, H.N. and Muly, E.C. (1980) "Scattered light measurements using a photodiode array." *Proc. ISA Conf. Advances in Instrumentation*, vol. 35, part 1, p. 544. The Instrument Society of America, Research Triangle Park, N.C.

Fruh, P. and Widmann, G. (1982) "Thermal analysis in routine and quality control." *Am. Lab.* **14**, 93.

Fuchs-Viniczay, G. and Huber, K. (1979) "Optical-electrical system for monitoring filaments." U.S. Patent 4,166,214.

Fujii, H. and Asakura, T. (1974) "A contrast variation of image speckle intensity under illumination of partially coherent light." *Opt. Comm.* **11**, 32.

Fujii, H. and Lit, J.W.Y. (1978) "Surface roughness measurement using dichromatic speckle pattern: An experimental study." *Appl. Opt.* **17**, 2690.

Fulton, J.C. (1984) "Sensors and controls for the steel industry". *Proc. ISA Conf., Advances in Instrumentation*, p. 735. The Instrument Society of America, Research Triangle Park, N.C.

Gadient, A.J. (1983) "Test results on Polaroid's acoustic range finder." *QualTest-2 Conference Proceedings*, p. 17-1. American Society for NDT, Columbus, OH.

Gagosz, R. M. (1974) "Pulsed holography." In: *Holographic NDT*, R.K. Erf, editor, p. 61. Academic Press, New York.

Gahinet, P., Mazer, E. and Peltier, M. (1986) "An intelligent vision controller for circuit card lead inspection." In: *Optical Techniques for Industrial Inspection*, P. Cielo, editor, vol. **665**, p. 135. SPIE, Bellingham, WA.

Galle, B., Olsson, T. and Svanberg, S. (1979) "Measurement of surface cleanness using laser techniques." *Proc. of the Nordic Industrial Symp.*, Voksenasen, Norway, December 1979.

Garg, A.O. and Claus, R.O. (1983) "Application of optical fibers to wide-band differential interferometry." *Mater. Eval.* **41**, 106.

Gaston, C.A. (1980) "Integrated autofocusing system and method of manufacture." *IBM Tech. Disc. Bull.* **23**, 1906.

Gathers, G.R., Shauer, J.W., Calder, C.A. and Wilcox, W.W. (1977) "Determination of sound velocity in liquid metals at temperatures greater than 4000K." *Proc. 7th Symp. on Thermophysical Properties*, Gaithersburg, MD.

Geverter, W.B. (1982) "An overview of computer vision." *Report NBSIR 82–2582*, National Bureau of Standards.

Giallorenzi, T.G., Bucaro, J.A., Dandridge, A., Sigal, G.H., Cole, J.H., Rashleigh, S.C. and Priest, R.G. (1982) "Optical fiber sensor technology.". *IEEE Journ. Quant. Electr.* **QE-18**, 626.

Giers, L. (1981) *Applications Report for Coated Paper and Board Laser Inspection.* Intec, Trumbull, CT.

Giglio, M., Musazzi, S. and Perini, U. (1979) "Surface roughness by means of speckle wavelength decorrelation." *Opt. Comm.* **28**, 166.

Gilbert, A.F. (1986) "Spectrophotometric measurement of chlorine dioxide residual in Kraft pulp bleaching." *6th Int. Conf. on Instrumentation and Automation in the Paper, Rubber, Plastics and Polymerization Industries*, Akron, Ohio, Oct. 27–29. The Int. Fed. for Automatic Control.

Gilbert, L.F. (1980) "CO control heightens furnace efficiency." *Chemical Engineering*, July 28, p. 69.

Godridge, A.M. (1983) "Television pyrometry in combustion systems." *The Transducer Tempcon Conf.*, Earls Courts, London. Trident, Tavistock, U.K.

Golden, W.G. and Saperstein, D.D. (1983) "FTIR reflection-absorption spectroscopy of surface speices." *J. Electron. Spectr. Rel. Phen.* **30**, 43.

Golden, W.G. (1985) "FTIR reflection-absorption spectroscopy." In: *Fourier-Transform Infrared Spectroscopy*, J.R. Ferraro and L.J. Basile, editors, vol. **4**, p. 315. Academic Press, Orlando.

Gonzalez, R.C. and Safabakhsh, R. (1983) "Computer vision techniques for industrial inspection and robot control: Tutorial review." In: *Tutorial on Robotics*, C.S.G. Lee, R.C. Gonzalez and K.S. Fu, editors, p. 300. IEEE Computer Society, Los Angeles.

Goodman, J.W. (1968) *Introduction to Fourier Optics*. McGraw-Hill, New York.

Goodman, J.W. (1975) "Statistical properties of laser speckle patterns.". In: *Laser Speckle and Related Phenomena*, J.C. Dainty, editor, p. 9–75. Springer-Verlag, Berlin.

Gorog, I., Goedertier, P.V., Knox, J.D., Ladany, I., Wittke, J.P. and Firester, A.H. (1976) "Information scanning technology: Applications of CW AlGaAs injection lasers." *Appl. Opt.* **15**, 1425.

Gotoh, T., Ojima, M., Arimoto, A. and Chinone, N. (1982) "Characteristics of laser diodes and picture quality." In *Optical Disk Technology*, R.A. Sprague, editor, vol. **329**, p. 56. SPIE, Bellingham, WA.

Gottlieb, M., Ireland, C.L.M. and Ley, J.M. (1983) *Electro-optic and Acousto-optic Scanning and Deflection*. Marcel Dekker, New York.

Green, D.R., Wandling, C.R., Gatto, F.B. and Rogers, F.S. (1984) "System for NDE of thermal-spray coating bonds." *Proc. 2nd Nat. Conf. on Thermal Spray*, Long Beach, CA, p. 125. American Society for Metals.

Greenler, R.G. (1966) "IR study of adsorbed molecules on metal surfaces by reflection techniques." *J. Chem. Phys.* **44**, 310.

Gregory, D.A. (1976) "Speckle photography in engineering applications." In: *The Engineering Uses of Coherent Optics*", E.R. Robertson, editor, p. 263. Cambridge University Press, London.

Griffiths, D.H. (1980) "Toward quantitative industrial thermography." In: *Contemporary Infrared Sensors and Instruments*, H. Kaplan and F.M. Zweibaum, editors, vol. **246**, p. 152. SPIE, Bellingham, WA.

Griffiths, P.R., and de Haseth, J.A. (1986) *Fourier Transform Infrared Spectrometry*. John Wiley & Sons, New York.

Grisar, R. (1983) "Spectroscopic analysis of liquids using IR light emitting diodes." *Proc. Int. Conf. on Optical Techniques for Process Control*, The Hague, June 14–16, p. 260. BHRA, Cranfield, U.K.

Groh, G. (1970) "Engineering uses of laser-produced speckle patterns." In: *The Engineering Uses of Holography*, E.R. Robertson and J.M. Harvey, editors, p. 483. Cambridge University Press, London.

Gumpertz, W.E. (1984) "Macro-viewing and lightening techniques for IC device inspection." *Electri-onics* **30**, 42.

Haberlin, M. M. and Campbell, N.T. (1983) "Temperature control of brazed joint repairs in the heat exchanger of the prototype fast reactor." *Proc. Transducer Tempcon Conf.*, p. 1. Trident, Tavistock, U.K.

Halioua, M. and Liu, H.C. (1986) "Optical sensing techniques for 3-D vision." In: *Optical Techniques for Industrial Inspection*, P. Cielo, editor, vol. 665, p. 150. SPIE, Bellingham, WA.

Halivi, G. (1980) *The Role of Computers in Manufacturing Processes*. John Wiley, New York.

Hall, E.L., Tio, T.B.K., McPherson, C.A. and Sadjadi, F.A. (1982) "Measuring curved surfaces for robot vision." *Computer* **15**, 42.

Hamilton, J.C. Nagelberg, A.S. (1984) "In-situ Raman spectroscopic study of zirconia attack by molten sodium vanadate." *J. Amer. Cer. Soc.* **67**, 686.

Hanma, K. and Saito, Y. (1983) "Fully automatic solid-state color video camera for consumer use." *Hitachi Rev.* **32**, 117.

Hanma, K., Masuda, M., Nabeyama, H. and Saito, Y. (1983) "Novel technologies for automatic focusing and white balancing of solid-state color video camera.". *IEEE Trans. Cons. El.* **CE-29**, 376.

Harada, N., Imai, K., Yamada, T. and Makabe, E. (1985) "New radiation thermometries using multiple reflections and their applications to color-coating line and continuous-annealing line." *5th Process Technology Conf. on Measurement and Control Instrumentation in the Iron and Steel Industry*, Detroit, Apr. 15–17, 1985.

Haralick, R.M., Shanmugam, K. and Dinstein, I. (1973) "Textural features for image classification." *IEEE Trans. Syst., Man., Cybern.* **3**, 610.

Harding, W.E. (1981) "Semiconductor manufacturing in IBM, 1957 to present: A perspective." *IBM Journ. Res. Dev.* **25**, Sept. 1981.

Harmer, A. L. (1984) "Optical fibre sensors for industrial applications." *Proc. 2nd Int. Conf. on Optical Fiber Sensors*, p. 17. VDE-Verlag, Berlin.

Harrick, N.J. (1971) "Determination of refractive index and film thickness from interference fringes." *Appl. Opt.* **10**, 2344.

Harrick, N.J. (1977) "Transmission and reflection spectroscopy, nature of the spectra." In: *Characterization of Metal and Polymer Surfaces*, L.H. Lee, editor, vol. **2**, p. 153. Academic Press, New York.

Harrick, N.J. (1979 *Internal Reflection Spectroscopy.* Harrick Scientific Corp., New York.

Harris, K.L., et al. (1983) "Wafer inspection automation: Current and future needs." *Solid State Technol.*, Aug. 1983, p. 199.

Harris, W.J. and Claus, F.J. (1971) "Inspecting bonded structures by laser holography." *Metal Progress* **8**, 63.

Harris, W.J. and Woods, D.C. (1974) "Thermal stress studies using optical holographic interferometry." *Mater. Eval.* **32** (3), 50.

Hartmann, A.W., Roseberry, F.W. and Simpson, J.A. (1973) "A noncontacting length comparator with 10 nm precision." *Opt. Eng.* **12**, 95.

Hassel, von A. (1985) "What's new in thickness gauging? Coextrusion and lots more." *Plastics Technology* **31**(2), 71.

Haugen, P.R., Keil, R.E. and Bocchi, C. (1984) "3-D active vision system." In: *Intelligent Robots and Computer Vision*, D.P. Casasent and E.L. Hall, editors, vol. **521**, p. 258. SPIE, Bellingham, WA.

Häusler, G. and Maul, M. (1985) "Telecentric scanner for 3-D sensing." *Opt. Eng.* **24**, 978.

Heikkinen, T., Ahola, R., Manninen, M. and Myllylä, R. (1986) "Recent results of the performance analysis of a 3-D sensor based on time-of-flight measurements." In:

Optical Techniques for Industrial Inspection, P. Cielo, editor, vol. **665**, p. 168. SPIE, Bellingham, WA.

Hensler, D.H. (1972) "Light scattering from fused polycrystalline aluminum oxide surfaces." *Appl. Opt.* **11**, 2522.

Herbst, H., Grassl, H.P. and Pfleiderer, H.J. (1981) "Experimental auto-focus system for lens shutter camera." *Proc. ESSCIRC*, Freiburg, Sept. 22–24, 1981, p. 184. VDE-Verlag, Berlin.

Herder, P., Olsson, T., Sjoblom, E. and Svanberg, S. (1982) "Monitoring of surface layers using fluorescence techniques." *Internal Report*. Institute for Surface Chemistry, Box 5607, S-114 86 Stockholm, Sweden.

Hewitt, F.C., Morris, K.S. and Rein, A.J. (1985) "Evolution of FTIR from the laboratory to the production environment." *Am. Lab.* **17**, 32 (Dec. 1985).

Higham, E.H. and Medlock, R.S. (1984) "The relevance of optical transducers to the process industry." *Optics and Lasers in Engineering* **5**, 193.

Hill, D.A. (1985) "The measurement of light for inspection." *Proc. 7th Int. Conf. on Automated Inspection and Product Control*, P.A. McKeown, editor. IFS/North-Holland.

Hill, J. and Eriksson, L. (1976) "Optical analysis of fibre suspensions." *Pulp and Paper Canada* **77**(4), 61.

Himmel, D.P. (1978) "A laser measuring system for automatic industrial inspection." *Proc. 4th Int. Joint Conf. on Pattern Recognition*, November 1978, p. 952.

Hirata, Y. and Tanaka, H. (1983) "Automatic visual inspection of semiconductor wafer." *Bull. Japan. Soc. of Proc. Eng.* **17**, Sept. 1983.

Hirschfeld, T., Callis, J.B. and Kowalski, B.R. (1984) "Chemical sensors in process analysis." *Science* **226**, 312.

Hirschfeld, T. (1986) "Remote and in-situ composition analysis: Have your cake, eat it too." *InTech* **33**, 45 (Feb. 1986).

Hobson, J.D. and Lomas, T.W. (1975) "Analysis of metals." U.S. Patent 3,909,133.

Hodkinson, J.R. (1966) "The optical measurement of aerosols." In: *Aerosol Science*, C.N. Davies, editor, p. 287. Academic Press, London.

Hodor, J.R., Schapp, T.J. and Buholz, N.E. (1984) "Experimental characterization of linear imaging arrays." In: *State-of-the-art Imaging Arrays and Their Applications*, K.N. Prettyjohns, editor, vol. **501**, p. 202. SPIE, Bellingham, WA.

Holik, A.S. (1986) "Optical microscopy." In: *Encyclopedia of Materials Science and Engineering*, M.B. Beaver, editor, vol. **5**, p. 3306. Pergamon Press, Oxford.

Holland, S.W., Rossol, L. and Ward, M.R. (1979) "Consight I: A vision-controlled robot system for transferring parts form belt conveyors." In: *Computer Vision and Sensor-based Robots*, G.G. Dodd and L. Rossol, editors, p. 81. Plenum Press, New York.

Holmsten, D. (1986) "Thermographic sensing for on-line industrial control." In *Optical Techniques for Industrial Inspection*, P. Cielo, editor, vol. **665**, p. 75. SPIE, Bellingham, WA.

Holownia, B.P. (1985) "Nondestructive testing of adhesive joints." *Proc. 2nd Int. Conf. on Lasers in Manufacturing*, M.F. Kimmit, editor, p. 151. IFS/North-Holland.

Holve, D.J., Tichenor, D., Wang, J.C.F. and Hardesty, D.R. (1981) "Design criteria and recent developments of optical single particle counters for fossil fuel system." *Opt. Eng.* **20**, 529.

Hopkins, R.E. and Buzawa, M.J. (1976) "Optics for laser scanning." *Opt. Eng.* **15**, 90.

Hopwood, R.K. (1980) "Design considerations for a solid-state image sensing system." In: *Minicomputers and Microprocessors in Optical Systems*, C.L. Koliopoulos and F.M. Zweibaum, editors, vol. **230**, p. 72. SPIE, Bellingham, WA.

Horwath, J.J. and Semerjian, H.G. (1986) "Laser-excited fluorescence studies of black liquor." In: *Optical Techniques for Industrial Inspection*, P. Cielo, editor, vol. **665**, p. 258. SPIE, Bellingham, WA.

Hsieh, C.K., Wang, X.A. and Yang, S.L. (1982) "IR scanning thermography for a quantitative detection of cavities in a plane slab and a rectangular prism." *J. Nondestr. Eval.* **3**, 99.

Hsieh, Y.Y. and Fu, K.S. (1980) "An automatic visual inspection system for integrated cirucit chips." *Computer Graphics Image Proc.* **14**, 293.

Hsu, T.R., Lewak, R. and Wilkins, B.J.S. (1978) "Measurement of crack growth in a solid at elevated temperature by holographic interferometry." *Exp. Mech.* **18**, 297.

Huang, C.C. (1983) "Optical heterodyne interferometer". In: *Precision Surface Metrology*, J.C. Wyant, editor, vol. **429**, p. 65. SPIE, Bellingham, WA.

Hudson, D.L. (1983) "A practical solution using a new approach to machine vision." *Qualtest-2 Conf. Proc.*, p.. 27-1. ASNT/ASQC/SME.

Hung, Y.Y. (1978) "Displacement and strain measurement." In: *Speckle Metrology*, R.K. Erf, editor, p. 51. Academic Press, New York.

Hung, Y.Y. (1982) "Shearography: A new optical method for strain measurement and nondestructive testing." *Opt. Eng.* **21**, 391.

Hunt, V.D. (1985) *Industrial Robotics Handbook*. Industrial Press.

Hunter, R.P. (1978) *Automated Process Control Systems: Concepts and Hardware*. Prentice-Hall, Englewood Cliffs, N.J.

Hutchins, D.A. (1986) "Ultrasonic generation by pulsed lasers." In: *Physical Acoustics*, W.P. Mason and R.N. Thurstone, editors (in press). Academic Press, New York.

Hutchinson, R.J. (1981) "An attenuated total reflection IR process analyser for liquids." *Proc. PROMECON '81 Conf.*, p. 31. The Instrument Society of America, Research Triangle Park, N.C.

Hyman, A. and Burns, N.D. (1985) "Automated optical inspection of knitted fabrics." In *Automated Inspection and Product Control*, P.A. McKeown, editor, p. 177. IFS/North-Holland, U.K.

Ibrahim, M.D. (1983) "Formation of asymmetric moiré fringes and their uses in moiré topography". *Opt. Lasers in Eng.* **3**, 65.

Idesawa, M. (1985) "A new type of optical range sensing method: RORS". In: *Optics in Engineering Measurement*, W.F. Fagan editor, vol. **599**, p. 419. SPIE, Bellingham, WA.

Idesawa, M., Yatagri, T. and Soma, T. (1977) "Scanning moiré method and automatic measurement of 3-D shapes." *Appl. Opt.* **16**, 2152.

Ikeuchi, K. and Horn, B.K.P. (1979) "An application of the photometric stereo method" *Proc. 6th Int. Joint Conf. on Artificial Intelligence*, Tokyo, p. 413.

Imai, M., Ohashi, T. and Ohtsuka, Y. (1980) "Fiber-optic Michelson interferometer using an optical power divider." *Opt. Lett.* **5**, 418.

Imhof, R.E., Birch, D.J.S., Thornley, F.R., Gilchrist, J.R. and Strivens, T.A. (1984) "Optothermal transient emission radiometry." *J. Phys. E: Sci Instrum.* **17**, 521.

Inasaki, I. (1985) "In-process measurement of surface roughness during cylindrical grinding process." *Precis. Eng.* **7**, 73.

Intrieri, A.J. (1977) "Optical fibers look around obstacles to measure temperature." *Control Eng.*, Dec. 1977.

Ishitami, A., Ishida, H., Soeda, F. and Nagasawa, Y. (1982) "FTIR reflection-absorption spectrometry for chemical analysis for surface analysis." *Anal. Chem.* **54**, 682.

Ito, K., Musha, T. and Kato, K. (1983) "Method and apparatus for detecting a focusing condition of an objective lens." U.S. Patent 4,422,168.

Iuchi, T. and Kusaka, R. (1982) "Two methods for simultaneous measurement of temperature and emittance using multiple reflection and specular reflection, and their applications to industrial processes." In: *Temperature, Its Measurement and Control in Science and Industry*, J.F. Schooley, editor, vol. **5**, p. 491.

Izutsu, M., Enokihara, A. and Sueta, T. (1982) "Optical-waveguide microdisplacement sensor." *Elecctron. Lett.* **18**, 867.

Jaarinen, J. and Luukkala, M. (1983) "Numerical analysis of thermal waves in stratified media for NDT purposes." *Journ. de Physique*, Coll. C6, Suppl. No. 10, vol. **44**, p. C6-503.

Jalki, J.A., Kim, R.C. and Case, S.K. (1985) "3-D inspection using multistripe structured light." *Opt. Eng.* **24**, 966.

Jarvis, J.F. (1980) "Automated visual inspection of printed wiring boards by local pattern matching." *IEEE Trans. Patt. Anal. Mahine Intell.* **2**, 77.

Jarvis, R.A. (1976) "Focus optimization criteria for computer image processing." *Microscope* **24**, 163.

Jarvis, R.A. (1983) "A perspective on range-finding techniques for computer vision." *IEEE Trans. P.A.M.I.* **5**, 122.

Jarvis, R.A. (1983a) "A laser time-of-flight range scanner for robotic vision." *IEEE Trans, P.A.M.I.* **5**, 505.

Jen, C.K., Farnell, G.W., Parker, M. and Cielo, P. (1984) "A fiber-optic damage monitor." *Review of Progress in Quantitative NDE*, D.O. Thompson, editor, La Jolla, July 8–13.

Jen, C.K., Cielo, P., Maldague, X. and El-Assal, K. (1985) "Noncontact ultrasonic characterization of peizoelectric ceramics." *J. Amer. Ceram. Soc.* **68**, C-146.

Johnson, P.M. and Burgess, T.J. (1968) "Free surface velocity measurement of an impacted projectile by optical Doppler shift." *Rev. Sci. Inst.* **39**, 1100.

Johnson, R.V. (1979) "Focus error detection in optical data storage systems." In: *Laser Recording and Information Handling*, A.A. Jamberdino, editor, vol. **200**, p. 73. SPIE, Bellingham, WA.

Johnsson, B. and Hill, J. (1983) "PQM technology and its application in pulp and stock process control." In: *Instrumentation and Automation in the Paper, Rubber, Plastics and Polymerization Industries*, A.R. Van Cauwenberghe, editor, p. 19. IFAC/Pergamon Press, Oxford

Jones, B.E. (1985) "Optical fibre sensors and systems for industry." *J. Phys. E: Sci. Instrum.* **18**, 770.

Jones, B.E. and Kvasnik, F. (1985), "Sense of touch for machines — an overview." *Measurement and Control* **48**, 55.

Jones, R. and Wykes, C. (1983) *Holographic and Speckle Interferometry*. Cambridge University Press, London.

Jones, R. (1985) "The on-line inspection of glass components." In: *Automated Inspection and Product Control* P.A. McKeown, editor, p. 331. IFS/North-Holland, U.K.

Judd, B.D. and Wyszecki, G. (1975) *Color in Business, Science and Industry*. John Wiley & Sons, New York.

Juran, J.M., Gryna, F.M., and Bingham, R.S. (1979) *Quality Control Handbook*. McGraw-Hill, New York.

Jutila, J. (1979) "Multicomponant on-stream analyzers for process monitoring and control." *InTech* **26**, 38 (July 1979).

Kamal, M.R. and Patterson, W.I. (1986) "Measurement and modelling problems in plastics processing of nonsheet processes." *6th Int. Conf. on Instrumentation and Automation in the Paper, Rubber, Plastics and Polymerization Industries*, Akron, OH. Oct. 27–29, 1986. The Int. Fed. for Automatic Control.

Kanade, T. and Asada, H. (1981) "Noncontact visual 3-D ranging devices." In: *3-D Machine Perception*, B.R. Altschuler, editor, vol. **283**, p. 48. SPIE, Bellingham, WA.

Kanstad, S.O. and Nordal, P.E. (1986) "Experimental aspects of photothermal radiometry." *Can. J. Phys* **64**, 1155.

Kapany, N.S. (1967) *Fiber Optics: Principles and Applications*. Academic Press, New York.

Kathyl, R.H. (1972) "Moiré screens with pseudo-random sequences." *Appl. Opt.* **11**, 2278.

Kaule, W. and Primbsch, E. (1983) "Apparatus for sensing ultrasonic waves by optical means." U.S. Patent 4,388,832, issued on 21 June, 1983.

Kay, L. (1964) "An ultrasonic sensing probe as a mobility aid for the blind." *Ultrasonics* **2**, 53.

Kay, L. and Crecraft, D.I. (1985) "Nonvisual perception by ultrasonic imaging of a robot workspace." In: *Automated Inspection and Product Control*," P.A. McKeown, editor, p. 289. IFS/North-Holland, U.K.

Kayaalp, A.E. and Jain, R. (1985) "A knowledge based automatic on-line wafer (IC) inspection system." *Vision '85 Conf. Proc.*, p. 5–117. MVA/SME, Dearborn, MI.

Keck, D.B. (1976) "Optical fiber waveguides." In: *Fundamentals of Optical Fiber Communications*, M. K. Barnoski, editor. Academic Press, New York.

Keitaaniemi, O. and Kujansivu, T. (1986) "Retention measurement in paper machines and its control using retention aid flowrate." *6th Int. Conf. on Instrumentation and Automation in the Paper, Rubber, Plastics and Polymerization Industries*, Akron, OH., Oct. 27–29. The Int. Fed. for Automatic Control.

Keller, B.E. (1967) "Noncontact surface contour analyzer." *Proc. Instn. Mech. Engrs.* **182**, Part 3K, 360.

Kelly, B.O. (1976) "Lateral-effect photo diodes." *Laser Focus*, March 1976, p. 38.

Kelsell, D. (1963) "Automatic emissivity — compensated radiation pyrometer." *J. Sci. Instrum.* **40**, 1.

Kendall, D.N. (1966) *Applied Infrared Spectroscopy*. Reinhold Publishing Co., New York.

Kennedy, C.W. and Andrews, D.E. (1977) *Inspection and Gaging*, Industrial Press, New York.

Kenny, L.C. (1983) "Automated focusing of an optical microscope." *J. Microsc.* **132**, 97.

Kerker, M. (1969) *The Scattering of Light and Other Electromagnetic Radiation*. Academic Press, New York.

Kersch, L.A. (1974) "Laminate structure inspection." In: *Holographic NDT*, R.K. Erf, editor, p. 303. Academic Press, New York.

Kessler, L.W. (1974) "Review of progress and applications in acoustic microscopy.". *J. Acoust. Soc. Am.* **55**, 909.

Kessler, L.W. and Yuhas, D.E. (1980) "Reliability assurance testing with the scanning laser acoustic microscope." *Proc. 30th Electronic Components Conf.*, p. 146, IEEE.

Khanna, S.M., Tonndorf, J. and Walcott, W.W. (1968) "Laser interferometer for the measurement of submicroscopic displacement amplitudes and their phases in small biological structures." *J. Acoust. Soc. Am.* **44**, 1555.

Kiedron, P. (1986) "Removing interference fringes to improve accuracy of film-thickness spectroscopic measurement." In: *Optical Techniques for Industrial Inspection*, P. Cielo, editor, vol. **665**, p. 110. SPIE, Bellingham, WA.

Kim, R.C., Case, S.K. and Schock, H.J. (1985) "Holographic optical system for aberration corrections in Doppler velocimetry." *Opt. Eng.* **24**, 786.

King, C. and Merchant, J. (1982) "A reflecting laser technique for noncontact measurement of molten glass and other liquids." *Proc. ISA Conf., Discover the New World of Instrumentation*, Columbus, OH. The Instrument Society of America, Research Triangle Park, N.C.

Kingslake, R. (1965) "Basic geometrical optics". In: *Applied Optics and Optical Engineering*, R. Kingslake, editor, vol. **1**, p. 201. Academic Press, New York.

Kingston, R.H. (1978) *Detection of Optical and Infrared Radiation*. Springer-Verlag, Berlin.

Kinsler, L.E. and Frey, A.R. (1962) *Fundamentals of Acoustics*. John Wiley and Sons, New York.

Kiparissides, C., MacGregor, J.F., Singh, S. and Hamielec, A.E. (1980) "Continuous emulsion polymerization of vinyl acetate, part 3: Detection of reactor performance by turbidity spectra and liquid exclusion chromatography." *Can. J. Chem. Eng.* **58**, 65.

Kirkbright, G.F., Narayanaswaing, R. and Welti, N.A. (1984) "A fibre-optic pH probe based on the use of immobilized colorimetric indicator." *Analyst* **109**, 1025.

Kishner, S.J. (1980) "An on-line spectrophotometer for color measurement of moving samples." In: *4th IFAC Conf.*, A. Van Cauwenberghe, editor, p. 67.

Kist, R. and Drope, S. (1983) "Fiber-optic Fabry-Pérot resonator as temperature sensor and optical spectrum analyzer." In: *Fiber Optic and Laser Sensors*, E.L. Moore and O.G. Ramer, editors, vol. **412**, p. 172. SPIE, Bellingham, WA.

Klappert, W.R. (1976) "The setting of double-exposure optical holographic NDT test parameters." *Mater. Eval.* **34**, 160.

Kleinknecht, H.P. and Meier, H. (1983) "Optical profilometer for monitoring surface contours of Si power devices." In: *Industrial Applications of Laser Technology*, W.F. Fagan, editor, vol. **398**, p. 266. SPIE, Bellingham, WA.

Kline, R.A., Green, R.E. and Palmer, C.H. (1978) "A comparison of optically and piezoelectrically sensed acoustic emission signals." *J. Acoust. Soc. Am.* **64**, 1633.

Kline, R.A., Green, R.E. and Palmer, C.H. (1981) "Acoustic emission from cracking steel: Experiment and theory." *J. Appl. Phys.* **52**, 141.

Knox, J.D., Goedertier, P.V., Fairbanks, D. and Caprari, F. (1977) "Inspecting IC masks with a differential laser scanning inspection system." *Solid State Technol.* **20**, 48.

Knudsen, J.K. and Sovik, R.E. (1984) "Vortex flowmeter or D/P?" *Proc. Pacific Cascade Instrumentation Conf.*, p. 171. The Instrument Society of America.

Koberstein, J., Russell, T.P. and Stein, R.S. (1979) "Total integrated light-scattering intensity from polymeric solids." *J. Pol. Sci.: Pol. Phys.* **17**, 1719.

Koch, I.D. (1983) "Machine vision applications in pressed metal parts manufacture." *Proc. Conf. on Applied Machine Vision*, p. 216. SME, Dearborne, MI.

Kochan, A. (1986) "Automatic inspection detects PCB faults." *Sensor Rev.* **6** (1), 29.

Kocher, D.G. (1983) "Automated Foucault test for focus sensing." *Appl. Opt.* **22**, 1887.

Koenig, J.L. and Antoon, M.K. (1978) "Recent application of FTIR spectroscopy to polymer systems." *Appl. Opt.* **17**, 1374.

Kogelnik, H. and Li, T. (1966) "Laser beams and resonators." *Appl. Opt.* **5**, 1550.

Koizumi, M. and Akiyama, N. (1985) "Automated visual inspection systems using optical detection." *Hitachi Rev.* **34**, 49.

Konczakowski, A. and Przybylski, W. (1983) "The rolling process effectivity evaluation method based on the spectral analysis of the surface geometrical structure." *Wear* **92**, 1.

Kopsala, A. and Paivarinne, J. (1984) "Process-control input/output modules." *Proc. of the Pacific Cascade Instrumentation '84*, p. 179. The Instrument Society of America, Research Triangle Park, N.C.

Köpf, V. (1973) "Correlation between speckle patterns of laser photographs recorded on different photographic plates". *Opt. Comm.* **9**, 374.

Korpel, A. (1980) "Acousto-optics." In: *Applied Optics and Optical Engineering*, R. Kingslake and B.J. Thompson, editors, vol. **6**, p. 89. Academic Press, New York.

Korpiun, P. and Tilgner, R. (1980) "The photoacoustic effect at first-order phase transitions." *J. Appl. Phys.* **51**, 6115.

Kortüm, G. (1969) *Reflectance Spectroscopy*. Springer-Verlag, Berlin.

Kramer, C. (1981) "Holographic laser scanners for nonimpact printing." *Laser Focus* **17**, 80 (June 1981).

Kramer, C.J. and Pizzutelli, P.A. (1985) "Hologon laser scanners." *Photonics Spectra* **19** (11), 89.

Krautkrämer, J. (1979) "Unconventional methods of generating, coupling and receiving ultrasound in NDT." *Proc. 9th World Conf. NDT*, Melbourne, Australia.

Krigman, A. (1984) "Turbidity instrumentation: Clearing muddy waters." *Intech* **31**, Feb. 1984, p. 31.

Krishnan, K. and Ferraro, J.R. (1982) "Techniques used in FTIR spectroscopy." In: *Fourier-Transform Infrared Spectroscopy*, J.R. Ferraro and L.J. Basile, editors, vol. **3**, p. 149. Academic Press, New York.

Krohn, D.A. (1984) "Fiber-optic displacement sensors." *Proc. ISA Conf., Advances in Instrumentation*, vol. **39**, p. 331. The Instrument Society of America, Research Triangle Park, N.C.

Kruger, R.P. and Thompson, W.B. (1981) "A technical and economic assessment of computer vision for industrial inspection and robotic assembly." *Proc. IEEE* **69**, 1524.

Kubiak, E.J. (1968) "IR detection of fatigue cracks and other near-surface defects." *Appl. Opt.* **7**, 1743.

Kuo, P.K., Favro, L.D., Inglehart, L.J. and Thomas, R.L. (1982) "Photoacoustic phase signatures of closed cracks." *J. Appl. Phys.* **53**, 1258.

Kwon, O., Wyant, J.C. and Hayslett, C.R. (1980) "Rough-surface interferometry at 10.6 μm." *Appl. Opt.* **19**, 1862.

Labutti, G., Frant, M.S. and Gilby, A. (1980) "New directions for the 80's in process infrared." In: *Advances in Instrumentation*, vol. **35**, part 2, p. 5. The Instrument Society of America, Research Triangle Park, N.C.

Lachambre, J.L. (1983) IREQ, Varennes, Québec, private communication.

Land, T. (1959) "Practical aspects of radiation pyrometry." *Trans. Soc. Instr. Techn.*, March 1959.

Lange, Y.V. (1985) "The local method of free elastic vibrations and its applications to testing in industry." *NDT Int.* **18**, 256.

Langlois, P., Lessard, R.H. and Boivin, A. (1985) "Real-time curvature radii measurements using diffraction edge waves." *Appl. Opt.* **24**, 1107.

Latter, T.D.T. (1983) "Measuring the thickness of paint and allied coatings." *Polymer Paint Colour Journal* **173**, 119.

Lauer, J.L., Vogel, P. and Seng, G.T. (1985) "Emission FTIR analyses of jet fuel deposits." *Appl. Spectr.* **39**, 997.

Laurendeau, D. and Poussart, D. (1986) "Unsupervised model building of 3-D objects using range information." In: *Optical Techniques for Industrial Inspection*, P. Cielo, editor, vol. **665**, p. 209. SPIE, Bellingham, WA.

Lavan, M.J., Van Damme, G.E., Cadwellender, W.K. and DeYoung, T.F. (1977) "Application of an optical heterodyne interferometer to the inspection of surfaces." *J. Test. Eval.* **5**, 382.

Lavigne, J.R. (1977) *An Introduction to Paper Industry Instrumentation.* Miller Freeman, San Francisco.

Lavigne, J.R. (1979) *Instrumentation Applications for the Pulp and Paper Industry.* Freeman, San Francisco.

Lawson, W.D. and Sabey, J.W. (1970) "Infrared techniques." In: *Research Techniques in NDT*, R.S. Sharpe, editor, vol. **1**, p. 443. Academic Press, London.

Lehureau, J.-C. and Bricot, C. (1978) "Optical focussing sensor." U.S. Patent 4,079,248.

Lemons, R.A. and Quate, C.F. (1975) "Acoustic microscopy: Biomedical applications." *Science* **188**, 905.

Lenef, A.L. and Gardner, C.S. (1985) "Optical emissions from weld arcs and their effects on the performance of welding robot vision systems." *Appl. Opt.* **24**, 2587.

Lenk, J.D. (1980) *Handbook of Controls and Intrumentation.* Prentice-Hall, New York.

Leonhardt, K., Kauffmann, E. and Tiziani, H.J. (1984) "Determination of average roughness and profile autocorrelation width of metallic surfaces with a white-light sensor." *Opt. Comm.* **51**, 363.

Lettieri, T.R. (1986) "Pulsed laser caliper for noncontact dimensional measurement." *Appl. Opt.* **25**, 1443.

Levine, M.D., O'Handley, D.A. and Yagi, G.M. (1973) "Computer determination of depth maps." *Computer Vision, Graphics Image Processing* **2**, 134.

Levine, S. (1983) "Applications of RVSI's 3-D vision system into industrial robotic manufacturing and inspection operations". In *Industrial Applications of Image Analysis*, Antwerp, Oct. 17–19. IRSIA/IWONL/CETA.

Levinstein, H. (1965) "Infrared detectors." In: *Applied Optics and Optical Engineering*, R. Kingslake, editor, vol. **2**, p. 311. Academic Press, New York.

Levy, R.L., Fanter, D.L. and Summers, C.J. (1979) "Spectroscopic evidence for mechano-chemical effects of moisture in epoxy resins." *J. Appl. Pol. Science* **24**, 1643.

Levy, R.L. (1986) "A new fiber-optic sensor for monitoring the composite-curing process." In: *Polymeric Materials: Science and Engineering.* Proc. ACS Spring Meeting, The American Chemical Society.

Lew, A. and Depeursinge, C. (1984) "Single-mode fiber evanescent wave spectroscopy". *Proc. 2nd Int. Conf. on Optical Fiber Sensors*, p. 71. VDE-Verlag, Berlin.

Lew, A., Depeursinge, C., Cochet, F., Berthou, H. and Parriaux, O. (1984) "Single-mode fiber evanescent wave spectroscopy." In: *Proc. 2nd Int. Conf. on Optical Fiber Sensors*, Stuttgart, p. 71. VDE-Verlag, Berlin. Proc. SPIE, vol. **514**.

Lewis, R.A. and Johnston, A.R. (1977) "A scanning laser rangefinder for a robotic vehicle." *Proc. 5th Int. Joint Conf. on Artificial Intelligence*, p. 762.

Liao, Y.T. and Koenig, J.L. (1984) "Applications of FTIR spectroscopy to the study of fiber-resin composites". *Dev. Reinf. Plast.* **4**, 31.

Lin, J.C. and Chi, Z.C. (1983) "Accuracy analysis of a laser/camera based 3-D measurement system." In: *Intelligent Robots: RoViSeC 3*, D.P. Casasent and E.L. Hall, editors, vol. **449**, p. 158. SPIE, Bellingham, WA.

Lin, S.C. and Giallorenzi, T.G. (1979) "Sensitivity analysis of the Sagnac-effect optical-fiber ring interferometer." *Appl. Opt.* **18**, 915.

Livingstone, F.R. and Rioux, M. (1986) "Development of a large field of view 3-D vision system." In: *Optical Techniques for Industrial Inspection*, P. Cielo, editor, vol. **665**, p. 188. SPIE, Bellingham, WA.

Livnat, Z., Kafri, O. ad Erez, G. (1980) "Hills and valleys analysis in optical mapping and its applications to moiré contouring." *Appl. Opt.* **19**, 3396.

Locke, D.H. (1975) "Lithographic plate surface inspection using fiber optics." In: *Quality Control*, vol. **60**, p. 39. SPIE, Bellingham, WA.

Lonardo, P.M. (1978) "Testing a new optical sensor for in-process detection of surface roughness." *CIRP Annals* **27**, 531.

Lord, H.C. and Maier, H.J. (1984) "A new NDIR analyzer incorporating a complete system calibration." In: *Instrumentation and Control in the 80's*, Des Plaines, Ill. May 1–3. The Instrument Society of America, Research Triangle Park, N.C.

Loree, T.R. and Radziemski, L.J. (1981) "Laser-induced breakdown spectroscopy: Time-integrated applications." *Plasma Chem. Plasma Proc.* **1**, 271.

Loree, T.R., Radziemski, L.J. and Cremers, D.A. (1982) "AES by LIBS may help you." *Electro-Optical Systems Design*, Oct. 1982. p. 35.

Lucas, J., Gracovetsky, S.A. and Marcus, J.J. (1978) "On-line paper machine sensor for roughness and dirt." *Pulp and Pap. Can.* **79**, 46.

Lucas, J.M. (1980) "Graininess sensor." U.S. Patent 4,213,708, issued on July 22, 1980.

Lukens, W.E. and Morris, R.A. (1982) "Infrared temperature sensing of cooling rates for arc welding control." *Welding Journal.* **61**, 27.

Luukkala, M., Lehto, A., Jaarinen, J. and Jokinen, M. (1982) "Photo-thermal imaging and thermal surface waves as an NDT tool for coatings." *Proc. IEEE Ultrasonics Symp.*, p. 591.

Lyle, J.H. and Pitt, C.W. (1982) "Vortex shedding fluid flowmeter using optical fibre sensor." *Electron. Lett.* **17**, 244.

Lyne, B. (1976) "Measurement of the distribution of surface void sizes in paper." *Tappi* **59**, 102.

Macfadyen, N. (1985) "Optical fibre sensors in industrial process control." *Proc. 2nd Int. Conf. on Lasers in Manufacturing*, M.F. Kimmit, editor, p. 173. The Laser Institute of America.

Macfarlane, J.W. and Woodruff, G. (1984) "Automated optical inspection of printed wiring boards." *Electro/84 Conf. Record.* Electronic Convention Management, Los Angeles.

Maddux, G.E. and Sendeckyi, G.P. (1979) "Holographic techniques for defect detection in composite materials." In: *NDE and Flaw Criticality for Composite Materials*, R.B. Pipes, editor, ASTM STP 696, p. 26. American Society for Testing and Materials.

Maggioli, V.J. (1985) "Fiber-optic sensors in petrochemical plants." *IEEE Trans. Ind. Appl.* **IA-21**, 1491.

Mairy, B. and Ramelot, D. (1981) "Descatherm, a new pyrometer for measuring strand surface temperature in continuous casting." *Ironmaking and Steelmaking* **No. 2,** p. 91.

Makosch, G. and Solf, B. (1981) "Surface profiling by electro-optical phase measurements." In: *High-resolution Soft X-ray Optics*, E. Spiller, editor, vol. **316**, p. 42. SPIE, Bellingham, WA.

Malacara, D. (1978) *Optical Shop Testing*. J. Wiley and Sons, New York.

Maldague, X., Cielo, P. and Jen, C.K. (1986) "NDT applications of laser-generated focused acoustic waves". *Mater. Eval.* **44**, 1120.

Maldague, X., Poussart, D. and Laurendeau, D. (1986) "3-D form acquisition apparatus." In: *Optical Techniques for Industrial Inspection*, P. Cielo, editor, vol. **665**, p. 200. SPIE, Bellingham, WA.

Mallow, A. and Chabot, L. (1978) *Laser Safety Handbook*. Van Nostrand Reinhold, New York.

Mannheim, L.A. (1981) "Pentax ME-F: The first TTL autofocus." *Brit. J. Phot.* **128**(48), 1214.

Mansour, T.M. (1983) "Nondestructive thickness measurement of phosphate coatings by infrared absorption". *Mater. Eval.* **41**, 302.

Mansour, S. and McGoldrick, P.F. (1985) "An image-processing system for automating the inspection of shot-blasting operations." In: *Automated Inspection and Product Control*, P.A. McKeown, editor. IFS/North Holland.

Margarinos, J.R. and Coleman, D.J. (1985) Holographic mirrors. *Opt. Eng.* **24**, 769.

Maron, S.H., Pierce, P.E. and Ulevitch, I.N. (1963) "Determination of latex particle size by light scattering." *J. Colloid Science* **18**, 420.

Marsh, K.A., Richardson, J.M., Schoenwald, J.S. and Martin, J.F. (1984) "Acoustic imaging in robotics using a small set of transducers." *Proc. 4th Int. Conf. on Robot Vision and Sensory Controls*, A. Pugh, editor, p. 261. IFS, U.K.

Marshall, G.F. (1980) "Scanning devices and systems." In: *Applied Optics and Optical Engineering*, R. Kingslake and B.J. Thompson, editors, vol. **6**, p. 203. Academic Press, New York.

Martin, B.G. and Becker, F.L. (1980) "The effect of near-surface metallic-property gradients on ultrasonic critical angle reflectivity." *Mater. Eval.* **38**, 92.

Martin, C., Rendl, J., Rochler, R. and Sieger, C. (1983) "Problems occurring with the application of holographic heterodyne interferometry." In: *Holographic Data NDT*, D. Vukicevic, editor, vol. **370**, p. 211. SPIE, Bellingham, WA.

Martin, Y., Wickramasinghe, H.K. and Ash, E.A. (1982) "Thermo and photo-displacement microscopy." *Proc. IEEE Ultrasonics Symp.*, p. 563.

Martin, Y. and Ash, E.A. (1984) "Imaging of micro-cracks by photo-displacement microscopy." *Proc. IEEE Ultrasonics Symp.*, p. 647.

Martinelli, M. (1982) "Optical fiber interferometer desensitized to the environmental perturbation acting on the fiber probe." *Proc. SPIE*, vol. **355**, p. 6.

Massey, G.A. and Carter, R.R. (1968) "Probable laser instrument for vibration analysis and transducer calibration." *Shock and Vibration Bull.* **37**(2), 1.

Matsubara, T., Toyota, T. and Fujiyama, A. (1986) "Hot slab surface inspection by laser scanning method." In: *Optical Techniques for Industrial Inspection*, P. Cielo, editor, vol. **665**, p. 17. SPIE, Bellingham, WA.

May, C.A., Hadad, D.K. and Browning, C.E. (1979) "Physicochemical quality assurance methods for composite matrix resins." *Polymer Engineering and Science* **19**, 545.

Maze, G., Cardin, V. and Poulain, M. (1984) "Fluoride-glass IR fibers for light transmission up to 5 μm." In: *IR Optical Materials and Fibers*, P. Klocek, editor, vol. **484**, p. 93. SPIE, Bellingham, WA.

McCamy, C.S. (1986) "Industrial color inspection." In: *Optical Techniques for Industrial Inspection*, P. Cielo, editor, vol. **665**, p. 290, SPIE, Bellingham, WA.

McCollum, A.J., Batchelor, B.G. and Cotter, S.M. (1985) "Three-dimensional optical sensing." In: *Automated Inspection and Product Control,"* P.A. McKeown, editor, p. 161. IFS/North-Holland.

McFarland, W.D. and McLaren, R.W. (1983) "Problems in 3-D imaging." In: *Intelligent Robots: RoViSec 3*, D.P. Casasent and E.L. Hall, editors, vol. **449**, p. 148. SPIE, Bellingham, WA.

McIntosh, W.E. (1983) "Automating the inspection of printed circuit boards." *Robotics Today*, p. 75, June 1983.

McKeown, P.A., Cooke, P. and Bailey, W.P.N. (1975) "The application of optics to the quality control of automotive components." In: *Quality Control*, vol. **60**, p. 77. SPIE, Bellingham, WA.

McLaughlin, P.V., McAssey, E.V. and Deitrich, R.C. (1980) "Nondestructive examination of fibre composite structures by thermal field techniques." *NDT Int.* **13**, 56.

McMahon, G.W. and Cielo, P.G. (1979) "Fiber-optic hydrophone sensitivities for different sensor configurations." *Appl. Opt.* **18**, 3720.

McMaster, R.C., editor (1959) *Nondestructive Testing Handbook*. American Society for NDT, Columbus, OH.

McVey, E.S. and Lee, J.W. (1982) "Some accuracy and resolution aspects of computer vision distance measurements." *IEEE Trans. PAMI* **4**, 646.

Mead, D.C., Kasdan, H.L. and Middleton, T.J. (1975) "Quality control by diffraction-pattern analysis." In *Quality Control*, vol. **60**, p. 57. SPIE, Bellingham, WA.

Meadow, D.M., Johnson, W.D. and Allen, J.B. (1970) "Generation of surface contours by moiré patterns." *Appl. Opt.* **9**, 942.

Medlock, R.S. (1987) "The present and future status of fibre optic sensors in industry" Meas. and Control **20**, 14.

Meyer-Arendt, J. (1972) *Introduction to Classical and Modern Optics.* Prentice-Hall, Englewood Cliffs.

Mezrich, R., Vilkomerson, D. and Etzold, K. (1976) "Ultrasonic waves: Their interferometric measurement and display." *Appl. Opt.* **15**, 1499.

Midgley, J.A. and Sander, K.F. (1971) "High-resolution homodyne interferometer." *Electron. Lett.* **7**, 117.

Mikami, T. Abe, F., Satoh, T. and Matsuda, T. (1983) "An ultra-high precision UV laser scanning system." *Proc. Int. Conf. on the Applications of Lasers and Electro-Optics (ICALEO '83)*, p. 36. Laser Institute of America.

Miles, R.O., Dandridge, A., Tveten, A.B., Taylor, H.F. and Giallorenzi, T.G. (1980) "Feedback-induced line broadening in CW channel-substrate planar laser diodes." *Appl. Phys. Lett.* **37**, 990.

Miles, R.O., Dandridge, A., Tveten, A.B. and Giallorenzi, T.G. (1981) "Limitations of diode-laser souces in fiber interferometer sensor systems." *Proc. CLEO '81*, p. 102.

Miller, C.S. (1984) "Improving productivity of ultrasonic thickness testing." *Proc. of the Pacific Cascade Instrumentation '84*, p. 39. Instrument Society of America, Research Triangle Park, N.C.

Miller, R.K. (1983) *Machine Vision for Robotics and Automated Inspection*, vols. **1** to **3**. Technical Insights, Fort Lee, N.J.

Miller, R.K. (1985) *3-D machine vision.* Report co-published by SEAI, Madison, GA and by Technical Insights, Ft. Lee, N.J.

Miller, R.W. (1983) *Flow-measurement Engineering Handbook.* McGraw-Hill, New York.

Miller, S.E., Marcatali, E.A.J. and Li, T. (1973) "Research toward optical-fiber transmission systems, part 1." *Proc. IEEE* **61**, 1703.

Mitsui, K. and Sato, H. (1978) "Frequency characteristic of cutting process identified by an in-process measurement of surface roughness." *Annals of the CIRP* **27**, 67.

Miyashita, T. and Manabe, T. (1982) "Infrared optical fibers." *IEEE J. Quant. El.* **QE-18**, 1432.

Molesworth, H. (1982) "An automated visual control system for glass-forming machines." In: *Automated Inspection and Product Control*, B. Batchelor, editor, p. 191. IFS, U.K.

Monchalin, J.-P. (1985) "Heterodyne interferometric laser probe to measure continuous ultrasonic displacements." *Rev. Sci. Instrum.* **56**, 543.

Monchalin, J.-P. (1985a) "Optical detection of ultrasound at a distance using a confocal Fabry-Pérot interferometer." *Appl. Phys. Lett.* **47**, 14.

Monchalin, J.-P. (1986) "Optical detection of ultrasound." *IEEE Trans. on Ultrasonics, Ferroelectrics and Frequency Control* **UFFC-33**, 485.

Montgomery, P.C. and Tyrer, J. (1985) "The use of ESPI as an inspection tool." In: *Proc. Snd Int. Conf. on Lasers in Manufacturing*, M.F. Kimmit, editor, p. 141. IFS/North-Holland.

Mooney, E.F. (1984) "Enhanced productivity by on-line analysis." *Proc. Pacific*

Cascade Instrumentation '84, Eugene, Ore. Sept. 11–13, p. 13. The Instrument Society of America, Research Triangle Park, N.C.

Mooradian, A. (1985) "Laser linewidth." *Physics Today* **38**(5), 45.

Moore, J.K. (1972) "The measurement of steel temperature during hot rolling." *The British Steelmaker*, Sept. 1972.

Morander, K.E. (1980) "The optocator. A high-precision, noncontacting system for dimension and surface measurement and control." *Proc. 5th Int. Conf. on Automatic inspection and Product Control*, p. 393.

Morey, W.W. (1986) "Diagnostics from imaging inside gas turbine combustors." In: *Optical Techniques for Industrial Inspection*, P. Cielo, editor, vol. **665**. SPIE, Bellingham, WA.

Mountain, D.S. and Webber, J.M.B. (1978) "Stress pattern analysis by thermal emission." *Proc. European Electro-Optics Conf.*, SPIE, vol. **164**, p. 189.

Müller, E., Hrdliczka, V. and Cuche, D.E. (1983) "Computer-based evaluation of holographic interferograms." In: *Industrial Applications of Laser Technology*, W.F. Fagan, editor, vol. **398**, p. 46. SPIE, Bellingham, WA.

Mundel, A.B., "Achieving productivity and manufacturing excellence." *Qualtest-2 Conference Proceedings*, Dallas, October 25–27, 1983.

Murata, M. and Kuroda, M. (1983) "Application of holographic interferometry to practical vibration study". In: *Industrial Applications of Laser Technology*, W.F. Fagan, editor, vol. **398**, p. 74. SPIE, Bellingham, WA.

Murphy, J.C. and Aamodt, L.C. (1981) "Optically detected photothermal imaging." *Appl. Phys. Lett.* **38**, 196.

Murphy, J.C. and Aamodt, L.C. (1983) "Reflective photothermal imaging." *J. Physique*, Coll. C6, p. 513.

Murphy, J.C., Maclachlan, J.W. and Aamodt, L.C. (1986) "Image-contrast processes in thermal and thermoacoustic imaging." *IEEE Trans. on Ultrasonics, Ferroelectrics and Frequency Control* **UFFC-33**, 470.

Musha, T., Itoh, K. and Shibata, T. (1982) "Optical head for digital audio disks." In: *Optical Disk Technology*, R.A. Sprague, editor, vol. **329**, p. 48. SPIE, Bellingham, WA.

Nadeau, F. and Bussière, J. (1984) "Acoustic sorting of grinder balls." *Material Evaluation* **42**, 1515.

Nadeau, F. and Hutchins, D.A. (1984) "A study of the interaction of surface waves with slots using noncontact laser generation and detection of ultrasound." *Proc. IEEE Ultrasonic Symp.*, p. 633.

Nakagawa, K. and Asakura, T. (1978) "Contrast dependence of white-light image speckles on surface roughness." *Opt. Comm.* **27**, 207.

Nakagawa, Y. and Rosenfeld, A. (1979) "A note on polygonal and elliptical approximation of mechanical parts." *Patt. Recogn.* **11**, 133.

Nakagawa, Y. (1982) "Automatic visual inspection of solder joints on printed circuits boards". In: *Robot Vision*, A. Rosenfeld, editor, vol. **336**, p. 121. SPIE, Bellingham, WA.

Nakaura, M. (1983) "Recent trend of on-line inspection techniques for hot slab surface and internal defects in Japan." In: *On-line Inspection of Steel Products*, p. 33. The International Iron and Steel Institute, Brussels.

Nakano, M. (1980) "Single-beam method nondispersive infrared gas analyzer." In: *Advances in Instrumentation*, vol. **35**, part 2, p. 337. The Instrument Society of America, Research Triangle Park, N.C.

Neiheisel, G.L. (1983) "Laser refractory wear sensor." *Proc. Int. Conf. Appl. Lasers and Electro-Optics* (ICALEO 183), p. 56. The Laser Institute of America.

Nervin, J.L. and Whitney, D.E. (1978) "Computer-controlled assembly." *Scientific American* **218**, 62.

Nevatia, R. (1982) *Machine Perception*. Prentice-Hall, Englewood Cliffs.

Newby, K., Reichert, W.M., Andrade, J.D. and Benner, R.E. (1984) "Remote spectroscopic sensing of chemical absorption using a single multimode optical fiber." *Appl. Opt.* **23**, 1812.

Nguyen, I.D. and Plaza, P. (1986) "Possibilités de mesures à distance et multi-sites par spectrométrie Raman-Laser avec association de fibres optiques." *Analusis*, **14**, 119.

Niepold, R. and Bruemmer, F. (1982) "Optical sensor system controls arc-welding process". *Proc. 2nd Int. Conf. on Robot Vivision and Sensory Controls*, SPIE, vol. **392**.

Nimrod, N., Margalith, A. and Mergler, H.W. (1982) "A laser-based scanning range finder for robotic applications." *Proc. 2nd Int. Conf. on Robot Vision and Sensory Controls*, SPIE vol. **392**, p. 241.

Nitzan, D., Brian, A.E. and Duda, R.O. (1977) "The measurement and use of registered reflectance and range data in scene analysis." *Proc. IEEE* **65**, 206.

Niwa, Y. and Kanie, S. (1981) "Active automatic focus detection through the taking lens system." *Terebijon* **35**, 637.

Nokes, M.A. Hill, B.C. and Barelli, A.E. (1978) "Fiber-optic heterodyne interferometer for vibration measurements in biological systems." *Rev. Sci. Instrum.* **49**, 722.

Nordal, P.E. and Kanstad, S.O. (1979) "Photothermal radiometry." *Physics Scripta* **20**, 659.

Norton-Wayne, L., Hill, W.J. and Brook, R.A. (1977) "Automatic visual inspection of moving steel surfaces." *Brit. J. NDT* **19**, 242.

Ogata, K. (1970) *Modern Control Engineering*. Prentice-Hall, Englewood Cliffs, N.J.

Ohi, M. (1980) "Frequency stabilization of a PbSnTe laser on a methane line in the I4 band." *Jpn. J. Appl. Phys.* **19**, L541.

Ohnuki, H., Fukui, Y., Ohnuki, N. and Inoue, M. (1982) "Three-dimensional actuator for optical head." In: *Optical Disk Technology*, R.A. Sprague, editor, vol. **329**, p. 103. SPIE, Bellingham, WA.

Ohtsuka, Y. (1982) "Optical coherence effects on a fiber-sensing Fabry-Pérot interferometer." *Appl. Opt.* **21**, 4316.

Okoshi, T. (1982) *Optical Fibers*. Academic Press, New York.

Oles, E.J. and Perfetti, B.M. (1980) "Effect of surface cleanliness on paintability of cold-rolled steel sheet." *SAE Congress*, Detroit, Feb. 25–29, paper 800148. The Society of Automotive Engineers, Warrendale, PA.

Olmstead, M.A., Awere, N.M., Kohn, S., Fournier, D. and Boccara, A.C. (1983)

"Photothermal-displacement spectroscopy: A new optical probe for solids and surfaces." *Appl. Phys. A* **32**, 141.

Olsen, E.D. (1975) *Modern Optical Methods of Analysis.* McGraw-Hill, New York.

Olsen, G.H. (1985) "Long-wavelength LEDs." *Photonics Spectra* **19**, 121.

Omini, M., Iazzi, F., Rastaldo, A. and Turco, C. (1981) "Potentiometric NDT of surface-hardened materials." *Rev. Sci. Instrum.* **57**, 1054.

Oomen, G.L. and Verbeek, W.J.P.A. (1983) "A real-time optical profile sensor for robot arc welding." In: *Intelligent Robots: RoViSeC3*, D.P. Casasent and E.L. Hall, editors, vol. **449**, p. 62. SPIE, Bellingham, WA.

Ortgies, H. (1985) "Automatic detection of surface defects on cables". *Wire-World Int.* **27** (Jan.–Feb. 1985).

Oshima, M. and Shirai, Y. (1979) "A scene description method using 3-D information." *Pattern Recognition* **11**, 9.

Oster, G. (1948) "The scattering of light and its applications to chemistry." *Chem. Rev.* **43**, 319.

Ott, E.R. (1975) *Process Quality Control.* McGraw-Hill, New York.

Ottesen, D.K. and Nagelberg, A.S. (1980) "Infrared spectroscopic studies of surface oxides." *Thin Solid Films* **73**, 347.

Ottesen, D.K. (1986) "Detection of contaminants in steel tubing using IR reflection spectroscopy." In: *Optical Techniques for Industrial Inspection*, P. Cielo, editor, vol. **665**, p. 226. SPIE, Bellingham, WA.

Ovren, C., Adolfsson, M., Hok, B. and Brogardh, T. (1985) "New opportunities with fiber-optic measurement." *Sensor Review* **5**, 199.

Owen, V.M. (1985) "Biosensors." *Sensor Review*, January 1985, p. 20.

Page, C.J. and Hassan, J. (1981) "Noncontact inspection of complex components using a rangefinder vision system." *Proc. Int. Conf. on Robot Vision and Sensory Controls*, Stratford-upon-Avon,U.K., p. 245.

Palmer, C.H. and Green, R.E. (1979) "Optical probing of acoustic emission waves." In: *Nondestructive Evaluation of Materials*, J.J. Barker and V. Weiss, editors, p. 347. Plenum Press, New York.

Palmer, C.H. (1984) "Circumventing laser relaxation oscillations in interferometers." *Appl. Opt.* **23**, 3510.

Pantzer, D., Politch, J. and Ek, L. (1986) "Heterodyne profiling for the angstrom region." *Appl. Opt.* **25**, 4168.

Pao, Y.H., editor (1977) *Optoacoustic Spectroscopy and Detection.* Academic Press, New York.

Papadakis, E.P. (1983) "Future growth of NDE." *Materials Evaluation* **41**, 1130.

Parker, W.J., Jenkins, R.J., Butler, G.P. and Abbott, G.L. (1961) "Flash method of determining thermal diffusivity, heat capacity and thermal conductivity." *J. Appl. Phys.* **32**, 1679.

Parker, J.G. (1973) "Optical absorption in glass: Investigation using an acoustic technique." *Appl. Opt.* **12**, 2974.

Parry, G. (1975) "Speckle patterns in partially coherent light." In: *Laser Speckle and Related Phenomena*, J.C. Dainty, editor, Springer-Verlag, Berlin.

Parthasarathy, S., Brick, J. and Dessimoz, J. (1982) "Laser rangefinder for robot control and inspection." In: *Robot Vision*, A. Rosenfeld, editor, vol. **336**, p. 2. SPIE, Bellingham, WA.

Paul, D.M. and Jackson, D.A. (1971) "Rapid velocity sensor using a static confocal Fabry-Pérot and a single-frequency argon laser." *J. Phys. E* **4**, 170.

Paul, D.R. and Barlow, J.W. (1980) "Polymer blends." *J. Macromol. Sci.*, Rev. Macromol. Chem. **C18**, 109.

Paul, G.R. (1983) "Image data synchronization techniques in rotating polygon scanners." In: *Advances in Laser Scanning and Recording*, L. Beiser, editor, vol. **396**, p. 204. SPIE, Bellingham, WA.

Paul, M., Betz, B. and Arnold, W. (1987) "Interferometric detection of ultrasound at rough surfaces using optical phase conjugation." *Appl. Phy. Lett.* **50**, 1569.

Peacock, G.R. (1979) "Infrared thermometers: A comparison of single band and ratio types." *Meas. Control* **13** (4).

Pentland, A.B. (1984) "Local shading analysis." *IEEE Trans. P.A.M.I* **6**, 170.

Pepper, G. (1985) "A fully automatic honing machine to correct bore geometry." In: *Automated Inspection and Product Control*, P.A. McKeown, editor, p. 263. IFS/North-Holland, U.K.

Perkins, W.A. (1978) "A model-based vision system for industrial parts." *IEEE Trans. Comput.* **C-27**, 126.

Pernick, B. and Kennedy, J. (1980) *Appl. Opt.* **19**, 3224.

Peterson, J.I., Goldstein, S.R., Fitzgerald, V. and Buckhold, D.K. (1980) "Fiber-optic pH probe for physiological use." *Anal. Chem.* **52**, 864.

Pfoutz, R.W. (1975) "Digital optics for dimensional gauging of fabricated parts and continuous webs." *Proc. SPIE, Quality Control*, vol. **60**, p. 110.

Piché, L. (1984) "Ultrasonic velocity measurement for the determination of density in polyethylene." *Polymer Eng. and Science* **24**, 1354.

Piché, L., Champagne, B. and Monchalin, J.P. (1987) "Laser ultrasonic measurement of elastic constants of composites." *Mater. Eval.* **45**, 74.

Piirainen, R.A. (1986) "Fiber length measurement in pulp and paper industry." In: *Optical Techniques for Industrial Inspection*, P. Cielo, editor, vol. **665**. SPIE, Bellingham, WA.

Pipitone, F.J. and Marshall, K.G. (1983) "A wide-field scanning triangulation rangefinder for machine vision." *Int. Journ. of Robotics Research* **2**, 39.

Pirlet, R., Mulder, J., Adrianensen, D. and Boelens, J. (1983) "The rometer, a noncontact system for measuring hot-strip flatness." In: *On-line Inspection of Steel Products*, p. 79. International Iron and Steel Institute, Brussels.

Pirlet, R., Boelens, J., Salmon, J., Balthazar, P. and Lückers, J. (1986) "Optical systems for dimensional control in hostile environments." In: *Optical Techniques for Industrial Inspection*, P. Cielo, editor, vol. **665**, p. 61. SPIE, Bellingham, WA.

Pirodda, L. (1982) "Shadow and projection moiré techniques for absolute or relative mapping of surface shapes." *Opt. Eng.* **21**, 640.

Podmancizk, A. (1983) "Diameter measurement of optical fibers and wires by laser." In: *Industrial Applications of Laser Technology*," W.F. Fagan, editor, vol. **398**, p. 274. SPIE, Bellingham, WA.

Popplestone, R.J., Brown, C.M., Ambler, A.P. and Crawford, G.F. (1975) "Forming models of plane-and-cylinder faceted bodies from light stripes." *Proc. 4th Int. Conf. on Artificial Intelligence*, p. 664.

Porter, G.B. and Mundy, J.L. (1982) "A noncontact profile sensing system for visual inspection." *Proc. 6th Int. Conf. on Pattern Recognition*, Munich, Oct. 19–22, p. 720.

Poulsen, P.D. (1979) "Electro-optical techniques for noncontact circuit probing." *Opt. Eng.* **18**, 335.

Powell, R.W. (1969) "Thermal conductivity determinations by thermal-comparator methods." In: *Thermal Conductivity*, R.P. Tye, editor, vol. **2**, p. 281. Academic Press, London.

Powers, S.R. and Somerford, D.J. (1979) "Error contour charts relevant to fibre sizing using light scattering." *J. Phys. D: Appl. Phys.* **12**, 1809.

Pratt, W.K. (1969) *Laser Communication Systems*. John Wiley & Sons, New York.

Pratt, W.K. (1978) *Digital Image Processing*. Wiley-Interscience, New York.

Prescott, L.J. and Van der Ziel, A. (1966) "Gas-discharge modulation noise in He–Ne lasers." *IEEE J. Quant. El.* **QE-2**, 173.

Pryor, T.R. and North, W. (1985) *Applying Automated Inspection*. SME, Dearborn, MI.

Puffer, L.G. (1984) "Optical detection of surface flaws on extruded cables." In: *Optical Systems Engineering*, P.R. Yoder, editor, vol. **518**, p. 120. SPIE, Bellingham, WA.

Pugh, J.F. (1980) "The infrared measurement of surface moisture in paper." *Tappi* **63**, 131.

Pugh, A., editor (1986) *Robot Sensors*. IFS/Springer Verlag.

Purll, D.J. (1978) "Automated surface inspection with solid-state image sensors." In: *Industrial Applications of Solid-state Image Scanners*, D.J. Purll, editor, vol. **145**, p. 18. SPIE, Bellingham, WA.

Purll, D.J. (1983) "Assessment of silicon imaging array performance." In: *Image Assessment: IR and Visible*, T.L. Williams, editor, vol. **467**, p. 122. SPIE, Bellingham, WA.

Pye, C.J. and Adams, R.D. (1981) "Detection of damage in fibre-reinforced plastics using thermal fields generated during resonant vibration." *NDT Int.* **14**, 111.

Raab, H. (1984) "Optical reflection sensors used in the chemical industry." *Process Automation* **6**, 47.

Radziemski, L.J. and Loree, T.R., (1981) "Laser-induced breakdown spectroscopy: Time-resolved spectrochemical applications." *Plasma Chem. Plasma Proc.* **1**, 281.

Radziemski, L.J., Loree, T.R., Cremers, D.A. and Hoffman, N.M. (1983) "Time-resolved laser-induced breakdown spectrometry of aerosols." *Anal. Chem.* **55**, 1246.

Rallison, R. and Lowe, R. (1983) "Hologram scanner design and fabrication in dichromated gelatin." In: *Industrial and Commercial Applications of Holography*, M. Chang, editor, vol. **353**, p. 47. SPIE, Bellingham, WA.

Randak, A. (1983) "On-line inspection of semi-finished and finished steel products for surface defects and internal quality." In: *On-line Inspection of Steel Products*, p. 1. The International Iron and Steel Institute, Brussels.

Ray, C. (1985) *Robots and Manufacturing Automation*. John Wiley & Sons, New York.

Reason, R.E. (1979) "Progress in the appraisal of surface topography during the first half-century of instrument development." *Wear* **57**, 1.

Reibold, R. and Molkenstruck, W. (1981) "Laser interferometric measurement and computerized evaluation of ultrasonic displacements." *Acoustica* **49**, 201.

Reid, G.T. (1984) "Moiré fringes in metrology". *Opt. Lasers in Eng.* **5**, 63.

Reid, G.T. and Rixon, R.C. (1986) "Automatic inspection of quasi-cylindrical objects by phase measuring Moiré topography." In: *Optical Techniques for Industrial Inspection*, P. Cielo, editor, vol. **665**, p. 162. SPIE, Bellingham, WA.

Reifsnider, K.L. and Williams, R.S. (1974) "Determination of fatigue-related heat emission in composite materials." *Exp. Mech.* **14**, 479.

Reynolds, W.N. and Wells, G.M. (1984) "Video-compatible thermograhy." *Brit. J. NDT* **26**, 40.

Rice, S.O. (1944) "Mathematical analysis of random noise." *Bell System Tech. J.* **23**, 282.

Richter, P., Lorincz, E., Peczeli, I. and Engard, F. (1983) "Coherent optical measurement of surface roughness and its application in the paper industry." In: *Industrial Applications of Laser Technoloy*, W.F. Fagan, editor, vol. **398**, p. 258. SPIE, Bellingham, WA.

Rider, G. (1975) "Measurement of weld-pool size by self-scanned photodiode arrays." *Proc. Int. Conf. on Low-Light and Thermal Imaging Systems*. IEE, London, U.K.

Riesen, R. and Sommerauer, H. (1983) "Curing of reaction molding resins studied by thermoanalythical methods." *Am. Lab.* **13**, 30.

Rines, G.A. (1981) "Fiber-optic accelerometer with hydrophone applications." *Appl. Opt.* **20**, 3453.

Rioux, M. (1984) "Laser range finder based on synchronized scanners." *Appl. Opt.* **23**, 3837.

Rioux, M. and Blais, F. (1986) "Compact 3-D camera for robotic applications." *Journ. Opt. Soc. Am. A*, **3**, 1518.

Roberts, D.W. (1977) "Particle sizing using laser interferometry." *Appl. Opt.* **16**, 1861.

Robeson, L.M. (1984) "Applications of polymer blends: Emphasis on recent advances." *Polym. Eng. Sci.* **24**, 587.

Rogers, L.M. (1978) "Applications of thermography in the steel industry." *Steel Times Ann. Rev.* **206** (8), 661.

Roney, J.E. (1982) "Steel surface temperature measurement in industrial furnaces by compensation for reflected radiation errors." In: *Temperature, its Measurement and Control in Science and Industry*, J.F. Schooley, editor, vol. **5**, p. 485.

Rosen, C.A. and Nitzan, D. (1977) "Use of sensors in programmable automation." *Computer* **10**, 12.

Rosenbaum, E.J., Steward, J.E. and Cleveland, F.F. (1960) "Raman spectroscopy." In: *The Encyclopedia of Spectroscopy*, G.L. Clark, editor, p. 675. Reinhold, New York.

Rosencwaig, A. and Gersho, A. (1976) "Theory of the photoacoustic effect with solids." *J. Appl. Phys.* **47**, 64.

Rosencwaig, A. (1980) *Photoacoustics and Photoacoustic Spectroscopy*. John Wiley and Sons, New York.

Rosencwaig, A., Opsal, J. and Willenborg, D.L. (1983) "Thin-film thickness measurements with thermal waves." *Appl. Phys. Lett.* **43**, 166.

Rosencwaig, A., Opsal, J. Smith, W.L. and Willenborg, D.L. (1985) "Detection of thermal waves through optical reflectance." *Appl. Phys. Lett.* **46**, 1013.

Rosenfeld, A. and Kak, A.C. (1976) *Digital Picture Processing*. Academic Press, New York.

Rosenfeld, A. (1983) "Computer vision research for industrial applications." *Proc. Trends and Applications 1983, Automating Intelligent Behaviour, Applications and Frontiers*, Gaithersburg, May 25–26, p. 43.

Ross, B. (1984) "Machines that can see: Here comes a new generation." *Business Week*, January 9, p. 118.

Ross, M. (1966) *Laser Receivers*. J. Wiley & Sons, New York.

Rostagni, A. (1952) *Ottica*, Chapter 2. Randi, Padova.

Roth, M.S. and O'Donnell-Leach, D. (1985) "FTIR analyzer for real-time multicomponent analysis of process streams." *Am. Lab.* **17**, 40 (Dec. 1985).

Rousset, G., Bertrand, L. and Cielo, P. (1985) "A pulsed thermoelastic analysis of photothermal surface displacements in layered materials." *J. Appl. Phys.* **57**, 4396.

Rudd, M.J. (1983) "Ultrasonic NDE using laser transducers." In: *Review of Progress in Quantitative NDE*, D.O. Thompson and D.E. Chimenti, editors, vol. **28**, p. 1763. Plenum Press.

Runge, E.F., Bonfiglio, S. and Bryan, F.R. (1966) "Spectrochemical analysis of molten metal using a pulsed laser source." *Spectr. Acta* **22**, 1678.

Rutledge, W.C. (1986) "Needs and expected developments in sensors for sheet products." *6th Int. Conf. on Instrumentation and Automation in the Paper, Rubber, Plastics and Polymerization Industries*, Akron, OH. Oct. 27–29, 1986. The Int. Fed. for Automatic Control.

Ryan, L.E., Gardner, M.P. and Vaughan, R.W. (1978) "Determination of moisture in polymers and composites." *Proc. 33rd Ann. Tech. Conf. Section 15-A*, p. 1. The Society of the Plastics Industry.

Saari, L.A. and Seitz, W.R. (1982) "pH sensor based on immobilized fluoresceinamine." *Anal. Chem.* **54**, 821.

Sakai, T., Matsui, K., Inade, K., Ohgaki, I., Suzuki, T., and Koyama, A. (1986) "Automatic detection of hot slab subsurface inclusion." In: *Optical Techniques for Industrial Inspection*, P. Cielo, editor, vol. **665**, p.91. SPIE, Bellingham, WA.

Saltzman, R.S. and Dell, C.G. (1984) "Photometric analyzer systems for monitoring low levels of H_2S." In. *Instrumentation and Control in the 80's*, Des Plaines, Ill., May 1–3. The Instrument Society of America, Research Triangle Park, N.C.

Samuels, R.J. (1971) "Small-angle light scattering from optically anisotropic spheres and disks." *J. Pol. Science*, part A-2, **9**, 2165.

Saridis, G.N. and Brandin, D.M. (1979) "An automatic surface inspection system for flat rolled steel." *Automatica* **15**, 505.

Saturday, K.A. (1983) "Absorption cell with fiber optics for concentration measurements in a flowing as stream." *Anal. Chem.* **55**, 2459.

Sawatari, T. (1972) "Surface-flaw dection using oblique-angle illumination." *Appl. Opt.* **11**, 1337.

Sawatari, T. and Müller, R.K. (1977) "Surface flaw detection using optical filters." *Int. Adv. NDT* **5**, 1.

Sawatari, T. and Zipin, R.B. (1979) "Optical profile transducer." *Opt. Eng.* **18**, 222.

Schliekelmann, R.J. (1982) "Interference holography in aerospace technology." In: *Industrial Applications of Holographic NDT*, J. Ebbeni, editor, vol. **349**, p. 91. SPIE, Bellingham, WA.

Schmieder, R.W. and Kerstein, A. (1980) "Imaging a conserved scalar in gas mixing by means of a linear spark." *Appl. Opt.* **19**, 4210.

Schneider, H., Gruner, H., Kohlhage, H., Deppe, G. J. and Ehlert, K.P. (1984) "Optical system for on-line detection of longitudinal surface cracks in hot continuously cast rounds." *Proc. 3rd European Conf. on Nondestructive Testing*, Florence, Oct. 15–18, p. 188.

Schoerner, J. and Rottenkolber, H. (1983) "Industrial applications of instant holography." In: *Industrial Applications of Laser Technology*, W.F. Fagan, editor, vol. **398**, p. 116. SPIE, Bellingham, WA.

Schooley, J.F., editor (1982) *Temperature, its Measurement and Control in Science and Industry*. American Institute of Physics, New York.

Schroeder, J. and Rosolowski, J.H. (1981) "Light scattering in polycrystalline materials." In: *Emerging Optical Materials*, Proc. SPIE, **297**, 156.

Schultz, A.W. (1968) "An infrared transient method for determining the thermal inertia, conductivity, and diffusivity of solids." *Appl. Opt.* **7**, 1845.

Schurink, H.D. (1983) "Vibration measurement by means of the laser Doppler vibrometer." *Journal A* **24**, 154.

Schwarzenbach, J. and Gill, K.F. (1978) *System Modelling and Control*. Edward Arnold, London.

Schweizer, M. (1977) "Sensors for program-controlled handling systems." *Foerdern und Heben* **27**, 22.

Scruby, C.B., Dewhurst, R.J., Hutchins, D.A. and Palmer, S.B. (1982) "Laser generation of ultrasound in metals." In: *Research Techniques in NDT", R.S. Sharpe, editor, vol.* **5**, p. 281. Academic Press, London.

Seitz, W.R. (1984) "Chemical sensors based on fiber optics." *Anal. Chem.* **56**, 16A.

Sell, J.A. (1985) "Gas velocity measurements by photothermal deflection spectroscopy." *Appl. Opt.* **24**, 3725.

Sevcik, A. (1982) "Methods and devices for scanning the surface of bearing rotating parts." In: *Automated Inspection and Product Control*, B. Batchelor, editor, p. 281. IFS, U.K.

Shafer, J.H. and Smith, V.N. (1982) "A low-cost, intelligent CO/CO_2 sensor for *in situ* combustion control measurements." *Proc. Symp. Discover New World Instrum.*, Columbus, Apr. 13–15. The Instrument Society of America, Research Triangle Park, N.C.

Shajenko, P. and Flatley, J. (1980) "Silicon rubber as elasto-optic material for optical hydrophones." *Appl. Opt.* **19**, 185.

Shannon, R.R. (1980) "Aspheric surfaces." In: *Applied Optics and Optical Engineering*, R.R Shannon and J.C. Wyant, editors, vol. **8**, p. 55. Academic Press, New York.

Sharp, H.J. (1976) "Technical and economic requirements for inspection." *Plastics and Rubber: Materials and Applications*, p. 98, Sept. 1976.

Sherman, P. (1970) *Industrial Rheology. Academic Press*, London.

Shibata, Y., Asano, Y., Kurita, K. and Shiozumi, M. (1986) "On-line glossmeter for stainless-steel sheets." In: *Optical Techniques for Industrial Inspection*, P. Cielo, editor, vol. **665**, p. 25. SPIE, Bellingham, WA.

Shiraiwa, T., Hiroshima, T., Sakamoto, T. and Matsui, K. (1979) "Development of inspection systems for hot slabs in a production line." *9th World Conf. NDT*, Melbourne, Australia.

Shiuh, J.C.H. (1972) "A method of determining one or more properties of asbestos fibers by turbidity measurement." U.S. Patent 3,695,763.

Sigel, G.H. and Tran, D.C. (1984) "Ultra-low loss optical fibers: An overview." In *IR Optical Materials and Fibers*, P. Klocek, editor, vol. **484**, p. 2. SPIE. Bellingham, WA.

Sigler, D. and Haworth, W.L. (1981) "Strain measurement by optical correlation." *J. Nondestr. Eval.* **2**, 125.

Simon, J. (1970) "New noncontact devices for measuring small microdisplacements." *Appl. Opt.* **9**, 2337.

Simonson, E.R., Reid, R.J. and Rosso, R.S. (1983) "Optical detection of surface flaws." *Proc. ISA Int. Conf.*, Houston, Oct. 10–13, 1983, p. 969. Instrument Society of America, Research Triangle Park, N.C.

Skolnik, M.I. (1962) *Introduction to Radar Systems*. McGraw-Hill, New York.

Slane, J.A., Clough, S.P. and Riker-Nappier, J. (1978) "Characterization of the surface of good and poor paintability cold-rolled steel". *Metall. Trans.* **9A**, 1839.

Sliney, D. and Wolbarsht, M. (1980) *Safety with Lasers and Other Optical Sources*. Plenum Press, New York.

Smeller, J.A. (1986) "Gaging breakthrough: Sensors that measure all coextrusion layers." *Modern Plastics*, May 1986, p. 52.

Smith, A.L. (1979) "Applied IR Spectroscopy: Fundamentals, Techniques and Analytical Problem-Solving." In: *Chemical Analysis*, vol. 54. Wiley-Interscience.

Smith, M.K.D. (1983) "The application of IR pyrometry to an industrial gas turbine for development and quality control purposes." *Transducer Tempcon Conference*, London, June 14–16. Trident, Tavistock, U.K.

Smith, T. (1976) "Effect of surface roughness on ellipsometry of aluminum." *Surface Sci.* **56**, 252.

Smith, W. (1966) *Modern Optical Engineering: The Design of Optical Systems*. McGraw-Hill, New York.

Smithgall, D.H., Watkins, L.S. and Frazee, R.E. (1977) "High-speed noncontact fiber-diameter measurement using forward light scattering." *Appl. Opt.* **16**, 2395.

Sneller, J.A. (1986) "Gaging breakthrough: Sensors that measure all coextrusion layers." *Modern Plastics*, May 1986, p. 52.

Snook, R. (1985) "Chemical sensors." *Lab. Pract.* **34** (7), 7.

Somerford, D.J. and Watson, D.J. (1983) "Fibre size distribution using optical scattering techniques." In: *Optical Techniques for Process Control*, p. 295. BHRA, Cranfield, Bedford, U.K.

Sommargren, G.E. (1979) "Heterodyne profilometer for measuring surface microstructure." *J. Opt. Soc. Am.* **69**, 1404A.

Sonner, R.H., Dailey, W.V. and Wright, D.E. (1984) "An open path IR hydrocarbon detector." In: *Instrumentation and Control in the 80's*, Des Plaines, Ill., May 1–3. The Instrument Society of America, Research Triangle Park, N.C.

Sontag, H. and Tam, A.C. (1985) "Optical monitoring of photoacoustic pulse propagation in silicon wafers." *Appl. Phys. Lett.* **46**, 725.

Spalding, I.J. (1971) "Laser systems development." *Phys. Bull.* **22**, 385.

Sparrow, E.M. and Cess, R.D. (1978) *Radiation Heat Transfer*. McGraw-Hill, New York.

Speake, J.H. (1978) "An absolute method of calibrating ultrasonic transducers using laser interferometry." *Proc. Conf. Eval. Calibr. Ultrason. Transd.*, p. 106. IPC, Guildford, U.K.

Sprague, R.A. (1972) "Surface roughness measurement using white light speckle." *Appl. Opt.* **11**, 2811.

Sridharan, G. (1985) "Optical reflectance sensor measures displacement in magnetic suspension." *IEEE Trans. Instrum. Meas.* **IM-34**, 31.

Stackhouse, G.E. (1966) "Control of beverages by continuous refractometry." *13th Ann. Meet. Soc. Soft-Drink Techn.*, Ashville, N.C.

Stapleton, T.T. (1983) "Applications of electro-optics in the factory" *Proc. ICALEO '83 Conf.*, vol. **39**, p. 1. The Laser Institute of America.

Stearns, R. G. and Kino, G.S. (1985) "Application of high-frequency transducers in air as photoacoustic detectors." *4th Int. Meet. on Photoacoustic, Thermal and Related Science*, Aug. 4–8, Ville d'Estérel, Québec.

Steel, W.H. (1967) *Interferometry*. University Press, Cambridge.

Stegeman, G.I. (1976) "Optical probing of surface waves and surface-wave devices." *IEEE Trans. Son. Ultrason.* **SU-23**, 33.

Steigmeier, E.F. and Auderset, H. (1983) "Optical scanner for dust and defect detection." *RCA Rev.* **44**, 5 (March 1983).

Stein, R.S. (1978) "Optical behaviour of polymer blends." In: *Polymer Blends*, D.R.Paul and S. Newmann, editors, **1**, 393. Academic Press, New York.

Sterling, W.M. (1979) "Automatic nonreference inspection of printed wiring boards." *Proc. IEEE Conf. on Pattern Recognition and Image Processing*, p. 93

Stetson, K.A. (1975) "A review of speckle photography and interferometry." *Opt. Eng.* **14**, 482.

Stetson, K.A. (1983) "A heterodyne optical strain sensor." In: *Instrumentation in the Aerospace Industry. Proc. 29th Int. Instrumentation Symp.*, p. 65. The Instrument Society of America, Research Triangle Park, N.C.

Stinebring, R.C. (1972) "Nondestructive testing of high-temperature coatings." *Materials Research and Standards* **12**, 18.

Stitch, M.L. (1972) "Laser rangefinding." In: *Laser Handbook*, F.T. Arecchi and E.O. Schultz-Dubois, editors, vol. **2**, p. 1745. North-Holland.

Stokely, G. (1983) "Combustion optimization, a key to efficiency." In: *Electro-optical Instrumentation for Industrial Applications*, F.M. Zweibaum and R.A. Carella, editors, vol. **411**, p. 62. SPIE, Bellingham, WA.

Stout, K.J., Obray, C. and Jungles, J. (1985) "Specification and control of surface finish: Empiricism versus dogmatism." *Opt. Eng.* **24**, 414.

Stowe, D.W., Moore, D.R. and Priest, R.G. (1982) "Polarization fading in fiber-interferometric sensors." *IEEE Journ. Quant. Electro.* **QE-18**, 1644.

Strand, T.C. (1985) "Optical three-dimensional sensing for machine vision." *Opt. Eng.* **24**, 33.

Straughan, B.P. and Walker, S. (1976) *Spectroscopy*, volumes 1 to 3. Chapman and Hall, London.

Streifer, W., Burnham, R.D., Paoli, T.L. and Scifres, D.R. (1984) "Phased array diode lasers". *Laser Focus* **20**, 100.

Strutt, J.W., Lord Rayleigh (1919) "On the problem of random vibrations and of random flights in one, two or three dimensions." *Phil. Mag.* **37**, 321.

Seumatsu, Y. (1985) "Advances in semiconductor lasers." *Physics Today* **38** (5), 32.

Sugden, E.A.N. (1986) "Optical approach for measuring delignification in Kraft pulping, part B: Using IR spectroscopy." In: *Optical Techniques for Industrial Inspection*, P. Cielo, editor, vol. **665**, p. 280. SPIE, Bellingham, WA.

Suresh, B.R., Funda Kowski, R.A., Levitt, T.S. and Overland, J.E. (1983) "A real-time automated visual inspection system for hot steel slabs." *IEEE Trans.*, vol. PAMI-5, p. 563.

Swift, K.J. and Dewhurst, R.J. (1983) "A flexible system handles engineering components." *Assembly Automation* **3**, 86.

Swing, D.M. (1975) "Dimensional gauging using a scanning laser." *Proc SPIE, Quality Control.*, vol. **60**, p. 118.

Tajima, A.T. (1982) "Monitoring cure viscosity of epoxy composite." *Polymer Composites* **3**, 162.

Takahashi, N., Sugimoto, i. and Sato, T. (1982) "GeO_2-glass optical fibre prepared by VAD method." *Electron. Lett.* **18**, 398 (1982).

Takasaki, H. (1970) "Moiré topography". *Appl. Opt.* **9**, 1467.

Takeyama, H., Sekiguchi, H. and Murata, R. (1976) "In-process detection of surface roughness in machining." *CIRP Annals* **25**, 467.

Talmi, Y. and Simson, R.W. (1980) "Self-scanned photodiode array: a multichannel spectrometric detector." *Appl. Opt.* **19**, 1401.

Tam, A.C. and Sullivan, B. (1983) "Remote sensing applications of pulsed photothermal radiometry." *Appl. Phys. Lett.* **43**, 333.

Tam, A.C. and Coufal, H. (1983) "Pulsed opto-acoustics: Theory and applications." *Journ. de Physique*, Coll. C 6, vol. **44**, suppl. to Nr. 10, p. C 6–9.

Tam, A.C. (1984) "Pulsed-laser generation of ultrashort acoustic pulses: Application for thin-film ultrasonic measurements." *Appl. Phys. Lett.* **45**, 510.

Tam, A.C. (1986) "Applications of photoacoustic sensing techniques." *Rev. Mod. Phys.* **58**, 381.

Tamura, Y., Tokuda, M., Tomono, H. and Nakamura, M. (1985) "Slag flow-out detection system in continuous casting process by optical method." *Trans. ISIJ.* vol. **25**, 109th ISIJ Meet., lecture S160.

Tanaka, T. and Benedek, G.B. (1975) "Measurement of the velocity of blood flow (*in vivo*) using a fiber-optic catheter and optical unmixing spectroscopy." *Appl. Opt.* **14**, 189.

Tankovsky, N.S. and Ivanov, D.V. (1985) "Photodisplacement imaging — A static regime", *Appl. Phys. Lett.* **46**, 639.

Taylor, R. (1983) "Thermal diffusivity of composites." *High Temp. — High Press.* **15**, 299.

Teele, R.P. (1965) "Photometry." In: *Applied Optics and Optical Engineering*, R. Kingslake, editor. Academic Press, New York.

Thomas, T.R. (1978) "Surface-roughnes measurements: Alternatives to the stylus." No. 4.2, *NELEX '78*. National Engineering Laboratory, Glasgow.

Thompson, B.H. (1984) "Magnetic flowmeters: Application and maintenance." *Proc. Pacific Cascade Instrumentation Conf.*, p. 169. The Instrument Society of America.

Thompson, B.J. (1978) "A review of Fourier plane signal processing for real-time measurement." *Proc. Int. Opt. Computing Conf.*, London, p. 139.

Thum, C. (1984) "Measurement of the entropy of an image with application to image focusing." *Opt. Acta* **31**, 203.

Tiikkaja, E., Makkonen, H., Piirainen, R. and Räsänen, S. (1983) "Instant fibre length determination." In: *Instrumentation and Automation in the Paper, Rubber, Plastics and Polymerization Industries.* A.R. Van Cauwenberghe, editor, p. 19. IFAC/Pergamon Press, Oxford.

Tilgner, R. (1976) "Testing plastic parts and foils by small-angle light scattering." *Siemens Forsch. Entwickl. Ber. Bd.* **5**, 217 (in German).

Tokida, M. (1981) "A new multipurpose fiber-optic sensor." *Trans. IECE Japan* **E-64**, 433.

Tolanski, S. (1973) *Introduction to Interferometry*. John Wiley and Sons, New York.

Tolanski, S. and Wood, A.F.B. (1958) "Multiple-beam interferometric studies on oscillating quartz crystals". *Physica* **24**, 508.

Torao, A., Kurita, K., Kitigawa, H., Nakamura, K. and Fujita, M. (1986) "On-line color measurement system for steel sheets." In: *Optical Techniques for Industrial Inspection*, P. Cielo, editor, vol. **665**, p. 39. SPIE, Bellingham, WA.

Toyen, G. (1976) "Generation of precision pixel clock in laser printers and scanners." In *Laser Scanning Components and Techniques*, L. Beiser and G.F. Marshall, editors, vol. **84**, p. 138. SPIE, Bellingham, WA.

Trolinger, J.D. (1985) "Automated data reduction in holographic interferometry." *Opt. Eng.* **24**, 840.

Truett, W.L., Dybwad, J.P., Propster, G.D,, Prozzo, C.D. and Szappanos, J. (1984) "A portable FTIR spectrometer using rotating refractor." *Amer. Lab.* **16**, 115.

Tsukada, T. and Yanagi, K. (1983) "An application of CCD image sensor to a measurement of surface roughness." *Bull. Japan Soc. of Prec. Eng.* **17**(3), 209.

Tyler, D.L. (1972) "Infrared moisture measurement experience with a press section application." *Tappi* **55**, 563.

Uchida, S., Sato, H. and Ohori, M. (1979) "Two-dimensional measurement of surface roughness by the light-sectioning method." *Annals of the CIRP*, **28**, 419.

Ueha, S., Shibata, N. and Tsujinchi, J. (1977) "Flexible coherent optical probe for vibration measurements." *Opt. Commun.* **23**, 407.

Van de Hulst, H.C. (1957) *Light Scattering by Small Particles*. John Wiley, London.

Van der Ziel, A. (1970) "Noise in solid-state devices and lasers." *Proc. IEEE* **58**, 1178.

Vanzetti, R., Traub, A.C. and Richard, A.A. (1980) "Laser inspection of solder joints." *Cleveland Electronics Conf. (CECON '80)*, Cleveland, May 21.

Vanzetti, R., Traub, A.C. and Ele, J.S. (1981) "Hidden solder joint defects detected by laser infrared system." *Proc. IPC 24th Annual Meeting*, p. 1.

Van Renesse, R.L. and Burgmeijer, J.W. (1985) "Application of pulsed reflection holography to material testing." *Opt. Eng.* **24**, 1086.

Varner, J.R. (1974) "Holographic and moiré surface contouring." In: *Holographic NDT*, R.K. Erf, editor, p. 106. Academic Press, New York.

Varennes, S., Charlet, G. and Delmas, G. (1984) "Use of the LCST for the characterization of polymer mixtures and the study of their compatibility: Application to PE, PP, and their copolymers." *Pol. Eng. and Science* **24**, 98.

Vavilov, V.P. and Taylor, R. (1982) "Theoretical and practical aspects of the thermal NDT of bonded structures." In: *Research Techniques in NDT*, R.S. Sharpe, editor, vol. **5**, p. 239. Academic Press, London.

Vest, C.M. (1974) "Crack detection." In: *Holographic NDT*, R.K. Erf, editor, Section 8.6. Academic Press, New York.

Vest, C.M. and Sweeney, D.W. (1977) "Applications of holographic interferometry to NDT." In: *International Advances in NDT*, W.J. McGonnagle, editor, vol. **5**, p. 17. Gordon and Breach, New York.

Vest, C.M. (1979) *Holographic Interferometry*. John Wiley and Sons, New York.

Vest, C.M. (1982) "Status and future of holographic nondestructive evaluation." In: *Industrial Applications of Holographic NDT*, J. Ebbeni, editor, vol. **349**, p. 186. SPIE, Bellingham, WA.

Vilkomerson, D. (1976) "Measuring picometer-displacement vibrations by optical interferometry." *Appl. Phys. Lett.* **29**, 183.

Von Gutfeld, R.J. (1980) "Thermoelastic generation of elastic waves for nondestructive testing and medical diagnostics." Ultrasonics **18**, 175.

Von Gutfeld, R.J., Vigliotti, D.R., Ih, C.S. and Scott, W.R. (1983) "Thermoelastic hologram for focused ultrasound.". *Appl. Phys. Lett.* **42**, 1018.

Vorburger, T.V. and Ludema, K.C. (1980) "Ellipsometry of rough surfaces.". *Appl. Opt.* **19**, 561.

Vorburger, T.V. and Teague, E.C. (1981) "Optical techniques for on-line measurement of surface topography.". *Precision Engineering* **3**, 61.

Vorburger, T.V., Teague, E.C., Scire, F.E., McLay, M.J. and Gilsinn, D.E. (1984) "Surface roughness studies with DALLAS-detector array for laser light angular scattering." *J. Res. Nat. Bur. Stand.* **89**, 3.

Vranish, J.M. (1984) "Magnetoresistive skin for robots." *Proc. 4th Int. Conf. on Robot Vision and Sensory Controls*, A. Pugh, editor, p. 269–284, IFS, U.K.

Wagner, G.G. (1982) "Vision methodology as it applies to industrial inspection applications." *Proc. Conf. on Applied Machine Vision*, p. 183. SME, Dearborne, MI.

Wall, D.L., Pearson, E.F. and Mantz, A.W. (1980) "Tunable diode lasers for industrial analysis and monitoring applications." *Proc. 4th IFAC Conf.*, A. Van Cauwenberghe, editor, p. 295. IFAC/Pergamon Press.

Wang, S.S.M. and Will, P.M. (1978) "Sensors for computer-controlled mechanical assembly." *The Industrial Robot* **5**, 9.

Wang, Y. and Chin, B.A. (1986) "On-line sensing of weld penetration using IR thermography." In: *Optical Techniques for Industrial Inspection*, P. Cielo, editor, vol. **665**, p. 314. SPIE, Bellingham, WA.

Ward, J.M. (1985) "A total quality improvement program." *Proc. 7th Int. Conf. on Automated Inspection and Product Control*, P.A. McKeown, editor, p. 1, IFS/North-Holland.

Ward, P.T., Hashim, A.A. and Schofield, B. (1985) "Expert systems in textile defect recognition." In: *Automated Inspection and Product Control*, P.A. McKeown, editor, p. 195. IFS/North-Holland, U.K.

Watrasiewicz, B.M. and Rudd, M.J. (1976) *Laser Doppler Measurements*. Butterworths, London.

Weill, M.E., Lhuissier, N. and Gouesbet, G. (1986) "Mean diameters and number densities in premixed flat flames CH_4–O_2 by DBS." *Appl. Opt.* **25**, 1676.

Weller, R. and Shepherd, B.M. (1948) "Displacement measurement by mechanical interferometry." *Proc. SESA* **6**, 35.

Wellman, R.J. (1980) "Laser system for the detection of flaws in solids." Harry Diamond Laboratories, Report HDL-TR-1902.

Wendlandt, W.W. and Hecht, H.G. (1966) *Reflectance Spectroscopy*. Interscience, New York.

Wertheimer, A.L. and Wilcock, W.L. (1976) "Light scattering measurements of particle distributions." *Appl. Opt.* **15**, 1616.

West, R.N. (1982) "Laser scanners for automatic inspection of strip products." In: *Automated Inspection and Product Control*, B. Batchelor, editor, p. 287. IFS, U.K.

Wetzel, D.L. (1983) "Near-IR reflectance analysis; sleeper among spectroscopic techniques." *Anal. Chem.* **55**, 1165A.

Wheeler, M.J. (1965) "Thermal diffusivity at incandescent temperatures by a modulated electron-beam technique." *Brit. J. Appl. Phys.* **16**, 365.

White, G. S. and Feldman, A. (1981) "Diffraction from a shallow rectangular groove." *Appl. Opt.* **20**, 2585.

White, R.M. (1963) "Generation of elastic waves by transient surface heating." *J. Appl. Phys.* **34**, 3559.

Whitefield, R.J. (1975) "Noncontact optical profilometer." *Appl. Opt.* **14**, 2480.

Whitehouse, D.J. (1985) "Instrumentation for measuring finish, defects and gloss." In: *Measurement and Effects of Surface Defects and Quality of Polish*, L.R. Baker and H.E. Bennet, editors, vol. **525**, p. 106. SPIE, Bellingham, WA.

Wickersheim, K. and Alves, R. (1979) "Optical temperature measurement." *Asea Research and Development*, December 1979.

Wiese, D.R., Ohlson, L.B., Hammerstron, H. and Broman, H. (1980) "Continuous on-line measurement of hot strip flatness." *Iron and Steel Engineer*. Available from Selective Electronic, Valdese, N.C.

Wilburn, D.K. (1977) "Temperature profiles observed in tensile specimens during physical test." *Mater. Eval.*, March 1977, p. 28.

Wilks, P. (1985) "On-line monitoring system for beverage packing lines." *Am. Lab.* **17**, 76 (Dec. 1985).

Will, P.M. and Pennington, K.S. (1971) "Grid-coding: A processing technique for robot and machine vision." *Proc. 2nd Int. Joint Conf. in Artificial Intelligence*, p. 66. See also *Proc. IEEE* **60**, 669.

Williams, J.H., Mansouri, S.H. and Lee, S.S. (1980) "One-dimensional analysis of thermal nondestructive detection of delamination and inclusion flaws." *Brit. J.NDT* **22**, 113.

Williams, R.V. (1975) "Applied optics in the European coal and steel community — its applications to quality control." *Proc. SPIE, Quality Control*, vol. **60**, p. 52.

Wilson, D.W. and Charles, J.A. (1981) "Thermographic detection of adhesive-bond and interlaminar flaws in composites." *Exp. Mech.* **21**, 276.

Winkler, T., Erdmann-Jesnitzer, F. and Haferkamp, H. (1981)" Holographic evaluation of disbondings in sandwich plates." *36th Annual Conf. Reinforced*

Plastics/Composite Institute, Feb. 16–20, 1981, session 10-E. The Society of the Plastics Industry.

Wirick, M.P. and Diede, H.A. (1980) "Infrared testing of printed circuit boards and hybrids." In: *Optics in Metrology and Quality Assurance*, H.L. Kasdan, editor, vol. **220**, p. 124. SPIE, Bellingham, WA.

Wolf, (1980) "On-line inspection of nonwoven fabrics." In: *Optics in Metrology and Quality Assessment*, H.L. Kasdan, editor, vol. **220**, p. 183. SPIE, Bellingham, WA.

Wolfe, W.L. and Zissis, G.J. (1978) *The Infrared Handbook*. IRIA, Environmental Research Institute of Michigan, Ann Arbor, MI.

Wong, Y.H., Thomas, R.L. and Hawkins, G.F. (1978) "Surface and subsurface structure of solids by laser photoacoustic spetroscopy." *Appl. Phys. Lett.* **32**, 538.

Wyant, J.C. (1975) "Optical gauging principles." *Proc. SPIE, Quality Control*, vol. **60**, p. 22.

Wyant, J.C., Koliopoulos, C.L., Bhushan, B. and George, O.E. (1984) "An optical profilometer for surface characterization of magnetic media." *Am. Soc. Lubrication Engrs. Trans.* **27**, 101.

Yakimovsky, Y. and Cunningham, R. (1978) "A system for extracting 3-D measurements from a stereo pair of TV cameras." *Computer Vision, Graphics Image Processing* **7**, 195

Yamada, J., Okamoto, M. and Toyonaga, T. (1985) "Fiber-optic sensor for temperature and pressure using polarization-maintaining single-mode fiber." *Trans. Inst. Electr. Comm. Eng. Jpn* **J68C**, 959.

Yamada, T., Harada, N. and Koyanagi, M. (1982) "Temperature distribution measurement with a silicon photodiode array." In: *Temperature, its Measurement and Control in Science and Industry*, J.F. Schooley, editor, vol. **5**, p. 395.

Yamada, T., Harada, N., Imai, K., Kawahata, S., Tezuka, H. and Terao, S. (1984) "New methods for temperature measurement and their applications to steel process." *ISIJ*, p. 23.

Yamaguchi, I. (1981) "A laser-speckle strain gauge." *J. Phys. E: Sci. Inst.* **14**, 1270.

Yamamoto, N., Nishi, H., Nishizawa, K. and Kitano, I. (1982) "Selfoc microlens with a spherical surface." *Appl. Opt.* **21**, 1021.

Yanagi, K., Miyoshi, T. and Saito, K. (1980) "Development of the displacement sensor by fiber optics." *Bull. Japan Soc. of Prec. Eng.* **14**, 175.

Yariv, A. (1976) *Introduction to Optical Electronics*. Holt, Rinehart and Winston, New York.

Yatagai, T., Nakadate, S., Indesawa, M. and Saito, H. (1982) "Automatic fringe analysis using digital image processing techniques." *Opt. Eng.* **21**, 432.

Yeh, J.T.Y. (1986) "On-line composition analyzers." *Chem. Eng.*, Jan. 20, 1986, p. 55.

Yoshino, T., Kurosawa, K., Itoh, K. and Ose, T. (1982) "Fiber-optic Fabry-Pérot interferometer and its sensor applications." *IEEE Journ. Quant. Electron.* **QE-18**, 1624.

Young, A.W. (1967) "Optical workshop instruments." Section 7.2. In: *Applied Optics and Optical Engineering*, R. Kingslake, editor, vol. **4**, p. 217. Academic Press, New York.

Yu, F.T.S., Zhuang, S.L. and Wang, N.H. (1982) "IC chip inspection with incoherent optical processing." *Proc. SPIE*, vol. **360**, p. 310.

Zanoni, C.A. (1977) "Dimensional gauging using a scanning laser beam." *Proc. SPIE, Clever Optics*, vol. 126, p. 104.

Zapka, W. and Tam, A.C. (1982) "Noncontact optoaoustic determination of gas flow velocity and temperature simultaneously." *Appl. Phys. Lett.* **40**, 1015.

Zeller, H. and Doemens, G. (1983) "Applications of pattern recognition in semiconductor and printed circuit board production." *Signal Processing* **5**, 399.

Zhao, C.J., Monchaud, S., Marcé, L. and Jullière, M. (1983) "Location of a vehicle with a laser range finder." In: *Intelligent Robots: RoViSeC 3*, D.P. Casasent and E.L. Hall, editors, vol. **449**, p. 82. SPIE, Bellingham, WA.

Ziegler, G. and Hasselman, D.P.H. (1981) "Effect of phase composition and microstructure on the thermal diffusivity of silicon nitride." *J. Mater. Sci.* **16**, 495.

Ziesemer, M. (1983) "Noncontact velocity-sensing and length-measurement of paper using the cross-correlation technique with optical sensors". In: *Instrumentation and Automation in the Paper, Rubber, Plastics and Polymerization Industries*, A.R. Van Conwenberghe, editor, p. 51. IFAC/Pergamon Press, Oxford.

Index

G